MECHANISMS OF EGG ACTIVATION

BODEGA MARINE LABORATORY
MARINE SCIENCE SERIES

Series Editor: James S. Clegg

Bodega Marine Laboratory
University of California, Davis
Bodega Bay, California

INVERTEBRATE HISTORECOGNITION
Edited by Richard K. Grosberg, Dennis Hedgecock, and Keith Nelson

MECHANISMS OF EGG ACTIVATION
Edited by Richard Nuccitelli, Gary N. Cherr, and Wallis H. Clark, Jr.

MECHANISMS OF EGG ACTIVATION

Edited by

Richard Nuccitelli
University of California, Davis
Davis, California

and

Gary N. Cherr and
Wallis H. Clark, Jr.
Bodega Marine Laboratory
University of California, Davis
Bodega Bay, California

Plenum Press • New York and London

Cover photograph: The wave of increased free Ca^{2+} in an aequorin-injected, zona-free hamster egg. A single sperm fertilized this egg at the top of the photographs. Continuous accumulation of light emitted by aequorin using the Hamamatsu photon-counting image intensification system: upper left, 0.5 s accumulation; upper right, 2.5 s accumulation; lower left, 4.5 s accumulation; lower right, 6.5 s accumulation. The wave of Ca^{2+} release starts at the sperm attachment site and spreads across the egg in about 7 min (see chapter by Miyazaki). *Photograph by Shun-ichi Miyazaki.*

ISBN 0-306-43245-5

Proceedings of a colloquium on the Mechanisms of Egg Activation,
held September 7–11, 1988, in Bodega Bay, California

A Division of Plenum Publishing Corporation
233 Spring Street, New York, N.Y. 10013

Printed in the United States of America

DEDICATION

Edward (Ted) Chambers

This book is dedicated to Ted Chambers who has just completed 50 years of scientific inquiry of the highest quality. He has made many important contributions to this field, and his work is highly regarded by colleagues as extremely reliable and thorough. Even after 50 years of research he still works in the laboratory and remains at the forefront of this field as evidenced by the lively discussion stimulated by his most recent work on the timing of sperm-egg fusion and egg activation. Ted's gentle manner, inquisitive intellect, and dedication to research are an inspiration to us all. We thank his wife, Eleanora, for her critical role in helping Ted write a brief biography and encouraging him to share with us some private moments and recollections. She is a professional in her own right.

CONTRIBUTORS

Edward L. Chambers

Department of Physiology and Biophysics
University of Miami School of Medicine
P.O. Box 016430
Miami, FL 33101

Ian Crossley

Department of Physiology
University College London
Gower Street, London WC1E 6BT
UNITED KINGDOM

Yoshihiro Endo

Division of Reproductive Biology
Department of Obstetrics and Gynecology
University of Pennsylvania
Philadelphia, PA 19104-6080

David Epel

Hopkins Marine Station
Stanford University
Pacific Grove, CA 93950

James Ferguson

Department of Zoology
University of California, Davis
Davis, CA 95616

Henri Goudeau

Departement de Biologie
Service de Biophysique
CEN-SACLAY
91191 GIF-sur-YVETTE Cedex FRANCE

Marie Goudeau

Departement de Biologie
Service de Biophysique
CEN-SACLAY
91191 GIF-sur-YVETTE Cedex FRANCE

Meredith Gould

Escuela Superior de Ciencias Biologicas
Universidad Autónoma de Baja California
A.P. 2921, Ensenada, B.C.N., MEXICO

Jin-Kwan Han

Department of Zoology
University of California, Davis
Davis, CA 95616

Laurinda A. Jaffe

Department of Physiology
University of Connecticut Health Center
Farmington, CT 06032

Raymond T. Kado

Centre National de la Recherche Scientifique
Laboratoire de Neurobiologie Cellulaire
et Moleculaire
GIF-sur-YVETTE 91198 FRANCE

Gregory S. Kopf

Division of Reproductive Biology
Department of Obstetrics and Gynecology
School of Medicine
University of Pennsylvania
Philadelphia, PA 19104-6080

Shigeaki Kurasawa

Division of Reproductive Biology
Department of Obstetrics and Gynecology
University of Pennsylvania
Philadelphia, PA 19104-6080

John W. Lynn

Department of Zoology and Physiology
Louisiana State University
Baton Rouge, LA 70803-1725

Peter Mattei

School of Medicine and
Department of Biology
University of Pennsylvania
Philadelphia, PA 19104-6080

David H. McCulloh

Department of Physiology and Biophysics
University of Miami School of Medicine
P.O. Box 016430
Miami, FL 33101

Shun-ichi Miyazaki

Department of Physiology
Tokyo Women's Medical College
Kawada-cho, Shinjuku-ku,
Tokyo, 162, JAPAN

Richard Nuccitelli

Department of Zoology
University of California, Davis
Davis, CA 95616

Lyanne C. Schlichter

Department of Physiology
Medical Sciences Building
University of Toronto Medical School
Toronto, Ontario CANADA M5S 1A8
present address:
Merck-Frosst Canada Inc.
P.O. Box 1005
Pointe Claire-Dorval
Québec CANADA H9R 4P8

Richard M. Schultz

School of Medicine
and Department of Biology
University of Pennsylvania
Philadelphia, PA 19104-6080

Sheldon S. Shen

Department of Zoology
Iowa State University
Ames, IA 50011

José Stephano

Escuela Superior de Ciencias
Universidad Autónoma de Baja California
A.P. 2921, Ensenada, B.C.N., MEXICO

Karl Swann

Department of Physiology
University College London
Gower Street, London WC1E 6BT
UNITED KINGDOM

Michael Whitaker

Department of Physiology
University College London
Gower Street, London WC1E 6BT
UNITED KINGDOM

PREFACE

The Second Annual Bodega Marine Science Colloquium was held at Bodega Marine Laboratory of the University of California, Davis in September of 1988. Many prominent investigators in the field of egg activation presented their most recent results and most of them contributed chapters to this volume. This was a very lively meeting with stimulating discussions that revealed a number of problems that are ripe for investigation in this area. Here I would like to introduce the reader to this work as well as point out some of the more lively topics of discussion at the meeting. We asked the authors to emphasize the description of useful methodology and tricks of the trade and the reader will find many helpful hints within these chapters. For example, Lyanne Schlichter has written down many of the details concerning optimizing electrophysiological recording from frog oocytes and David McCulloh and John Lynn have done the same for sea urchin eggs. Ray Kado has explained in detail all of the techniques for measuring membrane capacitance in a uniquely comprehensible manner. Henri and Marie Goudeau described the handling of gametes of several marine species that merit further investigation.

Jim Clegg, Director of the Bodega Marine Laboratory, first proposed an annual Marine Science Colloquium and this is the second of the series. There are many people who worked long hours to organize the meeting and assure that it ran smoothly. The initial organizing committee was composed of Jim Clegg, Wally Clark, Gary Cherr, Trisha Pedroia and myself. However, the meeting would never have happened without the dedication and hard work of the staff at Bodega, including Kitty Brown, Vicki Milam, and Alberta Doyle. Special thanks to Eleanor Uhlinger for her library and reference support as well as her E-Mail and FAX exertise. Thanks also to Vic Chow for his computer guidance and assistance. Finally, the single most important person in this whole project is Diane Cosgrove who created the camera-ready manuscript you see before you. She has done a fantastic, meticulous job and we are all indebted to her for her skill and care.

Richard Nuccitelli
University of California, Davis

**WE GRATEFULLY ACKNOWLEDGE
THE FOLLOWING SOURCES OF FINANCIAL FUNDING**

major support for this colloquium was provided by:

* **The Sonoma County Foundation** *

* **California Sea Grant College Program** *

* **Bodega Marine Laboratory** *

additional support was provided by:

* **Society for Developmental Biology** *

* **Lowell Berry Foundation** *
Mr. Herbert W. Funk

CONTENTS

INTRODUCTION

Research over the past decade has yielded several breakthroughs into the mechanisms of egg activation and most of them are discussed in this volume. Due largely to the technical advances of ion-specific microelectrodes, nuclear magnetic resonance and low-light imaging we have learned much about the intracellular ion concentration changes occurring during activation. In the late seventies and early eighties changes in intracellular pH (pH_i) were found to accompany the activation of many invertebrate and vertebrate eggs (reviewed in Busa and Nuccitelli, 1984). This fairly stable increase of about 0.4 pH units that accompanies egg activation in several species is not sufficient for complete activation in the species that have been most thoroughly studied and this is discussed by David Epel in this volume. Shortly after the first pH_i measurements, aequorin, ion-specific electrodes and fura-2 were used to study intracellular Ca^{2+} concentration changes during activation. A wave of increased free Ca^{2+} has been found to accompany activation in all eggs studied and many of the chapters here include some of these data, including those by Michael Whitaker, Richard Nuccitelli, and Shun-ichi Miyazaki. In most species studied this $[Ca^{2+}]_i$ increase is independent of extracellular Ca^{2+}, but in the egg of *Urechis* extracellular Ca^{2+} is required as discussed by Meredith Gould.

A critical step in the activation mechanism involves the connection between the sperm-egg interaction and these ionic changes. During the mid-eighties, the inositol lipid cascade and GTP-binding proteins (G proteins) were implicated in this signal transduction in sea urchins, frogs and mammals and much of this work is discussed in the chapters by Laurinda Jaffe, Michael Whitaker, Sheldon Shen, Richard Nuccitelli, Shun-ichi Miyazaki and Gregory Kopf. Briefly, there is now convincing evidence that the injection of the non-hydrolyzable GTP analogue, GTP-γS, can immediately activate sea urchin, frog and hamster eggs, while GDP-βS inhibits their activation. This implies that a G protein is involved in the activation mechanism. The activated G protein is then thought to stimulate phospholipase C which cleaves the membrane lipid, phosphatidylinositol-1,4,-bisphosphate (PIP_2) producing Ins(1,4,5)P_3 and diacylglycerol (DAG). Ins(1,4,5)P_3 has been shown to release Ca^{2+} in these eggs and is thought to be involved in the natural wave of increased free Ca^{2+}. DAG is known to stimulate protein kinase C and has been implicated in the activation of the Na^+/H^+ exchanger. However, here Sheldon Shen shows that the activation of this exchanger may not occur through this pathway during normal activation. Rather, it appears to be regulated by a calmodulin-dependent pathway. Another effect of DAG is described by Greg Kopf who points out that in the mouse egg, DAG also prevents sperm penetration of the zona pellucida by causing a modification of the zona pellucida protein, ZP2.

The involvement of a G protein naturally leads to the comparison with other systems in which G proteins are involved in signal transduction and in many of these a receptor activates the G protein when a hormone or growth factor binds to it. Thus, it is natural to envision a sperm receptor that plays a similar role of activating G proteins upon sperm-egg binding. This idea is also consistent with two other observations reported here: 1) In *Urechis* Meredith Gould finds that the sperm acrosomal protein can activate the egg upon contact with the plasma membrane, presumably by interacting with a membrane protein to open a Na^+ channel; 2) In the *Xenopus* egg Laurinda Jaffe discusses work from her lab indicating that the neurotransmitters, acetylcholine (ACh) and serotonin, can activate the frog egg when their respective receptors are inserted into the oocyte membrane (Kline, et al., 1988). Therefore, there is little doubt that receptors which stimulate G proteins can activate eggs, but does the sperm act in the same way?

One of the most controversial steps in the mechanism of activation discussed at the September meeting at Bodega Marine Laboratory is the initiation of the activation cascade by the sperm. All of this discussion hinged on the existence of a "latent period" in sea urchin fertilization. If the sperm stimulates a G protein when it binds to the egg in a manner analogous to the G protein stimulation resulting from ACh binding to its muscarinic M1 receptor, for example, one would expect a relatively rapid initiation of the inositol cascade. However, in the sea urchin egg there is a delay of approximately 12s between the first electrical event associated with sperm-egg binding and the initiation of the wave of Ca^{2+} release. This delay or "latent period" was first reported by Allen and Griffin in 1958 and has been most carefully studied by Chambers and his colleagues. Chambers describes it in his chapter in this volume as a shoulder in the activation current of a voltage-clamped egg. During this shoulder period of about 12s the egg's conductance increases slowly, in contrast to the rapid increase following the shoulder and coinciding with the Ca^{2+} wave. A second observation by Chambers' group that sperm can fuse transiently with the egg for many seconds without activating it also raises questions about why a G protein mechanism would not be activated under these conditions. This has led both Chambers and Whitaker to suggest that perhaps this latent period reflects the time required for a molecule to diffuse from the sperm into the egg. Whitaker presents his reasons for favoring such an hypothesis in his contribution to this volume and suggests that the diffusing activator molecule may be cGMP. He also rightly points out that Ehrenstein has identified a higher molecular weight candidate as part of a long-term investigation (Dale *et al.* 1985). The lively discussion surrounding these presentations made it clear that there are insufficient data to be sure of either mechanism. The latent period could indeed be due to a diffusible activator, but I would like to propose that it could just as easily be due to a slow sperm-G protein interaction. Since the bound sperm only interacts with receptors in a restricted region of the egg's surface and cannot exhibit the lateral mobility of a receptor such as that of ACh, G proteins have to be activated by diffusing to the sperm and then must diffuse away to activate the phospholipase C. Thus, it may well take some time for this process to release sufficient $Ins(1,4,5)P_3$ to activate the Ca^{2+} wave. We have found in the *Xenopus* egg that there is a small Ca^{2+} release for every pulse of $Ins(1,4,5)P_3$ injected and the wave of free Ca^{2+} is not triggered until a threshold Ca^{2+} level is exceeded (Busa *et al.* 1985). Thus, the shoulder of the voltage clamp activation current may reflect a slow buildup of $Ins(1,4,5)P_3$-induced Ca^{2+} release which triggers the Ca^{2+} wave when the threshold is reached. This interpretation is consistent with fura-2 data shown in Whitaker's chapter and would predict that the initial slow $[Ca^{2+}]_i$ increase observed during the latent period should be independent of extracellular Ca^{2+}. Clearly there are many more experiments to be done before we can explain all the steps in egg activation, but we have come a long way over the past decade.

Richard Nuccitelli
University of California, Davis

REFERENCES

Allen, R. D. and J. L. Griffin. 1958. The time sequence of early events in the fertilization of sea urchin eggs. 1. The latent period and the cortical reaction. *Exp. Cell Res.* 15:163-173.

Busa, W. B., J. E. Ferguson, S. K. Joseph, J. R. Williamson, and R. Nuccitelli. 1985. Activation of frog (*Xenopus laevis*) eggs by inositol trisphosphate. I. Characterization of the Ca^{2+} release from intracellular stores. *J. Cell Biol.* 101:677-682.

Busa, W. B. and R. Nuccitelli. 1984. Metabolic regulation via intracellular pH. *Am. J. Physiol.* 246:R409-R438.

Dale, B., L. J. DeFelice, and G. Ehrenstein. 1985. Injection of a soluble sperm fraction into sea-urchin eggs triggers the cortical reaction. *Experientia* 41:1068-1070.

AN AUTOBIOGRAPHICAL SKETCH

Edward L. (Ted) Chambers

FORMATIVE YEARS

I was born in Manhattan, New York City, January 12, 1917 to Bertha Inez (*née* Smith) and Robert Chambers. Since the age of 6 months, and all through my happy childhood the family summered at Woods Hole. The favorite family home was Bobtucket, off Gardiner Road, on the shore of Buzzards Bay barely a dozen feet from the high water mark. My father was working at the Marine Biological Laboratory, developing his microsurgical instrumentation and carrying out microdissection and microinjection studies on living marine eggs.

As a small boy I tagged along with the older students on the collecting trips. These were a major component of the courses taught at the MBL at this time, in conformity with Louis Agassiz' admonition "Study Nature, not books." My oldest brother, Bob, was an avid sailor. As his constant companion I learned to sail as if it was my second nature.

From time to time my father and my brothers and I would sail from Woods Hole on overnight cruises, camping on the Elizabeth Islands. The deer would often peer into the tent

Ted Chambers holding a horseshoe crab, on the beach at Woods Hole, Mass.

Bobtucket, the Chambers' family home on Buzzards Bay,
Woods Hole, Mass., about 1921.

as night fell. When it rained the water would collect on the underside of the roof of the tent. For all of us crowded into the tent it was extremely important not to touch the underside of the tent's roof, since if anyone did, the water would come pouring in.

Winters in New York City were barely tolerated. After leaving Woods Hole we counted the days until we were back again. However, I did enjoy walking every day across Central Park, to and from Trinity School, an Episcopalian day school.

An occasional summer was spent on Mount Desert Island, Maine, where my father worked at the Mount Desert Island Biological Station. I vividly recall fetching my father from the laboratory at night time. An intense light shown from the open doorway of his laboratory room, streaming out into the dark evergreen forest. My father was using an oil lamp as a lightsource for microscopic observations. Many years later I learned that the brightness of the

At Mount Desert Island, Maine, hiking in the Bubble Mountains. Ted's father to the right, Ted front center with blanket roll over shoulder. Ted's brother, Bill, behind, and his oldest brother, Bob, to the left.

On the freighter, American Farmer, *en route* to Europe.
Ted's mother, center; Ted to the left, Ted's younger brother, Brad to the right.

light was because he was using the flat, incandescent surface of the oil lamp flame to focus it on the object under the microscope. By this means he was able to achieve superb resolution (critical illumination). He was at the time manipulating with microneedles the asters and the astral rays in the transparent sand-dollars eggs.

My mother, who as a young woman had traveled extensively in Europe, determined that her sons in their formative years should spend a substantial period living in Europe. Consequently, in the fall of 1928, when I was eleven, my mother embarked with her sons on a small freighter for a ten day crossing to Plymouth, England. On the voyage we encountered a great storm. For three days waves smashed over the ship. The forward hatches were torn open, and then the steamer had to run with the storm until the seas subsided.

After disembarking at Plymouth, and a few weeks in Paris, my mother entered my brother Bill and myself in a pension, presided over by Monsieur and Madame Emile Cornu, in Lausanne, Switzerland. We were enrolled at Ecole Nauvelle, where we were placed in the regular classes with the French speaking boys. We were unable to comprehend the language sufficiently to grasp the subject matter of the courses. Monsieur Cornu complained to my mother that we should first spend our time learning French. His advice was followed that fall.

For the summer we joined our father at Roscoff in Brittany where he was working and measuring the intracellular pH of marine eggs at the Marine Laboratory there by micro-injecting pH indicators.

In Lausanne, Switzerland, at the Pension run by Monsieur and Madame Cornu, seated in the front. Ted is in the back row, just behind Madame Cornu.

In the fall we returned to the pension in Lausanne. Monsieur Cornu had been French tutor to the children of the Czar Nicholas II in Russia. Monsieur Cornu incited my sense of adventure with his stories of life in the Czar household, and with accounts of his hair breadth escapes from the revolutionary armies. He recounted these experiences to his guests at the evening meals. Monsieur Cornu worked at the prefecture of police. When a thunder and lightning storm threatened he would turn white, and "escape" to the basement of the prefecture to hide from the sound of the thunder claps over Lac Leman. He was unable to accept the reality that the thunder was not a bombardment by canon. In accordance with Monsieur Cornu's advice we spent most of our time that year learning French. At the end of the school year I could speak and understand French so well that I passed for a French boy.

In the fall of 1930 my father came to Cambridge, England, to work at the Strangeways Laboratory with Dr. Honor B. Fell. I was registered as a day student at the Perse School (a school established in the 1600's) in Cambridge. At this time my interest was concentrated on steam locomotives and model railroads. I spent an afternoon weekly at the home of Dr. Joseph Needham, (the prominent British embryologist and author of the 3 volume work: Chemical Embryology, 1931), operating with him his model trains in a room he had specially designed and equipped for this purpose.

My father and mother returned to the United States shortly after Christmas, and I was enrolled at Perse School as a boarding student. One day in the biology classroom the teacher drew the classes' attention to a section in the biology text which referred to my father, Robert Chambers, and which described his pioneering work in microsurgery. The teacher pointed to me as the son of this famous scientist. I, the new (and only) American student, had invariably been addressed by my British classmates as Al Capone, a notorious American gangster of the time. My teacher's reference to my father's achievements had a profound effect on me, rendered more intense by my homesickness. (This was the first time I had been separated from all my family members.) Rather than continuing to attempt the swaggering role of an American gangster surrounded by my retinue of protectors, that included a giant Sikh, I chose to live up to my father's name. I applied myself diligently to the Biology classwork. (Although this was four years before college, the material was presented at a level that I did not encounter until I later took a college level course for Biology majors after return to the USA.) Delving deeply into the subject matter of the course, interesting material emerged, whetting my interest. My identity with Biological Science began.

Ted (14 yr old) at Perse School, Cambridge, England, wearing the school blazer.

After returning to the USA in the spring, I joined my family at Woods Hole. My father helped me to set up a laboratory in the attic of our Bobtucket home. This was used for mounting insects, pressing plants, and for the identification of all manner of fauna and flora, including the teeming microscopic life of ponds and brackish swamps. For the early morning hours I set forth just before dawn nearly every morning to discover what birds and nests I could find. This devotion to Natural History preoccupied me during the summers of my high school years.

For the winter months, the family returned to New York City, where I was enrolled in Friends Seminary at Stuyvesant Square, N.Y.C. Among my after school activities, during the winter, my father's colleagues at N.Y. University (where my father was then chairman of the Department of Biology) arranged for me to take the biology course for college sophomores. This was a remarkably rigorous course, and I took all the laboratories and examinations. Nonetheless, I managed to come out at the top of the class. Meanwhile, at Friends Seminary I was taking the biology class there. I walked about as a know-it-all. The science teacher, Mr. Hinman, irked to some extent by this behavior decided to put me in charge of running the laboratory experiments for a week. The result was a disaster: none of my experiments worked. The virtue of humility was learned. Becoming restless with the tempo of the classwork, I completed high school in three years.

In accordance with the family tradition I entered Princeton University in the fall of 1933. (My grandfather, Robert Chambers, and his two brothers were graduates of the Princeton Seminary. My father had not followed this pattern, but graduated from Robert College in Istanbul, Turkey. This was because my grandfather was a Presbyterian minister and educator in Erzroom, in the eastern or Armenian sector of Turkey, where my father was born and spent his youth.)

I was particularly fortunate that when I went to Princeton it was still possible to attend the lectures delivered in the upper class biology course by Dr. Edwin G. Conklin (1863-1952) who must be included among the most eminent figures in American Biology. I recall sitting spellbound listening to his lectures on embryology and evolution. I spent the better part of the following summer at Woods Hole meticulously sketching each stage in the sequence of the mosaic development of the egg of the Ascidian, *Styela*, repeating the observations and following the segregation of the yellow crescent described by Conklin in his monumental monograph on mosaic development in the Ascidian egg (Conklin 1905).

At Princeton a series of courses entitled "Birds of the World" was taught by Mr. Rogers, the professor of ornithology. Through all of my years at Princeton I attended each of these courses together with one or two other interested students. We accompanied Mr. Rogers on frequent field trips through the swamps, fields, and woodlands near Princeton including trips to the New Jersey pine barrens and coastline. When it came time to select a topic for senior thesis Mr. Rogers was intensely disappointed that I did not elect to do a thesis with him. Instead I elected to do my thesis with Dr. E. Newton Harvey (discoverer of the centrifuge microscope and the luciferin-luciferase system) and his former student, Dr. Frank Johnson, studying the mechanism of the effects of narcotics on bacterial luminescence (Johnson and Chambers 1939).

The summer following my junior year at Princeton (1937), my father offered me a job as assistant to him and his associate, Dr. Milan J. Kopac, to work with them at the Dry Tortugas Laboratory of the Carnegie Institution of Washington. The laboratory was reached by an all day trip, 90 miles southwest from Key West, Florida, on a steam-powered launch, the Anton Dohrn (named after the illustrious first director of the Naples Marine Biological Station). The laboratory was located on Loggerhead Key, and was surrounded by coral reefs. This was my first exposure to the teeming marine life of the tropics, and the tropical birds-frigate birds, noddy terns, boobies.

My role in the laboratory was to denude sea urchin eggs of their extraneous coats at intervals after insemination. The fertilization membranes were removed as the membranes rose, and the hyaline layer washed off the surfaces of the eggs using isotonic salt solutions (of which 0.5 M KCl proved the most effective). Dr. Kopac and my father sitting at their microdissection instruments, mounted samples of the denuded eggs under the microscope repeatedly testing their ability to coalesce with expanding oil drops expelled from a micropipette. These experiments fascinated me, because the ability of the naked egg to fuse with an oil drop indicated the liquid nature of the protoplasmic surface film (or plasma membrane) (Chambers, R. and Kopac 1937a,b).

As a respite from the tropical heat, at midday everyone disrobed and walked about 200 feet to the opposite side of the key to swim. Needless to say at this time it was an all male lab. A somewhat discordant note was that the wife of the lighthouse keeper, a quarter of a mile away, complained about the scientists' skinny dipping. The ritual dip, secluded by a grove of palms, could only have been observed with the aid of field glasses from the top of the lighthouse.

For my spare time my father assigned to me the problem of mapping the paths of the male and female pronuclei in the highly transparent egg of the sea urchin, *Lytechinus variegatus*. A camera lucida had been found in the laboratory's storeroom. The "camera" superimposes the image of a sheet of drawing paper on the table top adjacent to the microscope on the image of the egg in the field of view of the microscope. Using this device, the positions of the pronuclei could be plotted. The path generated by the egg pronucleus, moving toward the center of the sperm aster, which itself is moving toward the center of the egg, resembles that of a celestial body moving under the influence of a gravitational force. The work resulted in my first paper (Chambers, E.L. 1937, 1939) After returning from the Dry Tortugas, I spent August at the Mount Desert Island Biological Station, Maine, working with the eminent cytologist and histologist, Dr. Ulric Dahlgren, in an attempt to isolate the luminescent organs of fireflies. The idyllic summer ended in late August by a phone call from Woods Hole. My mother and father were involved in a serious automobile accident, from which my mother never fully recovered.

After graduating from Princeton in 1938 I spent the summer working in my father's laboratory. In the oil coalescence studies mentioned earlier, but using *Arbacia punctulata* eggs at Woods Hole, it was noted that when the inseminated eggs were immersed in the 0.5 M KCl wash solution between 3 and 10 min after insemination, the eggs would develop to advanced cleavage stages, but when immersed before or after this period their development was completely arrested, as if the zygotes had been thrown into a fixative at the stage attained when immersed. My next project was to characterize this phenomenon. The experiments involved examining the effects of an almost infinite variety of different ionic media. Much of the inspiration for these experiments came from the classical experiments of Jaques Loeb (1859-1924) on the effects of ions on the development of the *Fundulus* egg.

The experiments that summer came to an abrupt halt. I was working on the 3rd floor, NE corner of the Crane wing of the Lillie building overlooking the Eel Pond. For some hours after the start of the experiment the wind seemed to be rising until it was blowing with uncommon force. I was obliged to keep to my counting of the eggs. Finally, when a break occurred, I glanced out the window and was astonished to see that the level of the water in Eel Pond had risen well over the top of the wall. I have in my notes for this experiment, dated September 21, 1938: "Experiment interrupted at 4:30 PM, due to Hurricane." This was the Hurricane which devastated the southern New England coast and caused the loss of several lives at Woods Hole.

MEDICAL SCHOOL AND WORLD WAR II YEARS

In the fall of 1939 I entered the first year class of New York University Medical School, where I was awarded the Haydn scholarship. The basic science courses at N.Y.U. were rigorous, heavily laden with laboratory work and were designed both for M.D. students and Ph.D. candidates. My working plan was to complete the first two years of medical school and then complete a research thesis. As the sophomore year progressed, war being imminent, all students in medical school were registered as members of the military forces. There were no options except to complete medical school, internship and enroll as an officer in one of the military services. The curriculum was accelerated, vacations minimized and I had little opportunity for research. I did manage some experimentation in the evenings studying irreversible shock in dogs subjected to a regime of bleeding, and the capillary bed in the mesoappendix of rats, in conjunction with Dr. Benjamin W. Zweifach and my father. After graduating from Medical School in April, 1943, I completed a nine-months' rotating internship in medicine, pathology and surgery, then a six months' residency in medicine, all at Bellevue Hospital, N.Y.C. By the time of my residency the enrollment of doctors in the military services had markedly depleted the hospital attending staffs, and I found myself having to assume responsibilities on the wards which strained to the limit the capabilities of my training.

I was thankful for having experienced such responsibilities. Several months after completing my residency, I found myself commandant of a 100-bed hospital at a camp near Phillippeville (named after the Roman Emperor, Augustus Phillippe, now called Skikda), Algeria. (I was a second lieutenant in the Navy assigned to the U.S. Public Health Service, and the United Nations Rehabilitation Administration.) One of my many functions was to treat prisoners from concentration camps in Germany, who were sent through occupied France and across the Mediterranean in shiploads of 500-600 at a time. For medical supplies I had to depend almost entirely on items captured from the defeated Nazi armies, and the day-to-day functions of the hospital were conducted largely in French.

During the push of the Allied forces northward through Italy I was put in charge of an even larger hospital in Apulia, situated at Santa Maria di Leuca at the extreme tip of the heel of Italy. This was a hospital for refugees of every conceivable nationality. During the closing days of the war we took care of many Russians, who had retreated with the Nazi forces, and then sought escape from the Red armies by fleeing south, in the hopes of obtaining passage to Africa. My nurses at Santa Maria di Leuca were mainly Italian nuns, most of whom spoke some French, luckily for me. I was assisted in surgical operations by a Russian and a Turk doctor, neither of whom spoke a word of English, but each could speak a few words of French.

The hospital staples were flour, corn meal, and dried beef. To alleviate the tedium of the resulting menu, every few days I would climb into the camp ambulance with the Apulian driver, Gino, and we would drive out into the nearby fields to forage for what vegetables or fruits could be found. Gino would negotiate the transactions with the peasants by bartering an exchange using sacks of flour from the hospital commissary.

In November, 1946, I was furloughed to Washington, D.C., where I was reassigned as hospital commandant to a refugee camp in Greece, where a communist rebellion threatened. These assignments were challenging in the extreme and all consuming, hence I found it difficult to pull myself out to follow my intended career of research in biology. Unable to make a sensible decision, in Washington, I finally flipped a coin–tails back to research, heads to Greece. The coin came up tails.

RETURN TO RESEARCH

Returning home, I consulted with my father, and his associate, Milan J. Kopac, who had become a close friend and mentor. Dr. Kopac, who urged me to strengthen my background in quantitative biology, advised me to obtain a fellowship with Dr. Sumner C. Brooks (1888-1948) at Berkeley, California. Dr. Brooks, a physical-chemical biologist, had pioneered in the use of radioactive tracers to study ion movements across cell membranes and cell permeability. He welcomed the proposal that I come to his laboratory, and he sponsored me for a Porter fellowship of the American Physiological Society, which in due time was awarded to me. At Berkeley, I stored the sea urchins I used, *Stronglylocentrotus purpuratus*, and *S. franciscanus*, in cages I built and hung under a dock beneath the Golden Gate Bridge in the Presidio in San Francisco. It was a memorable occasion to collect the *S. Purpuratus* on the reefs along the rugged shoreline south of San Francisco. The animals were obtained by rushing out to the surf line as a wave receded, chopping the animals out of their burrows in the rock with a geological pick, then clambering back up the reef before the next wave broke.

My first project was to examine the effect of fertilization on the rate of exchange of sodium across the egg membrane. At this time (1946) the $^{24}Na^{+}$ isotope was prepared by placing the non-radioactive precursor in a large aluminum chamber several feet across, and then bombarding the chamber in the cyclotron. After the bombardment was completed we would wait for the radioactivity of the chamber to diminish to what was then considered to be a reasonable level. I would then hoist the chamber to my shoulder and carry it across the campus to Dr. Brooks' laboratory. The chamber was then washed out and heavy metal contaminants removed. After adding an aliquot of the presumed $^{24}Na^{+}$ to a suspension of sea urchin eggs, and then inseminating them, we were astonished to find that the radioactive tracer rapidly accumulated in the fertilized eggs to levels greater than in sea water. Since the intracellular concentration of sodium in marine species is virtually always less than in the external medium, this result was entirely unexpected. The mystery was soon solved by the finding that the half-life of the tracer that had accumulated within the eggs far exceeded that of $^{24}Na^{+}$. The half-life corresponded to that of ^{32}P for the phosphate ion, a radioactive isotope which must have been created during the cyclotron bombardment and was present as a contaminant with the $^{24}Na^{+}$. This unexpected finding led to a series of papers on the accumulation of phosphate following fertilization in sea urchin eggs, and resulted in an extraordinarily fruitful collaboration with Dr. Arthur H. Whiteley, at first at the William B. Kerckhoff Marine Laboratory of Cal Tech, and later at the Friday Harbor Marine Laboratory of the University of Washington (this work and publications in this area date from 1948 to 1966). An extension of these investigations involved measurement of the content of the high energy phosphate compounds in collaboration with Dr. Thomas J. Mende, whose superb grasp of organic chemistry spelled the success of these endeavors (publications in 1953).

The phosphate carrier system in the fertilized egg is capable of extracting the orthophosphate anion from sea water to levels below 0.06 μM. This is a capability which is shared by among the most effective phosphate concentrating organisms known, the diatoms of the ocean plankton. Yet activation of the egg effects complete differentiation of this remarkable system, absent in the unfertilized egg, by 15 to 30 min following insemination. Further investigation of how this system is evolved could provide important insights into the mechanism of egg activation.

During the years at Berkeley from 1946 to 1949 (curtailed by the death of Dr. S.C. Brooks in 1947 while he was working at the Bermuda Biological Station) an enormous effort was devoted to measuring the rates of exchange of the K^{+} ion before and after fertilization using $^{42}K^{+}$ as a tracer. We showed that the rate of exchange of K^{+} in the unfertilized sea urchin egg occurs exceedingly slowly, while following fertilization the rate increases dramatically (10- to 16-fold after a brief lag period). The same result was obtained whether we measured the loss of $^{42}K^{+}$ from prelabelled eggs, or its uptake. Initially (Chambers *et al.* 1948) we proposed that the slow exchangeability of intracellular K^{+} in the unfertilized egg was due to the presence of a large inexchangeable K^{+} fraction, but that following fertilization this fraction became exchangeable. This proposal was based on our early finding that in unfertilized eggs further uptake of $^{42}K^{+}$ ceased after 20% of the intracellular K^{+} had exchanged (equilibrium had seemingly been attained). However, later we found that the presumed state of equilibrium observed in the unfertilized eggs after about 18h incubation with the isotope, was the consequence of a simultaneous slow loss of total intracellular K^{+}, as $^{42}K^{+}$ continued to enter the eggs. Consequently, the concept of a non-exchangeable intracellular K^{+} fraction in the unfertilized egg was abandoned (Chambers, E.L., and Chambers, R. 1949, see also Steinhardt *et al.* 1971; Robinson 1976).

However, in the light of the work I have done using $^{42}K^{+}$ over the years, using both single and mixed batches of sea urchin eggs (Chambers 1949, 1975), further investigation is needed to rule out the possibility that K^{+} containing compartments may exist within the egg which exchange K^{+} with the external medium at significantly different rates (presumably first exchanging with the continuous aqueous phase of the cytoplasm).

After leaving Berkeley, I spent a year at the MBL, Woods Hole, Massachusetts, working on the biochemistry of high energy phosphate compounds in echinodem eggs in conjunction with Dr. T. J. Mende (referred to earlier). We were able to do this because in the fall we brought down from Boothbay Harbor, Maine, a load of the circumpolar sea urchin, *Strongylocentrotus dröbachiensis*, and stored the animals in cages under the MBL dock in Great Harbor. The warmer winter waters at Woods Hole permitted ripening of the gonads of these animals.

Subsequently, after two years in the Anatomy Department at Johns Hopkins Medical School, I spent a year with my father writing the book, *Explorations into the Nature of the Living Cell* (Chambers and Chambers 1961); then a year in the Anatomy Department, University of Oregon, Portland, Oregon. When the opportunity presented itself to shift my discipline from Anatomy to Physiology, an area closer to my research, I accepted an opening in the Department of Physiology at the recently established School of Medicine, University of Miami, Florida. A further attraction was the closeness of the sea, the abundant marine life of the coral reefs, and the availability of several different sea urchin species (the same as those with which I had previously become acquainted at the Dry Tortugas Laboratory), one or another of which was ripe with gametes throughout the year.

The experience in Oregon, however brief, was particularly eventful, since it was here that I met and married the lovely Oregonian, Eleanora (*née*) Strasel. As a painter Eleanora understands the demands of time and energy involved in creative work. We left Oregon for Miami in October, 1954, and spent the month *en route* camping in the mountains of eastern California and the deserts of the southwest.

In Miami, although continuing work on phosphate uptake by the eggs, the emphasis was shifted from studies on K^+, to Na^+. The exchange of Na^+ was found to greatly increase following fertilization, even to a greater extent than for K^+. I was puzzled, however, by the finding that in $^{24}Na^+$ uptake studies, instead of seeing a steady uptake following fertilization, after an initial increase for the first 5 to 10 min, the $^{24}Na^+$ content then decreased transiently, following which the steady increase resumed. I was skeptical of this result because, to make the measurements, it was necessary to wash the eggs free of external isotope (a consequence both of the 20 fold lower intracellular Na^+ concentration compared to the external medium, and the small extent of isotope exchange in the eggs during the early uptake). My concern was that the washing procedure might have altered the intracellular $^{24}Na^+$ content.

Analyses of the total Na^+ content of the eggs were then carried out. The dangers incumbent on washing the eggs free of external Na^+ were minimized by rapidly diluting the egg samples with choline$^+$ (Na^+ free) sea water, immediately centrifuging the eggs, followed by decanting the supernatant as rapidly as possible (this experiment taxed one's manual dexterity to the limit!). The Na^+ content of the eggs was then calculated by subtracting the Na^+ content of the interstitial fluid included with the eggs. The results clearly showed that the intracellular content of Na^+ increased 20% following fertilization, but then decreased by 10 to 15 min to values close to that of the unfertilized eggs. This was a change consistent with the findings obtained using $^{24}Na^+$. I felt this was an important finding, and sought to publish the results in the *J. Cellular Physiology*, also in the *J. General Physiology*, but the reviewers rejected the manuscript as being of no significant interest (this is why my data is summarized only in an abstract: Chambers 1972).

The next step in the "Na^+ saga" was to find if the transient Na^+ increase might have a functional role. Accordingly the effect of immersing fertilized eggs at different stages of development in choline$^+$ sea water containing little or no Na^+ was examined. An important reason I felt impelled to go forward in this direction was my early experiments using KC1 (see above) where it was deeply impressed on me that the effect of an altered ionic environment depends on the developmental stage of the fertilized egg. In this way the essentiality of external Na^+ during the first mins after fertilization was demonstrated (Chambers 1975b, 1976; Chambers and Dimich 1975). Within a matter of weeks, after the first abstract appeared, Dr. Dave Epel's laboratory had carried out experiments which established that the extracellular Na^+ is required for exchange with intracellular H^+, accompanied by alkalinization of the cytoplasm (Johnson *et al.* 1976).

Studies using the isotope $^{45}Ca^{2+}$ were also carried out together with my graduate student, Roobik Azarnia (Azarnia and Chambers 1969, 1970, 1976; Chambers *et al.* 1970). From these investigations we were early on convinced that fertilization in the sea urchin egg is accompanied by the release of Ca^{2+} ions from an intracellular store of bound or sequestered calcium. This was further supported by the demonstration of the activation of sea urchin eggs by Ca ionophores (Chambers and Rose 1974, Chambers *et al.* 1974), a finding made coincidentally with Steinhardt and Epel 1974). A noteworthy feature of the Ca ionophore results is that virtually alone in all the work I've done, the experiment worked to perfection the very first time! At the peak of these investigations came the elegant, direct demonstration by Rick Steinhardt's laboratory of the intracellular release of free Ca^{2+} ions following fertilization in the sea urchin egg (Steinhardt *et al.* 1977).

An interesting extension of the studies on the Ca^{2+} ion is the subsequent experimental work which showed that external Ca^{2+} is not required for activation of the egg by the sperm

(Chambers 1980; Chambers and Angeloni 1981), or by microinjection of inositol trisphosphate (Crossley *et al.* 1988).

Several years after I came to the University of Miami, the University authorized for the first time, in 1959, the offering of the Ph.D. degree in qualifying departments. A group of us, in particular Dr. Walter B. Dandliker, of the Biochemistry Department, Dr. Ray Iverson and Dr. Howard B. Lenhoff of the Biology Department, and myself had conceived of the idea that a strong Ph.D. program could be offered by pooling those faculty members conducting active research in the areas of Cellular and Molecular Biology. At this time, the departments in the Biological Sciences, considered individually, lacked the resources to offer an interesting program, but on a University wide basis the excellence of the faculty group that could be amassed was remarkable. In due time (1959) the Ph.D. in Cellular and Molecular Biology (the Cell Program) was offered through the departments of Biochemistry, Biology, and Physiology, with the above mentioned committee of four in charge and myself, as director. The Cell Program recruited, admitted and supervised the progress of the students, and ran the courses in Cellular and Molecular Biology that the student took. For the first year a full time course was taught with extensive laboratory exercises, which covered the chemistry, structure, and function of the cell. This was followed by a series of single topic seminars which rotated over a three year period. The faculty, which numbered about thirty members, applied themselves to the challenge of this interdisciplinary endeavour with an astonishing *esprit de corps*. For the students the Cell Program proved to be highly attractive.

To orchestrate these proceedings, as cooperative and supportive as were the faculty members, directing the Cell Program proved immensely time-consuming, and activity I continued for 12 years (until 1971). In regard to my research, I could have regretted the time I spent on the Cell Program, but unquestionably it proved stimulating to the participating faculty, and played an important role in attracting to the basic biological sciences in the Medical School an active and gifted group of research scientists. Moreover, a review of Ph.D. programs offered in the U.S.A. revealed that our Cell Program was a prototype for others that were later developed elsewhere in the country.

I now devoted myself with additional vigor to experimental work. One of the shortcomings of studying ion fluxes during fertilization by means of tracer methodology is the limited time resolution achievable, accurate only to within several minutes by the methods I used. To secure improved time resolution and a new perspective on the activation of the egg, I resolved to extend my work on ion fluxes to studies on the electrophysiology of the egg. (Additionally it had been frustrating to attempt to handle the kinetics of ion fluxes using tracers in multicompartmental systems, a prominent feature for both the Na^+ and Ca^{2+} ions). This shift of emphasis was, of course, a natural extension of my long background working on the role of ions during fertilization and early development. It was my good fortune at this time that Dr. Werner Loewenstein came (in 1971) as the new Chairman of the Department of Physiology and Biophysics. Dr. Loewenstein, eminent for his discovery of permeable junctions between cells in tissues, and his extensive investigations on cell communication, brought to the department investigators thoroughly versed in electrophysiology and related areas. It was of inestimable value to have a consulting relationship with these faculty members. An early and most fruitful collaboration in the department was with Dr. Birgit Rose, involving the activation of eggs by Ca ionophores (referred to earlier).

The reasons why I decided to concentrate my efforts on voltage clamping the membrane potential of eggs during fertilization, and the early efforts in this direction, are summarized in the introduction to my article which follows. I am deeply indebted to Dr. John Lynn, who spent three years in my laboratory, and Dr. Dave McCulloh who has been working here since 1983. The unique contribution of each, working in collaboration with me (see my bibliography from 1982 on), and individually, made possible the advance of the work unraveling the phenomena associated with voltage clamping the membrane potential of eggs and oocytes during fertilization, together with the related aspects of sperm entry and activation: the subject matter of the first three chapters.

REFERENCES (For references to E. L. Chambers, see my bibliography below)

Chambers, R. and M. J. Kopac. 1937a. The coalescence of living cells with oil drops. I. *Arbacia* eggs immersed in sea water. *J. Cell. Comp. Physiol.* 9:331-343.

Chambers, R., and M. J. Kopac. 1937b. The coalescence of sea urchin eggs with oil drops. *Annu. Rep. Tortugas Lab. Carnegie Inst. Washington* 36:88-90.

Conklin, E. G. 1905. The organization and cell lineage of the *Ascidian* egg. *J. Acad. Nat. Sci.*, Second series, 8, Part 1:1-119.

Crossley, I., K. Swann, E. L. Chambers, and M. Whitaker. 1988. Activation of sea urchin eggs by inositol phosphates is independent of external calcium. *Biochem. J.* 252:257-262.

Johnson, J. D., D. Epel and M. Paul. 1976. Intracellular pH and activation of sea urchin eggs after fertilization. *Nature* 262:661-664.

Robinson, K. R. 1976. Potassium is not compartmentalized within the unfertilized sea urchin egg. *Dev. Biol.* 48:466-472.

Steinhardt, R. A. and D. Epel. 1974. Activation of sea urchin eggs by a calcium ionophore. *Proc. Nat. Acad. Sci. USA.* 71:1915-1919.

Steinhardt, R. A., L. Lundin, and D. Mazia. 1971. Bioelectric responses of the echinoderm egg to fertilization. *Proc. Natl. Acad. Sci. USA.* 68:2426-2430.

Steinhardt, R. A., R. Zucker and G. Schatten. 1977. Intracellular Ca^{2+} release at fertilization in the sea urchin egg. *Dev. Biol.* 58:185-196.

BIBLIOGRAPHY - Edward L. Chambers

Chambers, E. L. 1937-1938. Movement of the egg nucleus in relation to the sperm aster in the eggs of *Lytechinus variegatus. Annu. Rep. Tortugas Lab. Carnegie Inst. Washington* 36:86.

Chambers, E. L. and R. Chambers. 1938. The resistance of fertilized *Arbacia* eggs to immersion in KCl and NaCl solutions. *Biol. Bull.* 75:356.

Chambers, E. L. and R. Chambers. 1939. The suspension of activity in the fertilized *Arbacia* egg by early immersion in KCl. *Anat. Rec.* 75:44.

Chambers, E. L. 1939. Movement of the egg nucleus in relation to the sperm aster in the echinoderm egg. *J. Exp. Biol.* 16:409-424.

Johnson, F. H. and E. L. Chambers. 1939. Oxygen consumption and methylene blue reduction in relation to barbital inhibition of bacterial luminescence. *J. Cell. Comp. Physiol.* 13:263-267.

Chambers, R. and E. L. Chambers. 1940. Interrelations between egg nucleus, sperm nucleus and cytoplasm of the *Asterias* egg. *Biol. Bull.* 79:340.

Duncan, G., W. C. Hyman, and E. L. Chambers. 1943. Indirect method for the determination of blood pressure in rats. *J. Lab. Clin. Med.* 28:886-890.

Chambers, E. L. 1946. The effect of cations on the resistance of fertilized *Arbacia* eggs to the potassium ion. *Anat. Rec.* 94:372.

Chambers, E. L., W. White, N. Jeung, and S. C. Brooks. 1948. Penetration and effects of low temperature and cyanide on penetration of radioactive potassium into the eggs of *Strongylocentrotus purpuratus* and *Arbacia punctulata. Biol. Bull.* 95:252-253.

Brooks, S. C. and E. L. Chambers. 1948. Penetration of radioactive phosphate into the eggs of *Strongylocentrotus purpuratus, S. franciscanus*, and *Urechis caupo. Biol. Bull.* 95: 262-263.

Chambers, E. L., A. Whiteley, R. Chambers, and S. C. Brooks. 1948. Distribution of radioactive phosphate in the eggs of *Lytechinus pictus. Biol. Bull.* 95:263.

White, W. E. and E. L. Chambers. 1949. Le determination quantitative et l'isolement du phosphore labile sous forms d'adenosinetriphosphate dans les oeufs d'oursin. *Rev. Pathol. Comp. Hyg. Gen.* 613:647-648.

Chambers, E. L. and R. Chambers. 1949. Conference on problems in general and cellular physiology related to fertilization. III. Ion exchanges and fertilization in echinoderm eggs. *Am. Nat.* 83:269-284.

Chambers, R. and E. L. Chambers. 1949. Nuclear and cytoplasmic interrelations in the fertilization of the *Asterias* egg. *Biol. Bull.* 96:270-282.

Chambers, E. L. and W. E. White. 1949. Accumulation of phosphate and evidence for synthesis of adenosine triphosphate in the fertilized sea urchin egg. *Biol. Bull.* 97:225-226.

Chambers, R., E. L. Chambers, and L. M. Leonard. 1949. Rhythmic alterations in certain properties of the fertilized *Arbacia* egg. *Biol. Bull.* 97:233-234.

Chambers, E. L. 1949. The uptake and loss of ^{42}K in the unfertilized and fertilized eggs of *S. purpuratus* and *A. punctulata. Biol. Bull.* 97:251-252.

Kao, C., R. Chambers, and E. L. Chambers. 1951. The internal hydrostatic pressure of the unfertilized *Fundulus* egg activated by puncture. *Biol. Bull.* 101:210.

Chambers, R., E. L. Chambers, and C. Kao. 1951. The internal hydrostatic pressure of the unfertilized activated *Fundulus* egg exposed to various experimental conditions. *Biol. Bull.* 101:206.

Chambers, E. L. and T. J. Mende. 1953. Alterations of the inorganic phosphate and arginine phosphate content in sea urchin eggs following fertilization. *Exp. Cell Res.* 5:508-519.

Chambers, E. L. and T. J. Mende. 1953. The adenosine triphosphate content of unfertilized and fertilized eggs of *Asterias forbesii* and *Stronglocentrotus drobachiensis. Arch. Biochem. Biophys.* 44:46-56.

Chambers, R. and E. L. Chambers. 1953. Die Wirkung von Na-, K- und Ca- Salzen auf den physikalischen Zustand des Protoplasmas. *Arzneim. - Forsch.* 3:322-325.

Mende, T. J. and E. L. Chambers. 1953. The occurrence of arginine phosphate in echinoderm eggs. *Arch. Biochem. Biophys.* 45:105-116.

Kao, C. Y., R. Chambers, and E. L. Chambers. 1954. Internal hydrostatic pressure of the *Fundulus* egg. II. Permeability of the chorion. *J. Cell. Comp. Physiol.* 44:447-461.

Brooks, S. C. and E. L. Chambers. 1954. The penetration of radioactive phosphate into marine eggs. *Biol. Bull.* 106:279-296.

Chambers, E. L. and W. E. White. 1954. The accumulation of phosphate by fertilized sea urchin eggs. *Biol. Bull.* 106:297-307.

Chambers, E. L. and D. Hancock. 1955. Reactions of ameba to the microinjection of adenine nucleotides. *Am. J. Physiol.* 183:602-603.

Mende, T. J. and E. L. Chambers. 1957. Reaction of tissues with Schiff's reagent after treatment with anhydrous chromyl chloride. *J. Histochem. Cytochem.* 5:606-610.

Mende, T. J. and E. L. Chambers. 1957. Distribution of mucopolysaccharides and alkaline phosphatase in transitional epithelia. *J. Histochem. Cytochem.* 5:99-104.

Mende, T. J. and E. L. Chambers. 1958. Studies on solute transfer in the vascular endothelium. *J. Biophys. Biochem. Cytology.* 4:319-322.

Chambers, E. L. 1960. The effect of fertilization on the exchange of potassium in the sea urchin egg. p. 163. *In: Xe Congres International de Biologie Cellulaire.* Expansion scientifique Francaise, Paris.

Whiteley, A. H. and E. L. Chambers. 1960. The differentiation of a phosphate transport mechanism in the fertilized egg of the sea urchin. p. 387-401. *In: Symposium on the Germ Cells and Earliest Stages of Development.* Fondazione A. Baselli, Milano (1961).

Chambers, R. and E. L. Chambers. 1961. *Explorations into the Nature of the Living Cell.* pp 1-352. Harvard University Press, Boston.

Chambers, E. L. 1963. Role of cations in phosphate transport by fertilized sea urchin eggs. Fed. Proc. 22:331.

Chambers, E. L. and A. H. Whiteley. 1966. Phosphate transport in fertilized sea urchin eggs. I. Kinetic aspects. *J. Cell. Physiol.* 68:289-306.

Whiteley, A. H. and E. L. Chambers. 1966. Phosphate transport in fertilized sea urchin eggs. II. Effects of metabolic inhibitors and studies on differentiation. *J. Cell. Physiol.* 68:309-323.

Chambers, E. L. 1966. The apparent second and third dissociation constants, pK_2 and pK_3 of phosphoric acid in sea water. *J. Cell. Physiol.* 68:306-308.

Chambers, E. L. 1966. Effects of chloromycetin, and of hydrolytic enzymes on phosphate uptake in the fertilized sea urchin egg. *Physiologist* 9:151.

Chambers, E. L. 1968. Exchange of Na^+ following fertilization of sea urchin eggs. Excerpta Med., Int. Congr. Ser. 166:42-43.

Azarnia, R. and E. L. Chambers. 1969. Effect of fertilization on the uptake and efflux of calcium-45 in the eggs of *Arbacia punctulata. Biol. Bull.* 137:391-392.

Krischer, K. N. and E. L. Chambers. 1970. Proteolytic enzymes in the sea urchin egg: characterization, localization, and activity before and after fertilization. *J. Cell Physiol.* 76:23-36.

Chambers, E. L., R. Azarnia, and W. E. McGowan. 1970. The effect of temperature on the efflux of ^{45}Ca from the eggs of *Arbacia punctulata. Biol. Bull.* 139:417-418.

Azarnia, R. and E. L. Chambers. 1970. Effect of fertilization on the calcium and magnesium content of the eggs of *Arbacia punctulata. Biol. Bull.* 139:413-414.

Chambers, E. L. 1972. Effect of fertilization on the Na^+ and K^+ content, and the flux of Na^+ in the sea urchin egg. *Physiologist* 15:103.

Jobsis, F. F. and E. L. Chambers. 1972. Respiratory chain in sea urchin eggs and embryos. *Physiologist* 15:182.

Chambers, E. L., B. C. Pressman, and B. Rose. 1974. The activation of sea urchin eggs by the divalent ionophores A 23187 and X-537 A. *Biochem. Biophys. Res. Comm.* 60:126-132.

Chambers, E. L. and B. Rose. 1974. The activation of sea urchin eggs by Ca^{2+} ionophores. *J. Gen. Physiol.* 64:2a.

Chambers, E. L. 1974. Effects of ionophores on marine eggs and cation requirements for activation. *Biol. Bull.* 148:471.

Chambers, E. L. and R. A. Dimich. 1975. A Na^+ requirement for cytoplasmic and nuclear activation of sea urchin (*Lytechinus variegatus*) eggs by sperm and by divalent ionophores. *J. Gen. Physiol.* 66:9a.

Chambers, E. L. 1975a. Potassium exchange in unfertilized sea urchin (*Arbacia punctulata*) eggs. *Biol. Bull.* 149:442-443.

Chambers, E. L. 1975b. Na^+ is required for nuclear and cytoplasmic activation of sea urchin eggs by sperm and divalent ionophores. *J. Cell. Biol.* 67:60a.

Chambers, E. L. 1976. Na^+ is essential for activation of the inseminated sea urchin egg. *J. Exp. Zool.* 197:149-154.

Azarnia, R. and E. L. Chambers. 1976. The role of divalent cations in activation of the sea urchin egg. I. Effect of fertilization on divalent cation content. *J. Exp. Zool.* 198:65-77.

Mackenzie, D. O. and E. L. Chambers. 1977. Fertilization of voltage clamped sea urchin eggs. *Clin. Res.* 25:643A.

Chambers, E. L. and J. de Armendi. 1977. Electrophysiological studies of fertilization in the sea urchin egg. *J. Gen. Physiol.* 70:3a-4a.

Chambers, E .L. and J. de Armendi. 1977. Membrane potential, activation potential and activation current of the sea urchin egg. *Physiologist* 20:15.

Chambers, E. L. and J. de Armendi. 1978. Cortical reaction, cytoplasmic and nuclear activation and activation potential of sea urchin eggs in low Na^+ media. *J. Cell Biol.* 79:162a.

Hinkley, R. E. and E. L. Chambers. 1978. Anaesthetic-induced inhibition of cell cleavage: reversible inhibition of contractile ring assembly. *J. Cell Biol.* 79:265a.

Chambers, E. L. and J. de Armendi. 1979. Membrane potential, action potential, and activation potential of eggs of the sea urchin, *Lytechinus variegatus. Exp. Cell Res.* 122:203-218.

Chambers, E. L. and R. E. Hinkley. 1979. Non-propagated cortical reactions induced by the divalent ionophore A 23187 in eggs of the sea urchin, *Lytechinus variegatus. Exp. Cell Res.* 124:441-446.

Chambers, E. L. 1980. Fertilization and cleavage of eggs of the sea urchin *Lytechinus variegatus* in Ca^{2+} free sea water. *Eur. J. Cell Biol.* 22:476.

Chambers, E. L. and S. V. Angeloni. 1981. Is external Ca^{2+} required for fertilization of sea urchin eggs by acrosome reacted sperm? *J. Cell Biol.* 91:181a.

Chambers, E. L. and S. V. Angeloni. 1981. Effects of diltiazem, a water soluble inhibitor of Ca^{2+} influx, on exocytosis of cortical granules in fertilized sea urchin eggs. *J. Gen. Physiol.* 78:14a.

Hinkley, R. E. and E. L. Chambers. 1982. Structural changes in dividing sea urchin eggs induced by the volatile anaesthetic, halothane. *J. Cell Sci.* 55:327-339.

Chambers, E. L. and J. W. Lynn. 1982. Sperm-egg interaction and development of voltage-clamped eggs of the sea urchin, *Lytechinus variegatus. J. Gen. Physiol.* 80:12a-13a.

Lynn, J. W. and E. L. Chambers. 1982. Sperm entry and development of voltage-clamped sea urchin eggs with or without vitelline envelopes. *J. Cell Biol.* 95:154a.

Chambers, E. L. and J. W. Lynn. 1983. Current patterns generated during sperm-egg interaction in voltage clamped eggs of the sea urchin, *Lytechinus variegatus. Biophys. J.* 41:130a.

Lynn, J. W. and E. L. Chambers. 1983. Ion substitution studies on inseminated voltage clamped eggs of the sea urchin, *Lytechinus variegatus. J. Cell Biol.* 97:25a.

Lynn, J. W., E. L. Chambers, and R. E. Hinkley. 1983. Studies on voltage clamped sea urchin eggs. *J. Cell Biol.* 97:25a.

Lynn, J. W. and E. L. Chambers. 1984. Voltage clamp studies of fertilization in sea urchin eggs. I. Effect of clamped membrane potential on sperm entry, activation, and development. *Dev. Biol.* 102:98-109.

McCulloh, D. H. and E. L. Chambers. 1984. Capacitance increase following insemination of voltage-clamped sea urchin eggs. *Biophys. J.* 45:73a.

McCulloh, D. H. and E. L. Chambers. 1985. Localization and propagation of membrane conductance changes during fertilization in eggs of the sea urchin, *L. variegatus. J. Cell Biol.* 101:230a.

Lynn, J. W. and E. L. Chambers. 1985. Sperm entry and fertilization envelope elevation are inhibited in sea urchin eggs voltage clamped at -90 and -100 mV. *J. Cell Biol.* 101:230a.

McCulloh, D. H., P. I. Ivonnet, and E. L. Chambers. 1985. Cooperativity of sperm related to sperm entry in immature oocytes of the sea urchin, *Lytechinus variegatus. J. Cell Biol.* 101:230a.

Chambers, E. L. 1985. Fertilization and activation in voltage clamped sea urchin eggs and oocytes. *Dev. Growth & Differ.* 27:177.

Longo, F. J., J. W. Lynn, D. H. McCulloh, and E. L. Chambers. 1986. Correlative ultrastructural and electrophysiological studies of sperm-egg interactions of the sea urchin, *Lytechinus variegatus. Dev. Biol.* 118:155-166.

McCulloh, D. H. and E. L. Chambers. 1986a. Changes of loose patch clamp seal resistance associated with exocytosis of cortical granules in relation to changes of patch membrane conductance in eggs of the sea urchin *Lytechinus variegatus. Biophys. J.* 49:178a.

McCulloh, D. H. and E. L. Chambers. 1986b. When does the sperm fuse with the egg? *J. Gen. Physiol.* 88:38a-39a.

McCulloh, D. H. and E. L. Chambers. 1986c. Fusion and "unfusion" of sperm and egg are voltage dependent in the sea urchin, *Lytechinus variegatus. J. Cell Biol.* 103:286a.

McCulloh, D. H., J. W. Lynn, and E. L. Chambers. 1987. Membrane depolarization facilitates sperm entry, large fertilization cone formation and prolonged current responses in sea urchin oocytes. *Dev. Biol.* 124:177-190.

McCulloh, D. H. and E. L. Chambers. 1987. Where does the activation wave of the sea urchin egg propagate? *J. Cell Biol.* 105:239a.

Lynn, J. W. and E. L. Chambers. 1987. Effects of cytochalasin B on egg activation currents in *Lytechinus variegatus* eggs voltage clamped at -20 mV. *J. Cell Biol.* 105:339a.

Crossley, I., K. Swann, E. L. Chambers, and M. Whitaker. 1988. Activation of sea urchin eggs by inositol phosphates is independent of external calcium. *Biochem. J.* 252:257-262.

Lynn, J. W., D. H. McCulloh, and E. L. Chambers. 1988. Voltage clamp studies of fertilization in sea urchin eggs. II. Current patterns in relation to sperm entry, nonentry, and activation. *Dev. Biol.* 128:305-323.

McCulloh, D. H., P. I. Ivonnet, and E. L. Chambers. 1988. Actin polymerization precedes fertilization cone formation and sperm entry in the sea urchin egg. *Cell Motil. Cytoskeleton* 10:345.

Chambers, E. L., P. I. Ivonnet, D. H. McCulloh. 1988. Inhibition of sperm entry in sea urchin eggs voltage clamped at negative membrane potentials is affected by $[Ca2+]_o$ but not by $[Na^+]_o$. *J. Cell Biol.* 107:175a.

McCulloh, D. H., P. I. Ivonnet, and E. L. Chambers. 1989. Do calcium ions mediate the voltage dependent inhibition of sperm entry in sea urchin eggs? *Biophys. J.* 55:155a.

Chambers, E. L. 1989. Fertilization in voltage clamped sea urchin eggs. *In: Mechanisms of Egg Activation.* R. L. Nuccitelli, G. N. Cherr, and W. H. Clark, Jr. (Eds.) Plenum Press, New York (in press).

FERTILIZATION IN VOLTAGE-CLAMPED SEA URCHIN EGGS

Edward L. Chambers

Department of Physiology and Biophysics
University of Miami School of Medicine
P.O. Box 016430
Miami, Florida 33101

ABSTRACT

Following insemination of the voltage-clamped sea urchin egg a characteristic component of the activation current is the initial shoulder with abrupt onset. This is the counterpart of the shoulder of the activation potential, and has a duration of ~12s, equal to that of the latent period. After attaining a maximum, the shoulder of the activation current is followed by a large increase in the inward current culminating in the major peak at ~31s. One of the most interesting findings in voltage clamp studies of fertilization is that the shoulder phase can be fully, or partially, dissociated from the subsequent phases of the activation current by holding the egg's membrane potential (V_m) in the neighborhood of the resting value (-70 mV). When the dissociation occurs, either complete or partial, the attached sperm fails to enter the egg. When the dissociation is complete, the isolated shoulder (duration ~11s, now termed a sperm transient current) terminates abruptly, subsequent phases of the activation current do not occur, and the egg otherwise remains in the unfertilized state. When the dissociation is partial, the same isolated shoulder (a step-like current profile, abrupt turn-on and turn-off, duration of ~12s) is observed, but from 5 to 25s after return of the current to the holding level, the delayed second or major current phase of a modified activation current occurs, accompanied by delayed elevation of the fertilization envelope. Cleavage fails to occur. Dissociation of the shoulder component from the subsequent phases of the activation current together with suppression of sperm entry is also observed in oocytes (germinal vesicle stage) when single sperm attach. Oocytes have a V_m of -70 mV, and, because of the 15- to 20-fold higher membrane conductance compared to that of eggs, single sperm can depolarize the oocytes' V_m by only 7 to 8 mV.

These data are consistent with the conclusion that unless depolarization to the neighborhood of 0 mV occurs, the shoulder component is dissociated from subsequent phases of the activation current and sperm entry will not occur. One possibility is that dissociation of the shoulder component accompanied by failure of sperm entry results from a suppressive effect of negative V_m on fusion of the sperm and egg plasma membranes. However, capacitance measurements carried out on eggs using a patch clamp method indicate that coincidentally with the abrupt turn-on of the shoulder current, cytoplasmic continuity between sperm and egg is attained. Nonetheless, the incipient stages of sperm incorporation were shortly terminated since simultaneously with the cutoff of current (which ends the dissociated shoulder component), the fusion event was reversed. Patch clamp measurements have established that the conductance increase which generates the inward shoulder current is not global, but localized to the immediate site of sperm attachment. A consequence of the restriction of the conductance increase to a localized site is that the density (current per unit surface area) could be large. This could result in alteration of the concentration of ions at the site of the entering sperm, and suppress its penetration. This possibility is currently under investigation in ion substitution experiments.

INTRODUCTION

My earlier work showing that following fertilization large changes in the fluxes of Na^+, K^+, and Ca^{2+}, as well as changes in their intracellular concentrations occur, coupled with the demonstration by several investigators of an initial large depolarization following fertilization, incited my interest to determine how voltage clamping the membrane potential (V_m ; see Table 1 in Notes On Methods section for list of abbreviations used in this chapter) might affect activation of the egg. Consequently, a voltage clamp was constructed. By the time some of the "bugs" had been worked out, Laurinda Jaffe (1976) published her paper demonstrating the block to sperm entry at positive V_m in *Arbacia* eggs. Using our 2-electrode clamp we were immediately able to confirm this strikingly interesting result for the eggs of our local species, *Lytechinus variegatus* clamped at +20 mV.

In our early studies, carried out at this time, we found that inseminated eggs clamped at V_m between +10 and -70 mV activated and raised full fertilization envelopes. However, we were very much perplexed by the finding that, for eggs clamped at V_m's in the neighborhood of 0 mV, the activated eggs cleaved regularly, but when clamped at the resting V_m (-70 to -80 mV), the great majority of the eggs failed to undergo first cleavage. Since the fertilization envelope elevated to the same extent for the two groups of eggs we assumed (erroneously!) that the sperm had entered all the eggs.

These earlier studies were hampered by the frequent occurrence of injury to the eggs and excessive leakiness of the egg membrane, which resulted from the necessity of having to insert two microelectrodes into the egg, especially when the microelectrodes had been filled with the conventional 3 M KCl. In addition, an upright microscope was used. This necessitated using objectives with a long working distance (designed for metallurgical work), since the microelectrodes had to be inserted into the egg from above. These objectives lacked sufficient resolution to enable us to observe the site of sperm attachment and sperm entry; little more than elevation of the fertilization envelope and cleavage could be detected.

The conundrum of why the failure of development in inseminated eggs clamped at V_m in the neighborhood of resting V_m continued to baffle us, until the problem was revived two years later, when Dr. John Lynn joined the laboratory.

Meanwhile, following the basic precepts that injury to the egg had to be minimized, and that our conditions for observing the eggs were less than optimal, we proceeded to revise our methodology. A single-electrode, switched clamp was substituted for the 2-electrode system; a 0.5 M K_2SO_4 mixture for filling the microelectrodes (Lynn and Chambers 1984) was substituted for the 3 M KCl; and an inverted microscope replaced the upright version. These revisions in the methodology enabled us to observe the egg using high resolution optics at the same time that the electrophysiological responses (EPR's; see Table 1 in Notes On Methods section for a list of abbreviations used in this chapter) were recorded from the voltage-clamped egg using only one microelectrode.

Not long after these revisions had been completed Dr. John Lynn was at the microscope observing (magnification 400 x) an egg clamped at a V_m of -70 mV to which a sperm had just attached at the periphery. I was at the controls of the voltage clamp looking out to suppress an incipient oscillation. Shortly after the onset of inward current, which signalled a sperm hit, John literally shouted: "But the sperm is lifting off the egg on the rising fertilization envelope!" This was the answer to the failure of development in eggs clamped at negative V_m : the sperm activated the egg, but failed to enter.

This paper summarizes salient features of the electrophysiological work carried out together with my associates, Drs. John Lynn and David McCulloh, as well as in collaboration with Dr. Frank Longo, and with the assistance of Mr. J. de Armendi and Mr. Pedro Ivonnet. The data which is reviewed, besides abstracts to which specific references are made, is presented in Chambers and de Armendi (1979), Longo *et al* (1986), Lynn and Chambers (1984), Lynn *et al* (1988), and McCulloh *et al* (1987).

NOTES ON METHODS

Besides the main outline of the improvements in methodology described in the Introduction, a major feature that contributed to unravelling the sequence of events following insemination of voltage-clamped eggs, oocytes, and zygotes was that we concentrated our attention on eggs we felt certain were monospermic. To this end, the eggs of *Lytechinus variegatus* (used almost exclusively in the work described in this paper) are particularly suitable, because of their remarkable transparency. To be sure only one sperm elicited the EPR, and that no more than one sperm penetrated the egg, we depend on multiple criteria. These are as follows: (1) Only one sperm could be found on the egg. This criterion alone, however, is not reliable, since if the sperm attaches just above or just below the periphery, it may not be visible. Moreover, often not every sperm which attaches elicits an EPR, especially if the "dry" sperm had been removed from the animal for more than several hours. (2) One fertilization cone forms. (3) A single sperm monaster develops. (4) The egg cleaves to 2 regular blastomeres at the time of first cleavage. Examination of the normality of cleavage

Table 1. Abbreviations Used in This Chapter

mV	millivolt
V_m	membrane potential
EPR	electrophysiological response
I	current amplitude[a]
I_{on}	the initial onset of current[a]
I_{sm}	the maximum current of the shoulder (phase one) of the activation current of eggs penetrated by sperm (Type I EPR), and
t_{sm}	the time of its attainment.
I_m	maximum current of sperm transients (Type II EPR), or of the initial step-like component (phase one) of the modified activation current of eggs not penetrated by a sperm (Type III EPR), and
t_m	the time of its attainment.
I_p	the major current peak of activation currents, and
t_p	the time of its attainment.
$I_{(10\%\ Ip)}$	current equivalent to 10% of the major current peak of activation currents, and
$t_{(10\%\ Ip)}$	the time when I_p has decreased to this value.

[a] This and the subsequent I's and t's are graphically identified in Figure 3.

is also important, to be sure the egg had not been injured during the voltage clamping procedure. (5) As the work progressed, we learned a particularly valuable additional criterion, which was to examine the current record of the voltage-clamped egg to see if more than a single I_{on} (see Figure 3) could be found. (Each individual sperm elicits an I_{on} at the onset of an EPR). If the sperm fails to penetrate the egg and an EPR occurs, then if the egg activates only criteria (1), (2) and (5), are applicable, but if the egg does not activate only criteria (1) and (5) can be used.

Eggs dejellied by gentle centrifugation several times using a hand centrifuge are mounted in sea water in plastic Falcon tissue culture dishes (Becton, Dickinson and Co., Oxnard, CA), either 60 X 15 mm or 35 X 10 mm. The microelectrode is inserted into an egg which attaches loosely to the bottom of the dish. (Eggs that attach firmly to the bottom of the dish are best avoided).

Particularly useful microscope objectives are the Zeiss 40 X, 0.60 NA planachromat long working distance (1.5 mm), or if the specimen is not thicker than 100 μm, even better resolution can be obtained using a Zeiss planachromat 40 X 0.65 NA with 0.7 mm working distance. If illuminating the object through plastic is a problem, or higher resolution objectives with shorter working distances are needed, the bottom of the plastic dishes can be bored out, and a No. 0 or No. 1 coverslip cemented in place.

Details on methods are presented in the paper by Dr. David McCulloh in this volume, see also Chambers and de Armendi (1979), Lynn and Chambers (1984), Lynn *et al* (1988), and McCulloh *et al* (1987).

RESULTS AND DISCUSSION

For sea urchin eggs clamped at different V_m we can distinguish three different categories of voltage dependence. These are: (1) The inhibition of sperm entry at V_m more positive than +10 mV accompanied by complete suppression of the occurrence of an electrophysiological response (EPR) and the absence of any other kind of response following sperm attachment. (2) The inhibition of sperm entry at V_m more negative than -20 mV but without inhibition of EPR's following attachment of the sperm, accompanied or not by activation of the egg. (3) The inhibition of activation of the egg at clamped negative V_m more negative than -30 mV, following EPR's induced by a non-penetrating sperm. As is to be expected the threshold values differ for different sea urchin species.

What is the Relationship Between the Occurrence of an Electrophysiological Response (EPR) and the Attachment of Individual Sperm to an Egg Clamped at Different V_m?

Figure 1 beautifully illustrates the first type of voltage dependence, namely, the inhibition of the ability of the egg to generate electrophysiological responses (EPR's) following attachment of the sperm to the egg at clamped positive V_m. Neither activation of the egg, nor sperm entry ever occurs in the absence of an EPR. Jaffe (1976) first described the inhibition of sperm entry at positive V_m. At +20 mV attachment of a sperm fails to elicit any response whatsoever. The egg can be covered with sperm, yet the egg remains completely inert. As the V_m is clamped at more negative values, over a range of only a few mV, a steep increase in responsiveness of the egg to the attached sperm occurs. From between 0 and -10 mV to -100 mV, the majority to all of the individual sperm attachments elicit an EPR. I will describe later the different types of EPR, suffice it to say now that an EPR may or may not be followed by activation of the egg. No significance can be given to the variations in the percentages at the negative V_m's, our impression is that these result from variability in the potency of different batches of sperm. It should be emphasized that at all the V_m's examined the sperm attached readily to the eggs, *i.e.*, there is no evidence of an effect of V_m from +20 to -100 mV on the binding of the sperm to the egg's surface.

Our data have shown that whenever an EPR occurs this signifies that fusion of the plasma membranes of the sperm and egg had occurred. Indeed our electrophysiological data are consistent with the conclusion that fusion occurs at the very onset of each EPR. The evidence

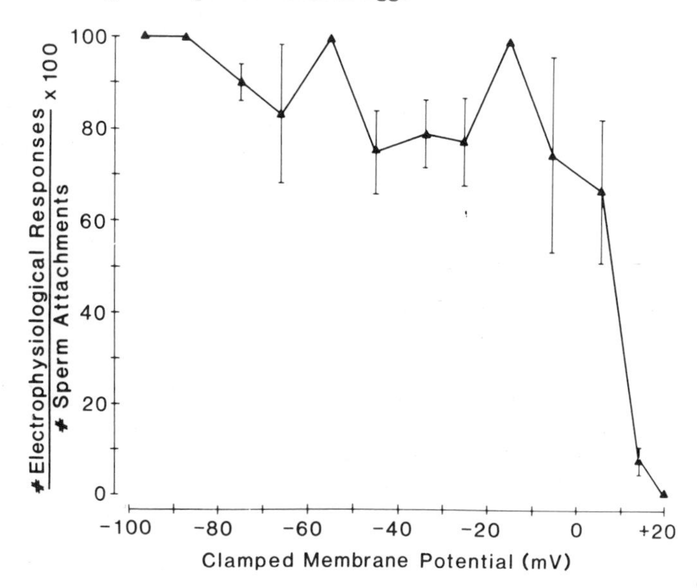

Figure 1. Effect of clamped V_m on the percentage of sperm attachments which elicited an electrophysiological response. Vertical bars indicate the standard errors. (From Lynn *et al.* 1988; by permission of the Acadeic Press.)

in support of this conclusion was obtained in patch clamp experiments designed by Dr. David McCulloh, in which sperm were introduced into the patch clamp pipette. Following attachment of a sperm to an isolated patch of egg membrane the capacitance of the membrane patch, the surface area of which approximated that of the sperm, increased simultaneously with the onset of an EPR (McCulloh and Chambers 1986a, b). We interpret the increase in capacitance as resulting from the establishment of electrical contact between the gametes at the moment when fusion of the membranes occurred. Conversely, a capacitance change was never observed when sperm attached to the patch, but failed to elicit an EPR, a situation uniformly observed at inhibitory positive V_m. Based on our patch clamp measurements, therefore, we conclude that Figure 1 depicts the effect of V_m on the fusion of sperm and egg plasma membranes. Negative V_m's do not inhibit the fusion event, but at positive V_m's fusion of the plasma membranes is suppressed, and then eliminated. That the block to sperm entry at positive V_m is due to inhibition of fusion of the gamete membranes had earlier been proposed by Jaffe and coworkers (e.g., Jaffe 1976; Gould-Somero and Jaffe 1984).

What is the Relationship Between Sperm Entry and the Occurrence of an EPR at Different Clamped V_m's?

Figure 2 (curve drawn with solid line) illustrates the second type of voltage dependence, the inhibition of sperm entry at clamped V_m more negative than -20 mV. From +15 to -15 mV the sperm enters the egg for every EPR that occurs, but at more negative V_m, a decreasing proportion of the EPR's is associated with the entry of a sperm. Only 10% of the EPR's are associated with sperm entry in eggs clamped at a V_m in the neighborhood of the resting value (-70 to -80 mV). At V_m's more negative than -80 mV none of the EPR's are associated with entry of a sperm. In Figure 2, data points at V_m's more positive than +17 mV are not plotted, because there were neither any EPR's nor any sperm entries (we chose not to plot 0 divided by 0). By combining the information of Figures 1 and 2, note that for V_m in the neighborhood of +15 mV, the attached sperm are highly ineffective in eliciting an EPR (Figure 1), but for every EPR which occurred, a sperm entered the egg (Figure 2, solid line). The situation is completely otherwise at the opposite end of the spectrum (at V_m's more negative than -80 mV). In this V_m range the sperm are highly effective in eliciting a response, but none of the sperm succeeded in entering the egg.

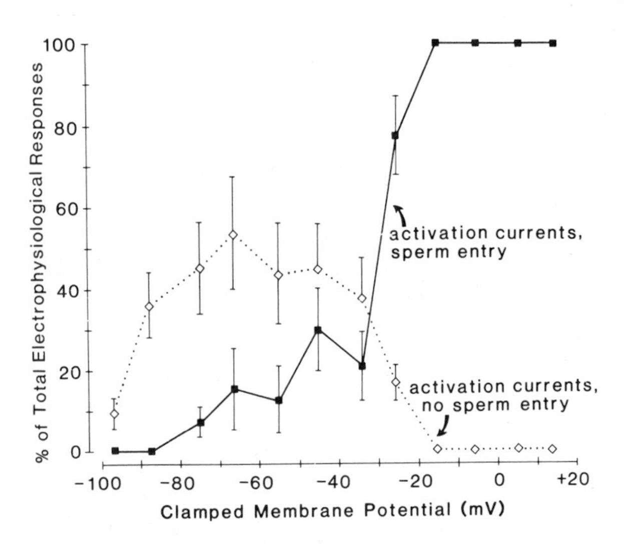

Figure 2. Effect of clamped V_m on the percentage of the total electrophysiological responses which are either: activation currents associated with sperm entry (solid line connecting filled squares), (type I responses); or modified activation currents not associated with sperm entry (dotted line connecting open diamonds), (type III responses). (From Lynn *et al.* 1988, by permission of Academic Press.)

These data obtained on voltage-clamped eggs show that depolarization is required for sperm entry to occur. For sea urchin eggs which have not been voltage-clamped and which in their normal physiological state have a resting V_m of -70 to -80 mV (Jaffe 1976; Jaffe and Robinson 1978; Hagiwara and Jaffe 1979; Chambers and de Armendi 1979), a successful sperm attachment induces a large depolarization of ~100 mV (see Figure 9, panel C; Chambers and de Armendi 1979) and entry of the sperm follows. It should be noted that because of the unfertilized egg's very low membrane conductance (~2.9 nS), electrical leakage at the site of insertion of the microelectrode is frequently sufficient to depolarize the V_m to values in the neighborhood of -10 to -20 mV (e.g., McCulloh *et al.* 1987). Attachment of a sperm to these eggs elicits a smaller depolarization (in Figure 9, compare panels B and C), yet sperm entry proceeds normally. The point I wish to make here is that for eggs with a small resting V_m, the sperm induced depolarization is not a requisite for penetration of the sperm, since the egg's V_m is in the range permissive for sperm entry even before the sperm attaches.

The block to sperm entry at negative V_m can also be demonstrated in sea urchin oocytes (McCulloh *et al.* 1987, and more fully discussed in the paper presented by Dr. David McCullloh in this volume), or in zygotes reinseminated from 30 to 60 min after the initial sperm had entered the unfertilized egg (unpublished), either voltage-clamped or not. Oocytes, or reinseminated zygotes are, figuratively speaking, spontaneously voltage-clamped at their resting V_m's of -70 to -80 mV. This is because of their high membrane conductance. Consequently, a successful sperm attachment can depolarize the cell by only a few mV, a depolarization which is insufficient to permit entry of the sperm.

Accepting that the occurrence of an EPR signifies that fusion of the gamete plasma membranes had occurred at the onset of the response (see above), we conclude that the inhibition of sperm entry at negative V_m is exerted after the gamete membranes have fused. Since the block to sperm entry at positive V_m occurs prior to the membrane fusion event, clearly the voltage dependent inhibitory mechanisms involved in the two cases are different.

What Kinds of EPR's are Observed in Eggs Clamped at Different Membrane Potentials?

Three types of responses are observed. The responses have either three phases (types I and III), or only the first phase (type II). The currents are inward at all V_m more negative than 0 mV.

Type I responses are activation currents seen only in eggs penetrated by a sperm. Their frequency of occurrence at different V_m's, therefore, is the same as for the percentage of EPR's that are associated with sperm entry (Figure 2). The first phase or shoulder (Figure 3A) is of abrupt onset, I_{on}, slowly increasing to the shoulder maximum, I_{sm}. This phase corresponds in duration (~12s) to the latent period (Allen and Griffin 1958). During the second phase (Figure 3A), or the phase of the major inward current, the current rapidly increases to a maximum at I_p (~31s after I_{on}). At the onset of phase three a rapid decrease of the current occurs until the amplitude attains a value approximating that of the shoulder maximum (I_{sm}). The end of phase three, and of the activation current, is arbitrarily defined as the time when the current approaches 10% of the peak of the major inward current, $I_{10\% Ip}$. Activation of the egg as evidenced by exocytosis of the cortical granules and elevation of the fertilization envelope occurs during phase two and is completed during the early part of phase three.

Type II are transient inward currents (Figures 3B and 11) and are associated with the attachment of a sperm to an egg, but without sperm entry. Following the occurrence of the transient, the egg otherwise remains in the unfertilized state as viewed by light microscopy, and the sperm swims away. Two types of transient currents are observed, those of long duration (over 3.5s), with an average of 10.6s, and those of short duration (less than 3.5s), with an average of 1.7s). Transient currents are not observed at all at a clamped V_m of -20 mV and more positive values. Long duration transient currents (Figure 4, line drawn with dashes) are the only type of transient observed at clamped V_m between -25 mV and -70 mV, and their frequency of occurrence decreases at a clamped V_m of -80 mV and more negative values. The

A. Activation I-sperm enters (V =-33mV)

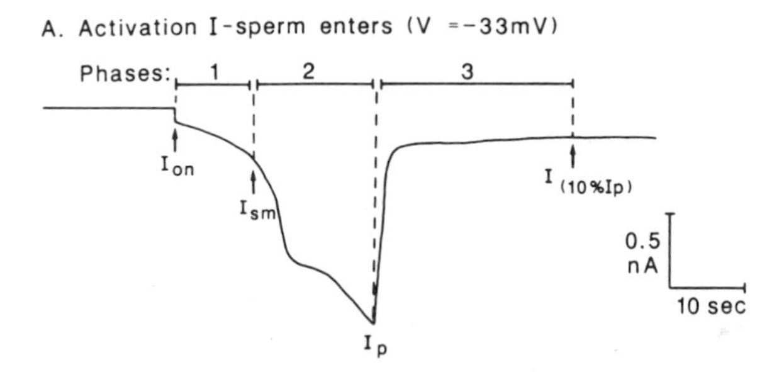

B. Transient I - no sperm entry (V =-58mV)

C. Modified Activation I- no sperm entry (V=-45mV)

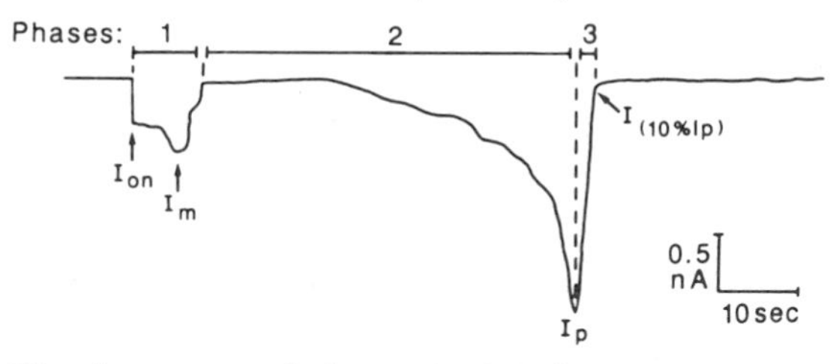

Figure 3. The three types of electrophysiological responses observed following insemination of voltage-clamped eggs. See text for description, and Table 1 for list of abbreviations.

short duration transient currents (Figure 4, line drawn with alternating dots and dashes) are not observed at clamped V_m more positive than -70 mV, but at more negative V_m their frequency increases steeply. The short duration transients will be further discussed in a later section of this paper. The long duration transient currents (Figure 3B and 11A) resemble phase one of an activation current (type I) except that, after attaining maximum amplitude at I_m, instead of undergoing a further increase the current abruptly turns off and returns, usually abruptly, to holding.

Type III responses, or modified activation currents occur only in eggs not penetrated by a sperm, but the non-penetrating sperm induces activation of the egg. These responses are not seen at a clamped V_m more positive than -20 mV. They are observed in the clamped V_m range between -25 mV and -90 mV (Figure 2, dotted line), a frequency distribution similar to that of the long duration transient currents (compare Figure 2, dotted line with Figure 4, line drawn with dashes). Modified activation currents have three phases (Figure 3C), but the first phase resembles a long duration transient current. Following the cutoff of current which terminates phase one, however, the current does not remain at holding. After a varying time interval (now phase two), the current starts to increase. The current increases to a peak, I_p, similar to phase two of type I responses. However, unlike type I responses, the time of occurrence of I_p is delayed, and the rapid decline of current of phase three occurs all the way to the holding value, and phase three is very short in duration. Activation of the egg as evidenced by exocytosis of the cortical granules and elevation of the fertilization envelope occurs during phases two and three, the same as during the corresponding phases in eggs penetrated by a sperm.

The characteristics of the three types of EPR's are analyzed in further detail in Lynn *et al.* (1988). Similar current profiles have recently been described by David *et al.* (1988) for the inseminated eggs of another sea urchin species.

The EPR's may be viewed as consisting of one elemental component, A, which is, or is not, followed by a second elemental component, B. Each component of current has its own unique characteristics, and is generated by a different conductance. Our data are consistent with the conclusion that component A is generated by a sperm-mediated conductance

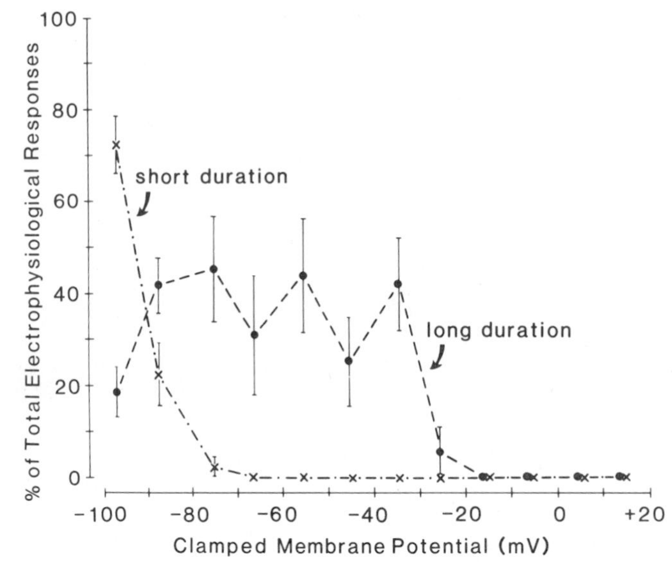

Figure 4. Effect of clamped V_m on the percentage of total electrophysiological responses which are either: long duration transient currents (line drawn with dashes connecting filled circles), or short duration transient currents (line drawn with alternating dots and dashes connecting x's). Both are Type II responses not associated with sperm entry. (From Lynn *et al.* 1988; by permission of Academic Press.)

increase, and component B by an increase in conductance of the egg membrane in response to a factor released by the sperm during the latent period.

Component A. Phase one of an activation current (type I) represents the early part of the component A current while the transient currents (type II) and phase one of the modified activation currents (type III) represent component A, terminated, by a cutoff of the component A conductance, shortly after the current has attained maximum amplitude (I_m).

That phase one of types I and III currents, and the long duration transient currents (type II) are indeed the same component is indicated by their identical characteristics with respect to amplitude at the onset (I_{on}), the maximum attained (I_{sm} or I_m), and duration (Lynn *et al* 1988).

For phase one of the activation currents (type I) the conductance responsible for generation of component A persists during the phase two major inward current, since when the peak current, I_p, cuts off, the current, as well as the conductance during the early segment of phase three revert to that observed at the maximum of the shoulder, I_{sm} (see Figure 3A). How long the component A current (or conductance) persists, however, could not be determined due to the turn-on of the late K^+ conductance.

An outstanding characteristic of component A is that the magnitude of the current is proportional to the number of successful sperm attachments. If two sperm attach, each sperm attachment results in an I_{on} of the same amplitude, and if they attach within several seconds of each other, the currents sum. An example of this summation is shown in Figure 5 for two long duration transient currents (type II EPR). Summation also occurs when more than one sperm attach to the egg during phase one of types I and III EPR's.

Component B. The conductance increase responsible for generation of the B component never turns on except following the A component, even though the A component may turn off completely. The B component invariably is associated with the occurrence of activation as indicated by the wave of exocytosis which sweeps over the egg during phase two and early phase three. The occurrence of the B component and activation of the egg are synonymous. For activation currents (type I) and for modified activation currents (type III) associated with elevation of a full fertilization envelope, the B component comprises the major inward current. After rising to a peak, the conductance increase responsible for generation of the B component turns off sharply at the beginning of phase three. The B component, without superimposition on the A component is represented by phases two and three of the modified activation currents (type III responses).

Unlike the A component, a proportional relationship between the number of successful sperm attachments and the amplitude of the component B current cannot be demonstrated.

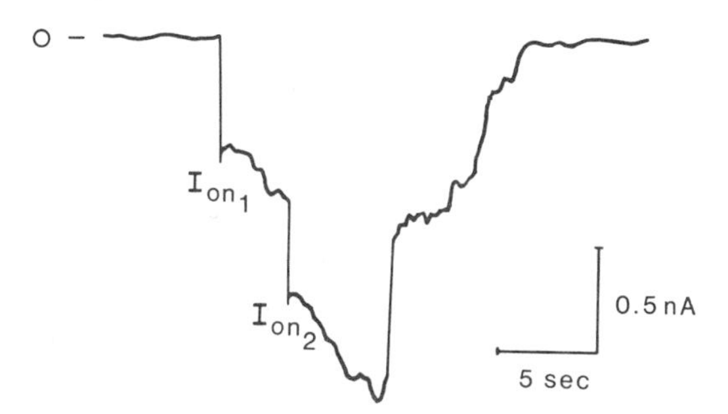

Figure 5. Summation of two long duration transient currents in egg clamped at -74 mV. The holding current was 0. The abrupt onsets of current following attachment of the first and then the second sperm are labelled $I_{on\,1}$ and $I_{on\,2}$, respectively. The cutoff of current following "unfusion" of the first, and then the second sperm are labelled $I_{off\,1}$ and $I_{off\,2}$. (From Lynn *et al.* 1988; by permission of Academic Press.)

When is the Decision Made Concerning Whether or Not the Sperm Will Enter the Egg?

Early in the work we became aware, by strictly concentrating our attention on eggs which we could conclusively show were monospermic, that the decision as to whether or not the sperm would enter the egg clamped at an inhibitory negative V_m must be made sometime before, or by the time the fertilization envelope starts to elevate. We based this conclusion on the observation that the non-penetrating sperm lifts off the egg as the fertilization envelope rises, while for the sperm which enters, the fertilization envelope rises around the sperm while the sperm remains attached to the egg's surface. Since, as is well known, the fertilization envelope starts to elevate at the site of sperm attachment, this meant the decision concerning sperm entry is made early, by 11 to 15s after the initiation (I_{on}) of the EPR, essentially during the latent period.

In addition we soon realized that the occurrence of an abrupt cutoff of current which terminates the transient currents (type II EPR) or phase one of the modified activation currents (type III EPR) signalled that sperm entry would fail. This cutoff of current occurred 11 to 15s after the onset of the EPR, the duration of the latent period. During this period electron microscopy (Longo *et al* 1986) shows that the plasma membrane at the tip of the acrosomal filament has fused with that of the egg, ~4s after I_{on} (the onset of the activation current). By the end of the latent period at 10 to 12s after I_{on} the sperm is still an intact entity at the surface of the egg, connected at its tip by a cytoplasmic bridge consisting of the widened acrosomal filament and an upwelling of egg cytoplasm, the incipient fertilization cone (Longo *et al* 1986). At this time, only several cortical granules in the vicinity of the membrane fusion site had undergone exocytosis.

In summary, the critical voltage-sensitive period when the decision is made whether or not the sperm will enter the egg, and whether or not the abrupt cutoff of current will occur is during the latent period or during the first ~12s after the occurrence of I_{on}. Additional data regarding the critical voltage-sensitive period for sperm entry is presented by Dr. John Lynn in this volume.

Genesis of the Activation and Transient Currents

Component A. I referred earlier to the capacitance measurements, which are consistent with the conclusion that the abrupt onset of current for the component A current, I_{on} (generated by an abrupt increase of the conductance) occurs simultaneously with the gamete membrane fusion event.

The occurrence of summation during phase one of both types of activation currents and of the transient currents also noted above (Figure 5) had earlier indicated that each sperm opens only a fraction of the total available channels. This suggested that the channels which opened were in the neighborhood of the attached sperm rather than being globally distributed. That this suggestion correctly explains the occurrence of summation has been demonstrated in patch clamp measurements, with sperm introduced into the patch clamp pipette. The entirety of the whole egg current of phase one of types I and III activation currents, as well as the transient type II currents (all representing the component A current) could be accounted for as inward current which was localized to the isolated sperm-attached patch (McCulloh and Chambers 1986a). Moreover, in patch clamp measurements where the sperm were added outside the patch clamp pipette, a conductance increase could not be detected in the patch until phase two of the activation current, although from the whole egg current the conductance increase which generates phase one was clearly observed (McCulloh and Chambers 1985). These findings are similar to those reported for the egg of the echiuroid worm, *Urechis*, for which evidence has been presented that the fertilizing sperm opens channels limited to, or in the immediate vicinity of, the site of sperm entry (Jaffe *et al* 1979; Gould-Somero 1981). Unlike the sea urchin egg, however, in *Urechis* eggs a second phase of channel openings, associated with an exocytotic response, does not occur.

Although we were well aware early on that the abrupt cutoff of current which terminated a sperm transient (type II EPR's) or phase one of modified activation currents (type III EPR's) signalled that sperm entry would fail, we could only guess what might be the cause of the cutoff. The patch clamp measurements on the isolated sperm-attached patch, referred to above, provided the answer (McCulloh and Chambers 1986b). Synchronously with the abrupt cutoff of current which terminated a sperm transient or phase one of modified activation currents, the capacitance of the patch also abruptly decreased and reverted to its original value (the capacitance of the patch had previously increased synchronously with I_{on}). These measurements are consistent with the conclusion that the sperm, which had initially fused with the egg's plasma membrane, then "unfused" when the cutoff of current occurred.

When experiments were done in which no cutoff of current was observed at the end of phase one of type I activation currents (associated with sperm entry), no decrease of the capacitance could be detected.

These data, taken together suggest that the component A current can be accounted for by the increase in conductance which would result from the insertion of channels of the acrosome reacted sperm into the egg's plasma membrane at the time of the fusion event. These channels would be introduced in parallel with those in the egg membrane. When the acrosome reaction occurs the sperm's V_m depolarizes (Schackman *et al* 1984). Consequently, the acrosome reacted sperm would be expected to have a high membrane conductance relative to that of the unfertilized sea urchin egg's exceptionally low membrane conductance. We do not know what accounts for the gradual increase in amplitude of the component A current until the maximum is attained at I_{sm} or I_m. The widening of the acrosomal process cannot, alone, account for the increase (D. H. McCulloh, unpublished). Possibly the increase in conductance could be related to the exposure of the internal surface of the sperm's membrane to the egg's intracellular environment, or a sperm-ligand-egg-receptor interaction might be involved (Gould and Stephano 1987). After phase two of the type I EPR's, the reversion of the activation current to the same amplitude as I_{sm} (~35s after I_{on}) can be accounted for by the persistence of channels in the sperm's membrane at the egg surface. Even at this time, when the fertilization envelope has risen over the egg surface, the fertilization cone only partially encloses the sperm head (McCulloh *et al* 1988). Moreover, proteins in the sperm's membrane become incorporated in the membrane of the fertilization cone (Longo 1986), and persist for considerable periods in the egg membrane (Gundersen *et al* 1986).

The "unfusion" event, which terminates the transient currents (type II EPR and phase one of the modified activation currents (type III EPR's), and which occurs before the sperm has been enveloped within the fertilization cone, would be achieved by severance of the cytoplasmic bridge (the acrosomal process fused at its tip with the incipient fertilization cone) between the sperm and the egg's surface. The possible involvement of depolymerization of cortical actin filaments in "unfusion" of the sperm is suggested in experiments where voltage clamped eggs were inseminated in the presence of cytochalasin (Lynn and Chambers 1987), more fully discussed in the paper presented by Dr. John Lynn in this volume.

Figures 6 and 7 illustrate diagrammatically the localized conductance increase (shown as the dark shaded area at the site of the sperm) responsible for generation of the A component of current, in relation to the evolving three types of EPR's previously illustrated in Figure 3. The component A conductance increase persists throughout the duration of the type I activation current (Figure 6, diagrams 1 to 5). On the other hand, the component A conductance cuts off abruptly for the type II transient response (Figure 7, diagrams 1 to 2A), as well as the type III modified activation current (Figure 7, diagrams 1 and 2B).

Figure 8 illustrates the localized component A conductance changes (dark shaded area as in Figures 6 and 7) elicited by two sperm which results in the summation of two transient currents. Following attachment of the first sperm a localized conductance A increase occurs which generates the abrupt onset of current, $I_{on\,1}$ (Figure 8, diagram 1). Before the component A conductance increase induced by the first sperm cuts off, a second sperm attaches, and elicits an additional localized conductance A increase, generating $I_{on\,2}$ (Figure 8, diagram 2). The first sperm then "unfuses", causing termination of the corresponding localized conductance increase, and the cutoff of current, $I_{off\,1}$ (Figure 8, diagram 3). Shortly thereafter the second sperm "unfuses" accompanied by $I_{off\,2}$ (Figure 8, diagram 4) and return of the current to holding.

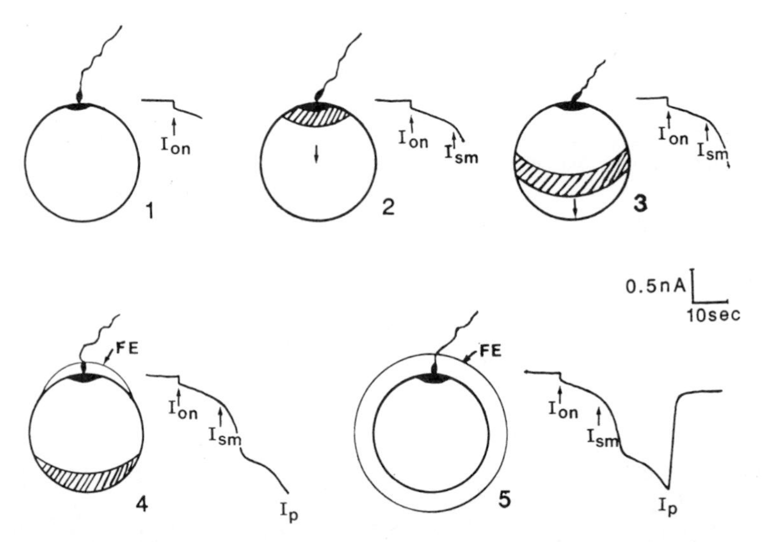

Figure 6. Diagrammatic representation of the conductance changes for an evolving activation current (type I) associated with sperm entry. The current profile is the same as shown in Figure 3A. The component A conductance is represented as the dark shaded area; the component B conductance as diagonal shading. While our data suggest that the component A conductance results from the insertion of sperm membrane channels into the egg membrane at the time the plasma membranes of the gametes fuse, for diagrammatic purposes the conductance A increase is represented as localized both in the membrane of the fused sperm, as well as in the egg membrane in the immediate vicinity of the attached sperm.

It has been stated that all the channels which open following sperm attachment and which generate the activation potential pre-exist in the egg membrane. This is based on the observation that a depolarization resembling a fertilization or activation potential can be evoked by exposing eggs to parthenogenetic agents, such as trypsin or the Ca ionophore A23187. These agents can certainly induce depolarizations associated with generation of action potentials and the further depolarization which accompanies activation (the counterpart of phase two of the activation current, or component B). However, I have yet to observe a well defined shoulder as a part of artificially induced activation potentials. It is the shoulder (Figure 9, panels B and C, segment a) which is the counterpart of phase one of the activation current (Figure 9, panel A) generated by the component A current, and which we propose results from the incorporation of the sperm's membrane channels into the egg's membrane at the time of the fusion event.

Component B. Our patch clamp data, with the sperm introduced outside the patch clamp pipette, showed no change in membrane conductance of the patch during phase one, as noted earlier. However, during phase two (the period of the major inward current), the conductance of the membrane patch increased for a much briefer period than the conductance increase measured during phase two for the whole egg. The conductance increase per cm^2 of the membrane patch, measured with the patch clamp, was much greater than the conductance increase per cm^2 of the egg surface measured for the whole egg current. Moreover, the time when the conductance increase in the patch occurred varied directly with the distance of the membrane patch from the site of sperm attachment (distance measured through the cytoplasm). From these data we concluded (McCulloh and Chambers 1985) that an increase in membrane conductance, which starts at the site of sperm attachment at the beginning of phase two, sweeps as an encircling band over the surface of the egg to the opposite pole, while behind the advancing band the conductance turns off. Our findings for the phase two conductance increase resemble those observed during fertilization of the amphibian egg (Kline and Nuccitelli 1985; Jaffe *et al* 1985; and Kline 1986). Using either a vibrating probe, or a patch

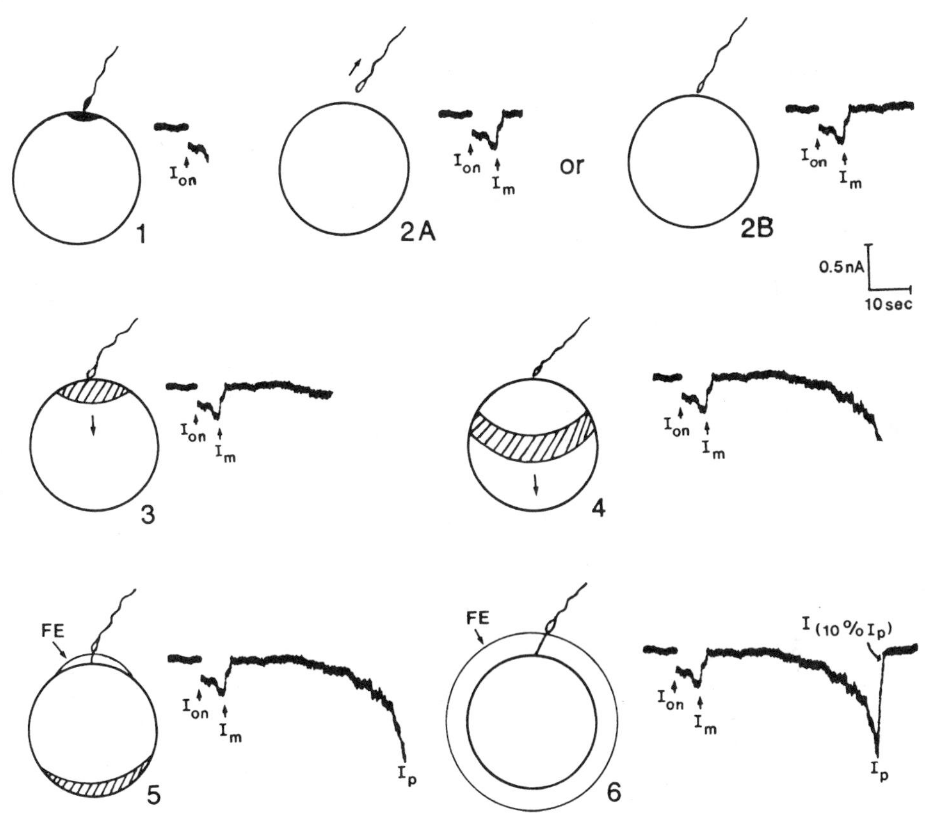

Figure 7. Diagrammatic representation of the conductance changes for an evolving transient current (type II), panels 1 and 2A, and an evolving modified activation current (type III), panels 1, 2B to 6. The current profiles are those shown in Figures 3B and 3C. The component A and B conductances are represented as described in the caption for Figure 6. For description see text.

electrode, a transient inward current localized to a ring shaped zone was found to progress over the entire egg surface as a wave starting from the point of activation.

Figures 6 and 7 illustrate diagrammatically the advancing zone of increased conductance (shown shaded with diagonal lines) which generates the component B current during phase two, and the cutoff of this conductance during phase three, in conjunction with an evolving activation current (type I, Figure 6, diagrams 2 through 5), and a modified activation current (type III, Figure 7, diagrams 3 through 6).

Voltage Dependence of Activation

Although the picture is dominated by the voltage dependence of sperm entry at positive V_m, and again at negative V_m, there is also a third category, the voltage dependence of activation. Recall that activation of the egg as indicated by the occurrence of an activation current (types I or III), exocytosis, and elevation of the fertilization envelope is observed whether or not the sperm enters the egg. On the other hand, when transient currents are recorded (type II EPR's), the egg otherwise remains in the unfertilized state.

If the sperm succeeds in entering the egg, which can occur only at clamped V_m more negative than +20 mV and more positive than -80 mV we cannot demonstrate an effect of clamped V_m on any parameter of the activation currents (type I EPR's) other than their amplitude.

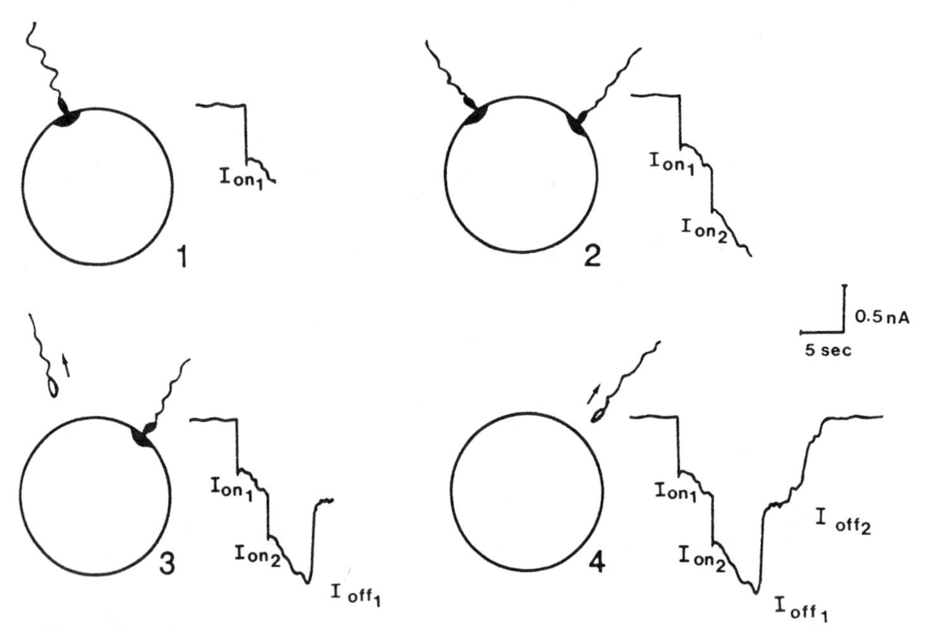

Figure 8. Diagrammatic representation of the conductance changes which occur during summation of two transient currents. The current profile is that shown in Figure 5. The localization of the component A conductance is represented as described in the caption to Figure 6. For description see text.

The voltage dependence of activation is best observed by considering only those EPR's which occur in eggs not penetrated by a sperm, as shown in Figure 10. That is to say, what is plotted is the number of responses of a particular type not associated with sperm entry, divided by the total number of responses observed which were not associated with sperm entry. The plot includes only V_m's more negative than -20 mV, simply because all of the EPR's between +17 mV and -20 mV are associated with sperm entry (and no EPR's occur at V_m more positive than +17 mV). Before considering this plot I will describe the short duration transients. These differ from the long duration transients in that they are not observed except at clamped V_m more negative than -80 mV, have a very short duration less than 3.5s as noted earlier, and frequently are much larger in amplitude. Examples are shown in Figure 11, B, in comparison to the previously described long duration transients, Figure 11A. Returning to Figure 10 note that at V_m's between -25 to -35 mV, where EPR's not associated with sperm entry are first encountered, approximately one half the responses are modified activation currents (Type III EPR's, Figure 10, solid line), accompanied by activation of the egg, and the other half are long duration transients (Figure 10, line drawn with dashes). This distribution persists throughout the range of clamped V_m to -75 mV. From -25 to -75 mV no significant difference could be detected in the mean duration of phase one of the modified activation currents, nor of the long duration transients.

We do not as yet know what determines, for the voltage range from -25 mV to -75 mV, whether sperm attachment will result in a long duration transient or a modified activation current. However, it is of interest that unless the inward current which culminates in I_p starts to increase by ~40s after the onset of phase one, no increase in the current will occur and the egg will fail to activate.

At clamped V_m more negative than -80 mV, Figure 10 shows that a sharp decline occurs in the percentage of responses which are modified activation currents (solid line), as well as long duration transients (line drawn with dashes). This decline coincides with a steep increase in the percentage of short duration transients (line drawn with dots and dashes), which supplant the other EPR's.

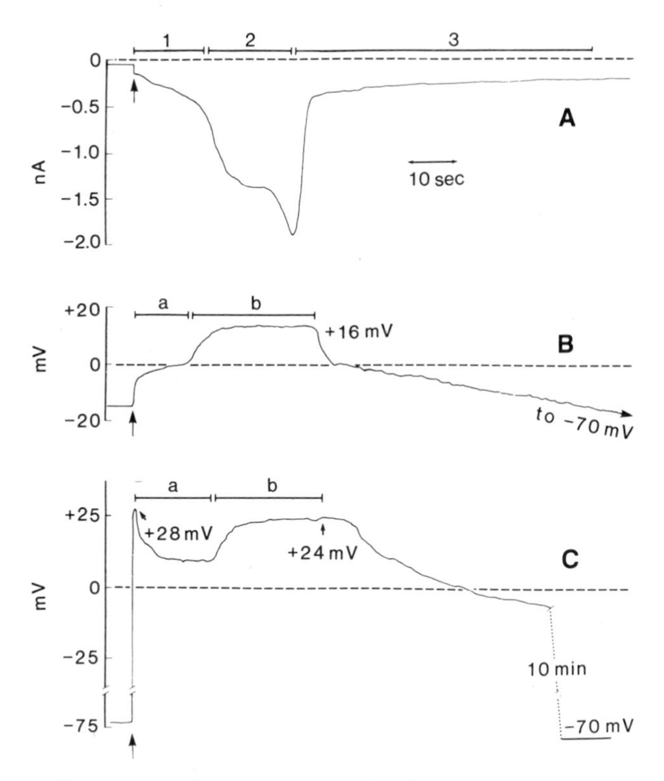

Figure 9. A: Activation current associated with sperm entry, type I response, egg clamped at -33 mV. B. and C: Activation (or fertilization) potentials. Resting V_m -18 mV for egg in panel B, -76 mV for egg in panel C. All records from monospermic eggs. The times of initiation of each response (vertical arrows) coincide. In panel A phases one and two of the activation current correspond in panels B and C with phases a and b, respectively, of the activation potentials. Clamping the V_m (panel A) prevents generation of the currents associated with the action potential, while in panel B the action potential has been inactivated since the resting V_m was positive to the action potential threshold. Note that the activation potential in panel B approximates the inverse image of the activation current in panel A. For C, a similar comparison cannot be made, since an action potential, which attains a peak of +28 mV is fired at the beginning of the first phase (a). (From Lynn *et al.* 1988; by permission of Academic Press.)

An interpretation of these data is that an effect of clamped V_m more negative than -80 mV to shorten the length of time with which the fused sperm maintains cytoplasmic contact with the egg (equal to the duration of the transient) diminishes the likelihood that activation will occur. The voltage dependence of activation evidently involves a voltage dependence of the length of time the sperm remains fused with the egg. This is consistent with the concept that a cytoplasmic factor contributed by the sperm initiates activation of the egg.

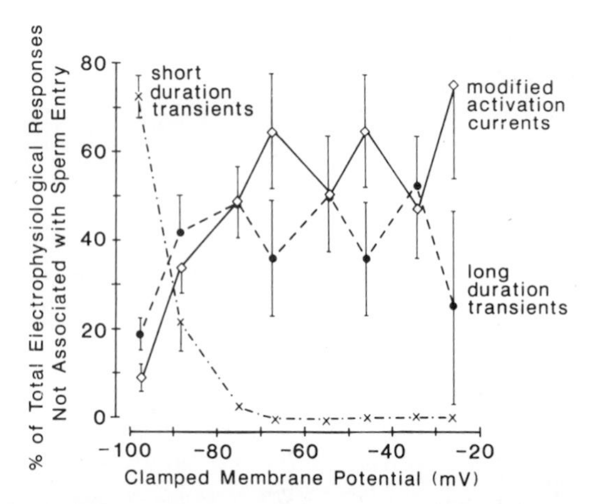

Figure 10. Effect of clamped V_m on the percentage of electrophysiological responses not associated with sperm entry which are: modified activation currents (solid line connecting open diamonds), long duration transient currents (line drawn with dashes connecting filled circles), or short duration transient currents (line drawn with alternating dots and dashes connecting x's).

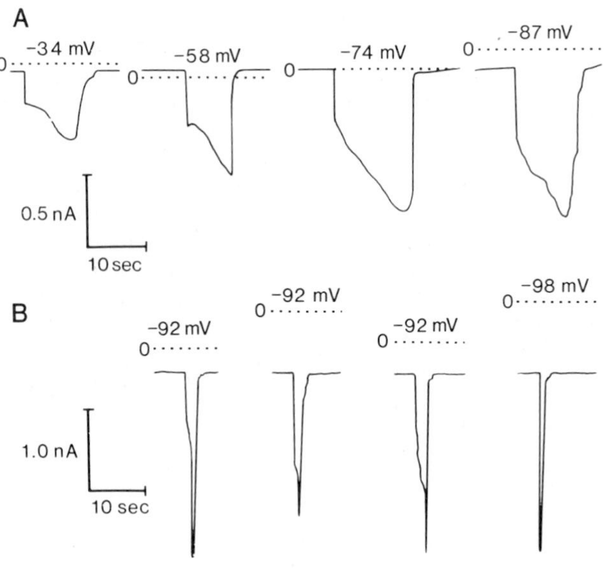

Figure 11. Sperm transient currents of the (A) long duration class, and (B) of the short duration class. The voltage shown above each current profile indicates the V_m at which the egg was clamped. Note that the nA scale for (B) is nearly twice that for (A). (From Lynn *et al.* 1988; by permission of Academic Press.)

ACKNOWLEDGEMENTS

This study was supported by NIH Research Grant HD 191267 and NSF Research Grants PCM 83-16864 and DCB-8711787.

REFERENCES

Allen, R. and J. L. Griffin. 1958. The time sequence of early events in the fertilization of sea urchin eggs. I. The latent period and the cortical reaction. *Exp. Cell Res.* 15:163-173.

Chambers, E. L. and J. de Armendi. 1979. Membrane potential, action potential and activation potential of eggs of the sea urchin, *Lytechinus variegatus*. *Exp. Cell Res.* 122: 203-218.

David, C., J. Halliwell, and M. Whitaker. 1988. Some properties of the membrane currents underlying the fertilization potential in sea urchin eggs. *J. Physiol. (Lond.)* 402:139-154.

Gould, M. and J. L. Stephano. 1987. Electrical responses of eggs to acrosomal proteins similar to those induced by sperm. *Science* 235:1654-1656.

Gould-Somero, M. 1981. Localized gating of egg Na^+ channels. *Nature* 291:254-256.

Gould-Somero, M. and L. A. Jaffe. 1984. Control of cell fusion at fertilization by membrane potential. p. 27-38. *In: Cell Fusion: Gene Transfer and Transformation.* R. F. Beers and E. G. Bassett (Eds.) Raven Press, New York.

Gundersen, G. G., L. Medill, and B. M. Shapiro. 1986. Sperm surface proteins are incorporated into the egg membrane and cytoplasm after fertilization. *Dev. Biol.* 113:207-217.

Hagiwara, S. and L. A. Jaffe. 1979. Electrical properties of egg membranes. *Annu. Rev. Biophys. Bioeng.* 8:385-417.

Jaffe, L. A. 1976. Fast block to polyspermy in sea urchin eggs is electrically mediated. *Nature (Lond.)* 261:68-71.

Jaffe, L. A., M. Gould-Somero, and L. Holland. 1979. Ionic mechanism of the fertilization potential of the marine worm, *Urechis caupo* (*Echiura*). *J. Gen. Physiol.* 73:469-492.

Jaffe, L. A., R. T. Kado, and L. Muncy. 1985. Propagating potassium and chloride conductances during activation and fertilization of the egg of the frog, *Rana pipiens*. *J. Physiol. (Lond.)* 368:227-242.

Jaffe, L. A. and K. R. Robinson. 1978. Membrane potential of the unfertilized sea urchin egg. *Dev. Biol.* 62:215-228.

Kline, D. 1986. A direct comparison of the extracellular current observed in the activating frog egg with the vibrating probe and patch clamp techniques. p. 1-8. *In: Ionic Currents in Development.* R. Nuccitelli (Ed.). Alan R. Liss, New York.

Kline, D. and R. Nuccitelli. 1985. The wave of activation current in the *Xenopus* egg. *Dev. Biol.* 111:471-487.

Longo, F. J. 1986. Surface changes at fertilization: Integration of sea urchin (*Arbacia punctulata*) sperm and oocyte plasma membranes. *Dev. Biol.* 116:143-159.

Longo, F. J., J. W. Lynn, D. H. McCulloh, and E. L. Chambers. 1986. Correlative ultrastructural and electrophysiological studies of sperm-egg interactions of the sea urchin, *Lytechinus variegatus*. *Dev. Biol.* 118:155-166.

Lynn, J. W. and E. L. Chambers. 1984. Voltage clamp studies of fertilization in sea urchin eggs. I. Effect of clamped membrane potential on sperm entry, activation, and development. *Dev. Biol.* 102:98-109.

Lynn J. W. and E. L. Chambers. 1987. Effects of cytochalasin B on egg activation currents in *Lytechinus variegatus* eggs voltage clamped at -20 mV. *J. Cell Biol.* 105:359a.

Lynn, J. W., D. H. McCulloh, and E. L. Chambers. 1988. Voltage clamp studies of fertilization in sea urchin eggs. II. Current patterns in relation to sperm entry, non-entry, and activation. *Dev. Biol.* 128:305-323.

McCulloh, D. H. and E. L. Chambers. 1985. Localization and proagation of membrane conductance changes during fertilization in eggs of the sea urchin, *Lytechinus variegatus*. *J. Cell Biol.* 101:230a.

McCulloh, D. H. and E. L. Chambers. 1986a. When does the sperm fuse with the egg? *J. Gen. Physiol.* 88:38a-39a.

McCulloh, D. H. and E. L. Chambers. 1986b. Fusion and "unfusion" of sperm and egg are voltage dependent in the sea urchin *Lytechinus variegatus*. *J. Cell Biol.* 103:236a.

McCulloh, D. H., P. Ivonnet, and E. L. Chambers. 1988. Actin polymerization precedes fertilization cone formation and sperm entry in the sea urchin egg. *Cell Motil. Cytoskeleton* 10:345.

McCulloh, D. H., J. W. Lynn, and E. L. Chambers. 1987. Membrane depolarization facilitates sperm entry, large fertilization cone formation, and prolonged current responses in sea urchin oocytes. *Dev. Biol.* 124:177-190.

Schackman, R. W., R. Christen, and B. M. Shapiro. 1984. Measurement of plasma membrane and mitochondrial potentials in sea urchin sperm. *J. Biol. Chem.* 259:13914-13922.

SPERM ENTRY IN SEA URCHIN EGGS: RECENT INFERENCES CONCERNING ITS MECHANISM

David H. McCulloh

Department of Physiology and Biophysics
University of Miami School of Medicine
P.O. Box 016430
Miami, Florida 33101

DEDICATION

This chapter is dedicated to Edward Lucas "Ted" Chambers in whose laboratory all of this work was done. Ted's love of science and his thorough approach toward experimentation have been and always will be inspirations for me. Even more impressive to me is his quiet, warm, and caring nature which I have come to know through more than five wonderful years of interaction. Thank you, Ted.

ABSTRACT

The mechanism by which a sperm enters the cytoplasm of an egg during fertilization is not known; but, in the sea urchin egg, membrane potentials in two different ranges are capable of precluding sperm entry. We have attempted to determine what steps of sperm entry are regulated by membrane potential in eggs and oocytes of *Lytechinus variegatus* with the objective of obtaining insight into the sperm entry process. Early sperm-egg interactions include: sperm-egg attachment or binding, sperm-egg fusion, and an abrupt increase of membrane conductance which leads to a depolarization of the egg's membrane potential. The positive-going activation potential of the sea urchin egg following attachment of the first sperm decreases the probability of entry for sperm which attach subsequently (rapid, voltage block to polyspermy). A sperm fails to enter the egg or oocyte unless it first causes a conductance increase. In voltage-clamped eggs and oocytes, the probability that an attached sperm causes a conductance increase is maximal at potentials more negative than -10 mV and decreases to 0 near +20 mV. It is possible that the lack of a conductance increase at potentials more positive than +20 mV results from a voltage-induced failure of sperm to fuse with the egg following attachment, hence precluding their entry. When sperm entry occurs at less positive potentials, the sperm's nucleus slowly migrates into the cytoplasm of the egg beginning roughly 60s after the initial conductance increase. However, if an egg or oocyte is voltage clamped at a potential more negative than -30 mV (their resting potentials are -75 mV), it is unlikely that a sperm which attaches and causes a conductance increase will be incorporated. A characteristic "cutoff" of the conductance increase occurs in association with the failure of these sperm to enter. In oocytes which are not voltage clamped, depolarizations associated with attachment of a single sperm also cut off prematurely and are too small to permit sperm incorporation. Sperm entry occurs in oocytes only when many sperm attach and involves a cooperative summation of their depolarizations. Cooperativity is not seen for eggs. In oocytes, sperm inçorporation, fertilization cone size, and the duration of the electrophysiological response are statistically associated and exhibit voltage dependences which are indistinguishable, suggesting that the three events are interrelated. We conclude that sperm entry involves two voltage-dependent, separable steps: sperm-egg fusion and nuclear incorporation. One common, voltage dependent mechanism may regulate nuclear incorporation and fertilization cone formation.

INTRODUCTION

The mechanism by which a sperm enters the cytoplasm of an egg during fertilization is not known; but, in the sea urchin egg, membrane potentials in two different ranges are capable of precluding sperm entry. Although this indicates that membrane potential can affect sperm entry, it is not clear how membrane potential regulates entry in these two distinct ranges. Understanding how membrane potential regulates sperm-egg membrane fusion and the subsequent incorporation of the sperm into the egg's cytoplasm will provide valuable insights into the mechanisms by which these two basic cellular processes occur.

Early sperm-egg interactions include: sperm-egg attachment or binding, sperm-egg fusion, and an abrupt increase of membrane conductance which leads to a depolarization of the egg's membrane potential (Figure 1). Voltage dependence is apparent for some step (as yet unknown) intervening between attachment and fusion. The positive-going activation potential of the sea urchin egg following attachment of a sperm reverses the polarity of the membrane potential from its resting membrane potential near -70 mV. Sustained positive membrane potentials act as a rapid, electrical block to polyspermy in eggs (Jaffe 1976). Oocytes are more susceptible to polyspermy (Runnstrom 1963; Franklin 1965; Longo 1978a). The activation potential in an egg initiates an action potential which is indistinguishable from action potentials elicited in eggs by a brief depolarizing current pulse (Chambers and de Armendi 1978). Action potentials attain positive membrane potentials near +25 mV. The activation potential subsides to a plateau level near +10 mV within 2-3s after the action potential. The plateau level is maintained for 15s while the sperm pivots about its point of attachment on the surface of the egg. During this latent period (Allen and Griffin 1958), the egg appears not to respond to the sperm. The latent period ends as the egg begins to respond with additional conductance increase (Lynn *et al.* 1988), exocytosis of cortical granules (McCulloh and Chambers 1984; McCulloh 1985), a wave of contraction which is observed at

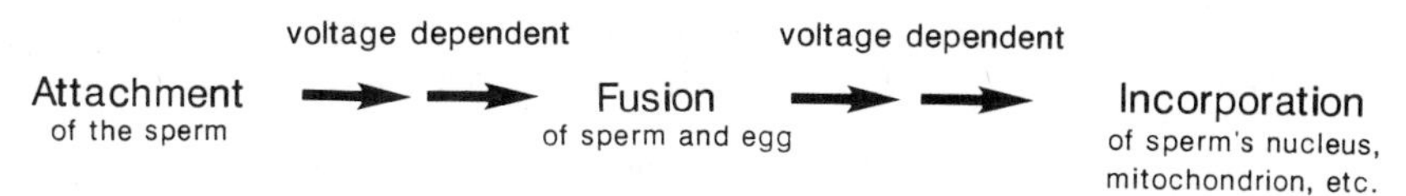

Figure 1. Two distinct voltage dependences during sperm entry.

the surface (McCulloh *et al.* 1988) and the formation of a fertilization cone which engulfs the head of the sperm. Beginning roughly one min after the initial conductance increase, the sperm is incorporated (Figure 1) as its nucleus and mitochondrion slowly migrate into the cytoplasm of the egg (McCulloh *et al.* 1988).

Recent evidence indicates that the voltage clamped membrane potential of the egg precludes sperm entry when it is maintained at values more positive than +17 mV (Lynn and Chambers 1984). The increase of membrane conductance which occurs in the egg following sperm attachment (Lynn *et al.* 1988) and the fusion of sperm and egg plasma membranes (McCulloh and Chambers 1986a; 1986b) are also precluded. While the egg's plasma membrane is clamped at potentials more negative than -20 mV, sperm entry fails; whereas fusion (McCulloh and Chambers 1986b) and the membrane conductance increase which immediately follows sperm attachment (Lynn and Chambers 1984; Lynn *et al.* 1988) occur. This second class of failure for sperm entry indicates that there is a step sensitive to membrane voltage present between fusion and incorporation of the sperm (Figure 1).

Ted Chambers, John Lynn, Pedro Ivonnet and I have attempted to gain insight into the sperm entry process in eggs and oocytes of *Lytechinus variegatus*. In this chapter, I have collected my thoughts concerning three basic issues: 1) characterization of the voltage dependence of both events in a more quantitative manner than has been done previously, 2) verification that membrane potential plays a role in eggs and oocytes which have not been voltage clamped, and 3) determination of what other readily observable phenomena are similarly voltage dependent and may either affect or be affected by sperm entry.

MATERIALS AND METHODS

Solutions

Seawater used for experiments was collected from the Gulf Stream near Miami, Florida. Large particulates were removed by filtration with filter paper within hours of arrival in the laboratory. Following filtration, seawater was stored for up to three months at room temperature. TAPS seawater was prepared by further filtration (0.80 μm filter, Millipore, Bedford, MA) and buffering to pH 8.3 by addition of 10 mM tris[hydroxymethyl]-methylaminopropane sulfonic acid (TAPS, SIGMA, St. Louis, MO), prior to use for experiments.

Animals

Collection and maintenance. Sea urchins, *Lytechinus variegatus*, were collected from either the ocean or Biscayne Bay near Miami, Florida. The animals were transported within several hours of collection to the laboratory and were placed in aquaria containing seawater (collected from Biscayne Bay) maintained at 20-22°C. Each aquarium was equipped with a recirculating pump which drew the seawater through a filter containing glass wool and gravel. A day or two after new animals were added to an aquarium, feces which had collected atop the gravel at the bottom of the tank were removed by siphoning them in seawater from the bottom of the tank. The siphoned seawater was replaced by fresh seawater. Roughly one seventh of the seawater in the aquaria was exchanged for fresh seawater and salinity and pH were adjusted weekly. If animals were maintained in the aquaria for several months, they were fed spinach leaves.

Obtaining gametes. The first step in obtaining gametes is to determine the sex of the sea urchin by observing the gonads. In our laboratory, this is done only after cutting away the oral portion of the test including Aristotle's lantern. Ovaries in females often appear more brilliantly orange and are more translucent than testes in males.

Eggs were collected by rinsing a female's coelomic cavity with 0.55 M KCl for 10-20s after which the KCl was discarded and the aboral surface was pointed down and suspended over a beaker so that the gonopores were placed in contact with the seawater filling the beaker. Yellow/orange eggs collected at the bottom of the beaker. After the shedding of eggs ceased from a female and all the eggs had settled to the bottom of the beaker, the supernatant seawater was decanted and replaced by fresh TAPS seawater. Eggs could be used for experiments immediately or could be maintained at 10°C for up to 12h.

"Dry" sperm were collected from males by removing the testes from the animal, blotting the coelomic fluid and seawater from them on filter paper and placing them in a dry Syracuse dish. Testes were broken in several places to release sperm and could be used within minutes or could be maintained at 4°C for up to 12h.

Females which shed oocytes along with eggs were selected as oocyte donors. This occurred most often when the female was collected just prior to a new moon or a full moon and its eggs were shed within one week thereafter. Oocytes were obtained from the ovaries of the selected females after completion of shedding because it was reasoned that removal of eggs from the ovaries by shedding enriched the ovaries for oocytes. The ovaries were minced in seawater and poured though silk bolting cloth (roughly 150 μm mesh) to remove as much ovarian debris as possible. Each oocyte was easily identified by its large germinal vesicle containing a nucleolus. Only large oocytes with diameters near that of mature eggs were used in these studies (McCulloh *et al.* 1987).

Selection of gametes for experiments. Sperm and eggs were collected each day of experimentation from several different males and females, respectively. Gametes from different animals of the same sex (batches) were never combined. Rather, the best batch of sperm and the best batch of eggs were selected. Usually, the largest batch of sperm was selected in order to assure that the same batch of sperm could be used throughout the test fertilizations used to select a batch of eggs and all subsequent inseminations for the day. Very seldom was a batch of sperm poor enough to warrant using a different batch.

Test fertilizations were performed to select a batch of eggs to be used for the remainder of the day. From each batch of eggs a small aliquot (roughly 20 to 40 μl) was removed (from the bottom of the collection beaker) and was suspended in a separate, small plastic petri dish containing 5 ml of seawater. Five to ten drops of sperm suspension (20 μl dry sperm diluted in 10 ml of seawater) were added to each Petri dish as the dish was swirled. Forty five s later, each Petri dish received another 5-10 drops of the same sperm suspension. Beginning roughly one min after the initial insemination, the batches were viewed carefully taking note of what percentage of the eggs elevated fertilization envelopes, how high the envelopes elevated, the clarity of the egg cytoplasm, and the size of the fertilization cones appearing at the site of sperm entry. Beginning about an hour after the initial insemination, the eggs were scored for division taking care to note what percentage of the eggs cleaved normally to the 2-cell stage, indicative of normal, monospermic fertilization.

No batch of eggs was used if fewer than 95% of the eggs elevated fertilization envelopes or if fewer than 95% of the eggs cleaved normally to the 2-cell stage. Among those batches which were not eliminated by these criteria, a batch was selected which raised the highest fertilization envelopes indicative of a robust cortical reaction, had the clearest cytoplasm enhancing the visibility of sperm anywhere on the surface of the egg, formed the largest fertilization cones indicating most clearly the site of sperm entry, or which provided the best combination of the above.

Removal of egg jelly. Most of the jelly coat which surrounds each egg or oocyte was removed to facilitate rapid access of sperm to the egg or oocyte surface and to aid in adherence of the eggs and oocytes to the bottoms of Petri dishes for microelectrode insertion. Sufficient jelly removal was accomplished for eggs and oocytes suspended in normal seawater within a 15 ml centrifuge tube as follows: The tube was filled with seawater to roughly 1.5 cm from the top and repeatedly inverted for several minutes waiting for the bubble to reach the end of the tube with each inversion. Eggs were pelleted gently using a hand centrifuge and the supernatant was discarded and replaced to the same level with fresh seawater. This procedure was repeated until the egg pellet which initially appeared as two bands (jelly free eggs on the bottom, eggs surrounded by jelly on top) became one homogeneous, jelly-free band. No attempt was made to determine whether trace amounts of jelly remained at the surface of the eggs or oocytes as long as they adhered to the bottoms of Petri dishes.

Electrophysiology

Microelectrodes. Micropipettes were pulled from glass tubing containing a small glass filament (#30-30-0, Frederick Haer, Brunswick, ME) and were filled with an electrolyte solution composed of 0.5 M K_2SO_4, 10 mM NaCl, 0.1 mM sodium citrate. The resistances of the microelectrodes used (indicative of the diameter at the tip) were roughly 20-40 Mohms for experiments in which the membrane potential was voltage clamped but were roughly 35-70 Mohms for experiments in which the membrane potential was recorded but not voltage clamped. When electrodes were used for voltage clamp, the micropipette's capacitance was decreased by shielding it with silver paint (GC Electronics, Rockford, IL) applied to within 500 μm of the tip. Subsequently, the silver paint near the tip was insulated from the bath by coating it with either Q-Dope (GC Electronics) or fingernail polish (Sally Hansen's Hard as Nails, clear, Del Laboratories, Inc., Farmingdale, NY) prior to filling the micropipette. The shield (silver paint) was maintained at the membrane potential by electrical contact with a 1X output from the headstage. Recently it has come to our attention that the insulation step alone decreases the microelectrode's capacitance sufficiently for voltage clamping (personal communication, R. E. Kane).

Recording membrane potential. Eggs or oocytes were placed in 2-3 ml of seawater within a Petri dish firmly fixed to the moveable stage of an inverted compound microscope (Invertoscope or IM35, Zeiss, Oberkochen, West Germany). Microelectrodes held in lucite microelectrode holders (E. L. Wright Co., Bridgeport, CN) were positioned over the dish with a micromanipulator (Ernst Leitz Wetzlar GMBH, Wetzlar, West Germany; or Narashige, Tokyo, Japan). A silver/silver chloride wire within the electrode holder placed the electrode solution in electrical contact with the headstage input of a voltage clamp (Model #8100 or #8800, DAGAN Corp., Minneapolis, MN). In addition, the tip of a Pasteur pipette containing seawater and plugged with a gelled solution of roughly 3% Agar in seawater was maintained in the solution within the Petri dish as a reference electrode. A wire attached to a silver/silver chloride pellet in contact with the seawater inside the reference electrode completed the circuit to the voltage clamp. The outputs of the voltage clamp permitted us to monitor the membrane potential as the difference of potential between the microelectrode and the reference electrode. Membrane potentials are reported in this paper as the potential inside the egg or oocyte relative to the reference in the bath. In addition, the voltage clamp, permitted us to monitor the amount of current passed through the electrode in an unclamped "bridge" mode, while allowing the membrane potential to respond to current. Alternatively, the voltage clamp permitted us to maintain the membrane potential constant while monitoring any currents which flowed through the membrane as previously described (Lynn and Chambers 1984; McCulloh *et al.* 1987). Inward currents are displayed as downward deflections.

Microelectrode insertion. The tip of a microelectrode was positioned near an egg or oocyte and advanced until a dimple was visible where it pushed against the surface. After the surface deformation of the egg or oocyte had been allowed several seconds to conform to the tip of the microelectrode, the negative capacitance circuit of the voltage clamp's headstage was oscillated for 5 to 200 ms. This oscillation was generally followed by an 2-10 mV negative shift of the potential recorded by the electrode immediately following the oscillation, indicative that the tip of the microelectrode had entered the cytoplasm of the egg. The recorded potential often decreased or reversed smoothly to a positive value within 100 ms after the oscillation and subsequently subsided to a value of -5 to -10 mV over the next two min. On a few occasions, the potential continued to become more negative and eventually came to rest at a value near -75 mV after 10 to 20 min. More commonly, the value rested near -10 mV until a small amount of hyperpolarizing current (< .125 nA) was applied. With time, the membrane potential of the egg could be maintained near -75 mV with a decreased amount of current. Roughly one-quarter of the eggs eventually rested at -75 mV using this technique, whereas the other three-quarters rested at potentials near -10 mV despite their greatly increased input resistance (McCulloh *et al.* 1987). Insertion of electrodes into oocytes often required repeated oscillations of the negative capacitance circuit. The membrane potential of oocytes most often rested spontaneously near -75 mV (McCulloh *et al.* 1987) and required no application of hyperpolarizing current.

Insemination. Sperm were added gently to eggs or oocytes so as not to dislodge the micro-electrode which had been inserted. In addition, sperm were added in low density to avoid polyspermy which otherwise results during voltage clamp experiments (Lynn and Chambers 1984; Lynn *et al.* 1988). Sperm were prepared no more than 1-2 min prior to insemination by adding 1 or 1.5 μl of dry sperm to 40 ml of seawater. Subsequently, 100 μl of this suspension was added to the dish within 1 cm of the egg or oocyte monitored by the micro-electrode. The density of sperm prior to addition to the Petri dish (amounting to 2×10^5 sperm/ml, above) estimates sperm density, because we can estimate the density near the egg no better than this after addition. For oocyte experiments, the sperm density used was elevated as stated in the text in order to obtain more than one sperm attachment in a brief period of time.

Determining the Number of Sperm Which Enter Eggs and Oocytes

Equal volumes of a mixed suspension of dejellied eggs and oocytes were dispensed into 15 ml centrifuge tubes, and gently pelleted to insure that equal numbers of eggs were present in all tubes (0.1 ml). The volume in each tube was adjusted with seawater so that upon sperm addition, the volume would be 10 ml.

An aliquot of dry sperm was diluted in seawater and mixed vigorously for 10s prior to a second dilution in seawater. Insemination was accomplished by the addition of various aliquots of the second dilution to the centrifuge tubes containing eggs and oocytes, 30s after initial sperm dilution. The sperm/egg/oocyte suspension was inverted for several seconds following sperm addition and then was capped and placed on its side on a rotary shaker.

Sperm entry was halted by addition of an equal volume of fixative (3.7% formalin in seawater, adjusted to pH 8.3, minutes prior to addition) to centrifuge tubes 2 1/2 min following insemination. After 5-15 min of fixation, the eggs and oocytes were washed 2-3 times with 10 ml of seawater and incubated for one hour in a solution of 0.5 mg/ml Pronase (SIGMA) in seawater to remove attached but unfused sperm (Vacquier and Payne 1973). Addition of 0.001 mg/ml bisbenzimide (Hoechst #33342, SIGMA, a gift from R. Wall) allowed visualization of sperm as punctate blue cones within the cytoplasm of eggs and oocytes when viewed with epifluorescence microscopy using a filter set specific for bisbenzimide (Zeiss #487702). The number of sperm inside each of 200 to 1200 eggs and oocytes at each concentration of sperm was counted and recorded.

Determining Whether Sperm Entered Independently

Histograms plotting the frequencies of occurrence for eggs containing 0, 1, 2, 3, etc. sperm (see Figure 6 in results) were generated for eggs and/or oocytes which were inseminated in the same tube (thus exposed identically to sperm). These histograms were compared with histograms generated according to the Poisson equation:

$$N_i = \frac{N_{tot}\, m^i \exp(-m)}{i!} \quad . \qquad \textbf{Eq. 1}$$

which yields the number of eggs or oocytes (N_i) containing i independently acting sperm in which N_{tot} is the total number of eggs or oocytes counted and m is the mean number of sperm per egg or oocyte.

Estimating the Relative Probability of Sperm Entry

The probability (P_n) of the nth sperm entering and the effect of (n-1) previous entries was estimated by the frequency of eggs which contained n sperm following staining with the Hoechst stain (described above). In order to estimate the probability of entry for the nth sperm the frequency of eggs containing ($\geq$ n) sperm (F_n) was divided by the frequency of eggs containing ($\geq$ n - 1) sperm ($F_{(n-1)}$),

$$P_n = \frac{F_n}{F_{(n-1)}} \quad . \qquad \text{Eq. 2}$$

Formally, this value estimates the probability that an egg containing (n-1) or more sperm will be found containing n or more sperm. Probabilities were estimated only when $\geq$ 5 eggs were observed containing n or more sperm. This arbitrary limit was chosen to decrease the error associated with small counts.

RESULTS

Membrane Potentials More Positive Than +20 mV in the Egg or Oocyte Preclude Sperm Entry and the Associated Conductance Increases

When an egg is activated by sperm, the egg's membrane potential reverses, attaining values as positive as +10 to +30 mV (Chambers and de Armendi 1979). Eggs were inseminated while the potential of the plasma membrane was maintained constant by voltage clamp in order to determine if sperm entry and egg activation occurred at potentials near those attained during the activation potential. Sperm entered eggs which were voltage clamped at potentials between -20 and +17 mV whenever the sperm attachment was followed by occurrence of an activation current. An activation current begins with an abrupt inward current which increases slowly over the following 10-15s and continues with a more rapid increase of inward current subsiding rapidly beginning 30s after the abrupt inward current to a level near that attained just prior to the more rapid increase (Lynn *et al.* 1988). Changes of inward current were accompanied by a concomitant change of membrane conductance. Sperm failed to enter eggs clamped at potentials more positive than +20 mV. In addition, neither current profiles nor conductance increases indicative of egg activation or of sperm-egg interaction nor formation of a fertilization cone were observed in eggs clamped at these positive potentials. Sperm entry never occurred unless a current response was recorded.

The strong association between the occurrence of an electrophysiological response and entry of the sperm led us to investigate more carefully the voltage dependence of whether or not an electrophysiological response occurred following sperm attachment. We observed eggs which were inseminated while the membrane potential was voltage clamped. Each sperm which attached to the surface of the egg was noted in addition to any electrophysiological response which occurred shortly after attachment. Attachment of a sperm was followed by an electrophysiological response in greater than 75% of the cases when the membrane potential of the egg was more negative than -10 mV (Figure 2a). At more positive potentials, this percentage was decreased such that no electrophysiological responses were observed at potentials of +20 mV or more positive. A smooth curve (Figure 2a) plots a variant of the Boltzmann equation:

$$B = \frac{B_{max}}{1 + \exp\left[\frac{zF(V_m - V_{50})}{RT}\right]} \qquad \text{Eq. 3}$$

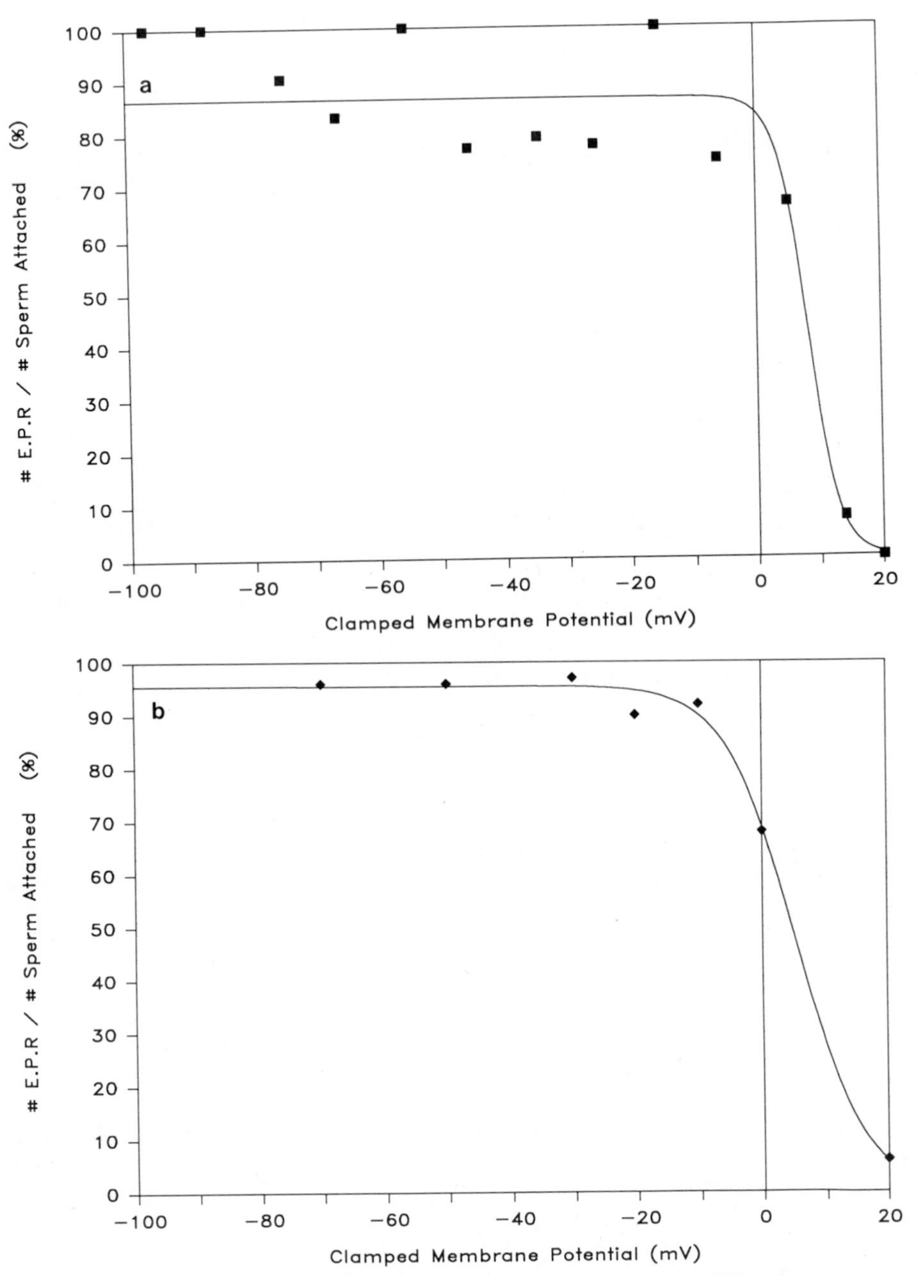

Figure 2. a = Eggs, b = Oocytes. Voltage dependence of the probability that a sperm will cause an electrophysiological response (E.P.R.) following its attachment to an egg **(a)** or oocyte **(b)**. Plotted points were determined by dividing the number of electrophysiological responses recorded from eggs clamped at each membrane potential by the number of sperm which attached to the same eggs at that membrane potential and by multiplying the quotient by 100 to obtain a percentage. Smooth curves were obtained by fitting Equation 3 to the data by a procedure which minimized the sum of the squared deviations between the data (weighted according to the number of sperm which attached) and the curve.

which was fitted to the data obtained using these eggs. The best fit occurred when the relationship between the probability of a sperm causing an electrophysiological response (B) as a function of the voltage clamped membrane potential (V_m) attained a maximum (B_{max}) at which 86% of the sperm were capable of causing an electrophysiological response following attachment. The relationship decreased to 50% of this maximum value at +8 mV (V_{50}) changing e-fold for a 2.4 mV shift of membrane potential (RT/zF; for which R = the universal gas constant, T = temperature in K, F = Faraday's constant and z, the valence is the only parameter free to vary within this ratio).

Sperm entered an oocyte during voltage clamp only if an electrophysiological response occurred. The potential at which the oocyte membrane was voltage clamped affected whether or not an electrophysiological response occurred. At membrane potentials more negative than -10 mV, an electrophysiological response occurred following over 90% of the sperm attachments. At increasingly positive membrane potentials, between -10 and +20 mV, fewer of the sperm which attached caused electrophysiological responses (Figure 2b). At +20 mV, two electrophysiological responses occurred following the attachment of 67 sperm in three oocytes. A Boltzmann equation (Equation 3) was fitted to this data (Figure 2b, smooth curve) obtained from oocytes. The best fit occurred when the relationship attained a maximum (B_{max}) at which 96% of the sperm caused an electrophysiological response following attachment. The relationship decreased to 50% of this maximum at +5 mV (V_{50}) changing e-fold for a 5.4 mV shift of membrane potential (RT/zF).

Membrane Potentials More Negative Than -20 mV Preclude Sperm Entry Despite the Occurrence of an Electrophysiological Response

Voltage clamp experiments. The membrane potential for eggs and oocytes prior to interaction with sperm is near -75 mV. The effect of the depolarization away from this membrane potential during activation by sperm was studied by voltage clamping eggs at potentials near -75 mV and at more positive potentials. Sperm failed to enter eggs clamped at -75 mV; although, an electrophysiological response was observed for more than 75% of the sperm which attached (Figure 2a, above). The electrophysiological response associated with failure of sperm to enter following sperm attachment at membrane potentials more negative than -20 mV, consisted of an abrupt onset of current sustained for 11 ± 3.6s (Lynn *et al.* 1988) prior to an abrupt cut-off of current which returned the current level to its pre-response baseline (Figure 3a). Increased inward current was accompanied by a concomitant increase of membrane conductance. The early, abrupt cut-off's of current and conductance observed in eggs clamped at large, negative membrane potentials were the primary features

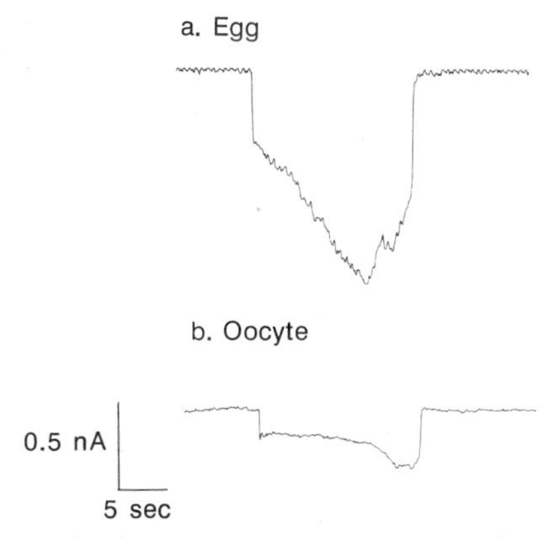

Figure 3. Transient current responses recorded from an egg **(a)** while its membrane potential was held at -70 mV by voltage clamp and from an oocyte **(b)** while its membrane potential was held at -70 mV by voltage clamp.

which distinguished the current response associated with failure of sperm to enter from the activation current, in which inward current and elevated conductance persisted for a longer time and diminished slowly (Lynn *et al.* 1988). When a second sperm attached during a transient current, a second transient inward current could superimpose upon the first, resulting in the summation of the two currents.

The disappearance of a sperm into the cytoplasm of an egg and the appearance of a sperm aster in the egg's cytoplasm (6-12 min after the abrupt inward current) indicated that the sperm had entered the egg. At membrane potentials more negative than -55 mV fewer than 15% of the sperm which caused electrophysiological responses entered (Figure 4a). At increasingly positive membrane potentials, between -55 and -15 mV, an increasing proportion of the sperm which attached and caused an electrophysiological response, subsequently entered (Figure 4a). At potentials more positive than -15 mV, 100% of the sperm which caused electrophysiological responses entered. A smooth curve (Figure 4a) plots a Boltzmann equation (Equation 3) with two free parameters adjusted for a best fit. The best fit was obtained when the relationship increased to 50% of its maximum at -31 mV (V_{50}) changing e-fold for a 7.6 mV shift of membrane potential (RT/zF). The maximum (B_{max}) was held constant at 100%.

Whenever a sperm entered an egg, a large fertilization cone formed surrounding the sperm at the site of sperm attachment. At first, the cone was small but it increased in size over the 2 to 10 min following the initial abrupt inward current until it was roughly 3-6 times the width of a sperm and appeared dome-shaped. When sperm failed to enter an egg but caused an electrophysiological response, either a thin, filament-like fertilization cone formed which elongated, radiating from the surface of the egg with the sperm located near the tip of the filament, or no cone formed at all. On a few occasions, sperm which caused electrophysiological responses but which failed to enter swam away from the egg tens of seconds after the interaction.

Voltage dependence for sperm entry was also observed in oocytes. At oocyte membrane potentials more negative than -20 mV, sperm never entered (Figure 4b) despite occurrence of a sperm transient current for more than 90% of the attached sperm (Figure 3b). At oocyte membrane potentials increasingly more positive than -20 mV, sperm entry occurred with higher and higher frequency (Figure 4b) and was accompanied by the occurrence of a prolonged electrophysiological response (McCulloh *et al.* 1987). Sperm entry occurred for 73% of the sperm which caused an electrophysiological response following attachment to an oocyte clamped at 0 mV. No data were sought at +10 mV. It was difficult to obtain data at potentials more positive than 0 mV because electrophysiological responses occurred seldom (see Figure 2b, above). A smooth curve (Figure 4b) plots a Boltzmann equation (Equation 3) which estimates the percentage of sperm which enter oocytes at each clamped membrane potential. Its two free parameters were adjusted for a best fit, while fixing the maximum (B_{max}) at a value of 100%. Sperm entry increased to 50% of this maximum at -11mV (V_{50}) changing e-fold for a 6.1 mV shift of membrane potential (RT/zF). The plot of sperm entry versus membrane potential for eggs (Figure 4a) is shifted toward more negative values than the comparable curve for oocytes (Figure 4b), whereas the values of RT/zF (mV's required for an e-fold change in the probability of sperm entry) are not significantly different. This suggests that the voltage sensor is similar in eggs and oocytes but that the voltage gradient may be different at the same membrane potential.

The sperm transient currents observed in oocytes when sperm failed to enter (Figure 3b) were similar to sperm transient currents in eggs (Figure 3a), although slightly smaller in amplitude for oocytes clamped at the same potentials (Table 1). The reversal potentials were not significantly different for currents in eggs and oocytes. Sperm transient currents in oocytes were characterized by an abrupt increase of inward current sustained for 27 ± 1.7s and terminated with an abrupt cutoff of current returning to the pre-interaction baseline. In oocytes, the inward current response associated with sperm entry lasted longer than the sperm transient current and thus is called a prolonged response (McCulloh *et al.* 1987). In addition, the inward current amplitude attained during the prolonged response was larger than the amplitude of sperm transient currents in oocytes clamped at the same potentials; but was smaller than the amplitude of activation currents in eggs clamped at the same potentials (Table 1) (McCulloh *et al.* 1984). The reversal potentials for prolonged responses in oocytes were not

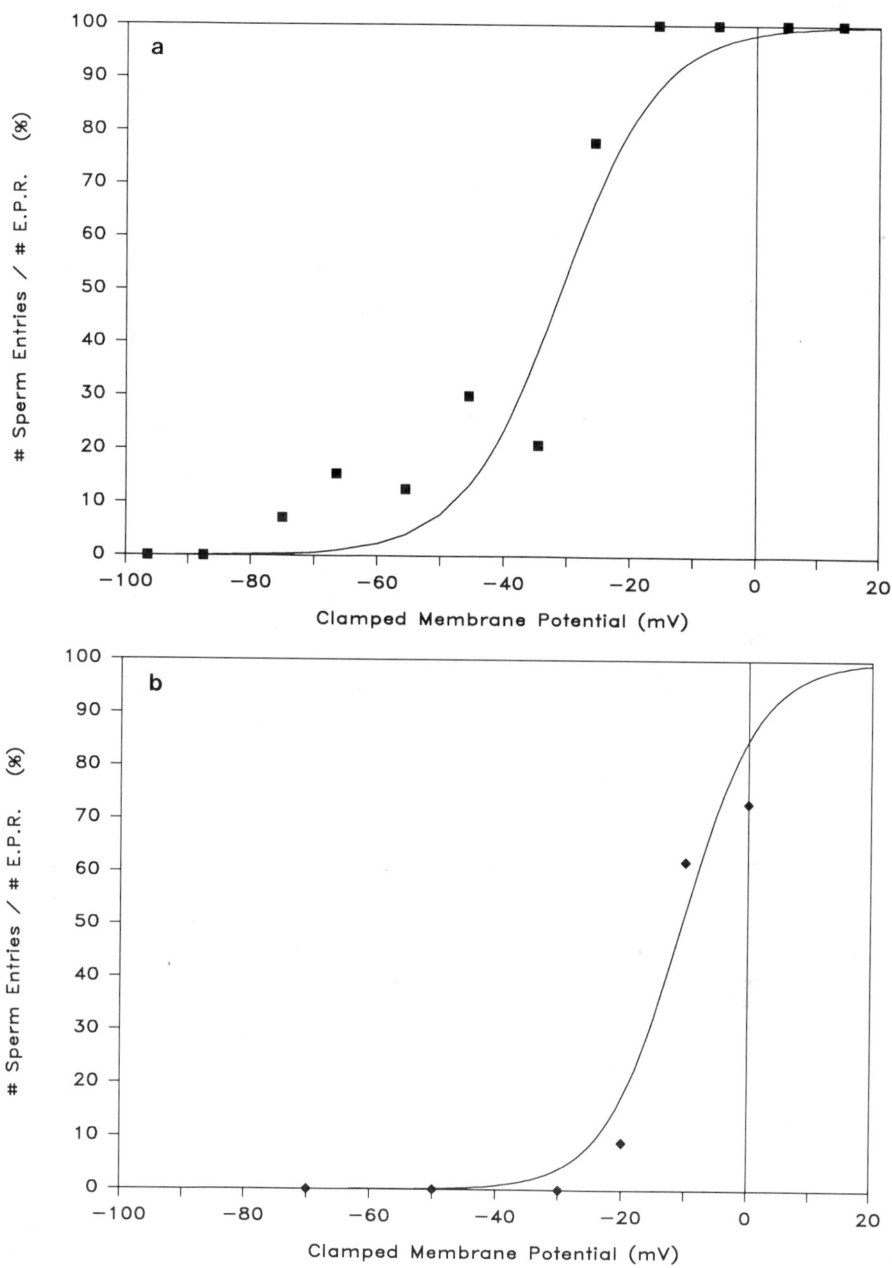

Figure 4. a = Eggs, **b** = Oocytes. Voltage dependence for the probability of entry for sperm which cause an electrophysiological response (E.P.R.) in an egg **(a)** or an oocyte **(b)**. Plotted points were determined by dividing the number of sperm which entered eggs or oocytes maintained at the indicated, clamped membrane potential by the number of sperm which caused an electrophysiological response in the same eggs or oocytes and then by multiplying the quotient by 100 to obtain a percentage. The smooth curves were obtained by fitting Equation 3 to the data by a procedure which minimized the sum of the squared deviations between the points (weighted according the the number of electrophysiological responses) and the curve.

Table 1. Comparison of Electrophysiological Responses in Eggs and Oocytes[a]

E.P. Response Type[b]:	Transient Currents		Prolonged Responses		
	I_{on}	I_{max}	I_{on}	I_{sm}	I_p
Conductance[c] (nS)					
Eggs	5.2±0.7	9±2	5.9±0.4	10.0±0.6	29±4
Oocytes	2.8±0.3	2.0±0.5	4.0±0.4	6±3	7±1
Ratio (oocyte/egg)	.54	.24	.68	.58	.24
Reversal Potential[d] (mV)					
Eggs	13±18	25±23	5±9	11±9	24±21
Oocytes	8±17	61±45	2±7	15±35	21±16
t test	p>.5	p>.5	p>.5	p>.5	p>.5

[a] The results from 90 responses in eggs and 67 responses in oocytes.

[b] The type of electrophysiological response: I_{on} = the amplitude of the initial inward current; I_{max} = the maximum current amplitude attained during a transient current; I_{sm}, I_p = the maximum current amplitude attained during the first and second phases, respectively, during a prolonged response in oocytes or an activation current in eggs.

[c] Conductance (slope ± standard error) determined as the slope parameter from linear regression of data plotted as current amplitude as a function of clamped membrane potential.

[d] Reversal potential (intercept ± standard error) determined by extrapolation to determine the V intercept from linear regression of data plotted as current amplitude as a function of clamped membrane potential. Data spanned potentials of -90 mV to +17 mV for eggs and -70 mV to 0 mV for oocytes.

significantly different from the reversal potentials of corresponding portions of the activation currents in eggs. These results suggest that sperm attachment causes an increase of conductance for the same ions in both eggs and oocytes but indicates that the conductance increase is smaller for oocytes than for eggs.

When a sperm entered an oocyte, a fertilization cone formed at the site of sperm attachment. Fertilization cones in oocytes were larger than the cones observed in eggs (Dale and Santanella 1985; McCulloh *et al.* 1987) and formed over a longer period of time. When a sperm failed to enter an oocyte despite the generation of an inward current, the cones were either much smaller consisting of a more filament-like protrusion from the oocyte surface, or no cone was observed.

Membrane potential records. Sperm transient currents which were associated with failure of sperm to enter either eggs or oocytes voltage clamped at membrane potentials more negative than -20 mV were characterized by an abrupt cutoff of conductance which was observed as an abrupt cutoff of inward current (Figure 3a, b). Voltage recordings were obtained from eggs and oocytes which were not voltage clamped in order to determine whether the failure of sperm to enter was caused by some epiphenomenon associated with the voltage clamp technique. In eggs, a constant hyperpolarizing current equal to that expected from a sperm transient current at -75 mV was passed through the microelectrode during insemination. This current maintained the egg's membrane potential at roughly -120 mV prior to sperm attachment. Upon attachment of a sperm, the membrane potential depolarized to between -90 and -50 mV for several seconds and then returned abruptly to roughly -120 mV (Figure 5a). Sperm which caused this type of response failed to enter the egg.

Voltage records were obtained from oocytes which were not voltage clamped and to which no exogenous currents were applied during interactions with sperm. The oocyte's membrane potential rests near -70 mV but the membrane resistance (26 ± 2.5 Mohm) is

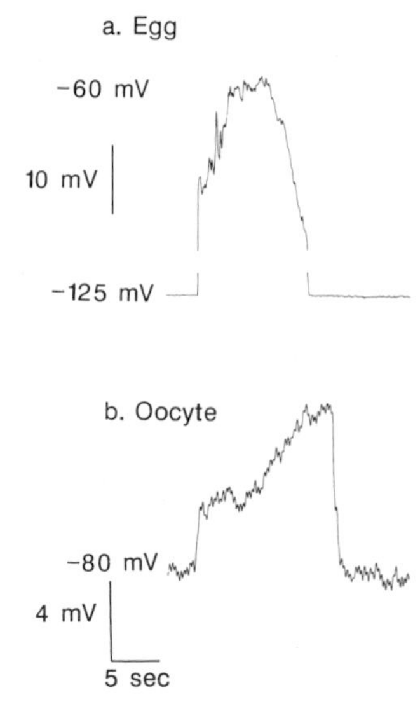

Figure 5. Transient depolarizations recorded from an egg **(a)** and an oocyte **(b)** which were not voltage clamped. Note the break in the voltage scale of trace a. Exogenous, constant current was passed through the membrane of the egg to maintain it at such negative potentials. The depolarization recorded from the oocyte (b) was obtained with no exogenous current passed through the oocyte membrane.

approximately 1/13 th that of the egg (340 ± 27 Mohm)(McCulloh *et al.* 1987). When a single sperm attached to an oocyte inseminated with 2 x 10^5 sperm/ml, a brief depolarization of 8 ± 1.4 mV lasting 19 ± 2.0s occurred (Figure 5b) (McCulloh *et al.* 1987). Sperm failed to enter the oocyte following such an interaction.

When several sperm attached to an oocyte (2 x 10^6 sperm/ml) within a short period of time several step-like depolarizations summed and were followed by several step-like cutoffs of current (McCulloh *et al.* 1987) as long as the membrane potential of the oocyte remained more negative than -30 mV. By summing the depolarizations, the oocyte attained more positive membrane potentials. Sperm failed to enter the oocyte following these interactions, in which 2-5 sperm cooperated to create a depolarization which was larger than the depolarization caused by a single sperm.

From these observations we conclude that the failure of sperm to enter eggs or oocytes voltage clamped at membrane potentials more negative than -20 mV can occur in eggs or oocytes which have not been voltage clamped and therefore is not an effect of the voltage clamp technique.

Cooperativity of Sperm-Induced Depolarizations and Sperm Entry in Oocytes but Not in Eggs

Currents summed in eggs and both currents and depolarizations summed in oocytes when several sperm attached to the same egg or oocyte within a brief period of time. This suggests that the small depolarizations caused by more than several sperm attached to an oocyte might sum to surpass the membrane potential necessary for entry of sperm in oocytes. If this suggestion is true, sperm would be expected to interact cooperatively in depolarizing the oocyte to a membrane potential permissive for sperm entry. To the contrary, no cooperativity would be expected in eggs because the depolarization caused by one sperm is sufficient to exceed the membrane potential required for sperm entry. The results of this would be observable in two types of experiment: 1) experiments to count the number of sperm which

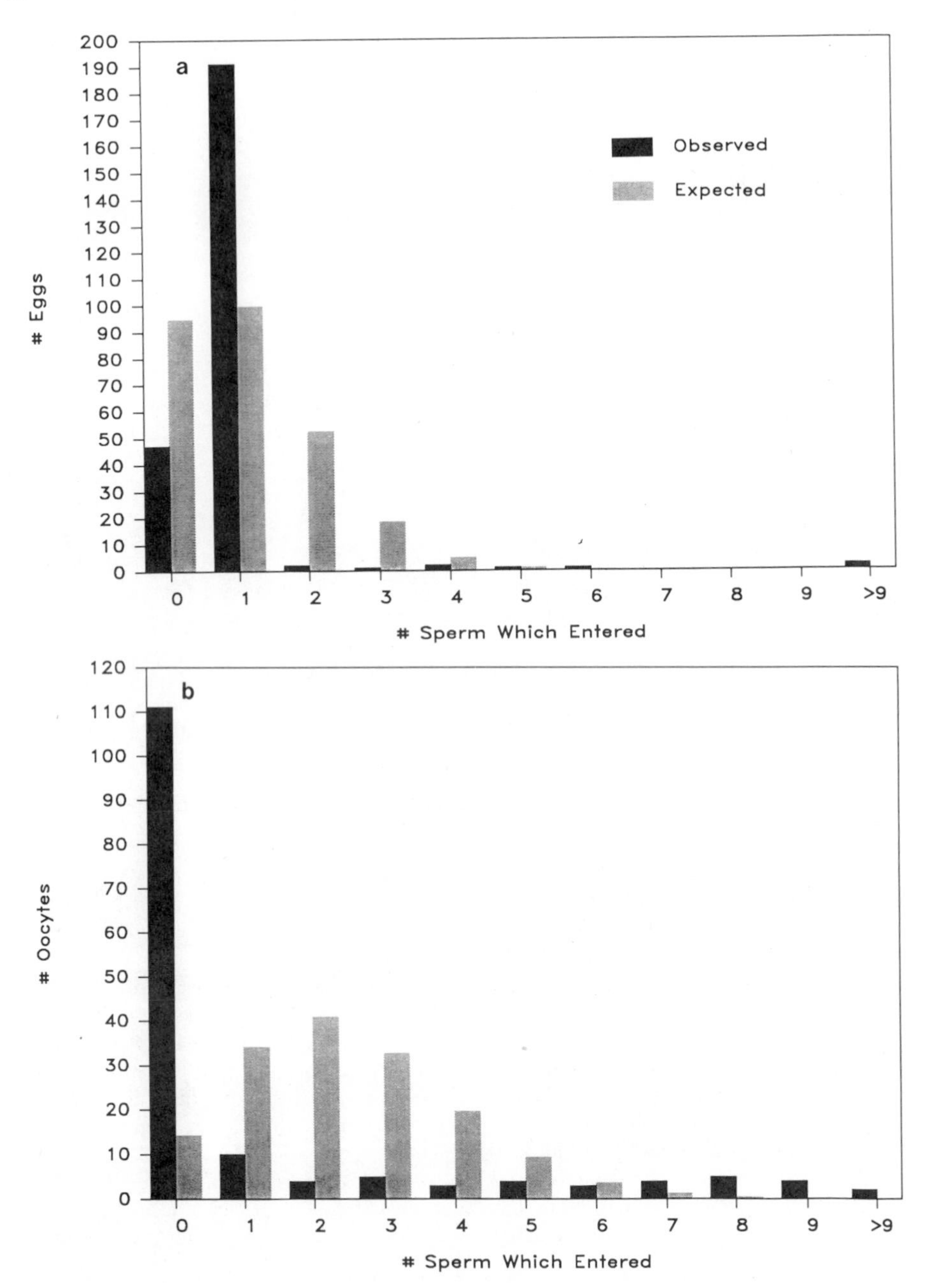

Figure 6. Examples of histograms indicating the numbers of sperm which entered eggs (**a**, black bars) and oocytes (**b**, black bars) in one, inseminated tube and the expectations (gray bars) determined using the Poisson distribution (Equation 1).

enter large numbers of eggs and oocytes, and 2) experiments in which single oocytes are observed both electrophysiologically and for sperm entry following insemination with a high density of sperm.

Experiments to count the number of sperm which enter eggs and oocytes. The number of sperm within each of 200 to 1200 eggs and oocytes which had been inseminated in the same tube at sperm densities between 1 and 10^5 sperm/ml were counted. Histograms (Figure 6a, b, black bars) display the number of eggs or oocytes which contained each number of sperm observed. This method was used previously by Presley and Baker (1970). The data indicate

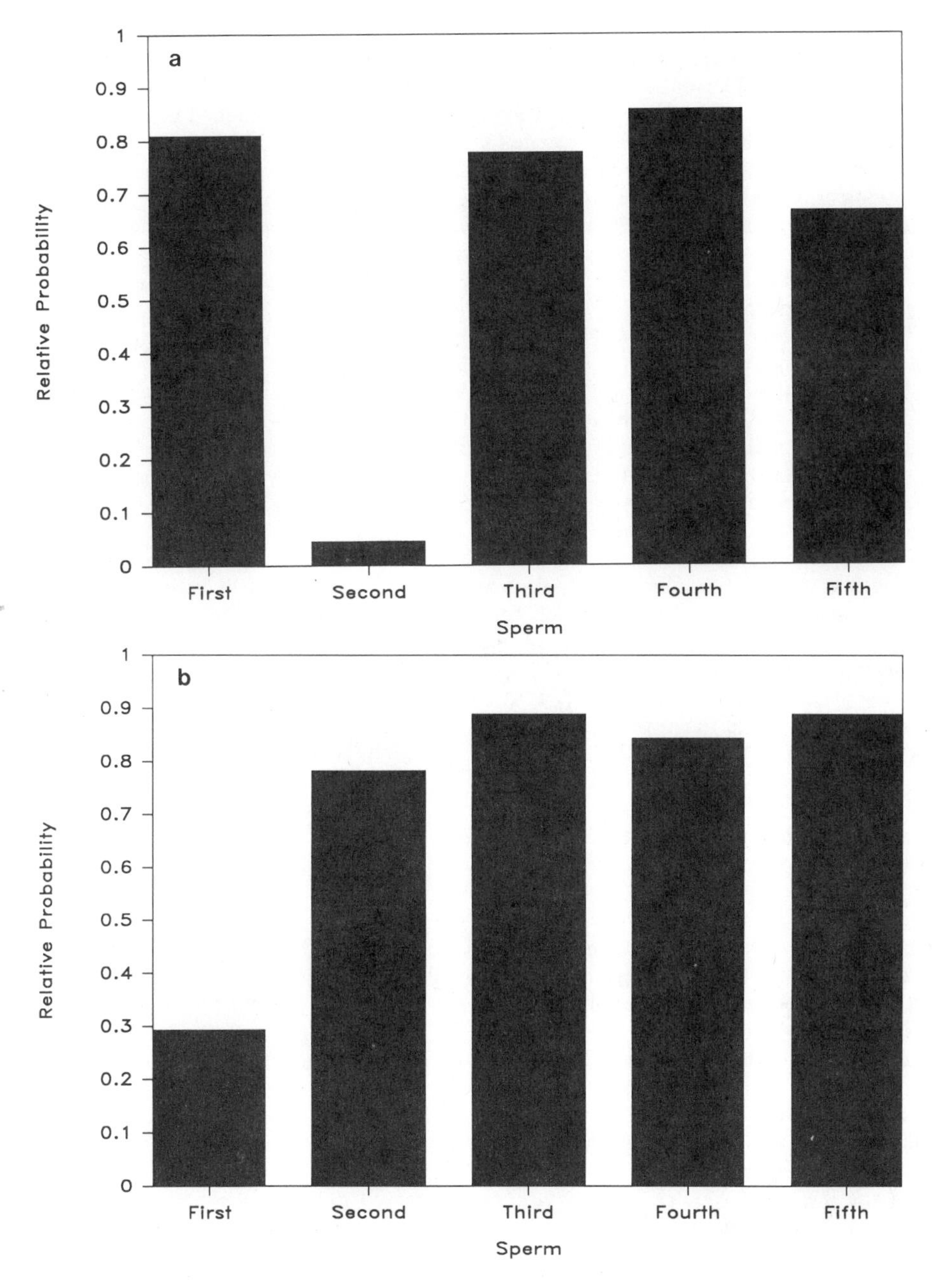

Figure 7. a = Eggs, **b** = Oocytes. Relative probability of entry for the first, second, third, fourth and fifth sperm in eggs **(a)** and oocytes **(b)**. Equation 2 and the data plotted in Figure 6 were used to generate the heights of the bars in this figure.

that the average number of sperm per oocyte (1.46) was greater than the average number of sperm per egg (1.00) when both were inseminated with 4000 sperm/ml.

In addition, the observed distributions (Figure 6, black bars) were compared with the Poisson distribution (Equation 1) expected if sperm acted independently during entry (Figure 6a, b, gray bars). Significant differences between the observed and expected distributions indicate that sperm do not act independently in entering either eggs or oocytes. The observed distribution for eggs is significantly different from the expected Poisson distribution determined from the mean number of sperm per egg ($X^2 = 746$ with four degrees of freedom;

$p \ll 0.005$). The observed distribution is much narrower than expected, clumping prominently at one sperm per egg (Figure 6a).

The observed distribution for oocytes (Figure 6b, black bars) is significantly different from the expected Poisson distribution (Figure 6b, gray bars) determined using Equation 1 and the mean number of sperm per oocyte ($X^2 = 210$ with 5 degrees of freedom; $p \ll 0.005$). The observed distribution is much more dispersed than the Poisson expectations. The observations in the center of the distribution, between 1 and 5 sperm per oocyte, are less than expected. The observations further from the center of the distribution, at 0 and at greater than 6 sperm per oocyte, are greater than expected. The sperm interactions in eggs and oocytes, indicated by the lack of independence for entries, must differ to cause clumping in eggs and dispersion in oocytes.

The difference of interactions between sperm in eggs and in oocytes was assessed further by estimating the probability of sperm entry (Equation 2) for the first, second, third, fourth and fifth sperm in eggs and oocytes (Figure 7). In eggs, the probability of entry was high for the first sperm but was precipitously lower for the second sperm at a value near 5% of the first sperm's probability (Figure 7a). This suggests that sperm interact negatively during entry in eggs such that the first sperm decreases the probability of entry for subsequent sperm. In oocytes, the probability of entry for the second and third sperm was greater than the probability for the first sperm (Figure 7b). This suggests that sperm interact positively (cooperatively) during entry in oocytes such that the entry of one or two sperm increases the probability of entry for subsequent sperm entries.

The mean numbers of sperm which entered eggs or oocytes were plotted as a function of the concentration of sperm during insemination as an independent check of the cooperativity of sperm interaction in oocytes (Figure 8, inset). If j sperm must cooperate to result in an entry, then the probability of entry (E) will be proportional to the concentration of sperm [S] raised to the power j:

$$E \propto [S]^{j} \quad . \qquad \textbf{Eq. 4}$$

The logarithm of each side of Equation 4 yields:

$$\log(E) \propto j \log[S] \quad , \qquad \textbf{Eq. 5}$$

indicating that a plot of log[S] vs log(E) will have a slope of j. In practice, the <u>steepest</u> slope at low sperm concentrations when plotted on double logarithmic axes is a <u>minimum</u> estimate of the number of sperm which cooperate in entering eggs or oocytes. We use the mean number of sperm per egg or per oocyte as an estimate of E (Figure 8). The steepest slope (j) for eggs at low sperm concentrations was 1.00 ± 0.069 (n = three experiments) and was not significantly different from 1 for eggs (Figure 8); whereas, the slope (j) for oocytes was 2.0 ± 0.20 (n = three experiments) and is significantly greater than 1 (Figure 8). These slopes indicate that sperm interact cooperatively during entry in oocytes but that no cooperativity is detected for sperm entry in eggs.

<u>Experiments in which single oocytes were observed during insemination with a high density of sperm</u>. Sperm cooperate in causing depolarizations in oocytes (see membrane potential records section, above) and interact cooperatively during sperm entry in oocytes. It was not certain that the cooperative depolarization was the same mechanism which was responsible for the cooperativity of sperm entry. In order to test this, oocyte membrane potential was monitored during insemination with high densities of sperm (2×10^7 sperm/ml). When many sperm attached to the oocyte in rapid succession, many small step-like depolarizations summed, attaining levels of membrane potential between -10 and 0 mV. Many of the sperm which had attached were observed as they entered the cytoplasm of the oocyte. Large mound-shaped fertilization cones formed at the sites where sperm entered. This suggested that the large depolarization resulting from the summed responses was sufficient to permit sperm entry.

In a similar experiment, oocytes, voltage-clamped to maintain membrane potentials constant near -70 mV were inseminated with high densities of sperm. Many sperm transient currents indicative of sperm-oocyte interaction were observed as sperm attached in close

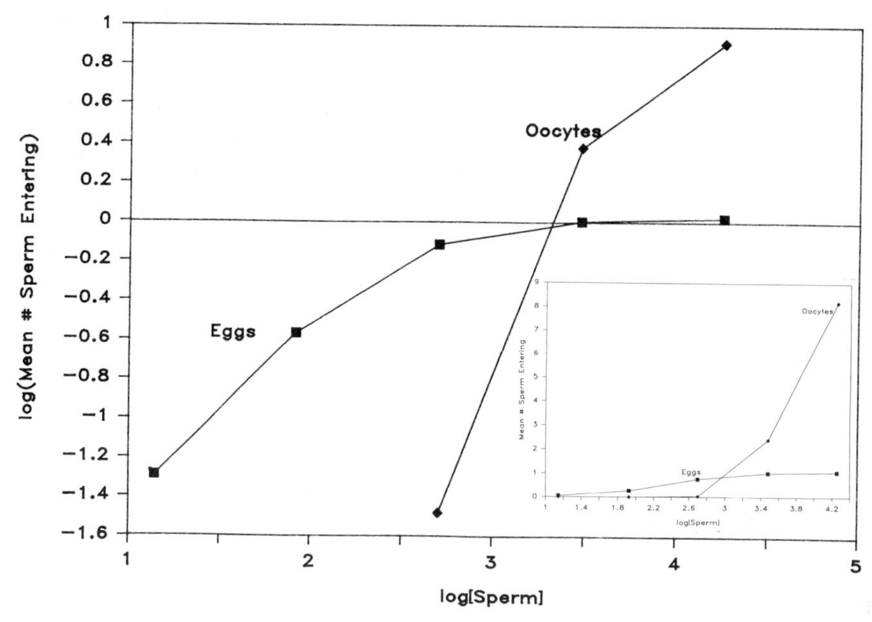

Figure 8. Double logarithmic plot of the mean number of sperm entering eggs and oocytes inseminated at different concentrations of sperm. The logarithm of the mean number of sperm/egg or sperm/oocyte is plotted in order to estimate the slopes of the relationships at low sperm concentrations. Data were obtained by inseminating five different tubes each containing both eggs and oocytes in the same proportion. The same data appear in the inset plotted as the mean number of sperm/egg and sperm/oocyte vs. the logarithm of the sperm concentration. The data used to obtain the means plotted at log[Sperm] = 3.5 in this figure were also used in Figures 6 and 7. (Note: the density of eggs and oocytes was maintained constant so that sperm/egg ratio and the sperm/oocyte ratio is in direct proportion to the sperm concentration.)

succession to the surface of the oocyte, however, sperm failed to enter. This indicated that the cooperative interaction of sperm leading to entry was eliminated when the depolarization was precluded. Therefore, we conclude that the depolarization is the critical, cooperative step which is required to permit sperm entry in oocytes.

Sperm Entry, Type of Electrophysiological Response, and Fertilization Cone Morphology Are Associated

Three phenomena are controlled by the clamped membrane potential of an egg or oocyte following sperm attachment and generation of a conductance increase. Sperm entry occurs only at membrane potentials between -30 and +17 mV (Figure 4, p. 29). The electrophysiological response occurs at all membrane potentials equal to or more negative than +17 mV (Figure 2, p. 26); but cuts off abruptly in more and more cases at membrane potentials increasingly more negative than -30 mV (Figure 4). A fertilization cone large enough to contain the sperm's head forms at potentials between +17 and -30 mV. The type of cone which forms at more negative membrane potentials when sperm fail to enter is much smaller and shaped like a filament or is not seen. The voltage dependences for activation currents which do not cut off and for formation of large fertilization cones in eggs are identical to the voltage dependence for sperm entry in eggs (Figure 4a). In oocytes the voltage dependences are very similar as shown in Figure 9, in which the points displaying the three different voltage-dependent phenomena at the same membrane potential are nearly identical. When sperm entry occurred in an egg, an electrophysiological response which did not cut off occurred, and a large fertilization cone formed (Table 2). The converse was also true: when

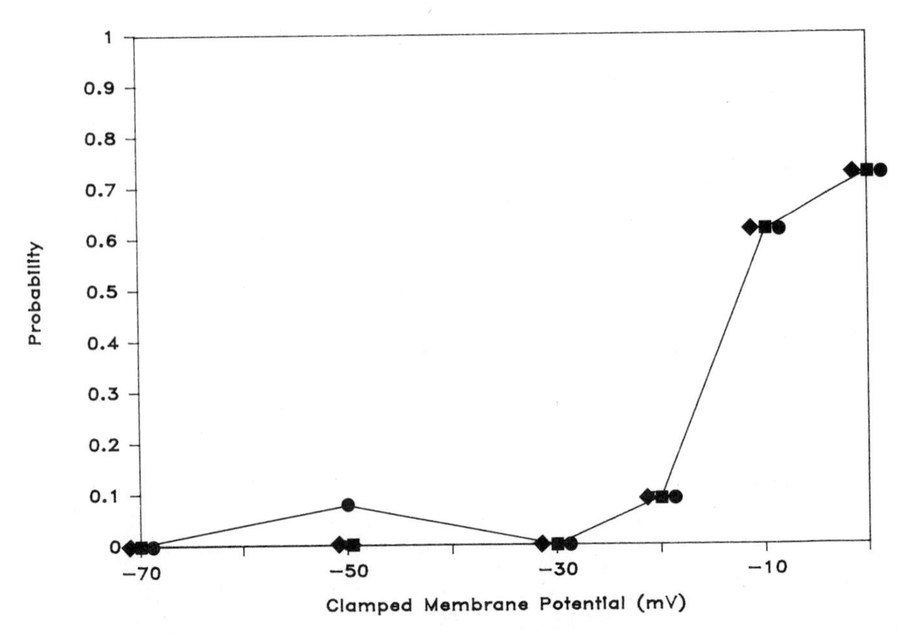

Figure 9. Similarity of voltage dependence for sperm entry (solid squares), formation of a large fertilization cone (solid diamonds), and occurrence of a prolonged electrophysiological response (closed circles) in oocytes.

the sperm did not enter an egg, an electrophysiological response which cut off abruptly occurred, and a small, filament-like fertilization cone or no fertilization cone was formed (Table 2). These three phenomena did not occur independently ($X^2 = 329$ with 3 degrees of freedom; $p << 0.005$). In oocytes, the association was also apparent because sperm entry was always accompanied by an electrophysiological response which did not cut off prematurely, and by formation of a large fertilization cone (Table 3). Most of the sperm which failed to enter oocytes caused electrophysiological responses which abruptly cut off (54/58 responses were brief transients) and nearly all of the fertilization cones at the site of sperm attachment either did not form or were tiny (57 cones/58 sperm which caused electrophysiologcial responses but failed to enter).

Table 2. Association[a] of Sperm Entry, Fertilization Cone Size and Type of Electrophysiological Response in Eggs

Entry:	Sperm Enters		No Sperm Enters	
Fert. Cone Size:	Large	Small/None	Large	Small/None
E.P. Response Type[b]				
Sustained	60	0	0	0
Abrupt Cut-off	0	0	0	191

a These data were tested for independence of occurrence of sperm entry, fertilization cone size, and type of electrophysiological response using a contingency Chi squared test. $X^2 = 82$ with 2 degrees of freedom. The probability that these events occurred independently is much less than 0.005.

b type of electrophysiological response

Table 3. Association[a] of Sperm Entry, Fertilization Cone Size and Type of Electrophysiological Response in Oocytes

Entry:	Sperm Enters		No Sperm Enters	
Fert. Cone Size:	Large	Small/None	Large	Small/None
E.P. Response Type[b]				
Prolonged	17	0	1	3
Sperm Transient	0	0	0	54

a These data were tested for independence of occurrence of sperm entry, fertilization cone size, and type of electrophysiological response using a contingency Chi squared test. $X^2 = 329$ with 3 degrees of freedom. The probability that these events occurred independently is much less than 0.005.

b type of electrophysiological response

DISCUSSION

Entry of sperm into sea urchin eggs and oocytes is controlled by two different voltage dependent steps: 1) a step sensitive to positive membrane potentials and which determines whether or not the egg's membrane conductance increases, and 2) a step sensitive to negative membrane potentials and which determines whether the conductance increase cuts off and whether a large or small fertilization cone forms.

Voltage-Sensitive Step at Positive Membrane Potentials

At membrane potentials more positive than 0 mV, decreasingly fewer of the sperm which attach cause a conductance increase. The fitted Boltzmann equations (Equation 3) provide estimates of the probability of the conductance increase occurring as a function of membrane potential. The membrane potential at which the relationship declines to 50% of its maximum was +8 mV for eggs and +5 mV for oocytes. The errors associated with measuring membrane potential (± 4 mV) and the interval between potentials tested (10 mV) are too great to permit us to determine whether these potentials or the corresponding estimates of voltage dependence are significantly different. Therefore we conclude that the voltage sensor is not distinguishable between eggs and oocytes for this voltage-dependent event. The steepness of the relationship (e-fold for 2.4 mV and for 5.4 mV, respectively, in eggs and oocytes) indicates that the voltage dependence of the conductance increase is mediated by the equivalent of between 5 and 10 electron charges traversing the entire field of the plasma membrane.

We have never observed sperm entry which was not preceded by the characteristic conductance increase. Sperm which cause the conductance increase in eggs held at more negative potentials but which are subsequently stepped to potentials more positive than this potential are not blocked from entering (Lynn 1985; Shen and Steinhardt 1984). Therefore the effect of positive membrane potential associated with precluding sperm entry occurs at a critical step which also precludes the step-like increase of conductance.

The membrane potential attains a potential near +25 mV during the activation potential in eggs but subsides during the latent period to a value near +10 mV. The probability of sperm causing an electrophysiological response following attachment to an egg is greatly reduced at such positive potentials (Figure 2a). This reduction of the probability for an event which is a prerequisite to sperm entry is consistent with the proposal that the depolarization during fertilization acts as a rapid block to polyspermy (Jaffe 1976).

The membrane potential depolarized no more than to -10 to 0 mV in oocytes even following attachment of many sperm to the surface. Potentials in this range do not significantly decrease the probability of causing an electrophysiological response (Figure 2b). This may provide an explanation for the observation that polyspermy occurs in sea urchin oocytes (Runnstrom 1963; Franklin 1965; Longo 1978a). Similar attenuation of fertilization or activation potentials resulting in a greater susceptibility to polyspermy has been reported for the oocytes of starfish (Miyazaki 1979) and amphibians (Schlichter and Elinson 1981).

We suggest that the conductance increase is a prerequisite for sperm entry; although, we cannot discount the converse based on our negative results. Our recent observation that a step-like increase of membrane capacitance occurs at the instant that the step-like, sperm-induced conductance increase occurs suggests that sperm-egg membrane fusion occurs at the instant of conductance increase (McCulloh and Chambers 1986a,b). The simplest interpretation for these simultaneous changes is that the plasma membranes of the sperm and the egg fuse at this early time and that the conductance increase in the egg is caused by the sudden presence of sperm plasma membrane channels in the membrane of the egg at the time of fusion. It is possible that the lack of a conductance increase at potentials more positive than +20 mV results from a voltage induced failure of sperm to fuse with the egg following attachment, hence precluding their entry. It is not surprising that a sperm which does not fuse will not enter.

Voltage-Sensitive Step at Negative Membrane Potentials

Sperm fail to enter either eggs or oocytes voltage clamped at membrane potentials more negative than -20 mV despite the occurrence of electophysiological responses which cut off abruptly (Figures 3, 4). Boltzmann equations (Equation 3) fitted to the data estimate the voltage dependence for the probability of sperm entry once an electrophysiological response occurs. The membrane potentials at which eggs and oocytes block 50% of the entries (V_{50} = -30 and -10 mV, respectively) are significantly different. This indicates that the control of sperm entry differs in eggs and oocytes. The slopes of the voltage dependence for eggs and oocytes (RT/zF = 7.6 mV and 6.1 mV for an e-fold change, respectively) cannot be distinguished because of the errors associated with measurement of membrane potential. Similarity for these estimates of how strongly the process depends on membrane potential suggests that the voltage sensors for processes which regulate these voltage dependent events are the same in both oocytes and eggs. In addition, it is estimated from these voltage dependences that whether the sperm enters or not is mediated by the equivalent of between three and four electron charges traversing the plasma membrane's field. This mechanism is less voltage dependent than the mechanism apparent at positive membrane potentials (above).

Membrane potential rather than some artifact associated with its measurement regulates whether sperm entry occurs. The phenomenon occurs in eggs whose membrane potential was maintained negative by application of a constant hyperpolarizing current rather than by voltage clamp. This rules out some artifact of the voltage clamp feedback. Sperm entry fails to occur in oocytes during membrane potential recording, as long as only one or a few sperm attach within a brief period of time, even when no hyperpolarizing current is applied. This failure of sperm entry in oocytes can be explained by the failure of the small, summed depolarizations to attain a potential more positive than -10 mV, the value necessary to permit sperm entry in voltage clamped oocytes (Figure 4b). From this observation and the similarity of the voltage dependence curves in eggs and oocytes (Figures 4a, b), we conclude that the blockade of sperm entry at negative membrane potentials is truly dependent on membrane potential and is not an artifact of passing exogenous current through the oocyte's membrane.

Summation of electrophysiological responses was observed when more than one sperm attached within a brief period of time during membrane potential recordings from oocytes and during voltage clamp experiments in either eggs or oocytes. Each sperm which attached was capable of generating a conductance increase of similar size (a quantum). The result of linear summation of conductances leads to a non-linear summation of depolarizations in the oocyte. The summation indicates that these responses are evoked by an interaction between sperm and egg or oocyte which does not saturate with the interaction of a single or even several sperm. The conductance increase is localized to the region near the sperm (McCulloh and Chambers 1986a). When many sperm attached to different locations on the oocyte surface, it was possible to sum many depolarizations and for the resulting depolarization to attain membrane potentials more positive than those required to permit entry of sperm in voltage clamped oocytes. When these potentials were attained, sperm entry occurred.

From the observation that sperm-induced currents and depolarizations can sum, it was predicted that sperm entry in oocytes would rely upon the cooperative interaction of sperm-induced depolarizations. The histograms of numbers of sperm which entered either eggs or oocytes indicated that sperm entries did not occur independently. It is concluded that sperm interacted during entry. The interaction of sperm in eggs and oocytes differed. The first sperm decreased the probability of subsequent sperm entries in eggs; whereas, in oocytes, the first sperm increased the probability of subsequent sperm entries. Cooperativity was suggested for sperm entry in oocytes by the values greater than one for slopes in double logarithmic plots of the mean number of sperm which entered oocytes versus the concentration of sperm. Cooperativity for sperm entry in eggs is not evident from the double logarithmic plots of data from eggs. All of these data, together, agree with the notion that sperm interact cooperatively to enter oocytes whereas they interact negatively in eggs. The results support the predictions made by assuming that depolarization is required for sperm entry. The agreement between results and predictions suggests that this assumption is correct in oocytes even when the membrane potential is not monitored with the invasive microelectrode technique. Although the cooperative interaction indicated by these results fulfills the prediction, it is not sufficient to demonstrate that the membrane potential is the means by which the cooperative interaction of sperm entry in oocytes is mediated.

Two experiments address the issue of whether summation of depolarizations is responsible for the cooperative interaction of sperm entering oocytes. Sperm entry occurred in oocytes during membrane potential recording only when many sperm attached in close succession and the membrane potential depolarized to values between -10 and 0 mV. These potentials fall within the range of membrane potentials permissive for sperm entry when single sperm attach to voltage clamped oocytes (Figure 4b). When oocytes voltage clamped at -70 mV (a non-permissive potential) were inseminated with a high density of sperm, sperm entry failed. This indicated that the cooperative interaction of sperm was eliminated by precluding the depolarization by membrane potential control. Therefore, we conclude that the summation of depolarizations caused by different sperm attaching to the oocyte is the means of interacting cooperatively employed by sperm during entry in oocytes.

Two Distinct Voltage-Regulated Events Involved in Sperm Entry

Two distinct voltage-dependent events regulate sperm entry. The events are distinguished by the different voltage ranges over which they are effective in precluding sperm entry, and by the observation that fusion and the conductance increase are permitted to occur at membrane potentials more negative than -20 mV, whereas fusion and the conductance increase occur rarely at membrane potentials more positive than +10 mV. This indicates that sperm entry procedes to a more advanced stage prior to halting at negative membrane potentials than at positive membrane potentials. The specific time during which the negative potential exerts its effect occurs several seconds after fusion and after the conductance has increased in eggs (Lynn 1985 and in this volume).

In eggs which are not voltage clamped, the potential depolarizes to a level near +25 mV following fusion of the first sperm and then subsides to a level near +10 mV. This depolarization has two roles: 1) to permit the entry of the first sperm (by depolarizing the membrane potential of the egg to a value more positive than - 30 mV, the potential at which 50% of the sperm fail to enter despite having caused an electrophysiological response) and 2) to block the fusion of sperm which attach to the egg after the first sperm (by depolarizing the membrane potential of the egg to a value more positive than +8 mV).

Voltage dependence can be sensed in at least two ways: 1) by a charge or dipole which resides in the membrane and responds to changes in the electrical field through the membrane by migrating or exerting the force of the field toward a conformation change of a macromolecule; or 2) by charges which reside in the bulk solutions on either side of the membrane but which can pass through the membrane resulting in flux and accumulation of the charged moiety on one side and/or its depletion on the other side. It cannot be determined which of these sensing methods is used in the two different voltage dependent phenomena described. The voltage-dependence term RT/zF (mV/e-fold change of V) obtained from fitting for both negative and positive events in eggs were indistinguishable from the respective voltage dependence terms in oocytes. This suggests that the charges or dipoles which sense the membrane potential during the two events are either not affected by development or are associated with the sperm.

Although the potential at which the positive block of sperm entry occurred was indistinguishable in eggs and oocytes, the negative block of sperm entry in eggs required more negative membrane potentials than the negative block in oocytes. Neither the slopes nor the charges on the voltage sensors were significantly different in eggs and oocytes. The shift of the potential at which the voltage dependence occurred is likely to be associated with a developmental change of the oocyte as it matures to become an egg. Several possible explanations for this exist. The shift of voltage dependence could be caused by an accumulation or depletion of an ion or ions in the oocyte as it matures. If sperm entry is sensitive to the cytoplasmic concentration of such an ion species, changes of this baseline concentration would alter the level of voltage dependent ion flux required to exceed the threshold for an effect on sperm entry. Alternatively, the observed shift could be caused by a change of the surface potential (charge density) at either the cytoplasmic or extracellular surface of the oocyte's plasma membrane. Surface potential alterations are capable of affecting the voltage gradient (field) through the membranes without any direct effect on membrane potential. Charges or dipoles located within the membrane respond to the voltage gradient. If both the positive and negative blocks were mediated by charges or dipoles in the membrane, it would be expected that a difference of surface potential in eggs compared to oocytes would shift both voltage dependence curves in the same direction and by the same amount. Therefore, it is unlikely that both positive and negative blocks sense voltage via charges or dipoles residing in the membrane. However, either block alone could sense voltage in this way.

Association of Sperm Entry, Size of Fertilization Cone and the Type of Electrophysiological Response

Sperm entry, the failure of the electrophysiological response to cut off, and the formation of a large fertilization cone were associated in both oocytes and eggs (Tables 2, 3). The determination that these phenomena do not occur independently implies that the mechanisms generating these phenomena are related. Either one event leads to the occurrence of the others, or some fourth event causes all three to occur. The common voltage dependence for each of these events further supports the notion that both the mechanism for fertilization cone formation and the mechanism for causing the conductance increases are linked to the same membrane potential sensing mechanism which regulates sperm entry.

Sperm entry and fertilization cone formation are blocked jointly by treatment of eggs with cytochalasins (Longo 1978b; 1980; Schatten and Schatten 1981; Cline *et al.* 1983; Cline and Schatten 1986). Shortly following sperm egg fusion, filamentous actin appears in the cortex of the egg (Tilney and Jaffe 1980; Yonemura and Mabuchi 1987) and at the site of sperm-egg fusion (McCulloh *et al.* 1987). It is expected that the cytochalasins act by disrupting the normal polymerization of actin filaments which accompanies fertilization. It is possible that the common blocking effects of negative membrane potential on sperm entry and fertilization cone formation may be mediated by an effect on cytoplasmic polymerization of actin filaments, perhaps in the region near the sperm.

CONCLUSION

Sperm entry involves two voltage dependent, separable steps: sperm-egg fusion and nuclear incorporation. The inhibition at membrane potentials more positive than +10 mV involves movement of many charges through the membrane's voltage gradient and may act as a rapid block to polyspermy in eggs by blocking fusion. The inhibitory effect of negative membrane potentials and the summation of depolarizations evoked in oocytes by sperm account for the cooperative interaction of sperm evident during entry in oocytes. No such cooperativity is evident in eggs. The inhibition of sperm entry at membrane potentials more negative than -20 mV requires the movement of fewer charges through the membrane's voltage gradient. Only one of the two voltage dependent steps (although which step is unclear at this time) could be mediated by a charge or dipole in the membrane. One common, voltage dependent mechanism affecting the membrane currents following sperm attachment may regulate nuclear incorporation, and fertilization cone formation perhaps by a mechanism which disrupts the formation of actin filaments.

ACKNOWLEDGEMENTS

I wish to thank my collaborators, Ted Chambers and John Lynn, who collected (prior to my arrival in the lab) much of the data analyzed in this chapter. In addition, I thank them for their open exchange of ideas and helpful discussions which directed the development of my thoughts. I thank Pedro Ivonnet for his careful technical assistance and especially his willingness to count zillions of sperm which had entered eggs and oocytes. I thank Bill Ferguson for providing the membrane potential records from eggs which were maintained at negative potentials by application of constant current. I thank Bob Wall of the USDA, for making me aware of the Hoechst staining technique and for generously supplying me with a quantity sufficient to perform many experiments. I also wish to thank Ellen McGlade McCulloh for critically reading the manuscript prior to submission. This work was supported by a National Research Service Award HD 06505, and by research grants to E. L. Chambers: NIH HD 19126, NSF PCM 8316864, and NSF DCB 8711787.

REFERENCES

Allen, R. D. and J. L. Griffin. 1958. The time sequence of early events in the fertilization of sea urchin eggs. I. The latent period and the cortical reaction. *Exp. Cell Res.* 15:163-173.

Chambers, E. L. and J. de Armendi. 1979. Membrane potential of eggs of the sea urchin, *Lytechinus variegatus*. *Exp. Cell Res.* 122:203-218.

Cline, C. A., and G. Schatten. 1986. Microfilaments during sea urchin fertilization: Fluorescence detection with rhodaminyl phalloidin. *Gamete Res.* 14:277-291.

Cline, C. A., H. Schatten, R. Balczon, and G. Schatten. 1983. Actin mediated surface motility during sea urchin fertilization. *Cell Motil.* 3:513-524.

Dale, B. and L. Santanella. 1985. Sperm-oocyte interaction in the sea urchin. *J. Cell Sci.* 74:153-167.

Franklin, L. E. 1965. Morphology of gamete membrane fusion and of sperm entry into oocytes of the sea urchin. *J. Cell Biol.* 25:81-100.

Jaffe, L. A. 1976. Fast block to polyspermy in sea urchin eggs is electrically mediated. *Nature (Lond.)* 261:68-71.

Longo, F. J. 1978a. Insemination of immature sea urchin (*Arbacia punctulata*) eggs. *Dev. Biol.* 62:271-291.

Longo, F. J. 1978b. Effects of cytochalasin B on sperm-egg interactions. *Dev. Biol.* 67:249-265.

Longo, F. J. 1980. Organization of microfilaments in sea urchin (*Arbacia punctulata*) eggs at fertilization: Effect of cytochalasin B. *Dev. Biol.* 74:422-433.

Lynn, J. W. 1985. Time and voltage dependent sperm penetration in sea urchin eggs. *Dev. Growth & Differ.* 27:177-178.

Lynn J. W. and E. L. Chambers. 1984. Voltage clamp studies of fertilization in sea urchin eggs. I. Effect of clamped membrane potential on sperm entry, activation, and development. *Dev. Biol.* 102:98-109.

Lynn, J. W., D. H. McCulloh, and E. L. Chambers. 1988. Voltage clamp studies of fertilization in sea urchin eggs. II. Current patterns in relation to sperm entry, nonentry, and activation. *Dev. Biol.* 128:305-323.

McCulloh, D. H. 1985. Cortical reaction of sea urchin eggs: Rate of propagation and extent of exocytosis revealed by membrane capacitance. *Dev. Growth & Differ.* 27:178.

McCulloh, D. H. and E. L. Chambers. 1984. Capacitance increases following insemination of voltage-clamped sea urchin eggs. *Biophys. J.* 45:73a.

McCulloh, D. H. and E. L. Chambers. 1986a. When does the sperm fuse with the egg? *J. Gen. Physiol.* 88:38a-39a.

McCulloh, D. H. and E. L. Chambers. 1986b. Fusion and "unfusion" of sperm and egg are voltage dependent in the sea urchin *Lytechinus variegatus*. *J. Cell Biol.* 103:236a.

McCulloh, D. H., E. L. Chambers, and J. W. Lynn. 1984. Membrane currents associated with sea urchin sperm-oocyte interactions. *J. Cell Biol.* 99:57a.

McCulloh, D. H., P. I. Ivonnet, and E. L. Chambers. 1988. Actin polymerization precedes fertilization cone formation and sperm entry in the sea urchin egg. *Cell Motil. Cytoskeleton* 10:345.

McCulloh, D. H., J. W. Lynn, and E. L. Chambers. 1987. Membrane depolarization facilitates sperm entry, large fertilization cone formation, and prolonged current responses in sea urchin oocytes. *Dev. Biol.* 124:177-190.

Miyazaki, S. 1979. Fast polyspermy block and activation potential: Electrophysiological bases for their changes during oocyte maturation of a starfish. *Dev. Biol.* 70:341-354.

Presley, R. and P. F. Baker. 1970. Kinetics of fertilization in the sea urchin: A comparison of methods. *J. Exp. Biol.* 52:455-468.

Runnstrom, J. 1963. Sperm-induced protrusions in sea urchin oocytes: A study of phase separation and mixing in living cytoplasm. *Dev. Biol.* 73:38-50.

Schatten, G. and H. Schatten. 1981. Effects of motility inhibitors during sea urchin fertilization. Microfilament inhibitors prevent sperm incorporation and restructuing of fertilized egg cortex, whereas microtubule inhibitors prevent pronuclear migrations. *Exp. Cell Res.* 135:311-330.

Schlichter, L. C. and R. P. Elinson. 1981. Electrical responses of immature and mature *Rana pipiens* oocytes to sperm and other activating stimuli. *Dev. Biol.* 83:33-41.

Shen, S. and R. A. Steinhardt. 1984. Time and voltage windows for reversing the electrical block to fertilization. *Proc. Natl. Acad. Sci. USA* 81:1436-1439.

Tilney, L. G. and L. A. Jaffe. 1980. Actin, microvilli, and the fertilization cone of sea urchin eggs. *J. Cell Biol.* 87:771-782.

Yonemura, S. and I. Mabuchi. 1987. Wave of cortical actin polymerization in the sea urchin egg. *Cell Motil. Cytoskeleton* 7:46-53.

Vacquier, V. D. and J. E. Payne. 1973. Methods for quantitating sea urchin egg binding. *Exp. Cell Res.* 82:227-235.

CORRELATIONS BETWEEN TIME-DEPENDENT AND CYTOCHALASIN B AFFECTED SPERM ENTRY IN VOLTAGE-CLAMPED SEA URCHIN EGGS

John W. Lynn

Department of Zoology and Physiology
Louisiana State University
Baton Rouge, LA 70803-1725

ABSTRACT

Sperm entry is only successful in approximately 15% of *Lytechinus variegatus* eggs voltage-clamped at -70 mV, where as 100% of eggs voltage-clamped at -20 mV are penetrated by sperm (Lynn and Chambers 1984). Suppression of sperm penetration at -70 mV is reversible if the clamped egg membrane potential is stepped to -20 mV (a permissive potential for sperm entry). Sperm incorporation occurred in 50% of the eggs when the clamped potential was shifted at approximately 7.5s with an increasing percentage of penetrations as the time period from the sperm-initiated conductance increase to the time of the step was decreased. In the reciprocal experiments where the membrane potential was first clamped at -20 mV and then stepped to -70 mV at specific time points following a sperm-egg interaction, 50% of the eggs were penetrated at an average time of 10.4s with an increasing percentage of sperm incorporation occurring as the time to the step was increased. A similar failure of sperm penetration is induced by treating the sea urchin egg with cytochalasin B (cyto B) (Longo 1978; Schatten and Schatten 1980). Experiments with cyto B-pretreated eggs clamped at -20 mV (an otherwise permissive potential for sperm entry) revealed that not only is sperm entry blocked, but voltage-clamp current profiles associated with the failure of sperm entry at -70 mV were diagnostic of nonpermissive voltage-clamped potentials. In addition, localized FE elevation lifting the sperm from the surface of the egg was frequently observed (an event normally seen only at clamped potentials of -90 and -100 mV). One extended interpretation of these experiments is that microfilament polymerization is interrupted in both the cyto B experiments and in experiments where the egg is clamped at -70 mV and more negative potentials without cyto B treatments.

INTRODUCTION

The phenomenon of electrophysiological events associated with sperm-egg interactions has been the subject of numerous investigations utilizing intracellular recording techniques over the past 30 years (Hagiwara and Jaffe 1979). These studies have included such diverse organisms as the echinoderms (e.g., Chambers and De Armendi 1979; Hagiwara and Jaffe 1979; Steinhardt *et al.* 1971), amphibians (e.g. Cross and Elinson 1980; Grey *et al.* 1982; Jaffe and Schlichter 1985), mammals (e.g. Miyazaki and Igusa 1981, McCulloh *et al.* 1983), annelids (Kline *et al.* 1985), crustaceans (Goudeau and Goudeau 1985), teleosts (Nuccitelli 1980), and echiuroids (Gould-Somero *et al.* 1979). The references presented for each of these groups are in no way complete, but should serve to direct the reader into the literature.

The significance of recordable electrical events at the time of fertilization is two-fold. First, they are the earliest recordable physiological change in the egg cell during the fertilization process and, as such, allow incredible time resolution for sequential physiological and morphological events associated with egg activation. A second important aspect is that, at least in some species, a shift in the membrane potential to a positive value during the earliest sperm-egg interactions acts as a very early, although possibly incomplete, block to polyspermy (Jaffe 1976; Jaffe and Gould 1985). In the sea urchin, the normal resting potential of the egg cell membrane is approximately -75 mV, and following contact of the sperm with the egg surface, the membrane potential is rapidly (<100 ms) depolarized and reverses with a positive overshoot to +20 mV (Chambers and De Armendi 1979). This is particularly important since some species (most notably the echinoderms and anurans) may be challenged simultaneously with large numbers of sperm and the so-called late, long lasting block to polyspermy (formation of the fertilization envelope) may require over 30s to become established. Therefore elevation of the fertilization envelope would be too slow for the observed degree of monospermy (Rothschild and Swann 1952; for review, Nuccitelli and Grey 1984). In contrast, the rapid depolarization would provide an immediate barrier to those sperm already present which are not the initial sperm activating the egg. The subsequent electrophysiological events including and following the initial egg membrane depolarization have since become referred to as the activation potential (Chambers and De Armendi 1979; Nuccitelli and Grey 1984). Demonstration of the effectiveness of positive membrane potentials in excluding supernumerary sperm has centered on the lack of sperm entry and absence of recordable electrophysiological events when the plasmalemma has been artificially maintained at positive values (Jaffe 1976; Lynn and Chambers 1984).

More recently, Lynn and Chambers (1984) demonstrated that sperm entry was also inhibited in the sea urchin *Lytechinus variegatus* when the membrane potential was voltage-clamped at values more negative than -30 mV. A major difference between the inhibition of sperm entry at clamped egg membrane potentials of +20 mV and those at potentials more negative than -30 mV is that no electrophysiological events are recorded at the positive potentials. The question was posed as to why sperm entry did not occur at either of these values (termed nonpermissive potentials), but electrophysiological events are recorded at the more negative potentials. The original hypothesis for the fast block to polyspermy predicted, in fact, that sperm entry should be easier at more negative potentials increasing the likelihood of polyspermic eggs when eggs are artificially clamped at those potentials (Jaffe 1976). In addition, a natural sperm-induced shift from the normal resting membrane potential (-75 mV) to the +20 mV value allows fertilization to proceed normally.

These observations suggest that an initial interaction between the sperm and the egg could occur at the negative membrane potentials, but that a second, voltage-sensitive event required a shift from the original resting potential to more positive values. In this paper, we will examine preliminary evidence suggesting that a second, voltage-dependent step occurs and that the time when the membrane potential is permissive for sperm entry is also time dependent. In addition, we will also present evidence that the electrophysiological events recorded in sea urchin eggs voltage-clamped at -75 mV can be mimicked when eggs are pretreated with cytochalasin B, voltage-clamped at -20 mV, and then inseminated.

MATERIALS AND METHODS

For a more complete description of the electrophysiological methods, egg selection criteria and analysis of currents detected during the experiments described below, the reader is referred to Lynn *et al* (1988), Lynn and Chambers (1984), and Chambers and De Armendi (1979). An extensive but not necessarily complete description will be given here.

Gamete Collection and Storage

Eggs. Specimens of *Lytechinus variegatus* were collected in 3-5 feet of water south of the Florida State University Marine Laboratory at Sopchoppy, Florida. Upon return to the laboratory, the animals were held in a refrigerated aquarium at 19-20°C. Maintaining the animals at this temperature aids in preventing premature spawning or reabsorption of the gonadal tissue. Gametes were collected by removing the mouth parts of the urchin and examining the gonads to determine if the animals were male or female. Females were inverted over a 100 ml beaker of natural or artificial seawater (see below) and the coelomic cavity was filled with 0.5 M KCl. When females had ceased spawning, they were removed from the beakers and the seawater overlying the settled gametes was decanted and replaced at least two times to remove excess K^+ from the media.

Selection of the eggs to use for the experiments was primarily based on the following criteria: 1) following insemination, 95-100% of the eggs must elevate fertilization envelopes; 2) large prominent insemination cones must develop; and 3) 98-100% of the fertilized eggs elevating FEs must cleave synchronously to the 2-cell stage. The selected group of eggs was further monitored during the experiments to insure continued synchrony of cleavage and normal development to the swimming blastula stage. Preference was given to those batches of eggs which met the above criteria and which displayed the greatest clarity of the egg cytoplasm. Eggs were stored at 14-16°C until the experiments. In all cases, eggs were used within 6-8h of the original spawning time. Within this time frame, no observable changes in the electrophysiological recordings from voltage-clamped eggs were detected.

Fresh eggs from the stored spawns were prepared just before use by washing repeatedly in fresh seawater until the jelly layer was removed as indicated by the eggs packing densely in the tip of a conical glass centrifuge. Eggs were sedimented gently on a hand crank centrifuge during washings to minimize damage.

Sperm. Sperm were collected by removing the whole testis from a male, blotting it dry on a large piece of filter paper, and transferring it to a glass syracuse dish. The testes were stored in a refrigerator at 10-12°C. Immediately before use, approximately 2-5 μl of the semen which had oozed out of the testis was diluted to 40 ml of seawater. The final sperm dilution was between 1.5×10^6 and 3.0×10^6 sperm/ml. Sperm were added to the culture dish at a distance of 1 to 1.5 cm from the egg to allow further dilution of the sperm before they reached the egg. Sperm dilutions were used within 45s of preparations. Subsequent inseminations were performed with freshly diluted sperm.

Seawater Preparations

Osmolarity of the artificial seawaters was adjusted to the level of the seawater in which the animals were collected (940 to 1020 mosm). Formulations of the artificial seawaters were based on the ratios of ionic components reported by Chambers and De Armendi (1979). The seawaters were buffered with 10 mM TAPS (Sigma) and the pH adjusted to 8.3.

Cytochalasin B Treatment

Eggs were treated 10 min in a 1 μg/ml solution of cytochalasin B (cyto B) prepared from a 1 mg/ml stock solution in DMSO. Before eggs were added to the dish, the cytochalasin B and DMSO were removed from the eggs by two rapid rinses in artificial seawater. Eggs were immediately added to the dish and the microelectrode inserted into the cytoplasm (see below). Insemination was performed as rapidly as possible after the voltage and current recordings had stabilized but in no case was longer than 15 min from the last rinse.

In an alternative procedure, untreated eggs were added to the dish, the microelectrode was inserted into an egg and the electrophysiological recordings allowed to stabilize. Cyto B was then added to the dish and remained during insemination. Five to 10 min following the observation of a fertilization envelope, the seawater in the dish was exchanged with several volumes of normal seawater not containing cyto B.

Control eggs were treated with 0.1% DMSO and rinsed as above for the cyto B treated eggs.

Electrophysiological Preparations

Dejellied eggs were placed in virgin Falcon 35 mm culture dishes containing approximately 2.5 ml of seawater. Sufficient settling time was allowed to permit the eggs to attach to the bottom of the culture dish. No treatment of the dish was used to increase cell attachment because many treatments may artificially activate the egg (Chambers, personal communication).

Unfertilized eggs were voltage-clamped using a switched, single-electrode clamp with a driven shield (Dagan Instruments, Inc; Wilson and Goldner 1975). Electrodes were prepared by pulling 1 mm OD thin wall (0.75 I.D., with an internal fiber) capillary tubing (WPI, Inc.) using a Narashige horizontal puller. Electrodes were painted with conductive silver emulsion paint (GC Electronics) to within 150 μM of the tip (Lewis and Wills 1980). After drying, the silver was covered with Liquid Tape (GC Electronics) to prevent shorting of the driven shield to the bath ground. A small area of silver paint on the shank of the electrode was left bare to allow connection of the Dagan driven shield. The bath was connected to the instrument ground through an Ag-AgCl agar bridge.

Microelectrodes were filled with a solution of 450 mM K_2SO_4, 30 mM NaCl, 30 mM KCl, and 1 mM NaCitrate. Electrode impedance averaged 25-35 MΩ with this filling solution. This solution was chosen since it more closely approximates the internal concentration of the ions in the sea urchin egg than the standard 3 M KCl. Potentially lethal effects on the egg from KCl leakage from electrodes filled with 3 M KCl are thereby avoided (Blatt and Slayman 1982). Tip potentials were zeroed upon entrance of the microelectrode into the bath, just before insertion of the microelectrode into the egg, and checked upon removal of the microelectrode from the egg.

Eggs were continuously voltage-clamped during the experiments before, during and up to 30 min after the eggs were inseminated. In the case of the stepped clamp experiments, command voltages were switched manually based on visual detection of the sperm-induced conductance increase on a Tektronix oscilloscope. For experiments on cytochalasin B treated eggs (see above), membranes were clamped at -20 mV for the duration of the experiment. Both current and voltage responses were recorded on a Vetter instrumentation recorder and a X-Y pen plotter (Soltec). Electrophysiological recordings are graphed as inward current (or negative potential) as down.

RESULTS

The eggs of *Lytechinus variegatus* are remarkably clear and allow easy observation of the formation of the sperm aster in the peripheral cytoplasm, the migration of the female pronucleus following sperm entry, and the formation of the large monaster during the centration of the female pronucleus. Observation of these morphological events is essential in following the effects of voltage-clamp procedures and treatment with various drugs. Further, the pattern of the inward currents in the voltage-clamped eggs is diagnostic of the success or failure of sperm entry into the egg cytoplasm (Lynn and Chambers 1984; Lynn, *et al* 1988). For a detailed description of these currents see Lynn *et al* (1988), David *et al* (1988), and Chambers (this volume). In addition, a brief description of the initial phase of the sperm-induced currents is presented in the discussion below.

For all experiments that follow, sperm concentrations were kept low where possible to increase the probability that only a single sperm would interact with the surface of the egg. This tactic allowed the identification of individual sperm on the surface of the egg so that the correlation of a sperm-egg interaction resulting in an electrophysiological event could be maximized (see Lynn and Chambers 1984).

Sperm Entry Is Time Dependent on the Clamped Voltage

Since voltage-dependent sperm entry has been previously demonstrated in the eggs of several species of sea urchins (Lynn and Chambers 1984), reversibility and time dependency of this phenomenon were investigated by stepping the clamped voltage of the egg at various intervals following the sperm-induced conductance increase in the egg cell membrane. Initial experiments comprised: 1) a step in the clamped membrane potential from -70 mV (the normal resting potential of the egg and a nonpermissive value for sperm entry (Lynn and Chambers 1984) to -20 mV (a permissive value for sperm entry) where the voltage was held constant for the remainder of the experiment; and 2) a step in the clamped membrane potential from -20 mV to -70 mV where the voltage was held constant for the remainder of the experiment. Evaluations of the resulting sperm-induced current profiles and the observance of the morphological events described above were used in determining the entrance of the sperm into the egg cytoplasm.

Control Steps in Unfertilized Eggs

When voltage-clamped eggs are stepped from -70 to -20 mV or more positive values, a transient inward current increase occurs (less than 0.5s, Figure 1). In the example shown, a step from -70 mV to 0 mV was made and the inward current transient rapidly obtains a maximum value of 2.0 nA. The inward current (I) quickly returns to a steady holding level very close to the original holding level. The transient inward current is presumably caused by the opening of the voltage dependent Ca^{2+} channels which underlie the action potential mechanism of non-voltage-clamped egg (Chambers and De Armendi 1979). Such transient currents can be diminished in reduced Ca^{2+} seawater or amplified by increasing the normal concentration of Ca^{2+} in the seawater (Lynn, unpublished data; see also David *et al* 1988). This transient current is also superimposed on all current recordings stepped from -70 to -20 mV following a sperm-induced conductance increase, but are not observed when eggs were stepped from -20 to -70 mV. A small compensating overshoot in the holding current is observed when stepping from -20 to -70 mV, but never obtains more than a small fraction of the value of the reverse step and is not affected by calcium concentrations. It should be stressed that these brief transient currents do not resemble in any way the sperm transients described previously by Lynn *et al.* (1988).

Time Dependent Steps from -70 to -20 mV

In eggs stepped from -70 mV to -20 mV within 6s following a sperm induced conductance increase, 100% of the clamped eggs were penetrated by the sperm initiating the conductance increase. When the step from -70 mV to -20 mV was delayed to 6-9s, only 50% of the sperm were incorporated into the egg. If the step was delayed to greater than 9s, sperm incorporation into the egg cytoplasm did not occur.

When the voltage-clamp step occurs before 6s, the current profile recorded is similar to that reported by Lynn *et al* (1988) for eggs voltage-clamped at -20 mV. Such currents comprise 1) a shoulder initiated by a sharp current increase, 2) a major current peak, and 3)

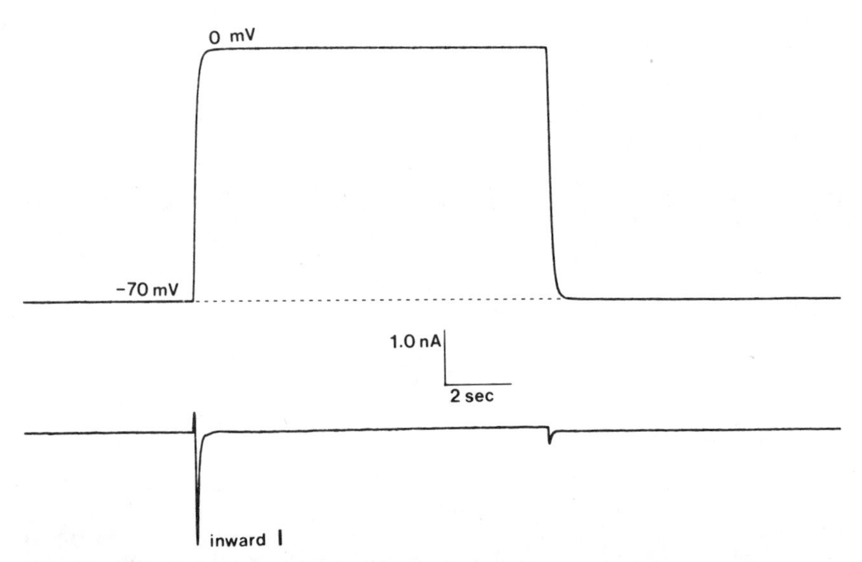

Figure 1. An electrophysiological recording from an unfertilized egg showing the clamped membrane potential (upper trace) and the holding currents of the clamp (lower trace). A step from -70 mV to 0 mV results in a transient inward current (inward I) which follows an initial outward current shift. This inward current transient is superimposed on the stepped potential currents of fertilized eggs. Note that the return step from 0 to -70 mV results in only a small current in the same direction as the step.

a phase of decreasing current which requires an extended period of time to return to the original holding level. In addition, these eggs will form normal insemination cones, and sperm asters. The female pronuclear centration will be followed by the formation of a large central monaster.

In contrast, when the voltage-clamp step occurs after 9s, the current profile recorded may resemble either that of a long-duration, sperm-induced transient or an activation current of an egg not penetrated by a sperm (Figure 2). In both cases, the holding current returns sharply to the original holding level with no subsequent inward currents recorded in the case of a sperm-induced transient, or near to the original holding level before the recording of the major inward current associated with the elevation of the fertilization envelope (see below for a more complete description of current characteristics or Lynn *et al.* 1988).

Time Dependent Steps from -20 to -70 mV

Eggs stepped from -20 to -70 mV before 9s following the sperm-induced conductance increase failed to incorporate the activating sperm in 90% of the eggs examined. Increasing the time delay to the voltage step to 9-12s or longer than 12s, allowed sperm entry into the egg cytoplasm in 50% and 100% of the observations, respectively.

Eggs incorporating sperm following a voltage step, had current profiles similar to those observed in continuously clamped eggs which incorporated sperm. In addition, failure of sperm incorporation into the egg was accompanied by a current profile characteristic of either a sperm-induced transient response, or an activation current distinctive of an egg elevating a full fertilization envelope without the entrance of a sperm nucleus (see for example Figure 3). Notably, these latter two diagnostic profiles were observed most frequently when eggs were stepped before 9s post sperm-induced conductance increase. The time relationships between those eggs stepped from -70 to -20 mV and those eggs stepped from -20 to -70 mV is summarized on a time line in Figure 4.

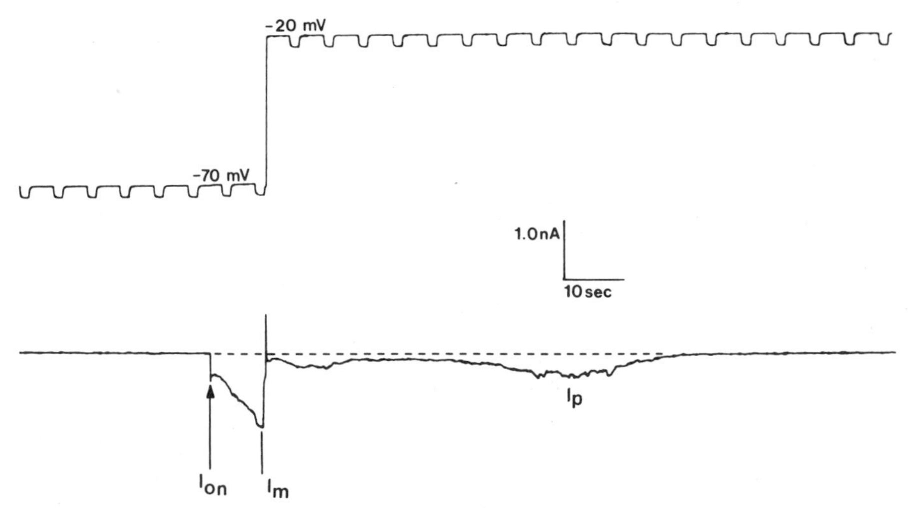

Figure 2. A stepped clamped potential recording. The clamped potential has small 3 mV hyperpolarizing voltage pulses superimposed so that membrane conductance changes can be followed. The egg is initially clamped at -70 mV and at approximately 10.5s following the detection of a sperm-induced conductance increase (I_{on}), the membrane potential is shifted to -20 mV. In this case, the step occurred just slightly after the beginning of the return of the holding current towards the original holding level. The step to a permissive clamped potential, however, could not rescue this egg and sperm entry was inhibited. The remainder of the holding current profile is diagnostic for an activated egg not penetrated by a sperm (Lynn *et al* 1988). I_m = Phase I maximum current, I_p = maximum inward current associated with cortical granule exocytosis.

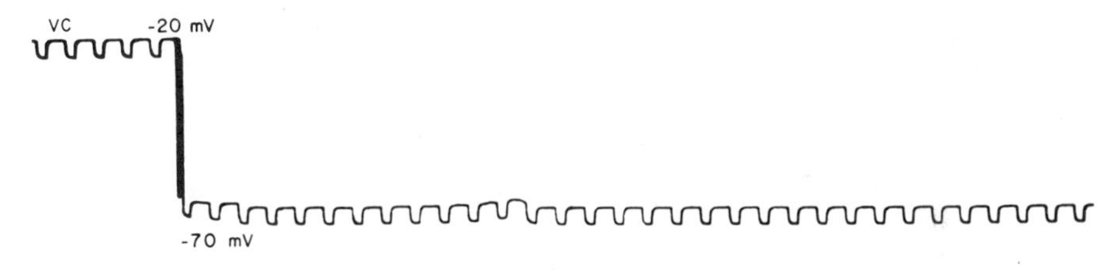

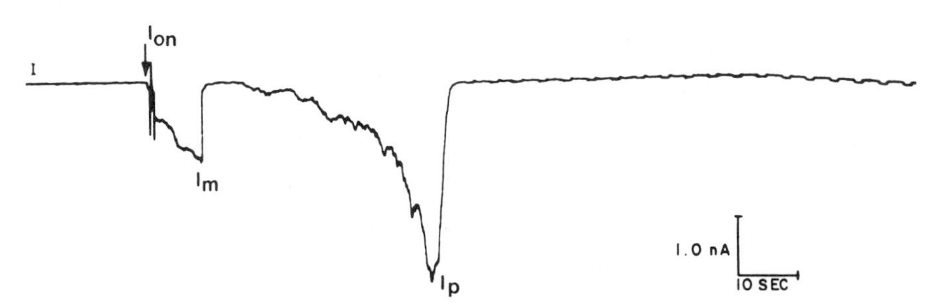

Figure 3. A recording of a stepped clamp experiment involving the shift of the membrane potential from -20 mV to -70 mV. In this case, the membrane potential was stepped within 2s of the sperm-induced conductance increase (arrow, I_{on}) to a nonpermissive potential. At 11s post-conductance increase, a rapid return to the original holding current is observed followed by a second conductance increase associated with the elevation of the fertilization envelope. At the end of I_p, the holding current returns to near the original holding current. This curve type is diagnostic for inhibited sperm entry (Lynn *et al* 1988). I_m = Phase I maximum current.

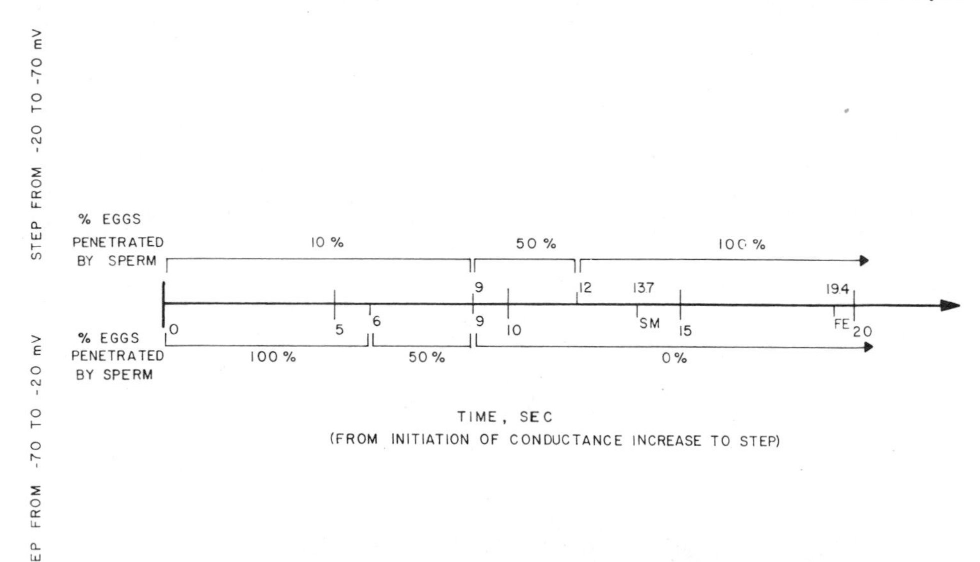

Figure 4. A summary of the results of stepping a voltage-clamped potential from either -20 to -70 mV or -70 to -20 mV following the detection of a sperm-induced conductance increase. A critical time is observed at 6-9s (brackets on lower half of the time line) for steps from -70 to -20 mV and beginning at 9s for steps from -20 to -70 mV. Specifically, the membrane potential must be at a permissive potential at 6s from the sperm-induced conductance increase and must remain at the potential until at least 9s following the conductance increase for 50% of the eggs to be penetrated by the activating sperm. SM = recorded time that sperm was observed to become immotile on the surface of the egg; FE = time of observation of the elevation of the fertilization envelope (see Lynn *et al.* 1988).

Cytochalasin B Allows Egg Activation But Prevents Sperm Entry at Permissive Voltage-Clamp Potentials

Pretreatment with cyto B greatly reduced the elasticity of the egg membrane as evidenced by the formation of a large indentation at the site of microelectrode insertion. In addition, the resting membrane conductance of these eggs was slightly, but not significantly greater than that of normal controls or controls treated with 0.1% DMSO (data not shown). Nevertheless, only those unfertilized eggs which attained a membrane conductance of less than 10 nS after the insertion of the microelectrode were utilized in this study (essentially the same criteria as used for untreated eggs in previous studies, Lynn and Chambers 1984).

DMSO treated eggs fertilized readily following the attachment of a sperm to the egg. The current profiles recorded (Figure 5) always resembled those of untreated controls and were diagnostic of voltage-clamped eggs penetrated by sperm. Eggs went through the normal sequence of morphological events including the full elevation of a fertilization envelope, the formation of an insemination cone at the site of sperm entry (Figure 5a) and cleaved regularly.

Eggs treated either continuously or pulse treated with cyto B were characterized by one of three current pattern profiles following insemination. Current profiles recorded included 1) full activation currents characteristic of eggs penetrated by sperm; 2) full activation currents characteristic of eggs not penetrated by sperm; 3) attenuated activation currents associated with a local or partial fertilization envelope elevation; and, 4) long duration sperm transients. [See descriptions below for more details of current profile characteristics or Lynn *et al.* (1988)].

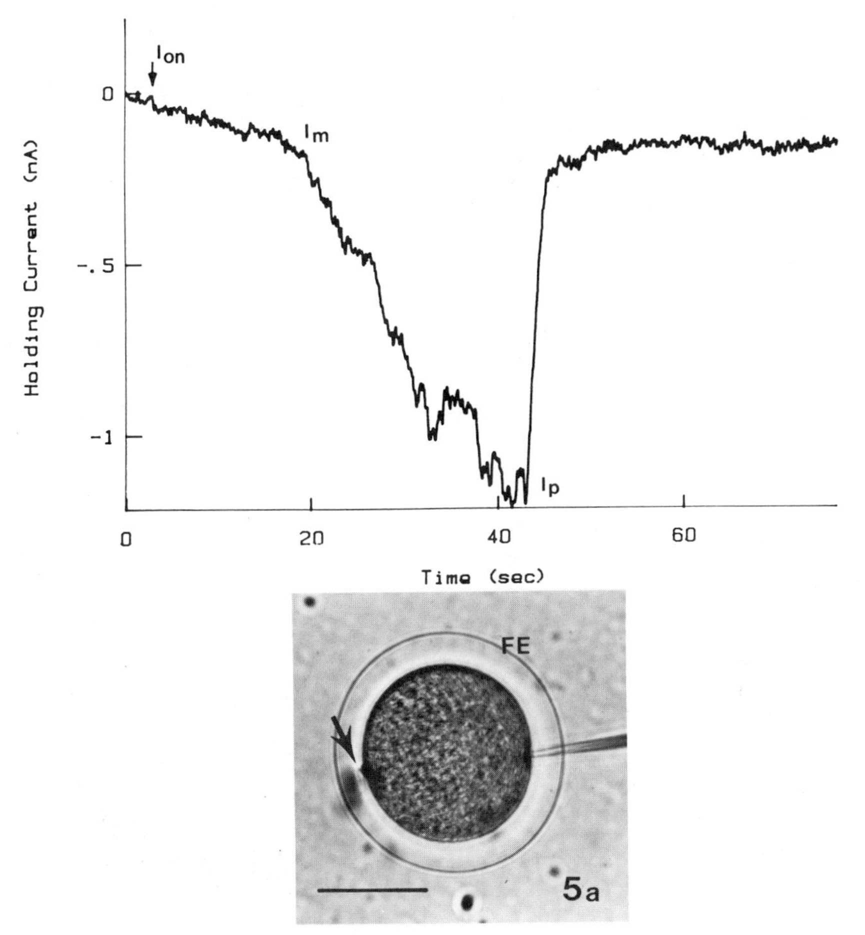

Figure 5. A current tracing of an egg voltage-clamped at -20 mV and inseminated following treatment with 0.1% DMSO. Note the sperm-induced conductance increase (arrow), followed by a shoulder current which reaches a maximum value of I_m, and eventually increasing to a maximum inward current (I_p). The holding current does not return to the originally holding level for a prolonged period after the I_p. This curve type is normally diagnostic of an egg penetrated by a sperm (also see text for possible exceptions). **Figure 5a** is a brightfield micrograph of the egg from which this recording was taken. Note the prominent insemination cone on the equator of the egg (arrow). FE = fertilization envelope. bar = 50 μM

Of the 11 full activation currents recorded, two closely resembled the current profiles characteristic of eggs penetrated by sperm. Curiously, neither of these eggs developed and in both cases the sperm were observed being lifted from the surface of the egg by the elevating fertilization envelope (see below). All eggs displaying a full activation current were accompanied by the elevation of a complete fertilization envelope which elevated to a normal height.

Attenuated activation currents were characterized by a significant and variable decrease in the magnitude of the major inward current phase (e.g. Figure 6). Otherwise, these currents followed a pattern similar but not identical to full activation currents associated with eggs not penetrated by sperm. Specifically, currents were initiated by a sharp inward current (I_{on}, see below) followed by a gradual increase of inward current forming a shoulder, a slight to dramatic increase in inward current associated in time with the formation of a fertilization

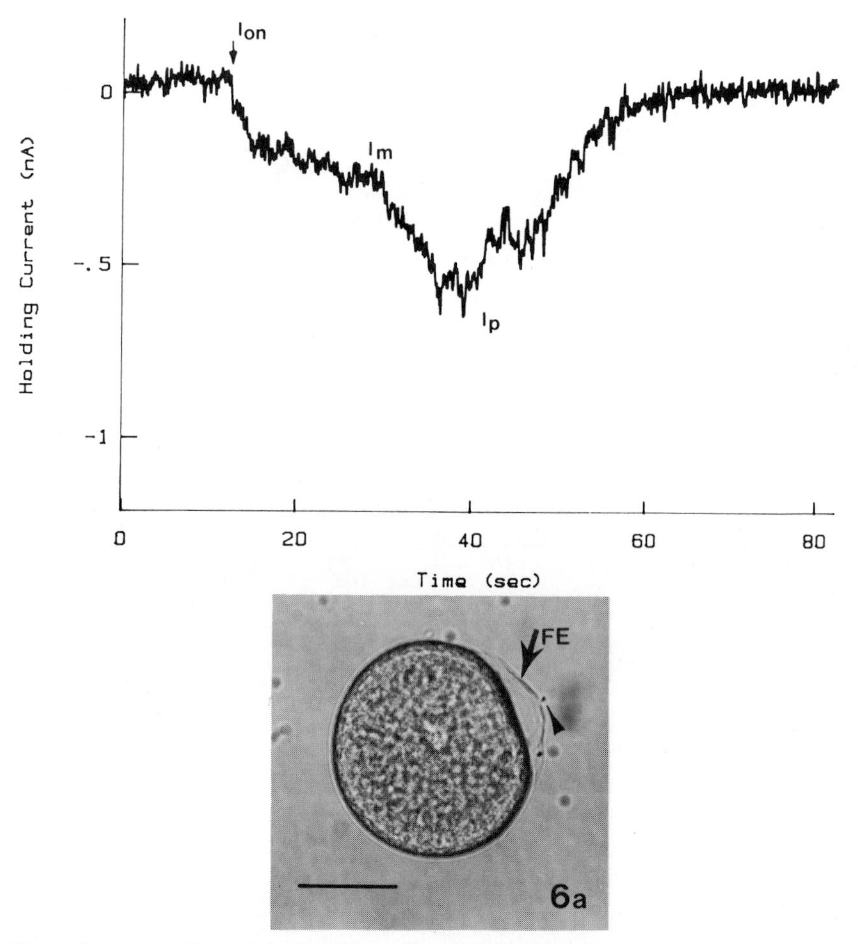

Figure 6. A recording of the holding current of an egg pretreated with cytochalasin B and voltage-clamped at -20 mV. After the holding current had stabilized, sperm were added to the dish. Attachment of a sperm to the surface of the egg resulted in the initiation of a conductance increase (arrow, I_{on}) within a few seconds consisting of a shoulder phase (see text, I_m = phase I maximum) followed by an attenuated major inward current (I_p). Notice that the holding current returns to near the original holding level of the unfertilized egg. **Figure 6a** is a brightfield micrograph of the egg from which this recording was taken. The sperm (arrowhead) initiating the conductance increase is lifted from the surface of the egg by the elevation of a partial fertilization envelope (FE, arrow). The envelope never surrounds the entire egg. This micrograph was taken approximately 5 min after the electrode was removed from the egg cytoplasm. Bar = 50 μM.

envelope, and terminated by a rapid return of the holding current close or equal to the original holding current of the preinsemination egg. Of notable absence in some but not all partial and attenuated activation currents was a sharp return towards the original holding level before the increase in current normally associated with the rise of the fertilization envelope.

Multiple, long-duration, sperm-induced transient currents were observed in the cyto B treated eggs (Figure 7) and followed the diagnostic pattern of such currents as described

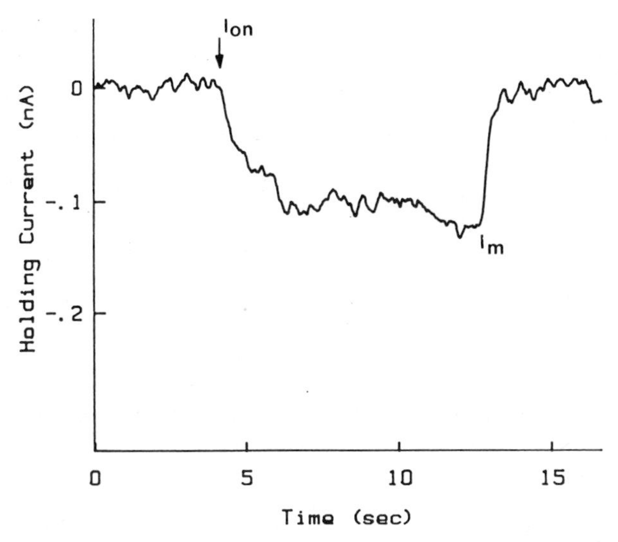

Figure 7. A recording of the holding current record of an egg pretreted with cytochalasin B and voltage-clamped at -20 mV. Following attachment of a sperm to the surface of the egg, a conductance increase was initiated (arrow, I_{on}) within a few seconds. No fertilization envelope was elevated and other than the appearance of a sperm-induced transient current, no further activation events were observed in the egg until a sperm sperm attached to the surface of the egg following addition of fresh sperm to the dish. Note the return of the holding current after the conductance increase back to the original holding level of the voltage-clamp. I_m = Phase I maximum current.

below. Following a sperm-induced transient, no fertilization envelope was detectable with either phase or brightfield microscopy and the holding current returned to the preinsemination holding level. These eggs essentially remained in an unfertilized state and required one or more additions of fresh sperm to the dish to activate the egg.

Partially activated, voltage-clamped eggs were characterized by an incomplete elevation of the fertilization envelope (Figure 6a). Similar partially activated eggs with incomplete fertilization envelopes were also observed in eggs which were not voltage-clamped but in the same dish as the experimental eggs. In those eggs which were partially activated (either voltage-clamped or simply residing in the dish) or which were characterized by sperm-induced transient currents, reinsemination was required for full activation to ensue. Regardless of whether fertilization envelope formation was partial or complete, the envelope elevation was always observed to proceed from the site of the sperm initiating a conductance increase in the egg membrane. Control eggs voltage-clamped at -20 mV never required reinsemination for complete egg activation once a sperm-induced conductance increase had been observed.

Analysis of the shoulder phase of the activation currents (designated Phase I by Lynn *et al* 1988) reveals that the initial conductance increase, I_{on}) ($x=0.07 \pm 0.04$nA, n=35), is more variable than observed in control eggs but is not significantly different from the means of I_{on} taken from a normal group of eggs voltage-clamped at values between -17 mV and -25 mV. A comparison of the duration of Phase I between controls and cyto B treated eggs, however, reveals that the mean duration of the Phase I current in the cyto B treated eggs is significantly longer ($t = 19.35 \pm 2.23$s, N=16) than that observed in untreated eggs (T test, $p<0.001$). Such comparisons were made between the Phase I current of sperm-induced transients, normally activated and fertilized eggs, and the Phase I portion of currents characteristic of eggs activated but not penetrated by the sperm.

Sperm were never observed to enter eggs treated with cyto B and were easily detected visually either on the surface of the egg or at the periphery of the elevated fertilization envelope (either local or complete; see for example Figure 6a). In all cases, insemination cones were not observed, sperm asters did not form in the egg cytoplasm, the pronucleus centered irregularly, if at all, and the eggs did not cleave regularly.

DISCUSSION

The preceding stepped voltage-clamped experiments demonstrate that the time at which the clamped voltage of the egg is at permissive versus nonpermissive values has a critical value between 6 and 9s in the sea urchin *Lytechinus variegatus*. Not only is the time at which the egg reaches these values critical, but the time that the egg remains at permissive values is also important, encompassing that time frame between 9 and 12s following a sperm-induced conductance increase. Further, the voltage-dependent sperm entry event can be mimicked and partial egg activation induced by pretreatment of the eggs with cytochalasin B.

Details of the inward currents recorded in eggs voltage-clamped at potentials between +17 mV and -75 mV are reported by Chambers (this volume), Lynn *et al* (1988), and David *et al* (1988). For the purpose of clarity, however, the inward currents associated with phase I (Lynn *et al* 1988) of the inward activation currents and sperm transient currents in eggs voltage-clamped at potentials more negative than -20 mV will be briefly repeated here. Phase I of activation currents in eggs either penetrated or not penetrated by a sperm and sperm transient currents observed at potentials more negative than -25 mV all have similar characteristics. Following sperm attachment, currents initiate abruptly and have the same magnitude at their onset (I_{on}) dependent on the clamped voltage. A slight increase in the current magnitude will occur until a maximum (I_{sm} or I_m) is reached. The time elapsed during this phase I current is identical at the same V_m for each of the cases indicated above. At the time of I_{sm} or I_m, subsequent currents may follow one of three patterns. First, in the case of eggs penetrated by a sperm, the inward current continues to increase toward a maximum inward current I_p associated with exocytosis of the cortical granules. Second, in eggs clamped at nonpermissive potentials (the sperm does not penetrate the egg), the inward current cuts off abruptly. If no further electrical events are recorded and no morphological events are observed, the currents are categorized as a sperm transient. Third, if the second case just described is followed by delayed current increases, sperm penetration does not occur but initial stages of egg activation characterized by the formation of the fertilization envelope will occur. Each of the currents observed is the result of an increase in the conductance of the egg membrane.

Stepped Clamped Potentials

The critical time for sperm entry as determined by both the stepped voltage-clamped experiments discussed here and the electron microscopy studies on eggs clamped at -20 mV (a permissive V_m)(Longo *et al.* 1986) correspond to the latent period of egg activation described by Allen and Griffin (1958) and is equivalent in duration to the length of phase I reported by Lynn *et al* (1988). The stepped voltage-clamp experiments indicate that the clamped voltage must be at a permissive value during the final period of the phase I current. For example, if the clamped voltage is not stepped to a permissive value within 6-9s of the I_{on}, sperm are not incorporated into the egg. In addition, if the clamped potential is stepped from a permissive value to the nonpermissive value before approximately 9-12s, sperm incorporation will not occur.

The magnitude of the step in itself does not appear to be detrimental to sperm incorporation for three reasons. First, unimpeded fertilization is characterized by a natural shift in the membrane potential of the sea urchin egg from approximately -70 mV to +20 mV, a potential shift of some 90 mV. Second, in experiments involving a step in the voltage from a clamped potential of 0 mV (a permissive voltage by definition of Lynn and Chambers 1984) to + 40 mV (a potential which would normally exclude all recordable events including sperm entry, therefore a nonpermissive potential) within two seconds of the detection of a sperm-induced conductance increase, sperm incorporation proceeds normally. The magnitude of the step is similar to that reported here for a shift from -70 mV to -20 mV and steps the potential from a permissive to nonpermissive value at an early time point. If the inhibition was correlated with step size and time of the step into a nonpermissive range, one might have predicted that this experiment would prevent sperm entry. Third, steps (or the natural shift in the membrane potential into positive, nonpermissive membrane potentials) may utilize a different mechanism to prevent entry of unsuccessful sperm in comparison to the present investigations which prevent the activating sperm from entering the cytoplasm. Shen and Steinhardt (1984) have demonstrated, however, that fertilization can proceed normally in eggs initially voltage-clamped at nonpermissive positive potentials. Following insemination and the observation of sperm attached to the egg surface, brief (30-60 ms) pulses to -60 mV (a negative nonpermissive potential) and rapid return to the original holding potential allows the entry of a single sperm as judged by a normal first cleavage. In these stepped clamped experiments, the absolute magnitude of the voltage shift was larger than in the experiments reported here, yet sperm entry was normal.

Other investigations concerned with the timing and visualization of critical early events in fertilization including sperm-egg fusion, insemination cone formation and cortical granule exocytosis also provide us with pertinent clues to the effects of negative potentials on sperm incorporation. For example in studies by Longo *et al* (1986), sperm-egg cytoplasmic continuity could not be detected by electron microscopy until 5-6s following the initiation of a sperm-induced conductance increase in the voltage-clamped egg. Hinkley *et al* (1986), using the transfer of the DNA dye Hoechst 33342, were in agreement with the findings of Longo *et al.* (1987). This time corresponds to the initial time (6s) reported here for stepping the clamped voltage from -70 to -20 mV to insure that 50% of the voltage-clamped eggs studied would incorporate the activating sperm. Electron microscopy studies further demonstrate that all morphological events associated with the initial activation of the egg are first confined to the site of sperm egg fusion and identification of the forming insemination cone could not be made until 8-9s following I_{on} (Longo *et al* 1986). These findings also correspond temporally to the detection of actin microfilaments in the forming insemination cone using phallacidin (McCulloh *et al* 1988).

The probability that the voltage-sensitive process interrupted by clamping the egg at negative potentials is a barrier to fusion is arguable, however, since the results of McCulloh and Chambers (1986) indicate that the initial conductance increase is a result of the fusion of the sperm and egg (see discussion below). Two alternative explanations include: 1) a rapid and sustained influx of ions at the negative clamped potentials destabilizes the sperm-egg fusion site; and/or, 2) the cytoskeletal elements of the membrane are adversely affected by the clamped potentials. Lynn and Chambers (1984) originally argued that the latter explanation was weak since some eggs clamped at -75 mV were still capable of forming normal insemination cones as long as sperm entry occurred (see Lynn *et al* 1988). More recent experiments, including those reported in this paper on the effects of cytochalasin B, support the theory that the negative potentials are affecting cytoskeletal functioning (see discussion below on cytochalasin B experiments). It is probable that in fact the cytoskeletal elements may be influenced by ion fluxes crossing the membrane at the negative potentials disrupting the assembly of microfilaments in the egg cortex. The timing of the visualization of actin microfilaments in the insemination cone (see above) are supportive of this argument but confirmation awaits further investigation.

Effects of Cytochalasin B

In the present study, eggs that were pretreated with cytochalasin B, subsequently voltage-clamped at permissive potentials, and inseminated failed to incorporate sperm into the egg cytoplasm. These findings are in agreement with previous investigations (Longo 1978; Gould-Somero *et al.* 1977; Byrd and Perry 1980; Schatten and Schatten 1981) on eggs of a variety of invertebrate species. In addition, the results show striking similarities to the failure of sperm incorporation at clamped membrane potentials near the egg's normal resting potential (Lynn and Chambers 1984).

Observations of egg activation in the presence of cytochalasins and accompanied by an electrophysiological response of the egg similar but not identical to untreated eggs have also been reported by Dale and De Santis (1981) and Hulser and Schatten (1982). The most significant alteration of the voltage-clamp currents seen in the present study was the increase in the duration of the phase I current. This corresponds with the delay between the initial sperm-induced step and the fertilization potential reported by Dale and De Santis (1981). The potential step reported by the latter investigators and the phase I currents in the voltage-clamp experiments described by Lynn *et al* (1988) and David *et al* (1988) most likely correspond to comparable changes in the properties of the egg membrane. The sperm-induced conductance increase became somewhat more variable in the voltage-clamped eggs in the experiments reported here. However, a significant difference from normal, untreated eggs was not detected in contrast to the report of Dale and De Santis (1981) in which the amplitude of the potential shift in cytochalasin treated eggs was altered. The exact role of cytochalasins in altering the membrane characteristics and the observed holding current patterns is not clear at this time.

Spudich and Lin (1972) reported that cytochalasin B inhibits the polymerization of actin. The present investigations, then, suggest that microfilaments are involved in the active migration of the sperm nucleus into the egg cytoplasm (see also Schatten and Schatten 1981; Schatten 1981). They do not on the other hand eliminate the possibility that cytochalasin B has inhibited sperm incorporation by secondary side effects on either the sperm or the early activation events of the egg itself (see below and Allemand *et al* 1987).

It is unlikely that failure of the sperm to be incorporated into the egg cytoplasm is a direct result of the effects of cytochalasin B on the sperm or sperm acrosome reaction itself for two reasons. First, sperm incorporation was inhibited in eggs pulse-treated with cytochalasin B. Since eggs were pretreated with the cytochalasin B and then washed before being inseminated, no cyto B would be continually present in the media unless it was leaking from the treated eggs. The effects of cyto B pulse-treated eggs are reversible, however, and may be attributable to the leakage of the cyto B from the egg. While such leakage may occur, it would be a slow process (Longo 1978; Byrd and Perry 1980) and the resulting concentration of the cytochalasin B in the media should be very low. Second, although the polymerization of actin is a critical event in the formation of the acrosomal process (Tilney 1973 *et al.*), previous investigators have found that the acrosome reaction is not affected by cyto B in *Lytechinus variegatus* (Sanger and Sanger 1975). In fact, pretreatment of sperm with concentrations of cytochalasin B that would normally inhibit sperm incorporation if eggs had been treated do not inhibit entry into untreated eggs (Byrd and Perry 1980; Allemand *et al.* 1987).

The role of microfilaments in the formation of the insemination cone in the sea urchin has been previously documented by several investigators (Tilney and Jaffe,1980; McCulloh *et al* 1988). In timed fixations and using video enhancement techniques, McCulloh *et al* (1988) detected microfilament formation in the insemination cone as early as 10s following sperm-egg contact. Other studies have further demonstrated that the movement of the sperm into the egg cytoplasm is dependent on the proper polymerization of the microfilaments (Longo 1978, 1980; Schatten and Schatten 1980, 1981; Schatten *et al* 1986). In many of these studies, the activating sperm was judged to have fused with the egg because the sperm suddenly became immotile (Schatten and Schatten 1981; Schatten 1981). However, the sperm was lifted from the surface of the egg by the elevating fertilization envelope. Byrd and Perry (1980) have suggested that fusion between the egg plasmalemma and the sperm plasmalemma may indeed have occurred, but the loss of microfilament integrity following cytochalasin B treatment resulted in a fusion bridge so weak, that it could not be maintained without cytoskeletal support from the cortex.

The likelihood that sperm-egg fusion has occurred, even in the presence of cytochalasin B, is supported by the following observations. When sperm are introduced to the egg through a pipette used to loose patch clamp the egg membrane, onset of a sperm-induced conductance increase is simultaneously accompanied by an increase in capacitance of the membranae patch (McCulloh and Chambers 1986). Since such a capacitance increase could only be logically explained under the conditions of the experiment as evidence that sperm-egg fusion had occurred, the conductance increase observed here in the presence of cytochalasin B would indicate that initial sperm-egg continuity had been achieved. Observations by Schatten and Schatten (1981) also suggest that sperm-egg fusion has occurred as judged by the attachment of the sperm to the egg surface by a thin cytoplasmic thread during the elevation of the fertilization envelope. Subsequent separation of the sperm and egg then would support the theory of Byrd and Perry (1980) suggesting that the contractile or cytoskeletal elements in the cortex of the egg were necessary to anchor the initial weak cytoplasmic bridge between the sperm and the egg. In contrast, however, following treatment of the egg with dithiothreitol to disrupt the FE precursor (the vitelline envelope) the sperm is still observed to detach from the egg surface despite the failure of the FE to elevate. A question then arises as to how (or why) the cytoplasmic bridge would break in the absence of a distinctive mechanical force.

Additional evidence of sperm-egg fusion is provided by the experiments of Gundersen *et al* (1982). These investigators reported the appearance of a patch of fluorescence on the surface of the eggs treated with cyto B or D and inseminated with FITC labeled sperm. Although nuclei were not detected in the cytoplasm of the eggs, detection of the fluorescent patches suggested that sperm-egg fusion had occurred to facilitate the dye transfer. Sperm in these experiments lie within the fertilization envelope, in contrast to the results reported here where the sperm are routinely lifted from the surface of the egg by the elevating FE.

A curious auxiliary effect of the treatment of eggs with cyto B was the observations of partial or incomplete elevation of the FE. While some investigators have reported that complete cortical discharge occurs in inseminated eggs treated with cyto B (Dale and De Santis 1981; Schatten and Schatten 1981), incomplete cortical reactions have been observed by Byrd and Perry (1980). Since at least partial cortical discharge occurs (usually restricted to the immediate vicinity of the sperm in the experiments described above) it seems unlikely that the cyto B is directly affecting the membrane fusion event between the cortical granule membranes and the oolemma. Two alternate explanations include; 1) a decoupling of the signal transmission from the sperm to the egg cytoplasm (whether through a receptor or by direct fusion) as discussed by Allemand *et al* (1987); or, 2) by the disruption of the cytoskeletal-mediated translocation of the cortical granules into apposition with the plasma membrane where fusion can take place. The latter explanation for partial cortical granule discharge is consistent with proposed mechanisms of vesicle translocation in other secretory systems (see for discussion Wade 1986). A decoupling of the sperm stimulus may also account for localized cortical granule discharge. For example, Dale (1985) has reported the association of a phosphatidylinositol 4-phosphate kinase with cytoskeletal elements in the human erythrocyte and suggested that they may play a key role in their physiological functioning. One might speculate, then, that in the sea urchin system, the cytoskeletal elements might also play a similar role by mediating the sperm-delivered stimulus through the membrane-bound G-protein, phosphatidyl kinase, and phospholipase C production of inositol 1,4,5 triphosphate and diacylglycerol, driving the intracellular free calcium release (Turner *et al* 1986; see also Allemand *et al* 1987 for discussion). Thus in the absence of a microfilament cytoskeleton, the sperm stimulus might be restricted to the immediate vicinity of the sperm or reduced by a variable transmission of the signal. In the present study, it is interesting to note that FE elevation still originated from the site of the activating sperm and proceeded in a wave-like fashion around the cortex. Such a mechanism is untested, however, and the observed results may actually reflect a secondary side effect of the cyto B on membrane processes (Lin and Spudich 1974; see also Longo 1978 for review).

ACKNOWLEDGEMENTS

I would like to thank the organizers of the Bodega Marine Laboratory Colloquium for the remarkable assemblage of speakers and topics. Thanks are also due to Dr. D. H. McCulloh and Dr. E. L. Chambers for their continued discussions and collaborations on the topics discussed in this paper. This work was supported in part by NIH Research Grant HD 19126 to E.L.C., NSF Research Grants PCM 83-16864 and DCB-8711787 to E.L.C. and a grant from the Louisiana State University Council on Research Grants to J.W.L.

REFERENCES

Allen, R. D., and J. L. Griffin 1958. The time sequence of early events in the fertilization of sea urchin eggs. II The latent period and the cortical reaction. *Exp. Cell Res.* 15:163-173.

Allemand, D., B. Ciapa, and G. De Renzis. 1987. Effect of cytochalasin B on the development of membrane transports in sea urchin eggs after fertilization. *Dev. Growth & Differ.* 29:333-340.

Blatt, M. R. and C. L. Slayman. 1982. KCl leakage from microelectrodes and its impact on membrane parameters of a non-excitable cell. *J. Gen Physiol.* 80:12a.

Byrd, W. and G. Perry. 1980. Cytochalasin B blocks sperm incorporation but allows activation of the sea urchin egg. *Exp. Cell Res.* 126:333-342.

Cross, N. L. and R. P. Elinson. 1980. A fast block to polyspermy in frogs mediated by changes in the membrane potential. *Dev. Biol.* 75:187-198.

Chambers, E. L. and J. De Armendi. 1979. Membrane potential of eggs of the sea urchin *Lytechinus variegatus*. *Exp. Cell Res.* 122:203-218.

David, C., J. Halliwell, and M. Whitaker. 1988. Some properties of the membrane currents underlying the fertilization potential in sea urchin eggs. *J. Physiol. (Lond.)* 402:139-154.

Dale, B. and A. DeSantis. 1981. The effect of cytochalasin B and D on the fertilization of sea urchins. *Dev. Biol.* 83:232-237.

Dale, G.L. 1985. Phosphatidylinositol 4-phosphate kinase is associated with the membrane skeleton in human erythrocytes. *Biochem. Biophys. Res. Comm.* 133:189-194.

Goudeau, H. and M. Goudeau. 1985. Fertilization in Crabs: IV. The fertilization potential consists of a sustained egg membrane hyperpolarization. *Gamete Res.* 11:1-17.

Gould-Somero, M., L. Holland, and M. Paul. 1977. Cytochalasin B inhibits sperm penetration into eggs of *Urechis caupo* (Echiura). *Dev. Biol.* 58:11-22.

Gould-Somero, M., L. A. Jaffe, and L. Z. Holland. 1979. Electrically mediated fast polyspermy block in eggs of the marine worm, *Urechis caupo*. *J. Cell Biol.* 82:426-440.

Grey, R. D., M. J. Bastiani, D. J. Webb, and E. R. Schertel. 1982. An electrical block is required to prevent polyspermy in eggs fertilized by natural mating of *Xenopus laevis*. *Dev. Biol.* 89:475-484.

Gundersen, G. G., C. A. Gabel, and B. M. Shapiro. 1982. An intermediate state of fertilization involved in internalization of sperm components. *Dev. Biol.* 93:59-72.

Hagiwara, S. and L. A. Jaffe. 1979. Electrical properties of egg membranes. *Annu. Rev. Biophys. Bioeng.* 8:385-416.

Hinkley, R. E., B. D. Wright, and J. W. Lynn. 1986. Rapid visualization of sperm-egg fusion using the DNA-specific fluorochrome Hoechst 33342. *Dev. Biol.* 118:148-154.

Hulser, D. and G. Schatten. 1982. Bioelectric responses at fertilization: Separation of the events associated with insemination from those due to the cortical reaction in sea urchin, *Lytechinus variegatus*. *Gamete Res.* 5:363-377.

Jaffe, L. A. 1976. Fast block to polyspermy in sea urchins is electrically mediated. *Nature (Lond.)* 261:68-71.

Jaffe, L. A. and L. C. Schlichter. 1985. Fertilization induced ionic conductances in egg of the frog, *Rana Pipiens*. *J. Physiol. (Lond.)* 358:299-319.

Jaffe, L. A. and M. Gould. 1985. Polyspermy preventing mechanisms. p. 223-92. *In: Biology of Fertilization*, Vol. 3. C. B. Metz and A. Monroy (Eds.). Academic Press, New York.

Kline, D., L. A. Jaffe, and R. P. Tucker. 1985. Fertilization potential and polyspermy prevention in the egg of the nemertean, *Cerebratula lacteus*. *J. Exp. Zool.* 236:45-52

Lewis, A. S. and N. K. Wills. 1980. Resistive artifacts in liquid ion-exchanger microelectrode estimates of Na^+ activity in epithelial cells. *Biophys. J.* 31:127-138.

Lin, S. and J. A. Spudich. 1974. On the molecular basis of action of cytochalasin B. *J. Supramol. Struct.* 2:728-736.

Longo, F. J. 1978. Effects of cytochalasin b on sperm-egg interactions. *Dev. Biol.* 67:249-265.

Longo, F. J. 1980. Organization of microfilaments in sea urchin (*Arbacia punctulata*) eggs at fertilization: Effects of cytochalasin *B. Dev. Biol.* 74:422-433.

Longo, F. J., J. W. Lynn, D. H. McCulloh, and E. L. Chambers. 1986. Correlative ultrastructural and electrophysiological studies on sperm-egg interactions of the sea urchin, *Lytechinus variegatus*. *Dev. Biol.* 118:155-166.

Lynn, J. W. 1985. Time and voltage dependent sperm penetration in sea urchin eggs. *Dev. Growth & Differ.* 27:177-178.

Lynn, J. W. and E. L. Chambers. 1984. Voltage clamp studies of fertilization in sea urchin eggs: I. Effect of clamped membrane potential on sperm entry, activation, and development. *Dev. Biol.* 102:98-109.

Lynn, J. W., D. H. McCulloh, and E. L. Chambers. 1988. Voltage clamp studies of fertilization in sea urchin eggs: II. Current patterns in relation to sperm entry, nonentry and activation. *Dev. Biol.* 128:305-323.

McCulloh, D. H. and E. L. Chambers. 1986. When does the sperm fuse with the egg? *J. Gen. Physiol.* 88:38a-39a.

McCulloh, D. H., P. I. Ivonnet and E. L. Chambers. 1988. Actin polymerization precedes fertilization cone formation and sperm entry in the sea urchin egg. *Cell Motil. Cytoskeleton* (in press).

McCulloh, D. H., C. E. Rexroad, and H. Levitan. 1983. Insemination of rabbit eggs is associated with slow depolarization and repetitive diphasic membrane potentials. *Dev. Biol.* 95:372-377.

Miyasaki, S. and Y. Igusa. 1981. Fertilization potential in golden hamster eggs consists of recurring hyperpolarization. *Nature (Lond.)* 290:702-704.

Nuccitelli, R. 1980. The electrical changes accompanying fertilization and cortical vesicle secretion in the medaka egg. *Dev. Biol.* 76:483-498.

Nuccitelli, R. and R. D. Grey. 1984. Controversy over the fast, partial, temporary block to polyspermy in sea urchins - a reevaluation. *Dev. Biol.* 103:1-17.

Rothschild, L. and M. M. Swann. 1952. The fertilization reaction in the sea-urchin. The block to polyspermy. *J. Exp. Biol.* 29:469-483.

Sanger, J. W. and J. M. Sanger. 1975. Polymerization of sperm actin in the presence of cytochalasin-B. *J. Exp. Zool.* 193:441-447.

Schatten, G. 1981. Sperm incorporation, the pronuclear migrations and their relation to the establishment of the first embryonic axis: Time-lapse video microscopy of the movements during fertilization of the sea urchin *Lytechinus variegatus*. *Dev. Biol.* 86:426-437.

Schatten, H. and G. Schatten. 1980. Surface activity at the egg plasma membrane during sperm incorporation and its cytochalasin B sensitivity: Scanning electron microscopy and time-lapse video microscopy during fertilization of the sea urchin *Lytechinus variegatus*. *Dev. Biol.* 78:435-449.

Schatten, G. and H. Schatten. 1981. Effects of motility inhibitors during sea urchin fertilization. Microfilament inhibitors prevent sperm incorporation and restructuring of fertilized egg cortex, whereas micro tubule inhibitors prevent pronuclear migrations. *Exp. Cell Res.* 135:311-330.

Schatten, G., H. Schatten, I. Specotr, C. Cline, N. Paweletz, C. Simerly, and C. Petzelt. 1986. Latrunculin inhibits the microfilament-mediated processes during fertilization, cleavage and early development in sea urchins and mice. *Exp. Cell Res.* 166:191-208

Shen, S. and R. A. Steinhardt. 1984. Time and voltage windows for reversing the electrical block to fertilization. *Proc. Natl. Acad. Sci. USA* 81:1436-1439.

Spudich, J. A. and S. Lin. 1972. Cytochalasin-B, its interaction with actin and actomyosin from muscle. *Proc. Natl. Acad. Sci. USA*. 69:442-446.

Steinhardt, R. A., L. Lundin, and D. Mazia. 1971. Bioelectric responses of the echinoderm egg to fertilization. *Proc. Natl. Acad. Sci. USA* 68:2426-2430.

Tilney, L. G., S. Hatano, H. Ishikawa, and M. S. Mooseker. 1973. The polymerization of actin: Its role in the generation of the acrosomal process of certain echinoderm sperm. *J. Cell Biol.* 59:109-126.

Tilney, L. G. and L. A. Jaffe. 1980. Actin, microvilli, and the fertilization cone of sea urchin eggs. *J. Cell Biol.* 87:771-782.

Turner, P. R., L. A. Jaffe, and A. Fein. 1986. Regulation of cortical vesicle exocytosis in sea urchin eggs by inositol 1,4,5-trisphosphate and GTP-binding protein. *J. Cell Biol.* 102:70-76.

Wade, J. B. 1986. Role of membrane fusion in hormonal regulation of epithelial transport. *Annu. Rev. Physiol.* 48:213-223.

Wilson, W. A. and M. M. Goldner. 1975. Voltage clamping with a single microelectrode. *J. Neurobiol.* 6:411-422.

ELECTRICAL RESPONSES TO FERTILIZATION AND SPONTANEOUS ACTIVATION IN DECAPOD CRUSTACEAN EGGS: CHARACTERISTICS AND ROLE

Henri Goudeau and Marie Goudeau

Departement de Biologie
Service de Biophysique
CEN/SACLAY
UA CNRS 686
STATION BIOLOGIQUE DE ROSCOFF, FRANCE

Mailing Address: Department de Biologie
Service de Biophysique
CEN/SACLAY 91191
GIF-sur-YVETTE Cedex, FRANCE

INTRODUCTION

Relatively little is known about egg activation in the Crustacea, with the exception of morphological observations on the resumption of meiosis and the cortical reaction. Morphological analyses have demonstrated that the resumption of meiotic maturation is initiated by seawater contact in some decapod species. In the prawn *Palaemon serratus*, meiotic resumption of oocytes which are spawned at first meiotic metaphase, depends on the presence of external Mg^{2+} but not on external Ca^{2+} (Goudeau and Goudeau 1986). In Penaeid shrimp, meiotic resumption of oocytes, which are spawned at late prophase or early metaphase, requires Mg^{2+} when the gametes are fertilized and both Mg^{2+} and Ca^{2+} if unfertilized (Pillai and Clark 1987). In crabs, spawned oocytes are in first meiotic metaphase and also resume meiosis upon exposure to seawater. The process requires only 30-50 μM Ca^{2+} in Mg^{2+} free artificial seawater (ASW) (unpublished results). With respect to the cortical reaction, cytological observations have permitted the detection of a sperm-dependent cortical vesicle exocytosis in barnacle eggs (Klepal *et al.* 1979). In Penaeid oocytes, a specific release of jelly components, initiated by spawning and originating from extracellular crypts formed by invaginations of the plasma membrane, has been observed. Jelly components undergo an enzyme-mediated transition from a heterogeneous to homogeneous state, which is dependent on both Mg^{2+} in seawater and a protease (Clark *et al.* 1974; Clark and Lynn 1977; Clark *et al.* 1980; Clark *et al.* 1985; Lynn and Clark 1975). Also in Penaeid shrimp, the formation of a "hatching" envelope is induced by contact with seawater, requiring only external Mg^{2+} when the eggs are fertilized, and both Ca^{2+} and Mg^{2+} when they are not (Pillai and Clark 1987). Finally, a complex cortical reaction has been described in the eggs of crabs (Goudeau and Lachaise 1980 a,b; Goudeau and Becker 1982), and lobsters (Talbot and Goudeau 1988). The cortical reaction in crab eggs is a two-step phenomenon that consists of 1) the exocytosis of a fine granular material, elicited by contact with seawater (Goudeau and Goudeau 1985), and 2) a slow and long-lasting exocytosis of ring-shaped elements, which is sperm-dependent and leads to the elaboration of a thick extracellular capsule (Goudeau and Lachaise 1980b; Goudeau and Becker 1982). In prawn eggs, the cortical reaction requires external Mg^{2+}, and is independent of fertilization (unpublished results).

At the present time there is no information concerning electrical potential changes in the membrane during crustacean egg activation. In many species, the first response of the egg to fertilization consists of modifications in membrane permeability leading to a change in egg membrane potential, which has been named the fertilization potential (reviewed by Hagiwara and Jaffe 1979). The role of this fertilization potential was unresolved until Jaffe (1976) presented evidence that the fertilization potential, a transient positive depolarization of the egg membrane, reduces the probability that additional spermatozoa will fuse with the egg plasma membrane, constituting a fast and transient electrical block to polyspermy. Indeed fusion of a single male pronucleus with the female pronucleus is a prerequisite for subsequent normal development. Similar evidence for an electrically-mediated block to polyspermy due to a transient depolarization of the egg membrane at fertilization has been demonstrated for other eggs from starfish (Miyazaki and Hirai 1979), the echiuroid worm *Urechis caupo* (Gould-Somero *et al.* 1979; Jaffe *et al.* 1982), the nemertean worm *Cerebratulus lacteus* (Kline *et al.* 1985) and anuran Amphibia (Cross and Elinson 1980; Grey *et al.* 1982). This fast and transient electrically-mediated block to polyspermy is followed by a permanent block that is mechanical and results from cortical granule exocytosis. This leads to the formation and concurrent elevation of the fertilization envelope in sea urchin, starfish and frog eggs. In the eggs of *Urechis caupo* and *Cerebratulus lacteus*, the establishment of the transient electrical block to polyspermy is followed by a permanent block of an unknown nature at the level of the egg plasma membrane (for a review concerning polyspermy-preventing mechanisms see Jaffe and Gould-Somero 1985). In some species such as elasmobranchs (Wourms 1977), urodele amphibians (Fankhauser 1948), reptiles (Rothschild 1956), birds (Romanoff 1960), some invertebrates like pulmonate molluscs (Raven 1966) and the ctenophore *Beroe ovata* (Carré and Sardet 1984), physiological polyspermy is the rule. In these cases more than one sperm penetrates the egg, however, only one male pronucleus fuses with the female pronucleus. In some of these examples it has been demonstrated that a fast electrical block to polyspermy does not exist (see review by Jaffe and Gould-Somero 1985).

Concerning crustacean eggs, several questions arise: Does an electrical response of the egg membrane to fertilization or to activation by seawater exist? Do changes in the reproductive strategy (internal versus external fertilization) involve any changes in the egg electrical response? Do changes in the preferred habitat of the decapodan Crustacea (Reptantia, including crabs and lobster, versus Natantia, including prawn and shrimps) involve changes in the egg electrical response? Are decapod crustacean eggs mono- or polyspermic? If they are monospermic, what is (are) the mechanism(s), electrical, mechanical or other preventing polyspermy?

In this chapter we compare crab, lobster and prawn eggs with respect to their electrical response to fertilization or to activation by seawater. We then demonstrate in morphological examinations of naturally spawned fertilized eggs, that crab and prawn eggs are physiologically monospermic. Finally, we present results concerning *in vitro* inseminations of voltage-clamped crab eggs that allow us to understand the nature of the mechanism preventing polyspermy.

MATERIALS AND METHODS

Reproductive Strategy Associated with Internal Fertilization in the Crab (*Maia squinado*), External Fertilization in the Lobster (*Homarus gammarus*) and Prawn (*Palaemon serratus*)

At mating, in crabs, spermatophores are introduced into the two female genital ducts. Spermatophores are stored for several weeks in the spermatheca, which is a large extension of the distal part of the female genital duct (Spalding 1942). The spermatheca is lined by a cuticular covering which forms numerous parallel folds that retain seminal plasm near the genital aperture. Between the time of mating and spawning, several weeks elapse during which the female crab does not molt. It is during this period that oogenesis takes place. Sperm and eggs are mixed just prior to the eggs exiting the female genital pores.

During mating, in lobsters, the male introduces spermatophores into a thelycum or female sperm receptacle that consists of a single pouch formed by an infolding of the ventral wall of the female thorax. The thelycum opens directly to the outside near the two female genital apertures (Herrick 1909). Oogenesis occurs during the months that elapse from mating to spawning. Females do not molt during this period. Eggs are fertilized externally (Bumpus, 1891; Herrick 1909; Talbot 1983) as they flow over the aperture of the thelycum where sperm are released.

In prawns, there is a special molt that takes place when ovarian development has ended (Nouvel and Nouvel 1937; Höglund 1943). This molt is usually followed by mating and spawning. At mating a male transfers spermatophores to the ventral thoracic wall of the female, near the female genital pores (Sandifer and Lynn 1980). Gamete mixing occurs externally as spawned eggs pass through the seminal plasm of the spermatophores located in the area of the female genital openings (Nouvel and Nouvel 1937; Höglund 1943).

Animal Handling

Female prawns were obtained during the early spring and autumn breeding seasons. Since the prawn's shell is translucent, ovarian development can be followed by looking through the cephalothoracic cuticle. The most gravid females were isolated from males and reared in individual aquaria supplied with natural seawater (NSW) at 12°C (the average seawater temperature during the breeding seasons). Female prawns were maintained in aquaria until they had completed the ovigerous molt that occurs at the end of ovarian development. Females spawned 24-48h after the ovigerous molt, if mating had occurred (Panouse 1946). If the female was unmated spawning did not occur for 36-44h after this ecdysis in the spring or 42-52h in the autumn (Goudeau and Goudeau 1986).

Female lobsters were matured in the laboratory and candled to evaluate ovarian development (Talbot 1981, 1983).

Female crabs, *Maia squinado*, undergo two successive spawnings during the breeding season (April to August), however it is impossible to visualize ovarian development through the cephalothoracic cuticle since the shell is not translucent. The timing of ovarian development can only be estimated in females that have released larvae from the previous brood.

Collection of Naturally-Spawned, Fertilized Eggs of Crabs and Prawns

Spawned, fertilized eggs of the crab *Maia squinado* were collected while still unattached within the brooding pouch, which is formed by the convex abdomen folded back to the sternal plates of the female thorax. Some of the fertilized eggs were also collected after they had attached via attachment stalks to the ovigerous setae of the female pleopods. Fertilized eggs were collected at various times after spawning, up to 5h.

Spawned, fertilized eggs of the prawn *Palaemon serratus* were collected about 20-60 min after egg laying, while they were still unattached in the brooding cavities formed by the lateral walls of the female abdomen and the bases of the pleopods.

Obtaining Unfertilized Ovulated Oocytes for *In vitro* Inseminations

Ovulated oocytes of *Maia squinado* are orange, very yolky and opaque, and measure 800 μM in diameter. Ovulated oocytes of lobsters and prawns are greenish and measure 1.6 mm and 650 μM, respectively. In this article, we use the term "oocyte" for the gametes not activated by NSW; these would include gametes maintained in cold (2°C) Mg^{2+}-free artificial seawater (ASW) for the prawn and cold ASW for crab and lobster. We use the term "eggs" for all gametes activated in ASW, or NSW during *in vitro* insemination, whether they were penetrated by a sperm or not. Ovulated oocytes of crabs, lobsters and prawns were obtained by dissecting ripe females just prior to or in the act of spawning. Ripe female crabs and lobsters were washed before dissection with Mg^{2+} and Ca^{2+}-free ASW (Table 1) to avoid contamination by these cations. Female crabs and lobsters were dissected dry and the ovaries were washed *in situ* rapidly with Mg^{2+}, and Ca^{2+}-free ASW. They were then collected and placed in cold ASW (Table 1) to inhibit spontaneous activation. Ripe female prawns were washed and dissected in Mg^{2+}-free ASW (Table 1). The ovarian wall was incised and ovulated oocytes were collected. The oocytes were then stored in cold Mg^{2+}- free ASW.

Electrophysiological Measurements

All experiments were performed in a 0.5 ml lucite chamber (Figure 1), perfused at a rate of 2-4 ml/min with 19°-21°C NSW or different ASWs used for substitution experiments, whose composition are given in Table 1.

Measurement of membrane potential and fertilization potential. A single glass fiber-filled microelectrode (1 mm O.D., 0.7 mm I.D.), with a resistance in the range of 30-50 MΩ in ASW when filled with 3 M KC1, was used. The microelectrode was inserted vertically into an oocyte under a binocular microscope (50-70 X magnification) with a step-motor driven Narishige type SM21 or hydraulic micromanipulator (type MO-103). A bridge circuit was used to pass current and measure potential with the same microelectrode (M707 amplifier, World Precision Instrument, New Haven Conn., or VF 180 amplifier, Biologic, Meylan France). In order to measure R_m, the resistance of the oocyte membrane, pulses of current lasting 5-10s

Table 1. Composition of the Artificial Seawaters Used (pH 8.2)

Artificial seawaters	Composition (mM)										Osmotic pressure (mOsm)
	Na^+	K^+	Ca^{2+}	Mg^{2+}	Cl^-	Tris	Me* SO_3	** NMG^+	EGTA	EDTA	
Standard ASW	475	12	12	30	581	10	-	-	-	-	1040
Magnesium-free ASW	475	12	12	-	561	50	-	-	-	-	1040
Low-magnesium ASW	475	12	12	10	571	40	-	-	-	-	1040
	475	12	12	15	581	30	-	-	-	-	1040
	475	12	12	20	591	20	-	-	-	-	1040
Calcium-free ASW	475	12	-	30	567	20	-	-	0.5	-	
Magnesium and calcium-free ASW	475	12	-	-	557	70	-	-	-	0.5	1040
Magnesium-free, high Ca^{2+}, ASW	475	12	42	-	497	10	-	-	-	-	1040
Hypo-osmotic ASW	360	12	12	30	446	10	-	-	-	-	760
Low-magnesium hypo-osmotic ASW	360	12	12	10	446	20	-	-	-	-	760
Magnesium-free hypo-osmotic ASW	360	12	12	0	446	50	-	-	-	-	760
120 Potassium ASW	367	120	12	30	581	10	-	-	-	-	1040
58 Chloride ASW	475	12	12	30	58	10	523	-	-	-	1040
48 Sodium ASW	475	12	12	30	581	10	-	427	-	-	1040

* Methane sulfonate; ** N-methyl-D-glucamine

were passed through the membrane to the external medium, grounded by a short 3 M KCl agar bridge. In order to measure E_m, oocytes were impaled by gently depressing the oocyte surface by 100-150 μM. Piercing of the vitelline envelope was indicated when the potential, which had became negative due to the contact of the microelectrode with the envelope, sharply returned to zero. The oocyte was then impaled by momentarily overcompensating the negative capacitance adjustment of the amplifier. We found that by further raising the microelectrode about 80-100 μM, we improved the seal between the microelectrode and the membrane, and obtained more negative values of E_m and higher values of R_m. This procedure was routinely followed to impale crab and lobster oocytes, but was less reliable with prawn oocytes because the tip of the microelectrode often broke during vitelline envelope penetration. To impale prawn oocytes we used fiber-filled microelectrodes, 1 mm O.D., made of thick-walled aluminosilicate glass with tips mechanically more resistant than those of conventional microelectrodes. We also used this procedure to impale fertilized eggs of crabs which are encased in a thick capsule (the fertilization envelope).

Estimating the membrane permeability. Usually to test the relative permeability of the membrane to K^+, Na^+ or Cl^-, ASW containing 10 X or 0.1 X the normal concentration of K^+, Na^+ or Cl^-, were used (Table 1). The values of the change in E_m, (ΔE_m), measured with the test solutions were corrected for the tip potential observed when the microelectrode was withdrawn from the oocyte. ΔE_m values were compared to 58 mV, the theoretical maximal change in membrane potential displayed in a membrane ideally selective for the tested ion.

Voltage-clamp measurements. Voltage-clamp measurements were done only on crab eggs. Eggs were impaled with two microelectrodes. The voltage microelectrode was as described for potential measurements. The current microelectrode was pulled from thin-walled glass (1.5 mm O.D.; 1.05 mm I.D.) and filled with 3 M KCl (resistance 3-6 MΩ in seawater). Impalement with the current microelectrode, the tip of which was less sharp than that of the E_m microelectrode, often required a light tap on the electrode holder. This introduced an additional leakage which usually decreased the R_m. A conventional two-electrode voltage-clamp was used. A current-to-voltage converter (1 mV = 0.1 nA) was connected to

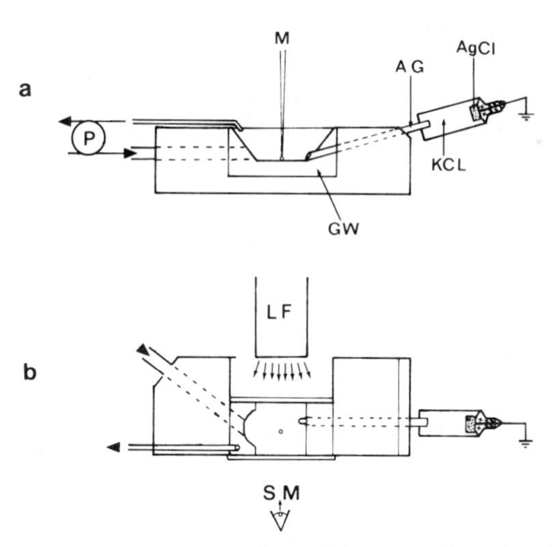

Figure 1. Front and upper view of the recording chamber. **a)** Front view showing the agar bridge made of polyethylene tubing (AG); a pellet of silver chloride (AgCl) is sealed in a syringe tip filled with KCl 3M (KCl). The egg is viewed through glass windows (GW) and is impaled with a vertical microelectrode (M). The recording chamber is perfused by a peristaltic pump (P). **b)** Upper view showing the impaled egg in the recording chamber, which is back illuminated with a light fiber (LF) and observed with a stereomicroscope (SM). X2.

the bath through an agar bridge (Biologic clamp amplifier, Meylan, France). During some measurements a grounded piece of copper foil, 0.1 mm-thick, was interposed between the two microelectrodes to minimize their coupling and improve the stability of the clamping device.

Sperm Collection and *In vitro* Insemination Procedures

Sperm were collected dry from dissected spermathecae of female crabs, thelyca of female lobsters, and from the spermatophores attached to the ventral thoracic wall of mated female prawns.

An attempt was made to *in vitro* fertilize crab and lobster eggs by pipetting diluted sperm onto the impaled egg; however this method was questionable since the sperm of lobsters and crabs are immotile and therefore may not reach the egg surface. We obtained successful *in vitro* fertilization of crab eggs by bringing a sticky pellet of sperm into contact with the egg vitelline coat using fine forceps. The number of spermatozoa deposited on the egg surface was to some extent regulated by two *in vitro* fertilization procedures. In the first, termed "heavy insemination", a sperm pellet was repeatedly brought to the surface of an egg in a measurement chamber in which the water circulation had been stopped. In the second procedure ("light insemination") a sperm pellet was touched once to the surface of an egg in a measurement chamber in which the water circulation was maintained. All inseminations were done at 19°C.

In vitro fertilization of prawn eggs was accomplished by introducing sperm to impaled eggs in Mg^{2+}-free hypo-osmotic ASW (osmolarity = 760 mOsm) or in low Mg^{2+}-hypo-osmotic ASW (10 mM Mg^{2+}). Immediately following sperm introduction the system was perfused with ASW. All *in vitro* fertilization procedures were carried out at 19°C.

Fixation Procedures for Voltage-Clamped Eggs

Voltage clamped eggs were usually kept in the recording chamber for 20 to 35 min post fertilization to be sure that complete sperm penetration of the egg cortex had occurred (see section 5). After this period, 1.5 to 2 ml of fixative for light microscopy (see below) was rapidly pipetted directly onto the voltage-clamped egg. At that moment, the fertilization current suddenly increased and the voltage clamp was quickly turned off. The perfusion was maintained until the seawater was completely replaced by fixative. The immediate increase in current, at the addition of fixative, indicated that fixation was instantaneous.

Light Microscopy

Since crab, lobster and prawn eggs are not translucent, the presence of sperm nuclei in the egg cortex of whole mounts was demonstrated using the technique of Zalokar and Erk (1977). Eggs were fixed with a mixture of 95% ethanol, 50% acetic acid and formaldehyde (40:10:2 by vol). After fixation the vitelline coat was removed with thin needles, to eliminate supernumerary spermatozoa trapped in the coat. Fixed eggs were hydrolysed with 2N HCl at 50°C for 5 min, quickly rinsed with distilled water, and stained with basic fuchsin (1% in 2.5% acetic acid). Eggs were then dehydrated with 100% ethanol, placed in xylene, and whole mounted in canada balsam. Eggs were observed on both sides. Whole mounts were examined with a bright field or phase contrast microscope. To determine the position of sperm nuclei in the egg cortex specimens were fixed in Carnoy's fixative with intact vitelline coats, embedded in historesin (LKB), and cut into 5 μM sections.

RESULTS

Electrical Properties of the Oocytes

1. Crab oocytes. Oocytes of the crab *Maia squinado* have an E_m of -53 ± 3 mV and R_m of 102 ± 10 M Ω (n = 37), giving a membrane-specific resistance of 2.05 MΩ.cm^2, assuming the oocyte to be a sphere, 800 μM in diameter. When bathed in ASW containing 120 mM

K^+, *i.e.* ten times the normal concentration of K^+ in ASW, the membrane of the oocytes depolarized by 35 ± 6 mV (n = 10) (as opposed to 58 mV for a K^+ electrode), while R_m decreased. Varying Cl^- and Na^+ also caused the potential to depolarize by 11 ± 1 mV and 8 ± 1 mV, respectively, for the 10 fold change in Cl^- and Na^+. In both situations R_m increased about 10%. We found that 20% of mature oocytes of *Maia squinado* displayed an average E_m of -42 ± 1 mV (n = 22) with a low R_m of 9 ± 2 MΩ (n = 22). In addition, ionic substitution experiments indicated that these oocytes were more permeable to Cl^- than to K^+, since they were depolarized by 30 ± 2 mV (n = 22) and by 7 ± 2 mV (n = 22), for the 10 fold change in Cl^- and K^+. These latter properties are likely due to impalement leaks. We found similar results for eggs of another crab, *Carcinus maenas*. In conclusion, it appears that the membrane of unfertilized ovulated crab oocytes is highly resistant and primarily permeable to K^+, with some underlying Cl^- permeability.

2. Lobster oocytes. We measured a resting potential of -32 ± 2 mV (n = 16, 2 females) in lobster oocytes. Ionic substitution experiments showed that the oocyte plasma membrane is more permeable to Cl^- than to other ions. We observed that its R_m is low (Goudeau and Goudeau 1986a); however, we found that impalement leaks were also responsible for these electrophysiological properties, since further measurements gave an E_m in the range of -50 mV which was primarily K^+ dependent, and higher R_m values than the previously recorded.

3. Prawn oocytes. When ovulated prawn oocytes were collected and immediately impaled in ASW, some displayed an average E_m of -42 ± 2 mV (n = 11), which is due largely to Cl^- permeability with some Na^+ and K^+ components. Their membrane resistance was relatively low, with an average value of 15 ± 5 MΩ (i.e. 170 KΩ.cm^2); however, some oocytes showed a highly resistant membrane (R_m = 160 ± 30 MΩ, n = 23) and it is likely that the presence of a leak around the microelectrode might explain this difference in R_m measurement (Goudeau and Goudeau 1986b). By contrast, when oocytes are collected and impaled in Mg^{2+}-free ASW, they display a high membrane resistance of 170 ± 20 MΩ (n = 23) (range 40-400 MΩ and mean specific resistance of 2.25 MΩ.cm^2) and an average E_m of -36 ± 4 mV (n = 23). In that situation, ionic substitution experiments indicated that the membrane is equally permeable to K^+, Cl^- and Na^+. We can conclude that unfertilized, ovulated prawn oocytes are highly resistant and consequently do not differ in this respect from unfertilized, ovulated crab oocytes.

Membrane Electrical Response to the Activating Effect of ASW, Independent of Fertilization, in Crab and Prawn Eggs

The electrical response of crab and prawn oocyte membranes were examined during sperm independent activation (meiotic resumption) by ASW. Lobster eggs were not examined because it is very difficult to obtain unfertilized, ovulated oocytes of this species.

1. Long term incubation in ASW hyperpolarizes the crab egg membrane. Figure 2 shows the time course of the membrane potential E_m and the membrane resistance R_m of an egg subjected to the effect of ASW for 100 min. E_m slightly decreased from -50 to -37 mV during the first 50 min of incubation, then became increasingly negative until it stabilized at -80 mV after 95 min of incubation. R_m remained unchanged at 170 MΩ for the first 15 min of incubation, then abruptly decreased to 36 MΩ after another 50 min. The large drop in R_m is correlated with the lifting of the vitelline coat due to the first step of the cortical reaction (Goudeau and Goudeau 1985). R_m then slowly decreased to reach very low values, in the range of 3-5 MΩ, after 80 min of incubation. This pattern was observed with 11 eggs: at the beginning of incubation E_m was -53 ± 3 mV (n = 11) and R_m = 149 ± 16 MΩ (n = 11). E_m became -54 ± 3 mV (n = 11), -61 ± 4 mV (n = 6) and -76 ± 2 mV (n = 10) at 50, 80 and 120 min of incubation, respectively. The correlative values of R_m were 21 ± 5 MΩ (n = 11), 6 ± 2 MΩ (n = 6) and 5 ± 2 MΩ (n =5), respectively. When incubation in ASW was terminated, the hyperpolarized membrane was primarily permeable to K^+.

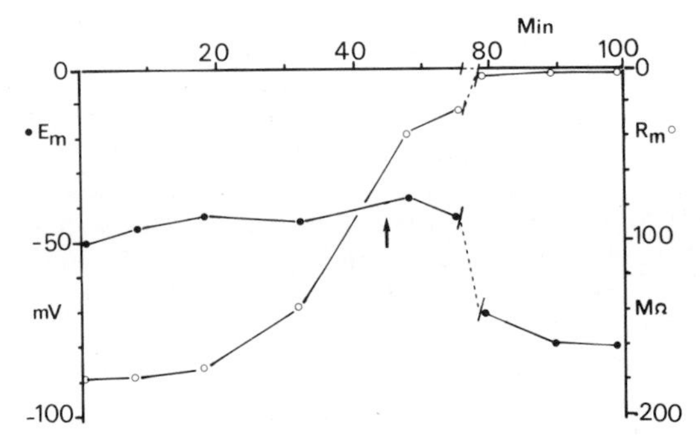

Figure 2. Time course of membrane potential, E_m (•) and resistance R_m (o) of a crab egg perfused with ASW for 100 min. The arrow marks the lifting of the vitelline envelope. Broken lines indicate that 15 min was omitted from the record.

2. Incubation in ASW also hyperpolarizes the prawn egg membrane, due only to external Mg^{2+}. Figure 3a shows a record of the time course of membrane potential of an egg incubated for 80 min in ASW. During the first 20 min E_m remained unchanged at -40 mV, thereafter becoming increasingly negative until it stabilized at -70 mV after one

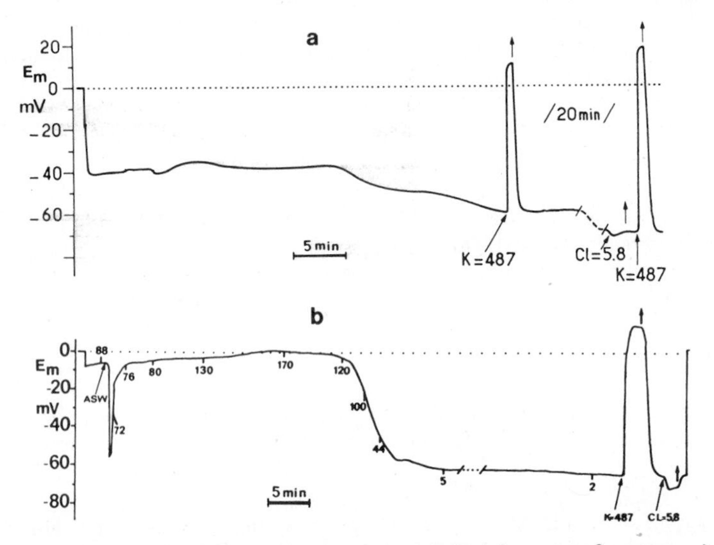

Figure 3. Record of membrane potential (E_m) in eggs of prawn. **a)** Time course of E_m of a prawn egg impaled in ASW at 21°C. The egg was impaled 10 min after having been collected in ASW. Cl = 5.8 and K = 487 indicate when ASW was replaced by ASW containing 5.8 mM Cl^- and ASW containing 487 mM K^+ respectively. Upward-directed arrows without other indications mark when ASW was reintroduced into the recording chamber. Broken line indicates that 20 min was omitted from the record. **b)** Initial transient hyperpolarization and delayed sustained hyperpolarization of egg membrane, induced by exchange of Mg^{2+}-free ASW by ASW at 21°C. This egg was collected in Mg^{2+}-free ASW at 6°C and incubated in that medium for 4h. Then the egg was impaled in Mg^{2+}-free ASW at 21°C. ASW indicates when ASW replaced Mg^{2+}-free ASW. Numbers under the record give the values of R_m.

hour of incubation. Ion substitution experiments showed that after this period the membrane, which until then was mainly Cl^- selective, became exclusively K^+ selective. This pattern was observed with 11 eggs; after 1 to 2h of incubation in ASW, E_m was -62 ± 2 mV. In eggs with a high initial membrane resistance, R_m markedly decreased to very low values (~ 2 MΩ) upon ASW incubation. These changes in membrane permeability and resistance were not induced by changes in osmotic pressure to which the egg is exposed during its passage from the interior of the female prawn (blood osmolarity = 780 mOsm) to ASW (osmolarity = 1040 mOsm). We also established that external Ca^{2+} was not required for hyperpolarization of the egg membrane and the decrease in its resistance. In addition, long-term incubation of eggs without external Mg^{2+}, resulted in E_m and R_m values that were not significantly different from their initial values (E_m = -36 ± 4 mV, R_m = 170 ± 20 MΩ, n = 23). By contrast, after preincubation in Mg^{2+}-free ASW (from 50 min to 24h), eggs bathed in ASW transiently hyperpolarized and then repolarized (Figure 3b) after 19.6 ± 1.5 min (n = 25, 12 females) to gradually attain an E_m value of -62 ± 1 mV (n = 16, 7 females). Concurrently R_m decreased to values of ~2 MΩ. We also found that the threshold Mg^{2+} concentration that induced secondary membrane hyperpolarization with a concurrent R_m decrease was around 20 mM (n = 30, 20 females) (Figure 4).

Membrane Electrical Response to Fertilization in Crab, Lobster, and Prawn Eggs

1. Crab eggs. Within 1s after the application of a small pellet of sticky sperm to the vitelline coat of the *Maia squinado* egg, we observed an abrupt and long-lasting hyperpolarization of the plasma membrane, from -54 ± 2 mV to -80 ± 1 mV (n = 16) (Figure 5). At the same time the membrane resistance decreased from 83 ± 7 MΩ to 3 ± 0.3 MΩ, and the membrane became mainly permeable to K^+. The hyperpolarization of the membrane occurred in less than 400 ms (Figure 6). We verified that the E_m of the eggs remained at -80 mV for at least 2h under *in vitro* conditions (n = 10, 6 females). 100% of the eggs were penetrated by sperm (n = 16). Of the 16 eggs studied, 15 showed one hyperpolarizing step and were monospermic. The remaining egg was polyspermic, exhibiting 4 hyperpolarizing steps and 4 sperm nuclei in its cortical cytoplasm (Figure 7). Moreover, polyspermy was usually associated with a fertilization potential value less negative than -75 mV, especially for eggs whose membrane resistance was low before fertilization (Goudeau and Goudeau 1985). Similar results were obtained from eggs of the crab *Carcinus maenas* (Figure 8). In a series of fertilization experiments with these crab eggs E_m hyperpolarized from -53 ± 7 mV to -76 ± 2 mV (n = 5), while R_m decreased from 104 ± 10Ω to 5 ± 2 MΩ (n = 5).

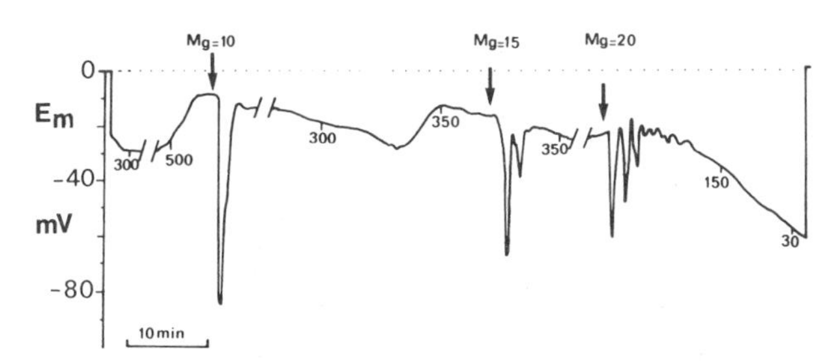

Figure 4. Determination of the threshold Mg^{2+} concentration required to obtain a sustained hyperpolarization of prawn egg membrane. This egg was collected in Mg^{2+}-free ASW and incubated in that medium for 8h, then the egg was impaled in Mg^{2+}-free ASW. Mg = 10, 15 and 20 mM indicate when ASW containing these concentrations in mM of Mg^{2+} replaced the preceding one. Note that at each increase of Mg^{2+} concentration, a transient hyperpolarization of the membrane occurs. Numbers under the record give the values of R_m. Separated bars on the record indicate that 10 min of the record has been omitted.

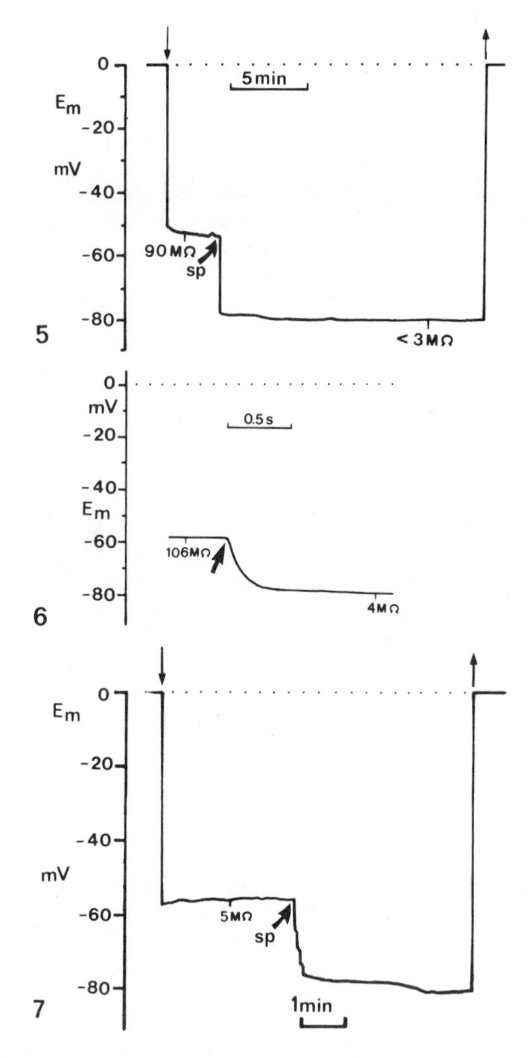

Figure 5. Fertilization potential of a crab egg, recorded at slow speed. The light and heavy arrows, respectively, indicate when the microelectrode penetrated the egg and when the egg was inseminated (sp). Numbers on the record indicate R_m.

Figure 6. Fertilization potential of a crab egg, recorded at high speed. The heavy arrow indicates insemination, and the numbers on the record indicate R_m.

Figure 7. Fertilization potential of a polyspermic crab egg. The light and heavy arrows, respectively, indicate when the microelectrode penetrated the egg and when the egg was inseminated (sp). Number on the record indicates R_m. The fertilization potential consists of four hyperpolarizing steps, four sperm nuclei were detected in the egg cortex.

We tested the E_m value of naturally spawned eggs of *Maia squinado* which had been collected from 30 min to 5h after laying. The E_m values measured did not differ significantly (average of -77 ± 0.5 mV, n = 66, 5 females). As such we can affirm that naturally fertilized eggs had an E_m remaining at the same negative value (-77 mV) for periods up to 5h after natural spawning. All the tested eggs were fertilized.

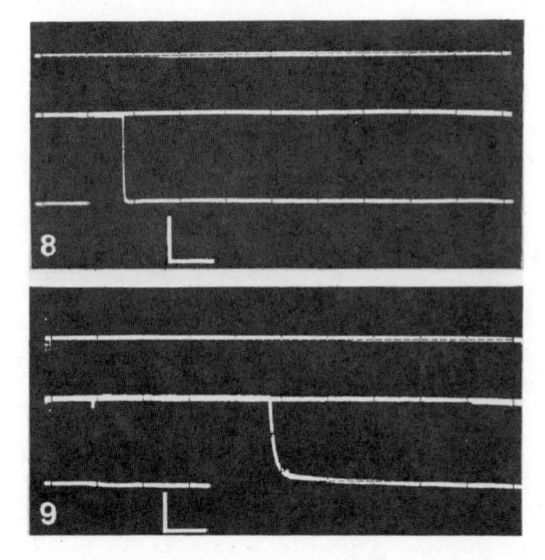

Figures 8 and 9. Oscilloscope records of fertilization potentials in *Carcinus maenas* and *Homarus gammarus*, respectively. The upper, middle and lower traces represent the zero potential level, one or several sweeps at the resting potential level and the membrane potential after the hyperpolarization response, respectively. Vertical and horizontal bars, correspond to 20 mV and 2s, respectively.

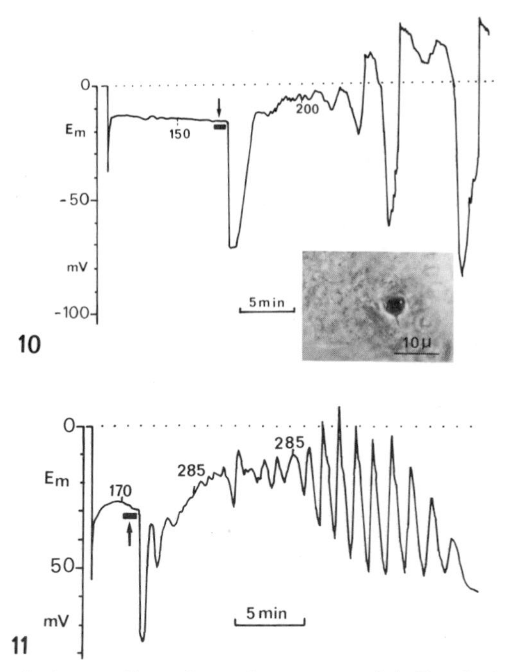

Figure 10. A typical recording of membrane potential, E_m, during insemination of a prawn egg incubated in Mg^{2+}-free ASW. This egg, inseminated at E_m = -17 mV, contained a sperm nucleus. The heavy bar indicates the period during which sperm were applied to the egg. Arrow marks the time when Mg^{2+}-free ASW was exchanged for ASW in the recording chamber. Numbers under the record give the values of R_m.

Figure 11. A recording of membrane potential, E_m, during insemination of a prawn egg incubated in Mg^{2+}-free ASW. This egg, inseminated at E_m = -27 mV, did not contain a sperm nucleus. Symbols have the same meaning as in Figure 10.

2. Lobster eggs. At fertilization, the egg membrane abruptly hyperpolarized within 400-600 ms (Figure 9). The negative shift averaged -32 ± 4 mV (n = 9) with the membrane potential remaining unchanged (E_m = -68 ± 1 mV, n = 9) for a period up to 70 min. The resistance of the membrane concomitantly decreased to about 1-2 MΩ, corresponding to an increase in K^+ permeability (Goudeau and Goudeau 1986a).

3. Prawn eggs. 12 of 49 eggs impaled in Mg^{2+}-free hypo-osmotic ASW were successfully fertilized (had sperm nuclei in their cortical cytoplasm). 11 eggs were monospermic and one contained two sperm nuclei. In contrast only 4 of 34 eggs impaled in low Mg^{2+} hypo-osmotic ASW contained sperm nuclei; three eggs were monospermic and one contained 4 sperm nuclei. We never detected a membrane response from eggs successfully fertilized. For example, Figure 10 illustrates a record of E_m from an egg collected and impaled in Mg^{2+}-free hypo-osmotic ASW, then exposed to sperm in ASW and concomitantly perfused with ASW. This record does not differ significantly from the record in Figure 11, which depicts the time course of E_m of an egg which failed to be fertilized by a sperm. There appears to be no correlation between successful fertilizations and E_m values (compare Figures 10-14). It should also be noted that raising the temp-

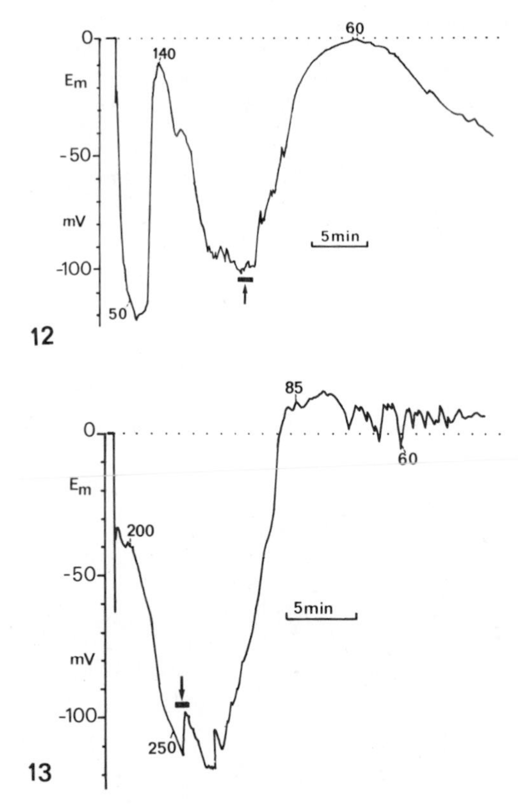

Figure 12. A recording of membrane potential, E_m, during insemination of a prawn egg incubated in Mg^{2+}-free ASW. This egg, inseminated at -100 mV, contained a sperm nucleus. Symbols have the same significance as in Figure 10.

Figure 13. A recording of membrane potential, E_m, during insemination of a prawn egg incubated in Mg^{2+}-free ASW. This egg, inseminated at -100 mV, did not contain a sperm nucleus. Symbols have the same significance as in Figure 10.

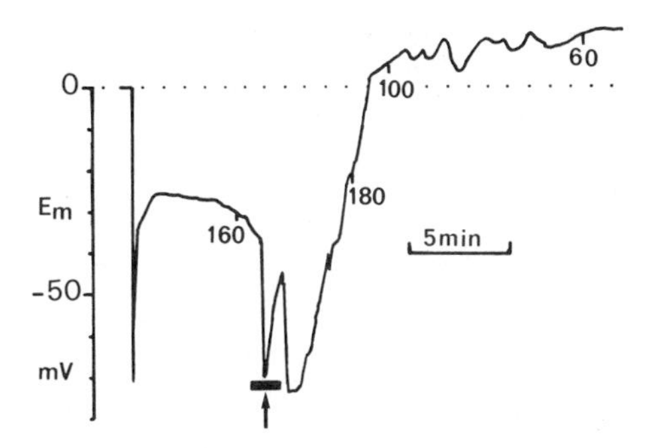

Figure 14. A recording of membrane potential, E_m, during insemination of a prawn egg incubated in Mg^{2+}-free ASW. This egg, inseminated at -70 mV, contained a sperm nucleus. Symbols have the same significance as in Figure 10.

erature of the ASW, which is normally kept relatively low to prevent egg activation prior to impalement, did not improve fertilization rates.

To conclude, four points emerge from the above results regarding *in vitro* fertilization of prawn eggs: i) eggs were penetrated by sperm when they were collected and impaled in Mg^{2+}-free hypo-osmotic ASW or low Mg^{2+} hypo-osmotic ASW; ii) in these trials the percent of eggs successfully fertilized was very low (12-24%); iii) most fertilizations were monospermic; iv) eggs penetrated by sperm never displayed an electrical change in membrane potential.

Are Fertilized Eggs of Crabs and Prawns Normally Monospermic or Polyspermic?

From 12 natural spawnings of *Maia squinado*, 510 eggs were individually examined. 30 to 90 spawned eggs were examined from each female. Eighty one percent of all the spawned eggs had one or more sperm nuclei in the cortical cytoplasm. The sperm nucleus is triangular in shape and contains condensed chromatin which stains strongly with fuchsin in whole mount preparations (Figure 15). 19% of the eggs examined did not contain a sperm nucleus. In 10 of the spawns, 100% of the eggs were monospermic (Table 2). In two spawns 3% and 12% of the eggs, respectively, were polyspermic and contained as many as 8 sperm nuclei.

347 naturally-spawned eggs collected from 7 *Palaemon serratus* females were individually examined. Thirteen to 75 spawned eggs were examined from each female. Eighty percent of all the spawned eggs had one or more sperm nuclei in the cortical cytoplasm (Table 2). The sperm nucleus looks like an inverted umbrella, as often described in natantian prawn species (Lynn and Clark 1983a) and contains chromatin that is easily visualized in fuchsin stained whole mounts when incorporated in the egg cortex (Figure 16). We also detected a thin, extended acrosomal process that resembles that of the reacted sperm of the penaeid shrimp *Sicyonia ingentis* (Clark *et al.* 1981, 1984, 1985; Shigekawa *et al.* 1982). The other 20% of the spawned eggs had no sperm nuclei in their cortical cytoplasm. In three natural spawnings 100% of the eggs were monospermic. In the 4 other spawnings, 2-31% of the spawned eggs were polyspermic, while the other 69 to 98% of the spawned eggs were monospermic. Monospermy was also previously reported for eggs of an another prawn species, *Macrobrachium rosenbergii* (Lynn and Clark 1983b).

To conclude, crab and prawn eggs are naturally monospermic, but the mechanism preventing polyspermy may be more efficient in crab than in prawn eggs (range 88-100% *versus* 69-100%). However, the lack of an electrical response to fertilization in prawn eggs, indicates that the mechanism(s) insuring monospermy is (are) different from the electrically-mediated block to polyspermy present in several other species.

Table 2. Monospermy in Eggs from Natural Spawnings of the Crab, *Maia squinado* and the Prawn *Palaemon serratus*

Female number	Time after spawning (min)*	Number of examined eggs	Percentage of eggs pene-trated by sperm	Percentage of monospermic eggs	Confidence limit (c.l.) **	Percentage of eggs resuming meiosis
			Maia squinado			
1	45	35	91	100	89-100%	70
2	30	41	85	100	90-100%	20
3	120	35	89	100	89-100%	100
4	20	35	97	100	90-100%	100
5	90	34	91	100	89-100%	100
6	90	33	94	100	89-100%	100
7	30	35	83	100	89-100%	100
8	30	30	57	100	79-100%	100
9	25	31	39	100	73-100%	100
10	90	68	91	100	94-100%	100
11	120	40	88	97	81-97%	100
12	45	93	65	88	78-95%	100
			(mean = 81 ± 5)			
			Palaemon serratus			
1	120	69	100	98		100
2	140	75	84	69		100
3	60	74	62	97		82
4	45	13	42	88		100
5	90	43	69	100		100
6	120	39	100	100		90
7	120	34	100	100		100
			(mean = 80 ± 9)			

* time elapsed between spawning and egg collection for fixation; ** confidence limits (c.l.) were calculated for a probability of 0.05.

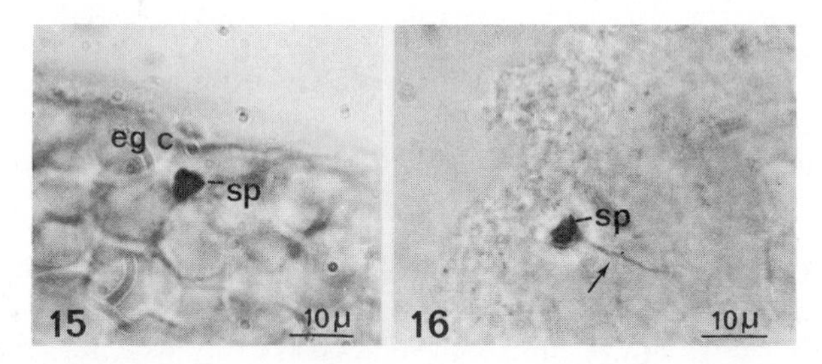

Figure 15. Light micrograph of a whole mount preparation of a naturally-spawned crab egg. Egg-laying started 30 min before fixation. One sperm nucleus (sp) is visible in the egg cortex (eg c). The location of the sperm nuclei in the egg is ascertained when, by changing the focus, the egg cortex is discernible above and below the sperm nucleus (x 1,000).

Figure 16. Light micrograph of a whole mount preparation of a naturally-spawned prawn egg observed with phase contrast. Egg-laying started 1h before fixation. One sperm nucleus (sp) is visible in the egg cortex. The nucleus has an inverted umbrella shape and contains strongly condensed chromatin. See also the acrosomal process (arrow) (x 1,000).

Voltage Clamp Studies of Fertilization in Crab Eggs

Since crab eggs are physiologically monospermic, we wondered if the unusually long-lasting membrane hyperpolarization displayed by the egg at fertilization might provide an effective block to polyspermy. To approach this question, unfertilized eggs were voltage clamped at various E_m, ranging from +20 to -90 mV, inseminated and examined morphologically.

1. Effect of clamped voltage on sperm penetration. Eggs clamped at +20 mV, -10mV, -40 mV, -50 mV, -60 mV, -80 mV, -90 mV or -100 mV (Table 3), and subsequently inseminated, had no sperm nuclei in their cortical cytoplasm as determined by whole mount or thick section observations (Figure 17). In contrast, penetration of sperm occurred in eggs clamped between -65 mV and -75 mV (Table 3). In whole mount preparations we often observed several sperm nuclei in the egg cortical cytoplasm. Sperm nuclei were completely incorporated in the egg cytoplasm 20 min after insemination. Figure 18 illustrates sperm penetrating an egg which was fixed 13 min after the electrical response had occurred. Figures 19 and 20 show sperm nuclei within the cytoplasm of eggs fixed 25 and 20 min, after the electrical response. Sperm nuclei were condensed, triangular in shape, and appeared no different than pronuclei in naturally fertilized eggs (Figure 15). The average number of sperm nuclei in these eggs was 3-6 (Table 3).

2. Effect of clamping the membrane potential on the fertilization current. All eggs clamped at potentials +20 to -60 mV displayed an outward current at fertilization. Figure 21 shows the fertilization current, I_f, obtained from an egg clamped at -40 mV before insemination. Each time sperm were applied to the egg, stepwise increments in outward current were seen. These increments are almost equal in amplitude but their number was variable, probably depending on the number of individual sperm contacting the egg. Figure 22 also depicts a fertilization current obtained from an egg clamped at E_m = -40 mV, but with a higher speed chart recording that shows five incremental steps. Usually, after the initial

Table 3. Effect of Egg Clamping Prior to Insemination

Clamped voltage (mV)	Percent of eggs⁺ with sperm nuclei in cortical cytoplasm	Average number of fertilizing sperm	Total number of eggs examined
+20	0	0	6
-10	0	0	16
-40	0	0	25
-50	0	0	21
-60	0	0	24
-65	58	5.9 ±1⁺⁺	26
-70	45	3.6 ± 0.8⁺⁺	29
-75	32	3.1 ± 1⁺⁺	22
-80	0	0	27
-85	0	0	18
-90	0	0	7
-100▼	0	6	6

▼ Current clamped eggs

⁺ Based on whole mount observations

⁺⁺ No significant difference between these values. Voltage clamped eggs were maintained 20 to 35 min after insemination and fixed.

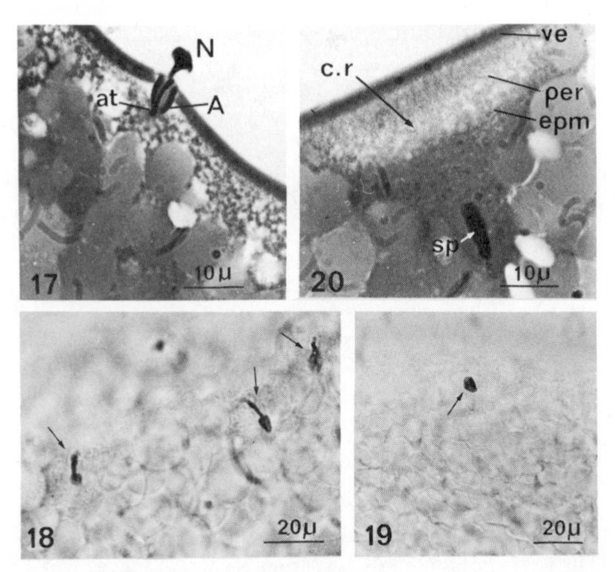

Figure 17. Thick section through a crab egg whose membrane was clamped at E_m 10 mV. The egg was inseminated, a fertilization current of 5 outward steps was recorded, the clamp was released and the egg membrane hyperpolarized to $E_m = -55$ mV. No sperm nuclei entered the egg cortex. We observed one partially reacted sperm with an acrosome tubule (at) and everted acrosomal vesicle (A) depressing the egg plasma membrane, but with its nucleus (N) was located outside the egg investment. Here sperm-egg contact did not initiatiate the release of cortical material (x 1,200).

Figure 18. Light micrograph of a whole-mounted crab egg, the membrane of which was clamped at -75 mV. At insemination 4 outward steps in the fertilization current were detected. The egg was clamped for 13 min before fixation. 4 sperm nuclei (three observable on the micrograph, see arrows) were being incorporated into the egg cortex. See the elongated shape of the nuclei (x600).

Figure 19. Light micrograph of a whole-mounted crab egg, whose membrane was clamped at -75 mV. At insemination 3 outward steps in the fertilization current were detected. The egg was clamped for 25 min before fixation. Three sperm nuclei were detected in the egg cortex, although only one is shown in the micrograph by the arrow (x500).

Figure 20. Thick section of a crab egg which was inseminated while clamped at -65 mV. 4 outward steps in current were recorded, and the egg was maintained clamped for 23 min before fixation. Four sperm nuclei were observed in the egg cortex; the micrograph shows one of them (sp), which had penetrated deep into the cortex. epm, egg plasma membrane, per, perivitelline space; ve, vitelline envelope; cr, cortical reaction at sperm-entry. (x1,200).

2. continued
phase consisting of stepwise increments, I_f decreased slowly, sometimes in step-like fashion, as shown in Figure 21. The egg illustrated in Figure 23 was clamped at -10 mV and, at insemination, displayed an I_f whose time-course was almost the same as that observed in eggs clamped at -40 mV. In all of these experiments, the conductance of the clamped membrane was maximal at the peak of the current and decreased with I_f (Figure 21). All eggs inseminated while clamped at +20 to -60 mV responded electrically, but none incorporated sperm nuclei (Figure 26).

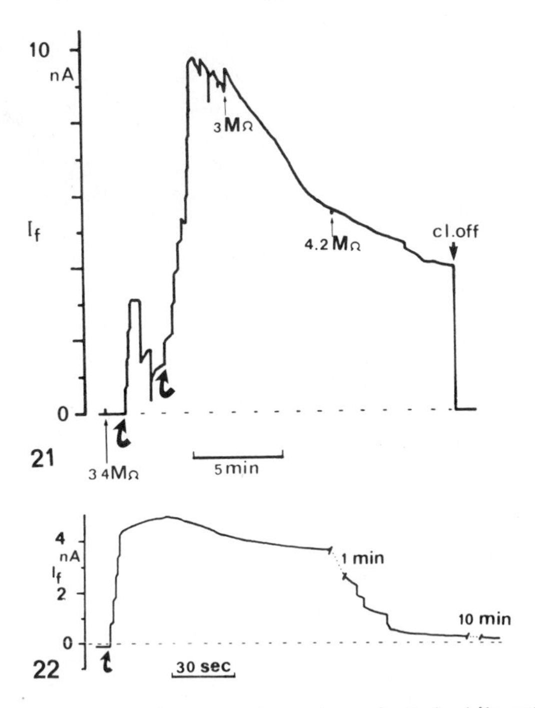

Figure 21. Fertilization current from a crab egg inseminated while voltage-clamped at -40 mV. The first contact of the sperm pellet with the vitelline envelope (curved arrow) induced four steps of current, and the second more than ten steps. Numbers over the record indicate R_m and cl. off indicates when the clamp was cut off. No sperm nuclei were detected in the egg cortex.

Figure 22. Fertilization current from a crab egg inseminated while voltage-clamped at -40 mV (high speed recording). The curved arrow indicates when the pellet of sperm was brought in contact with the vitelline envelope. Dotted lines indicate that sections were omitted from the record. No sperm nuclei were detected in the egg cortex.

2. continued

At clamped voltages from -65 to -75 mV, insemination of eggs also elicited a membrane outward current. Figure 24 shows the fertilization current from two eggs whose membrane was clamped at -65 mV. Both eggs were penetrated by several spermatozoa. Here again, I_f resulted from the summation of individual steps, but in contrast to eggs clamped at less negative potentials, two striking differences were seen: 1) After one or several initial steps the current, instead of decreasing, exhibited a delayed rise so that it reached its maximum in three or four minutes and remained almost at that level throughout the rest of the experiment (20 min). Thus the electrical response was prolonged in place of being transient as for more positive clamp voltages. 2) Whereas 100% of the eggs whose membrane was clamped at -65 mV (or at more positive potential) showed a fertilization current, only 77% of the eggs at -70 mV (n = 29) and 40% at -75 mV (n = 22) produced a fertilization current (Figure 26). 58%, 45% and 32% of the eggs clamped at -65, -70 and -75 mV respectively, exhibited sperm nuclei (Table 3 and Figure 26). Evidently, all eggs penetrated by a sperm gave an electrical response.

In eggs clamped at potentials from -80 to -90 mV, a fertilization current was observed only once in over 52 experiments. Figure 25 shows a record of an experiment in which an egg was clamped at -80 mV and inseminated. Repeated contact between the sperm mass and the vitelline envelope of the egg did not elicit a fertilization current, and sperm

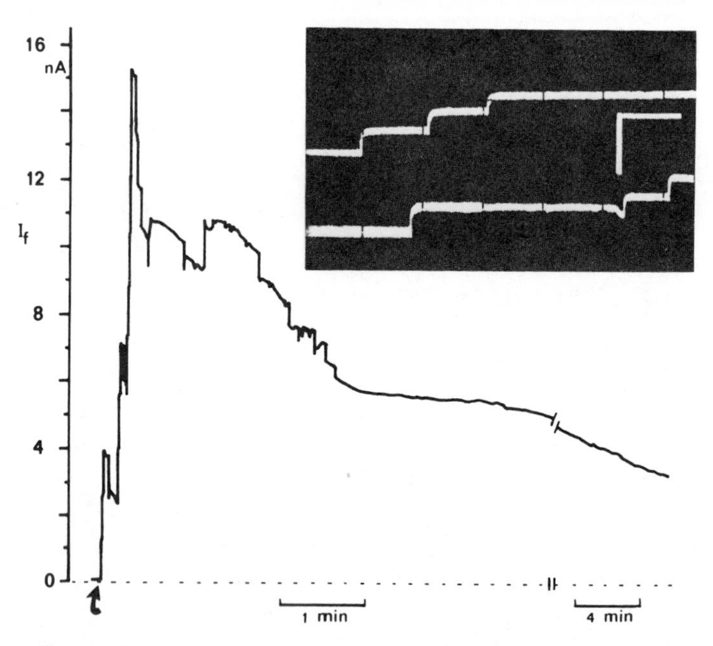

Figure 23. Fertilization current from a crab egg inseminated while voltage-clamped at -10 mV. Curved arrow indicates the time when the sperm pellet was applied. No sperm nuclei were detected in the egg cortex. Inset: Oscilloscope record of current representing the first steps of current; vertical and horizontal bars represent 5nA and 2s, respectively.

2. continued.
nuclei were not incorporated. None of the 27 eggs clamped at -80 mV, one of the 18 eggs clamped at -85 mV, and none of the 7 eggs clamped at -90 mV showed a fertilization current. Furthermore, none of the 6 eggs current clamped at -100 mV showed a fertilization potential. None of these eggs exhibited sperm incorporation (Table 3 and Figure 26).

The absence of a fertilization current at insemination in eggs voltage clamped at -80 mV and below cannot be due to the fact that the clamp potential is close to the reversal potential (~ -80 mV) of the fertilization current, for the following reasons: 1) Our recording apparatus was sufficiently sensitive to detect currents as small as 0.01 nA, and we never recorded any inward steps when we inseminated eggs clamped at -90 mV. 2) We inseminated current clamped eggs at -100 mV without ever obtaining a positive fertilization potential.

3. Effects of the release of clamped voltage on sperm entry. To determine whether the inhibition of sperm entry in voltage clamped eggs was reversible, we clamped eggs at various potentials, inseminated them and released the clamp. The clamp was released immediately after several steps of current were obtained (eggs clamped at -10 or -60 mV) or two minutes after insemination (eggs clamped at -80, -85 or -90 mV). Twenty minutes after the release of the clamp, eggs were fixed and examined for sperm penetration (Table 4).

Figure 27 shows an experiment where the initial clamp voltage of -10 mV was released after 6 steps of current were recorded (lower part of the figure). At release of the clamp, the potential went abruptly to -72 mV (inset of Figure 27). Morphological examination of this egg indicated the presence of 6 sperm nuclei in the egg cortex. Figure 28 shows an experiment where an egg was clamped and inseminated at -85 mV. After 2

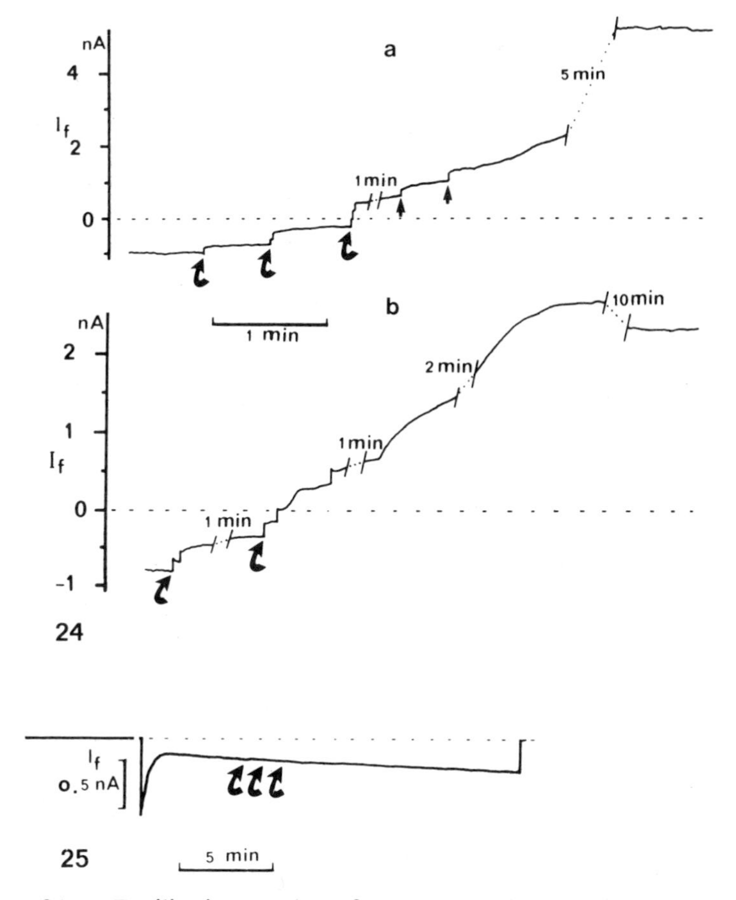

Figure 24. Fertilization current from two crab eggs inseminated while voltage-clamped at -65 mV. Curved arrows indicate when the sperm pellet was pressed against the vitelline envelope of the egg and straight arrows signal spontaneous steps. **a)** 8 steps of current were recorded and 6 sperm nuclei were found in the egg cortex; **b)** 5 steps of current were recorded, and 3 sperm nuclei were found in the egg cortex. Dotted lines indicate that sections of the record were omitted. Note that the fertilization current is maintained and not cut off in (a) and (b).

Figure 25. Insemination of a crab egg voltage clamped at -80 mV. The sperm mass was repeatedly brought in contact with the vitelline envelope (curved arrows), but no fertilization current was seen. No sperm nuclei were found in the egg cortex.

3. continued
min had elapsed without any electrical response, the clamp was released, leading abruptly to an E_m value of -72 mV. Figure 29 shows a high speed recording of eggs clamped at -80 mV and released. Figure 29a represents a control experiment without insemination. At the release of the clamp, the membrane potential went to a positive value which slowly became negative again with a time constant of 1.2s, indicative of the high R_m (50 MΩ) value. In Figure 29b, the egg was inseminated before releasing the clamp. When the clamp was released, E_m went immediately to -20 mV then shifted to -72 mV with a time constant in the range of that observed in fertilization potential records (compare Figures 29b and 6). At this time the R_m was low (4 MΩ). For all the clamp release experiments,

Table 4. Sperm Penetration after Release of Clamp Voltage in Eggs Clamped and Inseminated at Various Membrane Potentials

Initial clamp voltage (mV)	Percentage of eggs[+] with sperm nuclei in cortex	Average number of sperm entries seen after clamp release	Total number of eggs
-10	82	3.2 ± 0.6	11
-60	100	3.1 ± 0.6	12
-85	88	1.7 ± 0.4	16
-90	86	1	7

[+] Based on whole mount observations

cytological examinations showed that the eggs were fertilized: eggs whose records are given in Figure 28 and 29b, contained 5 and 2 sperm nuclei in their cortex. Conversely, for one egg voltage-clamped at -10 mV, the release of the clamp led to an E_m value of only -55 mV. Consequently there were no sperm nuclei in the egg cortex (Figure 17). Table 4 summarizes the results of these experiments; regardless of the clamp voltage during insemination fertilization occurred after release of the clamp. All eggs clamped at -90 mV and inseminated, were monospermic after the release of the clamp (Table 4).

4. Correlation between the number of sperm penetrating the egg and the number of incremental steps in the fertilization current or potential. At insemination of voltage-clamped crab eggs, the fertilization current usually consisted of summed equally-sized steps of current. It is likely that each step resulted from the interaction of a single sperm with the egg. As such the number of sperm nuclei in the egg was compared with the number of steps in the fertilization current displayed by eggs clamped at -60 mV and released. 100% of the eggs were penetrated by sperm (Table 4). Figure 30a shows that a correlation exists. We also observed fertilization potentials consisting of several negative steps in unclamped eggs (Figure 7). We determined the number of sperm nuclei found in the egg

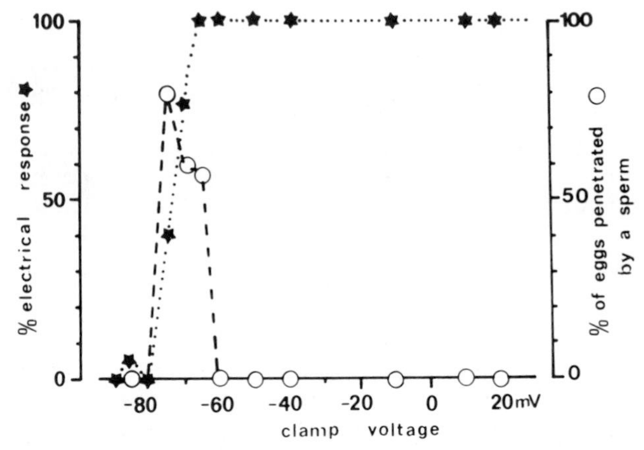

Figure 26. Percentage of inseminated crab eggs that produced a fertilization current, and that were penetrated by a sperm as function of the clamped voltage. Stars represent the percentage of eggs that responded at insemination by a fertilization current. Open circles represent the percentage of eggs that were penetrated by a sperm. Note that all eggs penetrated by a sperm gave an electrical response, but that all eggs which gave an electrical response were not necessarily penetrated.

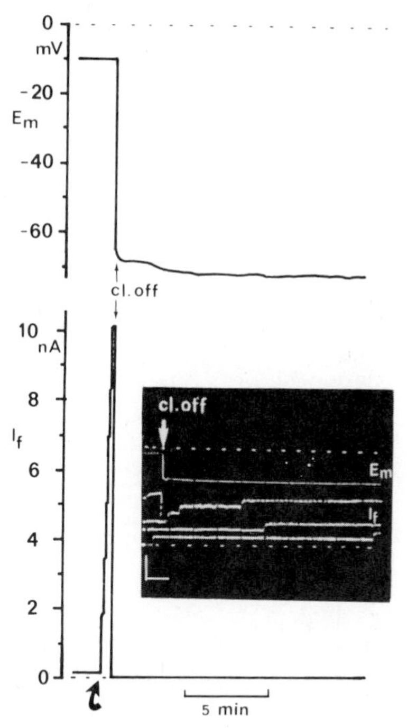

Figure 27. Insemination of a crab egg during voltage clamping at -10 mV followed by turning off the clamp after obtaining 6 steps of current (about 20s after insemination). The upper record shows voltage as a function of time; the lower record shows current as a function of time. The curved arrow indicates insemination, time 0, which caused 6 steps of current. cl. off indicates the time when clamp was turned off. E_m reached -72 mV.

4. continued.
cortex was correlated to the number of steps in the fertilization potential (Figure 30b). In summation these data show that the number of sperm nuclei is well correlated with the number of steps in the fertilization current of clamped (-60mV) and fertilization potential of unclamped eggs.

DISCUSSION

Electrical Response of Eggs to Fertilization in Decapod Crustacea

In vitro fertilization of crab and lobster eggs elicits an unusual electrical membrane response which abruptly (~ 400 ms) hyperpolarizes to about -30 mV. This hyperpolarization is sustained for at least 5h in, naturally spawned, fertilized eggs. This fertilization potential results in a dramatic increase in the K^+ conductance of the membrane, as evidenced by the decrease in the resistance of the membrane, which becomes selectively permeable to K^+. The low R_m value induces a correlative decrease in the time constant of the membrane, which accounts for the fast negative shift in E_m that characterizes the fertilization potential. A gain of K^+ conductance seems to be a general ionic mechanism related to fertilization and has been reported in the eggs of sea urchins (Chambers and De Armendi 1979), echiurians (Hagiwara and Jaffe 1979), amphibians (Jaffe and Schlichter 1985), fish (Nuccitelli 1980), hamsters (Igusa and Miyazaki 1983; Igusa *et al.* 1983), and nemerteans (Kline *et al.* 1985). It is clear, however, that there is a large variability in the onset of this increase of K^+ conductance. In sea urchin, echiurian, fish, amphibian and nemertean eggs, the increase of K^+ conductance follows an early, sperm-triggered increase in membrane conductance to Na^+ or Cl^-. This early depolarization may mask some early K^+ conductance, as in fact has been demonstrated in amphibians (Jaffe and Schlichter 1985) and recently in sea urchins (Obata and Kuroda 1987). In contrast, in the hamster egg, the increase of K^+ conductance occurs without any previous depolarization of the egg membrane and promotes transient hyperpolarizations superimposed on a slow hyperpolarization of the membrane, several seconds after gamete contact. In crab and lobster eggs, this increase of K^+ conductance arises without any early membrane depolar-

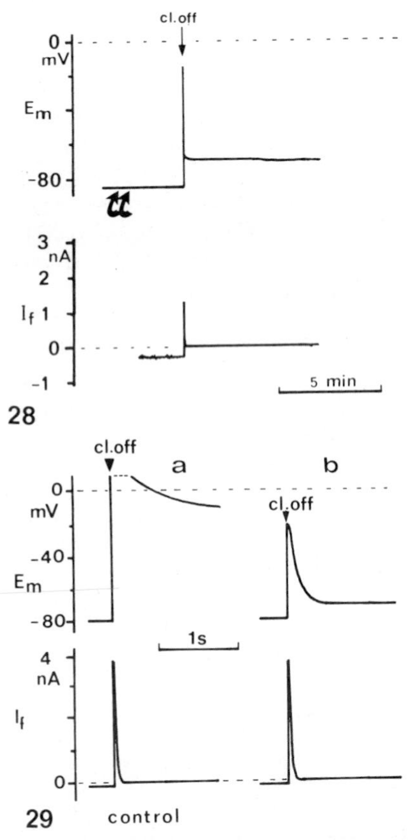

Figure 28. Insemination of a crab egg during voltage clamping at -85 mV followed by turning off the clamp 2 min later. The upper record shows voltage as a function of time; the lower record shows current as a function of time. Curved arrows indicate touches of the sperm mass to the vitelline envelope. At the release of the clamp (cl. off), E_m reached -72 mV.

Figure 29. Insemination of a crab egg during voltage clamping at -80 mV followed by turning off the clamp. Upper records show voltage as a function of time; lower records show current as a function of time. The time scale is greatly expanded compared with Figures 13 and 14. a) Control experiment without insemination. At the release of the clamp (cl. off), the potential shifted transiently to a positive level and then returned to -25 mV with a time constant of 1.2s. The shift of the potential to a positive level after the release of the clamp is due to an artifact. b) Insemination during voltage clamp. At the release of the clamp (cl. off), the potential shifted to -20 mV and then returned to -72 mV, with a time constant of 0.08s. Histological examination showed that 2 sperm nuclei were in this egg.

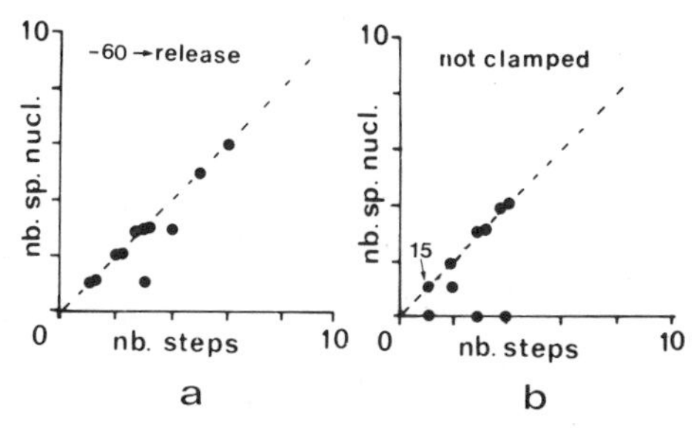

Figure 30. Number of sperm nuclei in the cortex of inseminated crab eggs as a function of the number of steps in the fertilization current (a) or fertilization potential (b). **a)** Eggs were clamped at -60 mV, inseminated and then released from the clamp. **b)** Eggs were inseminated while recording potential with one or two electrodes inserted. Fifteen of the eggs were impaled with one microelectrode, displayed only one step in the fertilization potential and were monospermic. These are indicated by a single point on the graph. Of the 9 other eggs, 3 were not penetrated by spermatozoa. Since their fertilization potentials were less negative than -65 mV. The 6 remaining eggs were polyspermic (5 were tested with 2 microelectrodes and one with only one microelectrode).

ization and, as such, is similar to what has been reported for hamster eggs. But in the eggs of crabs and lobsters, K^+ conductance increase requires only about 400 ms, and is sustained. Both the absence of the depolarizing step and the fact that hyperpolarization is sustained are intriguing features of the fertilization potential in lobster and crab eggs, the physiological significance of which are discussed below.

At the present time, we do not know the intrinsic membrane mechanisms that induce this increase in K^+ permeability. In eggs of the sea urchin (Slack *et al.* 1986), *Xenopus* (Busa *et al.* 1985; Kline 1988), the hamster (Igusa and Miyazaki 1983) and the nemertean, *Cerebratulus lacteus* (Kline *et al.* 1986), a sperm-triggered increase in cytosolic free calcium is responsible for the change in membrane permeability at fertilization. This is apparently not the case in crab and lobster eggs, whose membranes never display a fast hyperpolarization when treated with Ca^{2+} ionophores (A 23187 or ionomycin). We also do not know the nature of the membrane interactions at gamete contacts that elicit an electrical response by the egg membrane. Epel (1978), Longo (1978), Chambers and De Armendi (1979), and Longo *et al.* (1986) suggested that it might result either from an interaction between a ligand in or originating from the sperm membrane and a receptor on the egg membrane; alternatively, it might be due to the fusion of the sperm and egg membranes. The results of Longo *et al.* (1986) established that in inseminated sea urchin eggs, fusion could only be detected 4.5s after the onset of the fertilization current. Furthermore, it has been demonstrated that acrosomal proteins isolated from *Urechis caupo* sperm, were able to elicit an electrical response from homologous eggs (Gould *et al.* 1986). In addition, Swann and Whitaker (1987) suggested that the early electrical events at fertilization are not a consequence of sperm-egg fusion, but the result of the activation of a receptor coupled to an ionic channel. In contrast, McCulloh and Chambers (1986 a,b) found that in sea urchin eggs which were voltage-clamped and inseminated, gamete fusion was simultaneous with the onset of the fertilization current. Thus the conductance of the egg membrane increased at the instant of sperm-egg fusion. Regarding crab and lobster eggs, the question remains open.

During *in vitro* insemination of prawn eggs, we never detected an electrical response that could be recognized as a fertilization potential. This lack of electrical response at insemination cannot be due to an impairment of gamete fusion and further penetration of the sperm, since 24% of the *in vitro* inseminated eggs were penetrated by spermatozoa. This percentage of successful *in vitro* insemination appears, nevertheless, very low as compared to that obtained in crab eggs (24% *versus* 100%) or to that observed in naturally-spawned fertilized eggs of prawns (24% *versus* 80%). This reduced fertilization may be accounted for by experimental conditions that might be far from physiological.

The occurrence of a fertilization potential in the eggs of decapod crustaceans is not specifically related to the mode of fertilization. A fertilization potential occurs in both crab and lobster eggs, for which fertilization is respectively internal and external; however, it does not occur in prawn eggs which are fertilized in the external environment under natural conditions. Alternatively the occurrence of a fertilization potential in crab and lobster eggs might be associated with similar ecological conditions, since they are both benthic animals, while prawns may have a benthic or pelagic life.

Respective Roles of ASW and Fertilization in Eliciting a Membrane Hyperpolarization in Crab and Prawn Eggs

In crab eggs the membrane hyperpolarization, elicited by ASW due to an increase of K^+ permeability, begins after 50 min of incubation; this is compared to about 400 ms during fertilization. This dual membrane electrical response in crab eggs may be similar to the different time course in the acquisition of K^+ permeability in starfish eggs. These eggs gradually develop a high K^+ permeability if left unfertilized, or rapidly acquire a high K^+ permeability after their initial depolarization if fertilized (Miyazaki and Hirai 1979). Since the crab egg acquires the same E_m value (-80 mV), either gradually under the influence of seawater or rapidly at fertilization, it may be hypothesized that the same mechanisms operate in the membrane for the implantation or the activation of K^+ channels. This deserves further study.

For prawn eggs, it seems that the slow increase in membrane K^+ permeability occurring in seawater, is the only physiological response of these eggs. The effect of external Mg^{2+} on the K^+ permeability increase appears to be specific, as we never obtained the same effects when Ca^{2+} was used in place of Mg^{2+}. The mechanism(s) of action of external Mg^{2+} on the K^+, permeability increase, the resumption of meiosis and the cortical reaction, is (are) not yet understood in prawn.

Is Monospermy in Crab Eggs Electrically Mediated?

The results of *in vitro* insemination of voltage-clamped eggs showed that 100% of the eggs clamped, at potentials from +20 to -65 mV, displayed a fertilization current. The percentage of eggs that responded electrically decreased when they were clamped at potentials more negative than -65 mV and became nonexistant at -80 mV and at more negative E_m values. We observed that the penetration of sperm only occurred in eggs clamped at -65 to -75 mV. The fertilization current was transient in eggs voltage-clamped from +20 to -60 mV, while it was sustained in eggs voltage-clamped from -65 to -75 mV (these eggs contained one or more sperm nuclei in their cortex). Consequently, it is clear that the membrane processes occurring during the contact between gametes, which elicit an electrical response by the egg membrane, are voltage-dependent. Sperm entry is also voltage-dependent since it only occurs at voltages between -65 and -75 mV. The voltage dependence of the egg membrane electrical response may be due to the effect of the electrical field existing across the membrane. This may regulate the interactions of an acrosomal protein with an egg membrane receptor. Alternatively, the voltage may control the fusion of the gamete membranes, inducing transient fusions when the egg membrane is clamped from +20 to -60 mV, fusion followed by sperm entry for clamped potentials between -65 and -75 mV, and no fusion for clamped potentials of the egg membrane more negative than -80 mV. According to a scheme of voltage-dependent "fusion-unfusion" of gamete membranes, Lynn *et al.* (1988) and McCulloh *et al.* (1987) explain the occurrence of transient or prolonged fertilization currents in sea urchin eggs and oocytes when the membrane is clamped at voltages more negative than -30 mV or from - 10 to +20 mV, respectively. The transient response would be due to fusion and subsequent "unfusion" of a non-penetrating sperm, while the prolonged response one would correspond to fusion and penetration of the sperm.

In crab eggs as in sea urchin oocytes and eggs, the voltage dependence of the initiation of the current differs from the voltage dependence of sperm entry. In sea urchin eggs and oocytes, the sperm induces an electrophysiological response for membrane voltages between -70 and +17 mV. Sperm can only enter eggs, however, at membrane potentials between -25 and 17 mV and oocytes between -20 to +20 mV (Lynn and Chambers 1984; McCulloh *et al.* 1987). In crab eggs, the sperm induces an electrophysiological response for membrane voltages between +20 and -75 mV, but enters the egg at membrane potentials more negative than -60 mV and less negative than -80 mV. In this respect, the membranes of sea urchin eggs and oocytes and crab eggs respond in a similar way but with a voltage dependence of opposite sign. From these results, it can be suggested that the sustained value of E_m acquired by the crab egg at fertilization might constitute a permanent block to polyspermy, since we measured an average value of -77 ± 0.5 mV (n=66) in naturally spawned eggs. Because these eggs are already encased in a thick, tough envelope when impaled, we assume that some impalement leaks depolarize the egg membrane by a few mV. Consequently, in naturally spawned eggs, the true E_m might be more negative than -77 mV. Furthermore, our results for *in vitro* insemination of unclamped eggs gave an instantaneous E_m of -78 ± mV in less than 400 ms, which reached -80.5 ± 0.7 mV (n = 16) five minutes after fertilization. In the range of voltages attained by the egg membrane at fertilization, the probability of a second sperm entry is very small since our voltage clamp experiments indicated no sperm entry at E_m = -80 mV and 32% at E_m = -75 mV (*i.e.* a less negative value than that measured after natural fertilization or *in vitro* insemination). The hyperpolarization of crab eggs in response to fertilization may provide an effective block to supernumerary sperm. In addition, since the initial electrical egg response to fertilization lasted less than 400 ms, the narrowness (at most 20 mV) of the voltage window permitting sperm penetration into crab eggs might reduce the probability of these eggs being simultaneously fertilized by several spermatozoa during the E_m shift. This latter characteristic might constitute an additional mechanism improving the efficiency of the electrical block to polyspermy in crab eggs.

In eggs of sea urchins, starfish, anuran amphibia, the echiuroid worm *Urechis caupo*, and the nemertean *Cerebratulus lacteus*, the electrical block to polyspermy at fertilization consists of a fast positive shift of the egg membrane potential. This shift may act on a voltage-sensitive component of the sperm membrane (Gould-Somero and Jaffe 1984; Jaffe and Gould-Somero 1985). This electrically-mediated block in the above species is transient; to our knowledge the only exception to this is in *Chaetopterus* (Jaffe 1983) and in crabs. In the latter the electrical block to polyspermy is fast, negative and long-lasting since the hyperpolarized value of E_m remained for several hours after fertilization. Because the electrical block to polyspermy in crab eggs is long-lasting, it is unlikely that the lifting of the vitelline envelope is an effective block to polyspermy. This is due to the fact that the first step of the cortical reaction, elevation of the vitelline envelope, occurs 20 min after contact with seawater. The second step, the massive exocytosis of ring-shaped elements, starts 35 min after fertilization and is not completed for 6-8h. As in several other species, the mechanism preventing polyspermy in crab eggs involves an electrically mediated block, the novel aspect lies in the fast, negative and long-lasting character of the response producing that block.

ACKNOWLEDGMENTS

We would like thank Dr. Jo Ann Render for many helpful comments on the manuscript. We would also thank Nicole Guyard for typing the manuscript and the fishermen of the Station Biologique de Roscoff for crab and prawn collection.

REFERENCES

Bumpus, H. C. 1891. The embryology of the American lobster. *J. Morphol.* 5:215-252.

Busa, W. B., J. E. Ferguson, S. K. Joseph, J. R. Williamson, and R. Nuccitelli. 1985. Activation of frog (*Xenopus laevis*) eggs by Inositol Trisphosphate. I. Characterization of Ca^{2+} release from intracellular stores. *J: Cell Biol.* 101:677-682.

Carré, D. and C. Sardet. 1984. Fertilization and early development in *Beroe ovata*. *Dev. Biol.* 105:188-195.

Chambers, E. L. and J. De Armendi. 1979. Membrane potential, action potential and activation potential of eggs of the sea urchin, *Lytechinus variegatus*. *Exp. Cell Res.* 122:203-218.

Clark, W. H. and J. W. Lynn. 1977. A Mg^{++}-dependent cortical reaction in the eggs of Penaeid shrimp. *J. Exp. Zool.* 200:177-183.

Clark, W. H., H. Persyn and A. I. Yudin. 1974. A microscopic analysis of the cortical specializations and their activation in the egg of the prawn, *Penaeus* sp. *Am. Zool.* 14:1251.

Clark, W. H., J. W. Lynn, A. I. Yudin, and H. O. Persyn. 1980. Morphology of the cortical reaction in the eggs of *Penaeus aztecus*. *Biol. Bull.* 158:175-186.

Clark, W. H., A. I. Yudin, and F. G. Griffin. 1985. Gamete interaction in the Penaeidae, *Sicyonia ingentis*. *Dev. Growth & Differ.* 27:174.

Clark, W. H., A. I. Yudin, F. J. Griffin, and K. Shigekawa. 1984. The control of gamete activation and fertilization in the marine Penaeidae *Sicyonia ingentis*. p. 459-472. *In: Advances in Invertebrate Reproduction*, vol. 3. W. Engels (Ed.). Elsevier, New York.

Clark, W. H., M. G. Yudin, G. Kleve, and A. J. Yudin. 1981. An acrosome reaction in natantian sperm. *J. Exp. Zool.* 218:279-291.

Cross, N. L. and R. P. Elinson. 1980. A fast block to polyspermy in frogs mediated by changes in the membrane potential. *Dev. Biol.* 75:187-198.

Epel, D. 1978. Mechanisms of activation of sperm and egg during fertilization of sea urchin gametes. *Curr. Top. Dev. Biol.* 12:186-246.

Fankhauser, G. 1948. The organization of the amphibian egg during fertilization and cleavage. *Ann. N.Y. Acad. Sci.* 82:684-708.

Goudeau, H. and M. Goudeau. 1985. Fertilization in crabs: IV. The fertilization potential consists of a sustained egg membrane hyperpolarization. *Gamete Res.* 11:1-17.

Goudeau, H. and M. Goudeau. 1986a. Electrical and morphological responses of the lobster egg to fertilization. *Dev. Biol.* 114:325-335.

Goudeau, H. and M. Goudeau. 1986b. External Mg^{2+} is required for hyperpolarization to occur in ovulated oocytes of the prawn *Palaemon serratus*. *Dev. Biol.* 118:371-378.

Goudeau, M. and J. Becker. 1982. Fertilization in a crab. II. Cytological aspects of the cortical reaction and fertilization envelope elaboration. *Tissue Cell* 14:273-282.

Goudeau, M. and H. Goudeau. 1986. The resumption of meiotic maturation of the oocyte of the prawn *Palaemon serratus* is regulated by an increase in extracellular Mg^{2+} during spawning. *Dev. Biol.* 118:361-370.

Goudeau, M. and F. Lachaise. 1980a. Fine structure and secretion of the capsule enclosing the embryo in a crab (*Carcinus maenas*) (L.). *Tissue Cell* 12:287-308.

Goudeau, M. and F. Lachaise. 1980b. "Endogenous yolk" as the precursor of a possible fertilization envelope in a crab (*Carcinus maenas*). *Tissue Cell* 12:503-512.

Gould, M., J. L. Stephano, and L. Z. Holland. 1986. Isolation of protein from *Urechis* sperm acrosome granules that binds sperm to eggs and initiates development. *Dev. Biol.* 117:306-318.

Gould-Somero, M. and L. A. Jaffe. 1984. Control of cell fusion at fertilization by membrane potential. P. 27-38. *In: Cell Fusion, 14th Miles International Symposium*. R. F. Beers and E. G. Baddett (Eds.). Raven Press, New York.

Gould-Somero, M., L. A. Jaffe, and L. Z. Holland. 1979. Electrically mediated fast polyspermy block in eggs of the marine worm, *Urechis caupo*. *J. Cell Biol.* 82:426- 440.

Grey, R. D., M. J. Bastiani, D. J. Webb, and E. R. Schertel. 1982. An electrical block is required to prevent polyspermy in eggs fertilized by natural mating of *Xenopus laevis*. *Dev. Biol.* 89:475-484.

Hagiwara, S. and L. A. Jaffe. 1979. Electrical properties of egg membranes. *Annu. Rev. Biophys. Bioeng.* 8:385-416.

Herrick, F. H. 1909. Natural history of the American lobster. *Bull. U.S. Bur. Fish.* 29:149-408.

Höglund, H. 1943. On the biology and larval development of *Leander squilla* (L.) formatypica de Man. *Sven. Hydrogr. Biol. Komm. Skr. Ny Ser. Biol.* 2:2-44.

Igusa, Y. and S. I. Miyazaki. 1983. Effects of altered extracellular and intracellular calcium concentration on hyperpolarizing responses of the hamster egg. *J. Physiol.* 340:611-632.

Igusa, Y., S. I. Miyazaki, and N. Yamashita. 1983. Periodic hyperpolarizing responses in hamster and mouse eggs fertilized with mouse sperm. *J. Physiol.* 340:633-647.

Jaffe, L. A. 1976. Fast block to polyspermy in sea urchin eggs is electrically mediated. *Nature (Lond.)* 26:68-71.

Jaffe, L. A. 1983. Fertilization potentials from eggs of the marine worms *Chaetopterus* and *Saccoglossus*, p. 211-218. *In: The Physiology of Excitable Cells.* W.J. Moody and A.D. Grinnel (Eds.). Alan R. Liss, New York.

Jaffe, L. A. and M. Gould-Somero. 1985. Polyspermy preventing mechanism. p. 223-243. *In: Biology of Fertilization*, vol. 3. C.B. Metz and A. Monroy (Eds.). Academic Press, New York.

Jaffe, L. A., M. Gould-Somero, and L. Z. Holland. 1982. Studies on the mechanism of the electrical polyspermy block using voltage clamp during cross-species fertilization. *J. Cell Biol.* 92:616-621.

Jaffe, L. A. and L. C. Schlichter. 1985. Fertilization-induced ionic conductance in eggs of the frog *Rana pipiens*. *J. Physiol.* 358:299-319.

Klepal, W., H. Barnes, and M. Barnes. 1979. Studies on the reproduction of cirripedes. VII. The formation and fine structure of the fertilization membrane and egg case. *J. Exp. Mar. Biol. Ecol.* 36:53-78.

Kline, D. 1988. Calcium-dependent events at fertilization of the frog egg: injection of a calcium buffer blocks ion channel opening, exocytosis and formation of pronuclei. *Dev. Biol.* 126:346-361.

Kline, D., L. A. Jaffe, and R. T. Kado. 1986. A calcium-activated sodium conductance contributes to the fertilization potential in the egg of the nemertean worm *Cerebratulus lacteus*. *Dev. Biol.* 117:184-193.

Kline, D., L. A. Jaffe, and R. P. Tucker. 1985. Fertilization potential and polyspermy prevention in the egg of the nemertean, *Cerebratulus lacteus*. *J. Exp. Zool.* 236:45-52.

Longo, F. 1978. Effects of cytochalasin B on sperm-egg interactions. *Dev. Biol.* 67:249-265.

Longo, F., J. W. Lynn, D. H. McCulloh, and E. L. Chambers. 1986. Correlative ultra-structural and electrophysiological studies of sperm egg interactions of the sea urchin, *Lytechinus variegatus*. *Dev. Biol.* 118:155-166.

Lynn, J. W. and E. L. Chambers. 1984. Voltage clamp studies of fertilization in sea urchin eggs. I. Effect of clamped membrane potential on sperm entry, activation, and development. *Dev. Biol.* 102:98-109.

Lynn, J. W. and W. H. Clark. 1975. A Mg^{++} dependent cortical reaction in the egg of Penaeid shrimp. *J. Cell Biol.* 67:251a.

Lynn, J. W. and W. H. Clark. 1983a. The fine structure of the mature sperm of the fresh water prawn *Macrobrachium rosenbergii*. *Biol. Bull.* 164:459-470.

Lynn, J. W. and W. H. Clark. 1983b. A morphological examination of sperm-egg interaction in fresh water prawn *Macrobrachium rosenbergii*. *Biol. Bull.* 164:446-458.

Lynn, J. W., D. H. McCulloh, and E. L. Chambers. 1988. Voltage clamp studies of fertilization in sea urchin eggs: II. Current patterns in relation to spperm entry, nonentry and activation. *Dev. Biol.* 128:305-323.

McCulloh, D. H. and E. L. Chambers. 1986a. When does the sperm fuse with the egg? J. *Gen. Physiol.* 88:38a-39a.

McCulloh, D. H. and E. O. Chambers. 1986b. Fusion and "unfusion" of sperm and egg are voltage dependent in the sea urchin *Lytechinus variegatus*. *J. Cell Biol.* 103:236a.

McCulloh, D. H., J. W. Lynn, and E. L. Chambers. 1987. Membrane depolarization facilitates sperm entry, large fertilization cone formation and prolonged current responses in sea urchin oocytes. *Dev. Biol.* 124:177-190.

Miyazaki, S. I. and S. Hirai. 1979. Fast polyspermy block and activation potential. Correlated changes during oocyte maturation of a starfish. *Dev. Biol.* 70:327-340.

Nuccitelli, R. 1980. The electrical changes accompanying fertilization and cortical vesicle secretion in the medaka egg. *Dev. Biol.* 76:483-498.

Nouvel, H. and L. Nouvel. 1937. Recherches sur l'accouplement et la ponte chez les crustacés Décapodes Natantia. *Bull. Soc. Zool. Fr.* 208-221.

Obata, S. and H. Kuroda. 1987. The second component of the fertilization potential in sea urchin (*Pseudocentrotus depressus*) eggs involves both Na^+ and K^+ permeability. *Dev. Biol.* 122:432-438.

Panouse, J. B. 1946. Recherches sur les phénomiènes humoraux chez les Crustacés. L'adaptation chrom atique et la croissance ovarienne chez la crevette *Leander serratus*. *Ann. Inst. Oceanogr.* 23:65-147.

Pillai, M. C. and W. H. Clark. 1987. Oocyte activation in the marine shrimp, *Sicyonia ingentis*. *J. Exp. Zool.* 244:325-330.

Raven, C. P. 1966. Morphogenesis: The Analysis of Molluscan Development. Pergamon Press, Toronto.

Romanoff, A. L. 1960. The Avian Embryo. Macmillan, New York.

Rothschild, L. 1956. Fertilization. Wiley (Ed.), New York.

Sandifer, P. A. and J. W. Lynn. 1980. Artificial insemination of caridean shrimp. p. 271-288. *In: Advances in Invertebrate Reproduction*. W. H. Clark and T. S. Adams (Eds.). Elsevier, North-Holland.

Shigekawa, K., A. I. Yudin, and W. H. Clark. 1982. Separation and independence of the biphasic acrosome reaction events in *Sicyonia ingentis*. *J. Cell Biol.* 95:161a.

Slack, B. E., J. E. Bell, and D. J. Benos. 1986. Inositol-1,4,5-trisphosphate injection mimics fertilization potentials in sea urchin eggs. *Am. J. Physiol.* 250:C340-C344.

Spalding, J. F. 1942. The nature and formation of the spermatophore and sperm plug in *Carcinus maenas*. *Q. J. Microsc. Sci.* 83:399-422.

Swann, K. and M. J. Whitaker. 1987. Neomycin prevents the fusion of sperm with sea urchin eggs but not the early electrical events of fertilization. *J. Physiol.* 390:140p.

Talbot, P. 1981. The ovary of the lobster, *Homarus americanus*. I. Architecture of the mature ovary. *J. Ultrastruct. Res.* 76:235-248.

Talbot, P. 1983. Progress and problems in controlling reproduction in lobsters. *In: Abstracts of the Third International Symposium of Invertebrate Reproduction*. (Tübingen).

Talbot, P. and M. Goudeau. 1988. A complex cortical reaction leads to formation of the fertilization envelope in the lobster (*Homarus*). *Gamete Res.* 19:1-18.

Wourms, J. P. 1977. Reproduction and development in chondrychtyan fishes. *Am. Zool.* 17:379-410.

Zalokar, M. and I. Erk. 1977. Phase-partition fixation and staining of Drosophila eggs. *Stain Technol.* 52:89-95.

ION CHANNELS IN *RANA PIPIENS* OOCYTES: CHANGES DURING MATURATION AND FERTILIZATION

Lyanne C. Schlichter

Department of Physiology
Medical Sciences Building
University of Toronto Medical School
Toronto, Ontario
CANADA M5S 1A8

*Present address: Department of Pharmacology
Merck-Frosst Canada Inc.
P.O. Box 1005
Pointe Claire-Dorval
Québec CANADA H9R 4P8

Dedication: I would like to dedicate this chapter to the memory of Professor Bertrand Picheral, "l'homme à cent mille volts."

A. INTRODUCTION

1. Scope

This chapter represents a personal view of the frog oocyte during maturation and fertilization, as seen through the eyes of an electrophysiologist. "The frog" is, in this case, the northern leopard frog, *Rana pipiens*. To facilitate description of the electrical properties of the oocyte and to make this chapter of broader usefulness I begin with a description of the resting membrane (voltage and resistance) and give many technical tips on making and interpreting such measurements (Section B). Next, I describe the technique of voltage clamping and include many technical tips for voltage clamping frog oocytes (Section C). Then, the maturing oocyte (metaphase I to metaphase II) is considered in some detail; first by a description of the changes in action potential propagation and ionic basis (Section D), then a voltage-clamp study of the total membrane current and its ionic basis (Section E). Methods were developed to isolate each of the three voltage-dependent currents and some properties of each are treated in Section F (Cl^- current), Section G (K^+ current) and Section H (Na^+ current). Then, I will briefly describe electrical events associated with fertilization and activation of immature and mature oocytes and will speculate on their relationship to the ion channels present at the various stages of maturation (Section I). Finally, I will speculate about mechanisms of channel regulation and will suggest further studies (Section J). Throughout this chapter, I will intersperse figures drawn from my papers in press and redrawn from our published papers. The sections called Technical Tips will describe tricks and procedures that I have found useful in making electrical recordings from *Rana* eggs. Most of them will be applicable to eggs of other species.

2. Terms and Definitions

Membrane potential is defined with respect to the inside of the cell; that is, the bathing solution is taken as ground. Resting potentials are generally negative; hyperpolarization means a more negative potential and depolarization means a less negative (or even positive) potential. Injections of current pulses will be referred to as depolarizing if the resultant membrane potential change is in the depolarizing direction and hyperpolarizing if the membrane potential becomes more negative. For voltage clamp recordings inward current refers to the influx of positive charge (or efflux of negative charge) and outward current is the opposite flow. All current traces show inward current as downward and outward as upward, regardless of which ion species carry the current. Current-versus-voltage relations are plotted in the usual manner with the independent variable on the X axis; *i.e.* voltage in voltage-clamp experiments; current in current clamp, action potential studies. The term "oocyte" refers to oocytes at any stage of maturity, including fully mature and ready to be fertilized. Occasionally, the term "egg" will be used to refer to the fully mature, metaphase II oocyte.

B. THE RESTING MEMBRANE

1. Problems with Resting Potential Measurements

Until recently, amphibian eggs were thought to be electrically non-excitable since all-or-none action potentials had not been observed. However, the early membrane potential measurements of *Bufo* eggs by Maeno (1959) suggested to me that the toad egg had a voltage-dependent ion current. Although Maeno did not mention it, over a certain range of membrane potential, during a pulse of injected depolarizing current, the membrane first depolarized with an approximately mono-exponential time-course, then repolarized somewhat. As will become clear later, this is evidence for a time- and voltage-dependent outward current developing during a pulse to a permissive membrane potential.

For 20 years after Maeno's first recordings, electrophysiological studies of frog eggs consisted of membrane potential (V_m) and resistance (R_m) measurements at rest and during activation or fertilization (e.g. Maeno 1959; Morrill and Watson 1966; Ito 1972). There have been numerous discrepancies in the literature about the "true" resting potential of frog (and other) eggs. I would like to point out some of the problems and uncertainties inherent in measurements of resting potential and resistance of frog eggs.

a. The resting potential depends directly on the concentrations (actually activities) of ions in the bathing medium and on the relative permeabilities of the membrane to these ions. This can be seen from the Goldman-Hodgkin-Katz equation (1) for a membrane that is primarily permeable to Na^+, K and Cl^-; *i.e.*

$$V_m = \frac{RT}{F} \ln \left[\frac{P_K K_o + P_{Na} Na_o + P_{Cl} Cl_i}{P_K K_i + P_{Na} Na_i + P_{Cl} Cl_o} \right] \qquad \textbf{Eq. 1}$$

where V_m is membrane potential, R is the gas constant, T is temperature in degrees Kelvin, F is Faraday's constant, P_K, P_{Na}, P_{Cl} are membrane permeabilities to K, Na^+ and Cl^-, and K_o and K_i (etc.) are the concentrations outside and inside the cell.

If the membrane is primarily permeable to K (true for many cells at rest) then P_{Na} and $P_{Cl} \approx 0$ and V_m is close to the Nernst potential for K; *i.e.*

$$V_m = \frac{RT}{F} \ln \frac{K_o}{K_i} \qquad \textbf{Eq. 2}$$

which is about 58 mV per ten-fold difference between internal and external K.

b. For amphibian oocytes, we do not always know or choose the appropriate ion composition for the bathing medium for *in vitro* experiments. Normally, the immature oocyte is bathed in coelomic or oviducal fluid whose composition is not known. Moreover, the physiological Ringer's solutions used often differ in composition between labs. For fully mature oocytes the appropriate medium for fertilization is probably pond water; however the negatively charged jelly polymers might alter ion concentrations near the membrane in an unknown way. Most fertilization and activation experiments are done in 10% Ringer's or a similar solution which does not mimic pond water.

There is evidence that the composition of the bathing medium can alter ion permeabilities. For example, the relative permeabilities of frog oocytes to Na^+ and K^+, the fluxes of these ions, and the resting potential can be strongly affected by the external concentration of Ca^{2+} (Morrill *et al.* 1966, 1977; Ecker and Smith 1971; Bellé *et al.* 1977). The pH and amount of CO_2 in the medium can strongly affect ion permeabilities and these parameters may vary within the reproductive track and in the environment. The problem then is, what solution composition do we choose for measuring V_m or fertilization and activation potentials?

c. Perhaps the most significant source of error is the amount of damage caused by impalement. It is a mistake to think that frog eggs are easy to impale with microelectrodes because they are large. They are, in fact, rather fragile and have a high specific resistance (resistance per unit area). Damage will provide a parallel, low-resistance pathway for ions that is non-selective and tends to depolarize the cell toward the Donnan potential (probably 0 to -30 mV).

Figure 1 illustrates the damage problem in mature *Rana pipiens* oocytes, each impaled with one fine-tipped microelectrode. Part a shows that on impalement the electrode recorded -73 mV but the cell quickly depolarized to about -45 mV. R_m was periodically measured by injecting a small pulse of current. As the membrane healed V_m then hyperpolarized from -45 to -62 mV and R_m increased (*i.e.* $R_m = V_m/I$ injected) from 6 to 10 to 20 MΩ. Part b shows an oocyte that did not depolarize after impalement (V_m, -65 to -70 mV; R_m, 35 MΩ). Part c shows

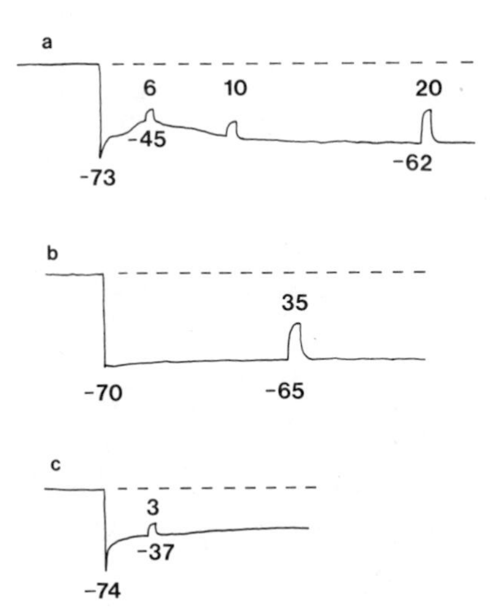

Figure 1. Resting potential and damage by impalement in mature *Rana pipiens* oocytes bathed in 10% Ringer's solution. Each oocyte was impaled with one microelectrode and occasional depolarizing current pulses were injected to monitor membrane resistance. (The electrode resistance has been compensated). For each trace the upper numbers are resistance (MΩ) and the lower numbers are membrane potential (mV). **a)** Depolarization and a low resistance resulted from impalement, but some healing occurred. **b)** Little or no depolarization occurred and the membrane resistance remained high. **c)** Depolarization occurred and no recovery took place. The membrane resistance was low.

a response typical of impalement damage; *i.e.* a rapid depolarization (from -74 to -37 mV) without recovery and a low R_m (3 MΩ). The true resting potential of the mature oocyte in this Ringer's solution is probably closer to -70 mV than the depolarized values often reported.

For mature *Rana* oocytes I have found a correlation between the resting potential (V_m) and resistance (R_m) that suggests depolarized values are a result of damage. Because of these problems, the resting potential is often an unreliable parameter. What is more interesting are changes in membrane potential brought about by various processes, such as maturation and fertilization, and most informative are studies of the ionic mechanisms responsible for these changes. The remainder of this chapter will focus on changes in membrane potential and on the ionic currents underlying these changes during maturation and fertilization or activation of *Rana pipiens* oocytes. But first, some technical tips for making electrical recordings from frog eggs.

2. Technical Tips

Impalement with one microelectrode is used for recording V_m and for passing pulses of current to monitor R_m or to construct current-versus-voltage relations. Several factors determine the amount of damage sustained during impalement; tip diameter, mechanical stability, position and method of impalement. For frog eggs, I found the following procedures to work best. Microelectrodes were pulled from thin-walled glass, usually containing a filament to aid in filling the electrode by capillarity. The ideal microelectrode has a very small tip diameter to reduce damage to the cell and a low resistance to reduce noise, tip potential and to allow larger currents to be injected (*i.e.* current passed = applied voltage / electrode resistance; $I = V/R_e$). Using thin-walled glass, resistances were in the range of 5-20 MΩ, which is much lower than for thick-walled glass with the same tip diameter (Jacobson and Mealing 1980). These values correspond with tips less than 1 μm in diameter; however, exact dimensions were not measured.

In choosing a filling solution, several factors should be considered. To reduce the tip potential the anion and cation should have about the same electrophoretic mobilities, hence the popularity of KCl as a filling solution. Because high ion concentrations reduce the electrode resistance and facilitate current injections 3 M KCl is often used; however, this solution could have adverse effects especially in small cells where ion injection (during current pulses) could change the intracellular composition. Many cells, including frog eggs, maintain a low intracellular Cl^- concentration and Cl^- injection could cause problems. I found that a 2:1 mixture of 3 M KCl and 0.6 M K_2SO_4 yielded small tip potentials, good current passing ability and stable long-term recordings with current pulses of either polarity. For small eggs, such as from echinoderms and mammals it would be best to fill the electrode with a solution resembling the cytoplasmic ion composition. This avoids artificial ion loading into the small egg volume and ensures that results are more relevant to the cells' normal physiology.

For impalement, the microelectrode is mounted in a holder in a very stable micropositioner (e.g. Leitz, Newport, Narishige) to avoid any vibrations or backlash that could damage the egg during penetration. I find it is easier to impale the animal hemisphere of frog eggs, regardless of the stage of maturity of the egg. The microelectrode is pushed lightly against the egg, just dimpling the surface then inserted by either sharply tapping the table (e.g. with a pen) or by overcompensating the capacitance adjustment on the recording amplifier to produce electrical oscillations (ringing) which cause the electrode tip to vibrate. The latter method is the least damaging for frog eggs and works well with small diameter microelectrodes. At this point successful impalement is noted by a sharply negative jump in recorded voltage (see Figure 1). Healing of the wound was often facilitated by passing small hyperpolarizing current pulses continuously through the electrode. Currents sufficient to hyperpolarize the membrane 10-30 mV seemed to work well. For the study of action potentials in frog eggs damage by impalement could be a serious problem. For example, spontaneously firing oocytes often stopped firing action potentials when a second electrode was inserted and the leak conductance effectively short circuited the voltage-dependent conductances. Therefore, for studying spontaneous firing a single, fine-tipped microelectrode was inserted.

3. Measuring Membrane Resistance

For monitoring R_m and constructing I-V relations there are several points to consider. Square pulses of current are injected by passing square voltage pulses through a large resistance (e.g. $10^7\Omega$ is good for large eggs). The amount of current is theoretically limited by the electrode resistance (R_e) and the maximum voltage available from the generator (V). In practice, R_e increases when large V pulses are applied (the electrode rectifies) and often in response to one polarity more than the other. I found rectification was reduced by using the KCl: K_2SO_4 solution mentioned above, by using thin-walled glass, and by adding millimolar amounts of EGTA to the filling solution. For R_m measurements, the current pulses must be long enough to charge up the membrane capacitance or V will not reach steady state and R_m will be underestimated. This can be seen by treating the membrane as a fixed resistance (R_m) in parallel with a capacitance (C_m). During an injected current pulse (I) the membrane potential change with time can be expressed as:

$$dV_m/dt = [I-(V_m - V_r)/R_m]/C_m \quad \textbf{(Eq. 3)}$$

The solution to this equation is a simple exponential function as illustrated in Figure 2a. For a positive current pulse that depolarizes the membrane (*i.e.* the beginning of the pulse):

$$V_m - V = IR_m [1 - \exp(-t/\tau_m)] \quad \textbf{(Eq. 4)}$$

where τ_m, the membrane time constant ($\tau_m = R_m C_m$), can be calculated from the slope of a semilogarithmic plot of voltage difference against time (Figure 2b). For a mono-exponential function, τ_m is the time at which V_m - V has reached 63.2%; *i.e.* [1 - exp(-1)] of its steady-state value. V_m is very nearly at steady state after 3 or 4 time constants. Suppose R_m is 20 MΩ and C_m is 0.1 μF (reasonable for frog eggs) then τ_m = 2s and current pulses should be 6 to 8s long to reach steady state. Figure 2a shows the membrane potential versus time and the total injected current separated into its resistive and capacitive components. This method of measuring R_m is only accurate for a passive membrane (*i.e.* R_m constant with V_m) or over very small regions of a current-versus-voltage relation where R_m is approximately constant. This point is considered further in the next section.

4. Constructing Current-Voltage Relations

Cell membranes rarely behave as simple, passive circuits in which V_m changes exponentially with time during pulses of injected current regardless of the amplitude of the V_m excursion. Most have time- and voltage-dependent changes in conductance ($G = 1/R_m$) outside a small range of voltages. For the frog egg (Figure 3, top) I will represent these changes as variable conductances for Na^+, K^+, Ca^{2+}, and Cl^-. The leakage conductance G_l is independent of voltage. To construct I-V relations, square pulses of current of increasing amplitude are injected, and the voltage excursion measured at steady state. The existance of voltage-dependent conductance (*i.e.* ion channels) is indicated by a non-linear I-V relation (such as that in Figure 3 (bottom), in which it is clear that R_m (slope) changes with V_m. Moreover, for small V_m excursions, R_m is approximately the slope of a tangent (point A), whereas if a single large current pulse were injected, one would inadvertantly measure R_m as the slope of a chord (A-B). Clearly, the two values of R_m can differ significantly.

To further illustrate the problem for *Rana* eggs, consider the I-V relations in Figure 4. Before activation (Δ) the I-V relation is strongly non-linear (*i.e.* it rectified). Around the resting potential (-30 mV in this case) membrane resistance (the slope of the line) is about 10 MΩ; however around 0 mV the resistance is less than 1 MΩ. Therefore, a simple shift in membrane potential would dramatically lower the resistance even in the absence of activation. At the peak of the activation potential (o) the resistance is about the same as in the resting cell at positive potentials. Suppose during an activation experiment R_m was continuously monitored by passing current pulses periodically. R_m would be smaller during the activation potential but, without knowing the I-V relations throughout activation, one could not safely conclude

a

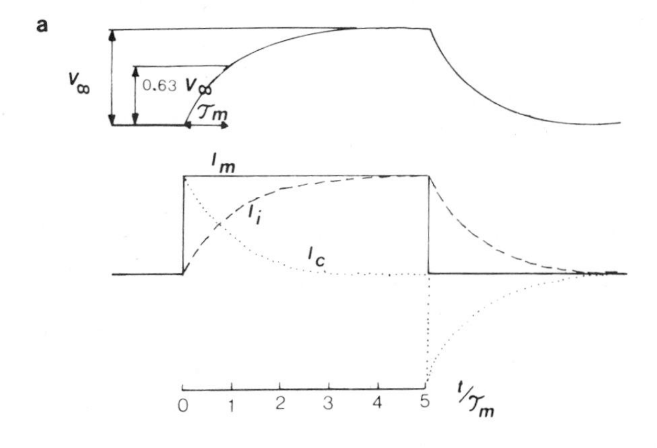

b

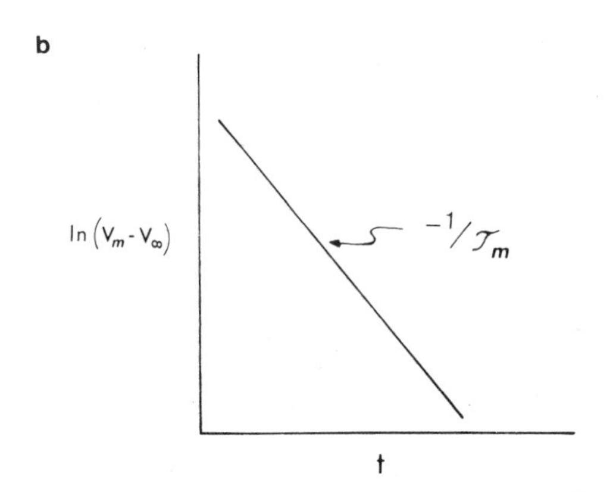

Figure 2. The membrane as a passive circuit with a resistance and battery in parallel with a capacitance. **a)** Membrane potential (V_m, upper trace) and membrane current (I_m, lower trace) as a function of time during and following injection of a square pulse of current. I_m is separated into the capacitive current (I_c) and ionic current (I_i). V_∞ is the change in membrane potential from the resting level to the steady-state level. τ_m is the time constant for the mono-exponential increase in V_m with time and is the time at which V_m has reached [1 - exp (-1)] or about 63% of its steady state value. V_m is nearly at steady state after 3 or 4 time constants. **b)** A semilogarithmic plot of the approach of V_m to the steady- state value, V_∞, as a function of time.

that activation opened new ionic pathways and thereby reduced R_m. In fact, the change in form of the I-V relations we observed during activation was a good indication that new channels were opening. The currents activated by fertilization or activation will be described in more detail in Section I.

C. VOLTAGE CLAMPING

1. Voltage Clamp Principles

As previously discussed, some information about membrane responses to injected current pulses can be obtained from constructing I-V relations. For example, non-linear I-V curves for frog oocytes suggested to me the existance of voltage-gated channels. However, detailed understanding of the channels is precluded by the simultaneously varying V_m and R_m in a system in which only I_m is fixed. A much more powerful approach is to control V_m, measure

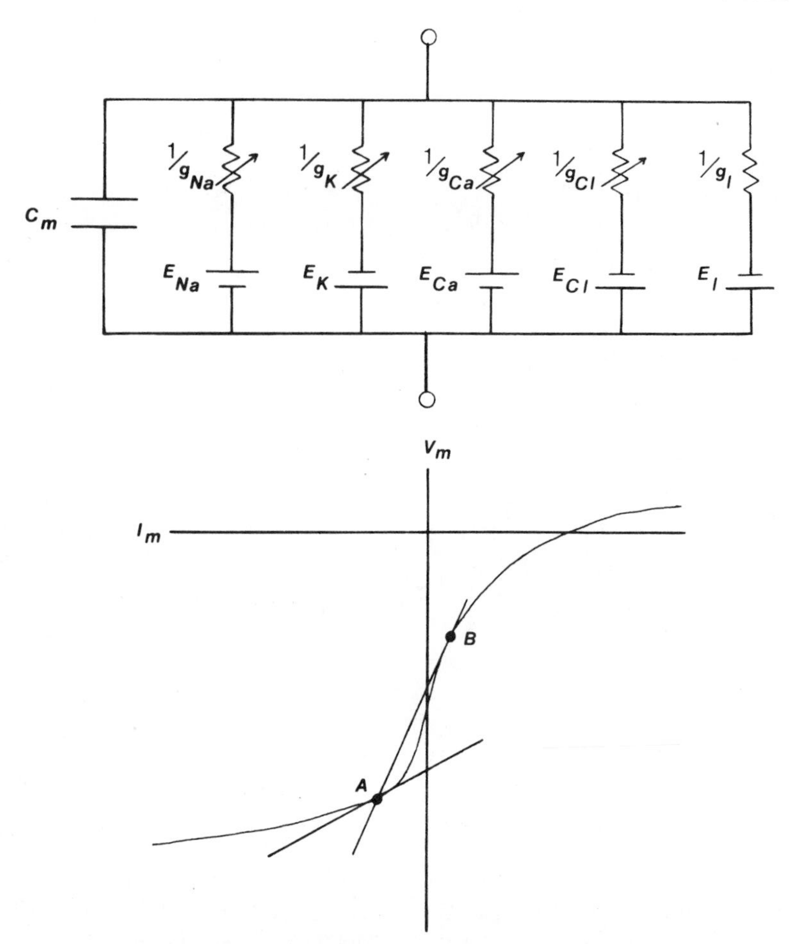

Figure 3. Current-versus-voltage (I-V) relations. Upper figure: Equivalent circuit for an excitable cell such as the *Rana* oocyte with several voltage-dependent conductances shown as variable resistors (1/g) and a fixed leakage resistance ($1/g_l$).

Lower figure: A hypothetical, steady-state I-V relation for a cell with voltage-dependent conductances. Current is injected and the resulting voltage excursion measured. Slope of tangent at A is slope resistance; slope of chord between A and B is chord resistance over this voltage range.

I_m exactly and study how R_m changes as ion channels open and close, for example, in response to voltage or to sperm. When V_m is held constant, the current flowing across the membrane is directly proportional to the number of channels that are open (N), the conductance of each individual channel (γ) and the driving force on the ions which, for each ion species, is the difference between the voltage and the equilibrium (Nernst) potential for the ion (*i.e.* $V-V_j$).

The general principle of voltage clamping is illustrated in Figure 5 which is a schematic of the home-made voltage clamp I used for frog eggs. The principle is to hold the membrane potential (V_m) at a desired level by means of a feedback circuit. V_m can be a constant DC level or transiently changed by square pulses, triangle waves, sine waves or any other command signal delivered to the input of a summing amplifier (SA) and from there to the positive input of a high gain (A, maximum 2500 for my clamp) differential amplifier (CA, clamp amplifier). Membrane potential was measured as the difference between an intracellular (VI) and an extracellular (VE) microelectrode connected to the inputs of a high-impedance differential

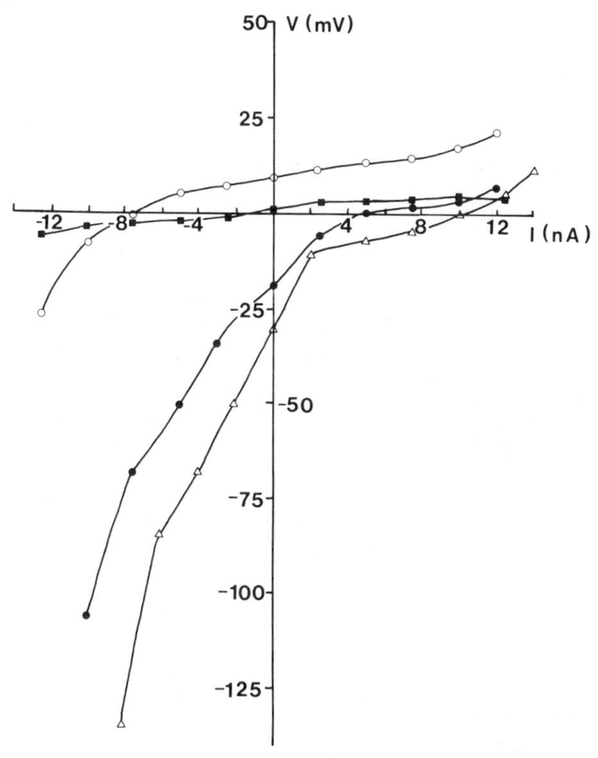

Figure 4. Actual current-versus-voltage relations obtained by injecting current pulses into a mature *Rana* oocyte; before artificial activation (Δ), at the beginning of the activation potential (■) when V_m reached its most positive value (○) and after the cell had repolarized (•).

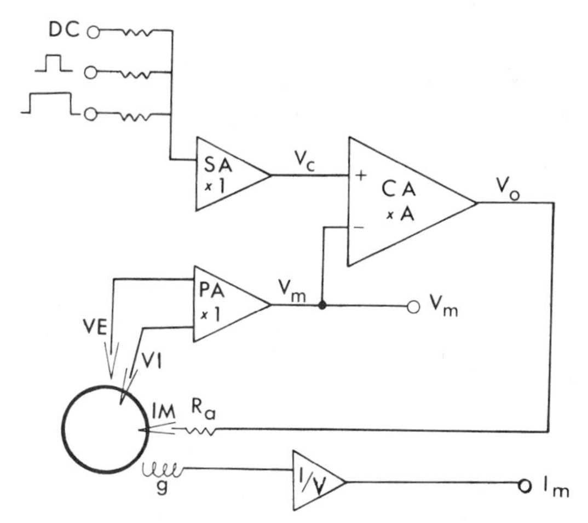

Figure 5. Voltage-clamp arrangement. SA, summing amplifier for command voltages (V_c); CA, voltage-clamp amplifier with gain = A ; PA, preamplifier with a high input impedance for measuring membrane potential (V_m); V_o, output voltage of clamp amplifier; R_a, access resistance; IM, current-injecting microelectrode; VE and VI, extracellular and intracellular voltage-recording microelectrodes for differential recording of V_m; g, Ag/AgCl ground wire in agar; I/V, current-to-voltage converter; I_m, injected current (current clamp) or membrane current (voltage clamp).

preamplifier (PA) with a gain of 1. The output of the PA (*i.e.* V_m) was taken to the negative input of the CA which compares the voltages on its positive and negative inputs. Any difference between V_m and V_c is amplified by the gain (A) and appears as an output voltage (V_o). V_o is then applied to an access resistance (R_a) that has at least three components: (1) a fixed resistor in the voltage-clamp circuit, (2) the resistance of the current-injecting microelectrode, and (3) the cytoplasmic resistance. In the frog oocyte (3) is much smaller than (1) or (2) and can be ignored. When V_o is applied to R_a a current is produced and injected into the egg through the intracellular current-injecting microelectrode (IM). This current crosses the cell membrane, carried by any ion species to which the membrane is permeable. To measure this current a chlorided silver ground wire (Ag/AgCl) in an agar bridge (g) is connected to a current-to-voltage converter (I/V). A voltage drop develops across a known resistance in the I/V converter and the current can be readily calculated from $I = V/R$. In summary, I_m is the current necessary to hold V_m equal to V_c and I_m is generated by ions passing across the cell membrane, through specific ion channels and non-specific leaks.

For good voltage clamping: (1) the membrane potential should be accurately measured and constant within a couple of millivolts; (2) the measurement of V_m and its feedback to CA should occur within microseconds to ensure electronic stability and (3) the current density across the membrane should be uniform within a few percent to ensure isopotentiality throughout the cell. A calculation of the clamp error, using arbitrary but reasonable values for the frog egg, will underline several important features of the clamp system. The clamp error ($V_c - V_m$) is a function of the access resistance (R_a), the magnitude of the membrane current (I_m) and the gain (A) of the clamp amplifier; *i.e.*,

$$V_c - V_m = (V_c + R_a I_m) / (A + 1) \qquad \textbf{(Eq. 5)}$$

R_a is dominated by the current-injecting microelectrode (IM) with a typical resistance of $5 \times 10^6\ \Omega$. For a maximum current of 2×10^{-6} A (*i.e.* $\Delta V_m = 100$ mV, membrane conductance = 2×10^{-7} S) and a gain of 2500 the clamp error is 4 mV. To increase the accuracy one can reduce the electrode resistance or increase the gain of the clamp. In practice, both are necessary but larger electrodes cause more damage to the egg, so a limit is reached. The speed of the clamp is also important so that the membrane potential stabilizes in a short time compared with the rate of development of the ionic currents. For the frog oocyte the currents activate quite slowly; nevertheless, slow clamping speeds can be a problem. To illustrate this point, consider the membrane as a simple, passive R_m in parallel with a C_m. It is the time required to charge the membrane capacitance that limits the speed of the clamp. The maximum current flowing through this capacitance I_c determines the rate of charging dV_m/dt; *i.e.*,

$$I_C = C_m\, dV_m/dt \qquad \textbf{(Eq. 6)}$$

The specific capacitance of frog eggs is 1-5 $\mu F/cm^2$ (Kado *et al.* 1979); the higher values presumably resulting from dense microvilli at certain stages of maturation. This yields 0.07 to 0.35×10^{-6} F for a frog oocyte (area 0.07 cm^2). For a clamp amplifier operating at ± 15 V and a microelectrode resistance (R_a) of $5 \times 10^6\ \Omega$ the maximum current is $\pm 3 \times 10^{-6}$ A. The maximum dV_m/dt is about 9 V/s or about 11 ms for a 100 mV change in membrane potential. Hence, for these parameter values the development of ionic currents that activate with time constants shorter than about 30 ms cannot be accurately followed. For oocytes (e.g. *Xenopus*) that can tolerate a much larger electrode and if a higher gain clamp amplifier is used, the time to achieve good clamping can be reduced to about a millisecond.

2. Technical Tips

For voltage clamping frog oocytes, I found the following procedure worked best. The voltage electrode (VI in Figure 5) was a 10-30 MΩ electrode pulled from thin-walled, filament glass as described in the previous Technical Tips section. It was inserted into the oocyte first and R_m was measured. The current microelectrode (IM) was about 5 MΩ, pulled from the same glass and filled with the same solution. Before inserting this electrode, small negative current pulses were continuously applied to it, sufficient to hyperpolarize the cell 10-20 mV; *i.e.* if R_m = 20 MΩ, then a -1 nA pulse would produce a -20 mV excursion of V_m. Then, the IM was inserted by touching and dimpling the cell and sharply tapping the table. Overcompensating the capacitance to vibrate the electrode (as previously described) did not usually work with these larger electrodes. Following this routine has two main advantages: (1) one can immediately detect impalement by the appearance of negative V_m excursions produced by the current pulses through IM, hence the temptation to push harder on the electrode is removed; and (2) damage can be assessed when the larger (IM) electrode enters, by the degree of cell depolarization and by the decrease in R_m; hence damaged cells can be discarded. After the current electrode is inserted, small hyperpolarizing pulses are passed less frequently for a few minutes. This seems to help the wound heal and I will speculate that Ca^{2+} driven into the wound by the negative voltage aids healing.

A typical voltage-clamp setup consists of the following. Inside the Faraday cage is a vibration isolation table, a perfusion chamber containing the cells, a microscope and micromanipulators for positioning the microelectrodes. Outside the cage are a perfusion pump for circulating the bathing solution, and the power source for the microscope lamp. On an equipment rack within easy reach are the voltage clamp controller, a pulse generator, an oscilloscope for monitoring voltage and current and devices for recording the data. Recording devices could be a fast strip-chart recorder, a video cassette recorder with pulse code modulator, an instrumentation tape recorder, a computer with analog-to-digital converters and appropriate software or even an old-fashioned camera to photograph the oscilloscope screen.

To reduce the electrical noise of the recording system; all apparatus in the Faraday cage are grounded to a common point, the headstage of the clamp and the current-to-voltage converter are placed inside the Faraday cage, and the perfusion system has grounded interruptions in both the inlet and outlet lines. It is often helpful to make a grounded shield around the current microelectrode or between the current and voltage electrodes. One method is to ground an aluminum foil strip and suspend it between the electrodes (not in contact with the bath solution!). A better method, though more laborious, is to coat the current electrode with conductive silver paint then cover all exposed paint with a clear insulating lacquer (Q-Dope, sylgard, clear nail polish) leaving a small patch of exposed silver near the shank of the electrode (top portion). A ground wire is attached to this patch of paint to make a continuous grounded shield around the current electrode. Finally, the current signal can be filtered to reduce high frequency noise, taking care that the lowpass frequency is high enough to allow faithful time resolution of the ionic currents.

D. ACTION POTENTIALS IN *RANA PIPIENS* OOCYTES

This section will describe the main features of action potentials in *Rana pipiens* oocytes, their ionic basis and changes during maturation. Because the basic data and conclusions have been published (see Schlichter 1983 a,b) I will treat this section briefly and will explain how the data were interpreted.

1. Action Potential Types

I discovered that *Rana pipiens* oocytes are electrically excitable (fire action potentials) when I realized that damage during impalement was reducing V_m and R_m (see Figure 1). My goal, then, was to obtain the most negative V_m and highest R_m possible by using fine-tipped microelectrodes and very gentle impalement procedures. To my surprise a series of immature oocytes bathed in Ringer's solution (to approximate the *in vivo* environment) produced the type of voltage oscillations shown in Figure 6. Part "a" shows an oocyte that spontaneously fired action potentials of a regular amplitude and duration. Each action potential can be divided into a gradual depolarizing phase, an all-or-none overshoot and a rapid repolarization. Occasionally an oocyte showed oscillations of increasing amplitude before each all-or-none spike (action potential) such as in Part b. I suspect these oocytes were more damaged by

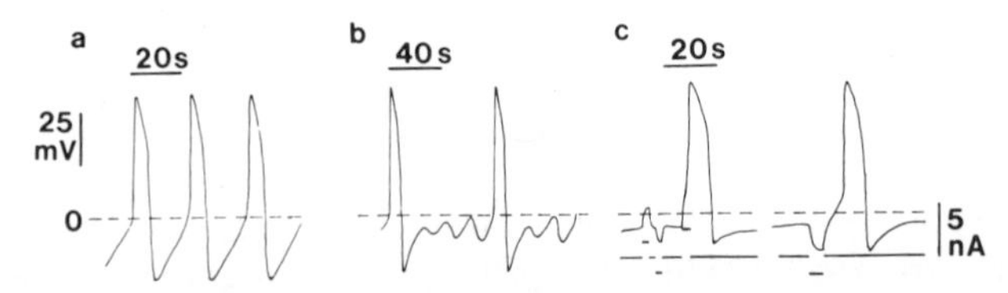

Figure 6. Examples of action potentials in three young oocytes (metaphase I) recorded in normal Ringer's solution. **a)** Spontaneous action potentials with one microelectrode. Spikes continued for 32 min until a second microelectrode was inserted and the oocyte leakiness increased. **b)** Increasing amplitude oscillations preceded each action potential for the first few minutes, then the action potentials resembled those in part a. **c)** Action potentials were not spontaneous with two microelectrodes in this oocyte. A small depolarizing or hyperpolarizing pulse failed to produce an action potential, whereas larger pulses of either polarity evoked action potentials.

(left) that a sufficiently large depolarizing current pulse could evoke an all-or-none response that outlasted the stimulus (*i.e.* the definition of an action potential). The right panel in Part c shows a so-called "off-response"; *i.e.* an action potential fires after a hyperpolarizing current pulse is injected. This behavior is also typical of many types of excitable cells; e.g. nerve and muscle.

I found that the form of the action potential changed in a characteristic way during oocyte maturation from metaphase I to the fully mature egg (Schlichter 1983 a,b) and I will briefly describe these changes.

2. Changes During Maturation

The youngest oocytes tested (early metaphase I) did not have true action potentials but showed the behavior illustrated in Figure 7a. In response to depolarizing current pulses V_m rapidly rose to very positive value (often +75 to +80 mV) then spontaneously fell toward 0 mV. The response did not outlast the current pulse. I will show later that the rapid depolarization is due to opening of Na^+ channels and the subsequent repolarization corresponds with the opening of K and Cl^- channels. Slightly more mature oocytes (around first polar body stage) showed spontaneous repetitive firing (Figure 7b) unless impalement damage was too large. No current injection was necessary. The appearance of spontaneous activity correlated with a decrease in K channel activity (K^+ conductance) so that V_m was not forced to remain as negative as in the less mature oocyte (explained further below). As the oocyte matured the action potentials became longer, reached a less positive potential and repolarized to a less negative potential (Figure 7c). These changes are due to a decrease in Cl^- conductance with time, and to an increase in the cellular Na^+ concentration (Schlichter 1983 a, and below). At a stage (early metaphase II) when action potentials had broadened into a continuous positive plateau, many oocytes had a positive V_m when first impaled (Figure 7d). In such cases strong hyperpolarizing currents could force V_m to a negative value, but as soon as current injection ceased V_m returned to the positive plateau. This behavior corresponds with loss of the Cl^- conductance and the continued presence of a Na^+ conductance forcing V_m toward the Nernst potential for Na^+ (Schlichter 1983 a,b and below). Finally, the fully mature, metaphase II oocyte could fire an all-or-none action potential only when depolarizing current was injected. Figure 7e shows a typical response to increasing amounts of positive charge injected (*i.e.* larger current amplitude or longer pulses). When sufficiently depolarized, threshold was reached (arrow at notch) and a sustained action potential fired. Oscillations during the plateau phase are simply responses to repetitive current pulses used to monitor R_m. The membrane of these mature oocytes could be restored to a stable negative V_m when a sufficiently large hyperpolarizing current pulse was injected.

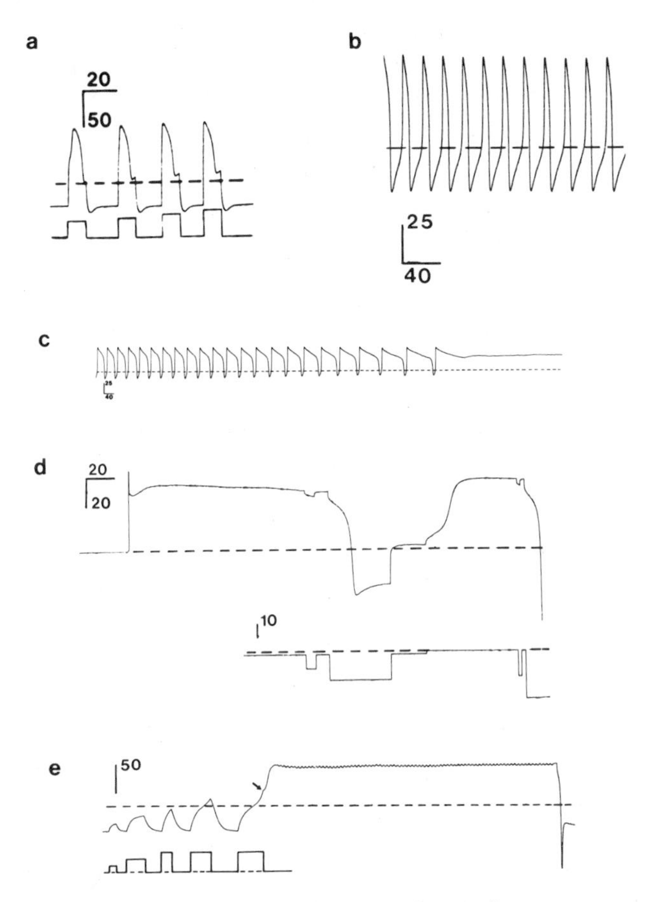

Figure 7. Changes in the form of the action potentials during oocyte maturation. As an indication of oocyte age the number of hours post-progesterone treatment (PPT) of the female frog are indicated. The timing of a given response varied. **a)** 17h PPT. The resting membrane potential (V_m) was quite negative (~ -35 mV). Current injection produced a regenerative depolarization which spontaneously repolarized during the pulse. The response never outlasted the current pulse. **b)** 18-19h PPT. Young, metaphase I oocyte with spontaneous recurrent action potentials lasting less than 10s each. Note slow depolarizing phase preceding each spike. **c)** 24-25h PPT. Spontaneous action potentials elongated, then remained positive for at least 20 min. Early metaphase II at end of recording. **d)** 26-27h PPT. V_m was positive upon microelectrode impalement. A large hyperpolarizing current ended the action potential but it spontaneously recurred when the current was turned off. **e)** 49h PPT. Mature, metaphase II oocyte with a resting V_m of about -35 mV. Characteristic all-or-none action potential in response to sufficient depolarizing current. Arrow shows threshold for firing; oscillations are due to injected current pulses and show that R_m is low during the positive phase. V_m remains positive until a sufficiently large hyperpolarizing current (not shown) is injected at the end of the recording.

3. Ionic Basis of the Action Potentials

The existance of action potentials requires an inward ionic current to depolarize the cell, and either an outward current to repolarize it or a mechanism to turn off the inward current and restore the membrane to its resting complement of ion permeabilities that produce a negative resting potential (according to Equation 1). The first approach to determine which ions carry these currents is to vary the external ion composition and observe the effect on the shape and amplitude of the action potentials. In this section I will very briefly outline the evidence that the action potentials are produced by an inward Na^+ current followed by an outward Cl^- current. (For a more complete treatment see Schlichter 1983 a,b.) Later, in the section on voltage clamp analysis of the ionic currents, I will show that young oocytes also possess a K^+ conductance that prevents the youngest from firing action potentials and helps to repolarize the membrane in slighly older oocytes.

To interpret the results of ion-substitution experiments it is useful to consider the direction of the driving forces on the ions moving across the membrane. In *Rana* oocytes the intracellular Na^+ activity increases from about 5 to 23 mM during maturation (Schlichter 1983b) whereas Na^+ in the normal bathing medium (Ringer's) was 115 mM. The Nernst potential for Na^+ (from Equation 2) would decrease from about +80 to +40 mV as the oocyte matures. This means that Na^+ will enter the cell if there is a permeation pathway and will tend to drive V_m to these very positive values; *i.e.* opening Na^+ channels will depolarize the oocyte. Internal K^+ activity in the mature oocyte is about 120 mM (Jaffe and Schlichter 1985) and the external K concentration was 1.9 mM in these experiments. If K^+ channels opened, K^+ efflux would occur, tending to hyperpolarize the cell toward the Nernst potential for K+ (about -100 mV in the mature oocyte). At positive potentials Cl^- would tend to enter the cell since internal Cl^- activity in the mature oocyte is about 44 mM (Jaffe and Schlichter 1985) and the extracellular concentration was 115 mM. Cl^- influx would tend to drive V_m toward the Cl^- Nernst potential of about -24 mV. These values of the K and Cl^- Nernst potentials are approximate since they are based on activities in the mature oocyte which might change during maturation. Overall then, Na^+ will tend to enter the cell and depolarize it, subsequently Cl^- will tend to enter and repolarize the cell and K^+ will always tend to leave the cell and hyperpolarize it.

Figure 8 shows typical V_m recordings from a young (metaphase I) oocyte that did not fire action potentials but showed a "regenerative" depolarization when injected with a current

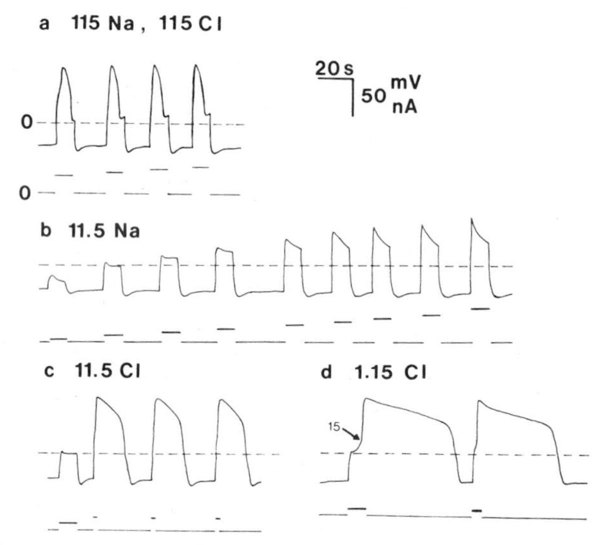

Figure 8. Ionic dependence of action potentials in a young, metaphase I oocyte evoked by injecting depolarizing current pulses. Na^+ was substituted by choline and Cl^- was substituted by methanesulfonate. **a)** Normal Ringer's. Regenerative depolarization and spontaneous repolarization occurred during each pulse. **b)** Reducing external Na^+ 10-fold eliminated the regenerative depolarization. Responses never outlasted the current pulse. **c)** Reducing external Cl^- 10-fold considerably broadened the regenerative depolarization and reduced the repolarization such that broad action potentials outlasted the current pulses. **d)** Reducing external Cl^- a further 10-fold broadened the spikes further.

pulse. Regenerative refers to the spontaneous rise in V_m after the initial rise resulting from the current pulse (see inflection in rising phase, part a). When external Na^+ was lowered ten-fold (part b) the regenerative depolarization was eliminated, strongly suggesting that an inward Na^+ current caused the depolarization. When external Cl^- was reduced (part c) an all-or-none response that outlasted the current pulse could be evoked. The action potentials were long (~20s) and became longer when Cl^- was reduced further (part d). This is strong evidence that an influx of Cl^-(outward Cl^-current) produces the repolarization in the young oocyte (see part a). Similarly, older oocytes that were spontaneously firing when impaled (Figure 9) stopped firing when external Na^+ was removed and resumed firing when Na^+ was restored. Increasing external Na^+ to 150% of normal broadened the spike and increased its peak amplitude. All of these results are best explained as an Na^+ influx causing the depolarization. Finally, removing external Cl^- prevented the membrane from repolarizing, by reducing the Cl^- influx. For the mature oocyte the sustained action potential evoked by a suprathreshold current pulse was insensitive to external Cl^- (down to 1% of normal) but very sensitive to external Na^+. This change corresponds with loss of the Cl^-current; *i.e.* the channel is lost or inactivated somehow (see figures and data in Schlichter 1983a).

The data to this point support the existence of Na^+ and Cl^- channels that are probably voltage dependent and can act in concert to produce action potentials. The section on voltage-clamp analysis of the ionic currents will show that immature oocytes possess three voltage- and time-dependent currents. (Na^+, K^+ and Cl^-) and a voltage-independent leakage current carried mainly by Na^+ entry. The currents will be examined separately and their voltage and time dependence and selectivity explored.

4. A Role for the Action Potentials?

This section will briefly describe conclusions from my published work and will speculate considerably on possible roles for the action potentials. For more details see Schlichter (1983b). I found that during the course of maturation the Nernst potential for Na^+ decreased, corresponding with an increase in intracellular Na^+ activity. Hence the first question was whether Na^+ influx during the action potentials raised internal Na^+. Three methods of preventing action potentials were used: (1) inserting a second microelectrode to increase the

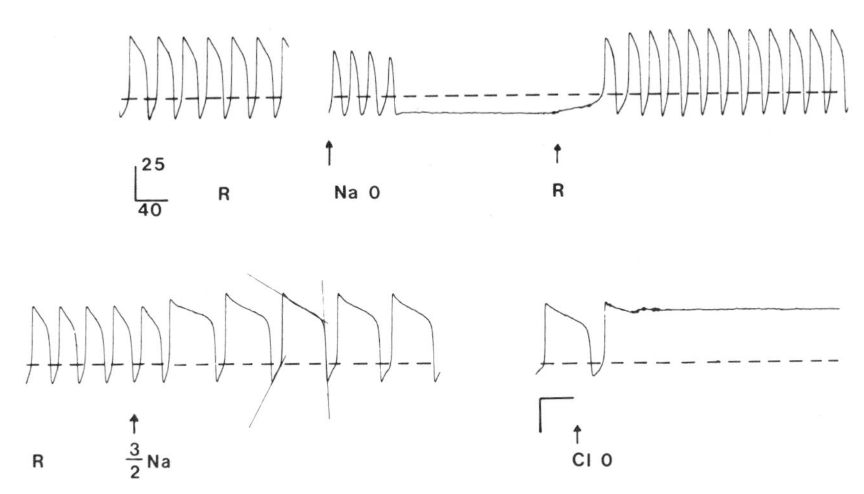

Figure 9. Ionic dependence of spontaneous action potentials. In normal Ringer's (R) the oocyte fired spontaneous action potentials but when all external Na^+ was substituted by choline (Na 0) the amplitude rapidly decreased and firing stopped. This effect was fully reversible on re-addition of Ringer's. The peak potential increased as external Na^+ was increased to 1.5 times normal and the spikes broadened. Finally, substituting all external Cl^- by methanesulfonate eliminated the repolarizing phase and the membrane potential remained positive.

leakage conductance; (2) voltage clamping the oocyte at a V_m just slightly too negative to allow firing, and (3) using 5 mM Co^{2+} to prevent the firing by blocking Na^+ channels. All three methods reduced the Na^+ accumulation by the cell, providing evidence that the action potentials are required for Na^+ loading. Conversely, evidence that action potentials fire spontaneously without microelectrode impalement was that Na^+ accumulation occurred in the absence of a microelectrode. This was determined by impaling many oocytes at different stages of maturation and observing that the Nernst potential for Na^+ decreased with age. An indication that the action potentials are involved (directly or indirectly) in maturation was that 5 mM Co^{2+} (which prevents firing) delayed, but did prevent, maturation. Maturity was judged by the ability to shock activate the oocytes.

There are many possibilities for how ion fluxes during the action potentials could affect oocyte maturation. All are, at present, highly speculative.

a. Na^+ accumulation could transiently activate the Na^+/K^+ ATPase to pump K^+ into and Na^+ out of the oocyte. Na^+/K^+ pump activity is thought to be down-regulated sometime before *Rana* oocytes become fully mature but the exact stage was not determined (Morrill *et al.* 1971).

b. If there is a Na^+/Ca^{2+} exchange mechanism at the plasma membrane or across an intracellular organelle membrane Na^+ uptake could cause Ca^{2+} to be released into the cytoplasm and this may be important for maturation. For example, on fertilization a wave of Ca^{2+} passes across the frog oocyte (Busa and Nuccitelli 1985) and opens Cl^- and K^+ channels (Jaffe *et al.* 1985). It may be important to charge up Ca^{2+} stores inside the cell or to increase the resting Ca^{2+} levels. A transient rise in Ca^{2+} has been observed immediately following progesterone treatment of *Xenopus* oocytes (Wasserman *et al.* 1980).

c. A rise in internal Na^+ might reduce Na^+/H^+ exchange across the plasma membrane or organelle membrane by decreasing the gradient for Na^+. This could result in an increase in intracellular pH as has been observed for *Xenopus* oocytes during maturation (Houle and Wasserman 1983).

d. The role of the action potentials may have nothing to do with Na^+ entry. For example, Cl^- will also enter during the action potential repolarization phase and might accumulate. One possibility is that Cl^- accumulation stimulates Cl^-/HCO_3^- exchange which could increase internal pH. Coincidentally, the voltage-dependent Cl^- channel in these oocytes is blocked by distilbene sulfonate drugs that traditionally block Cl^-/HCO_3^- exchange (Schlichter 1983b, and below).

e. It is possible that a small Ca^{2+} current is also present though undetected in my voltage-clamp experiments. If it is voltage sensitive the action potentials might increase Ca^{2+} influx and either charge up intracellular stores or increase the resting Ca levels. A tiny Ca^{2+} current has been detected in *Xenopus* oocytes, though at an earlier stage of maturity (Dascal *et al.* 1986).

f. Possibly the change in membrane potential directly regulates some process during maturation; e.g. the surface expression of a charged receptor molecule–the sperm receptor?

g. The reader can no doubt think of other possible roles, including no role at all! Perhaps the very specific protein molecules forming the various ion channels in the oocyte membrane are precursors of channels in the developing embryo. If so, one must wonder why the Na^+ channel is so different from other voltage-dependent Na^+ channels (no inactivation) and why the expression of the K and Cl^- channels comes and goes during maturation, to be replaced by Ca-activated Cl^- and K^+ channels at the moment of fertilization. (see Jaffe and Schlichter 1985; Jaffe *et al.* 1985).

E. IONIC CURRENTS IN MATURING *RANA* OOCYTES

In this section I will show mainly unpublished work on the Na^+, K^+ and Cl^- currents underlying the action potentials of maturing oocytes. Some of this work is in press (Schlichter 1989). Their voltage dependence will be analyzed in some detail as will the selectivity of the Na^+ channel. Some pharmacological data will be presented on agents blocking the channels.

Finally, evidence for two types of leakage current will be presented. First, it may be useful to examine the stylized voltage-clamp record below.

1. Components of the Membrane Current

Figure 10 is a highly schematic representation of the most complex current pattern observed in the *Rana pipiens* oocyte. Its purpose is to describe qualitatively the current seen during a voltage-clamp pulse and to dissect the components of the current to facilitate understanding of the voltage-clamp data presented in the remainder of this chapter. During a step in voltage from the resting level to a depolarized level (V_m not shown) three types of time-dependent currents develop; Na^+, K^+ and Cl^-. The figure shows the resulting membrane current: Part a shows the total current, divided into three phases A, B, C; Part b is an expanded view of phases A and B and Part C is a dissection of the three currents.

First there is a large, fast transient current through the membrane capacitance (phase A_1). The maximum amplitude of this capacitive current (I_c) is given by Equation 6, which also states that I_c flows only while the membrane potential is changing. For the frog oocyte, the membrane potential took 5 to 20 ms to stabilize after the command voltage was changed.

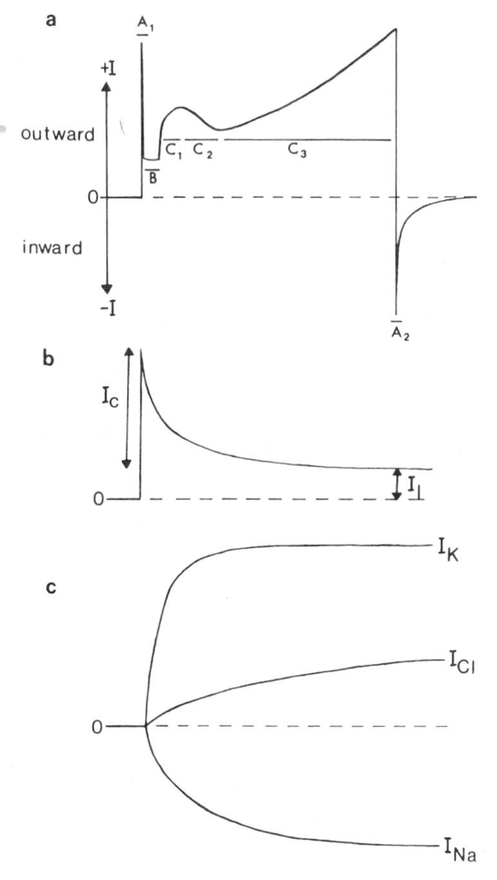

Figure 10. Schematic of components of membrane current from a young oocyte during a voltage-clamp step to a potential (e.g. +20 mV) at which K^+, Cl^-, and Na^+ channels open. **a)** Membrane current during a 10s long voltage-clamp pulse, divided into three phases: the capacitive currents (A_1 and A_2), the leakage current (B), and the ionic currents (C). **b)** An expanded view of phases A_1 and B, showing the transient capacitive current (I_c) and the time-dependent ionic currents, I_{Na}, I_K, and I_{Cl}.

The fast transient at the end of a clamp pulse (A_2) represents discharging of membrane capacitance. I_c is immediately followed by a component (B) that does not vary with time, called the leakage current (I_l). The magnitude of this current depends on the damage to the membrane during impalement and to the intrinsic non-selective ion permeability of the membrane. The third phase (C) is the sum of several time-dependent components. First, the outward current increases with time (C_1), then decreases (C_2), then slowly increases (C_3) until the end of the voltage-clamp pulse (A_2). Outward currents, conventionally considered as positive, correspond to an efflux of positive charges or an influx of negative charges. In principle, the decrease in outward current (C_2) could result from an influx of cations or an efflux of anions or from inactivation of the outward current. Part C shows phase C dissected into its three main components. (C_1) The first transient phase of outward current is carried by K^+) and when components 2 and 3 are eliminated, component 1 (called I_K) rises smoothly to a maximum after several seconds. (C_2) Superimposed on I_K is an inward current carried by Na^+ (called I_{Na}). When I_K and component 3 are eliminated, I_{Na} increases smoothly with time. In the absence of all other ion currents, I_{Na} can remain activated for many minutes. The direction of this current reverses at, or near, the equilibrium (Nernst) potential for Na^+. (C_3) The third time-dependent component is an outward current caused by an influx of Cl^- (called I_{Cl}).

Thus, during the voltage clamp step several currents flow through independent pathways. If the clamp is fast enough, the capacitive current can be separated, in time, from the ionic currents. To separate the three components of the ionic current drugs or blocking ions were used as described in the following sections.

2. Voltage-Clamp Analysis

Each ionic current is associated with a driving force resulting from concentration and voltage gradients, and a conductance that changes with time and voltage as the number of open channels changes. This can be simply described as:

$$I = \gamma NF (V_m - V_j) \quad \textbf{(Eq. 7)}$$

which states that the current flowing through a class of ion channels (e.g. K^+) is proportional to the conductance of each channel (γ), the number of channels in the membrane (N), the fraction of these channels that are open (F), and the driving force acting on the ion (V_m-V_j), *i.e.* membrane potential minus the Nernst potential for the ion. For voltage-gated channels such as the K^+, Cl^- and Na^+ channels in the frog oocyte, γ (which reflects the channel permeability) may vary with voltage (*i.e.* rectify), F depends on voltage and on time, N may be modulated by drugs or by the cell itself, and V_j depends on the internal and external ion activities and, weakly, on temperature.

The following simplified treatment of voltage-clamp analysis will concentrate on whole-cell currents since all my subsequent data will be from whole-cell recordings. Single ion channels can be studied using the patch-clamp technique (Simoncini *et al.* 1988). Single-channel analysis can be very powerful for dissecting the biophysical properties of the ion channels. However, I believe that most studies of currents in unexplored cell types should begin with the whole cell and proceed later to the single-channel level.

The description of a class of channels often begins with determining how the membrane conductance ($g = \gamma NF$) varies with voltage and with time and proceeds to characterizing the time and voltage dependence of each parameter; γ, F, and sometimes N. The current through an ensemble of open channels can be simply described as:

$$I = g (V - V_j) \quad \textbf{(Eq. 8)}$$

from Hodgkin and Huxley's formulation. In practice, g is often calculated from the slope of a tangent (slope conductance) or a chord (chord conductance) to the steady-state current-versus- voltage relation obtained from voltage clamp experiments such as those in the following section.

The rate at which voltage-gated channels open is a function of voltage and of how many "transition" states the channel goes through before opening. For simplicity, consider an ensemble of voltage-gated channels, each of which exists in either a closed state or an open state. If the membrane potential is abruptly changed to a new level at which more channels can open (in a typical step-wise, voltage-clamp experiment) then the number (n) of channels open at any time (t) is given by:

$$n(t) = n_{\infty} - (n_{\infty} - n_o) \exp (^{-t}/\tau) \qquad \textbf{(Eq. 9)}$$

where the time constant (τ = 1/rate of opening) is voltage dependent. For this simple model the ionic current should increase exponentially with time. The steady-state number of open channels (n) depends on the voltage (V) and the amount of energy (W) required to change the channel protein configuration from closed to open. As V-W becomes small the fraction of available channels open at steady state increases to a maximum according to a Boltzmann relation. Thus the proportion of channels open, ranges from 0 to 1 and depends both on time and on voltage. More complicated channels may have more than one closed or open state, inactivated state(s) and desensitized state(s). In such cases the current through an ensemble of channels will not rise mono-exponentially toward a steady-state value but might, for example, increase as an exponential raised to a power then decrease exponentially to a small value despite the maintained "permissive" voltage during a voltage-clamp step.

With this brief background to voltage-clamp analysis I will now show the currents in young *Rana pipiens* oocytes. First the effects of external ions on the total membrane current will be discussed.

3. Total Membrane Current

Throughout the remaining voltage-clamp analysis of the currents in young oocytes standard variations on Ringer's solutions were used as follows: Normal frog Ringer's contained (in mM): 11.1 NaCl, 1.9 KCl, 1.1 $CaCl_2$, 0.8 $MgSO_4$, 2.5 HEPES, 4 mM NaOH, pH 7.8, (*i.e.* total Na^+ 115, Cl^- 115). Low Cl^- solutions were made by substituting NO_3^- salts for KCl and $CaCl_2$ and methanesulfonate salts for NaCl. For low Na^+ solutions NaCl was replaced by cholineCl. K^+ concentration was increased 10-fold by simply adding KNO_3 (17.3 mM). The slight increase in osmolarity did not appear to harm the oocytes. Other solutions containing drugs will be described when the data are introduced.

Figure 11 shows action potentials in a metaphase I oocyte impaled with two microelectrodes. Action potentials could be evoked by a positive-going current pulse or as an off-response to a negative current pulse. The oocyte was bathed in normal Ringer's solution. Figure 12 shows total membrane currents during 10s long voltage-clamp pulses from the same oocyte bathed in Ringer's (part a), low Na^+ solution (part b) and decreasing Cl^- solutions (parts c and d). The corresponding current-voltage (I-V) relations in these four solutions are shown in Figure 13. In each solution the membrane was held at the resting potential (-32 to -38 mV) then stepped to a new potential. In each panel the currents in the left-hand set of traces correspond with hyperpolarizing pulses (below the zero-current mark) and depolarizing pulses above zero current. Traces in the other panels represent increasingly positive voltages. The numbers on some traces indicate currents at critical voltages that will be described below.

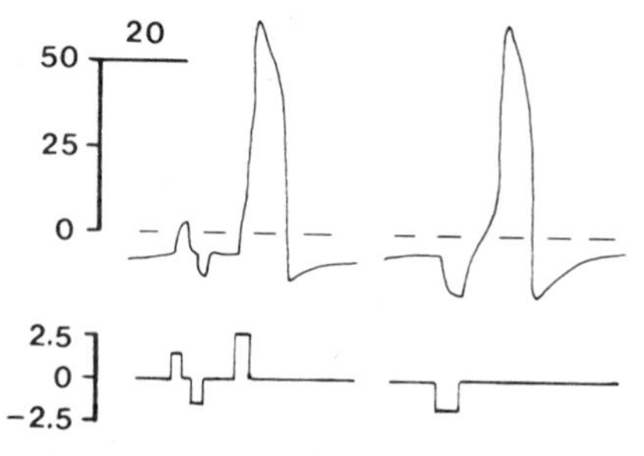

Figure 11. Action potentials evoked by current injection in a young metaphase I oocyte whose currents are studied under voltage clamp in Figures 12 and 13.

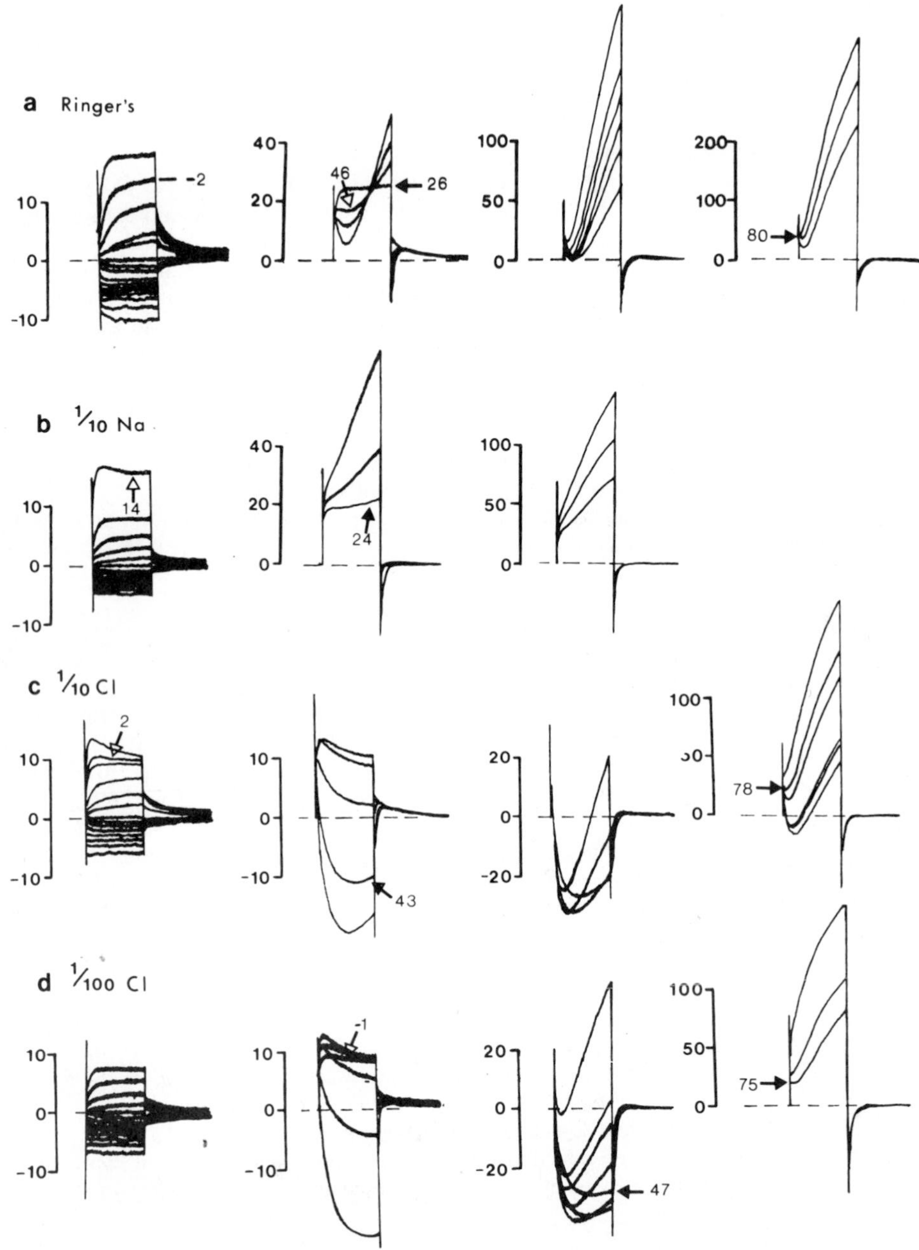

Figure 12. Effects of external Cl^- and Na^+ on the total membrane current from the oocyte in Figure 11. In this and all subsequent voltage-clamp records the current pulses are 10s long unless otherwise indicated. Leakage currents were not subtracted. In all cases the resting potential is the level at which the membrane current is zero. Hyperpolarizing voltage pulses are represented by currents below the zero current level in the first panel of a, b, c, d. All other currents are in response to depolarizing voltages increasing from the left to right panels. Voltage steps are at approximately 10-mV intervals; exact values can be read from the current-versus-voltage relations in Figure 14. **a)** Normal Ringer's; Na 115, Cl 115 mM. **b)** Low-Na Ringer's; 90% of the Na^+ replaced by choline **c)** Low-Cl Ringer's; 90% of the Cl^- replaced by methanesulfonate. **d)** Low-Cl Ringer's; 99% of the Cl^- replaced by methanesulfonate.

Ringer's. At hyperpolarized potentials a time-dependent inward current developed (described in more detail later in Figure 17). The amplitude of this current depended on the stage of oocyte maturity and decreased with age. The portion of the I-V relation corresponding with hyperpolarized potentials was nearly linear (Figure 13 a,b). At depolarized potentials the current was the sum of several time-dependent currents. With small depolarizations a smoothly developing outward current appeared which increased in amplitude and rate of rise with voltage (*i.e.* it is a voltage-dependent current). It will be shown below that this low threshold outward current is carried by K^+ through a non-inactivating channel. With further depolarization a second outwardly directed current appeared (see trace at +46 mV) which increased in amplitude and rate of rise with depolarization. It will be shown that

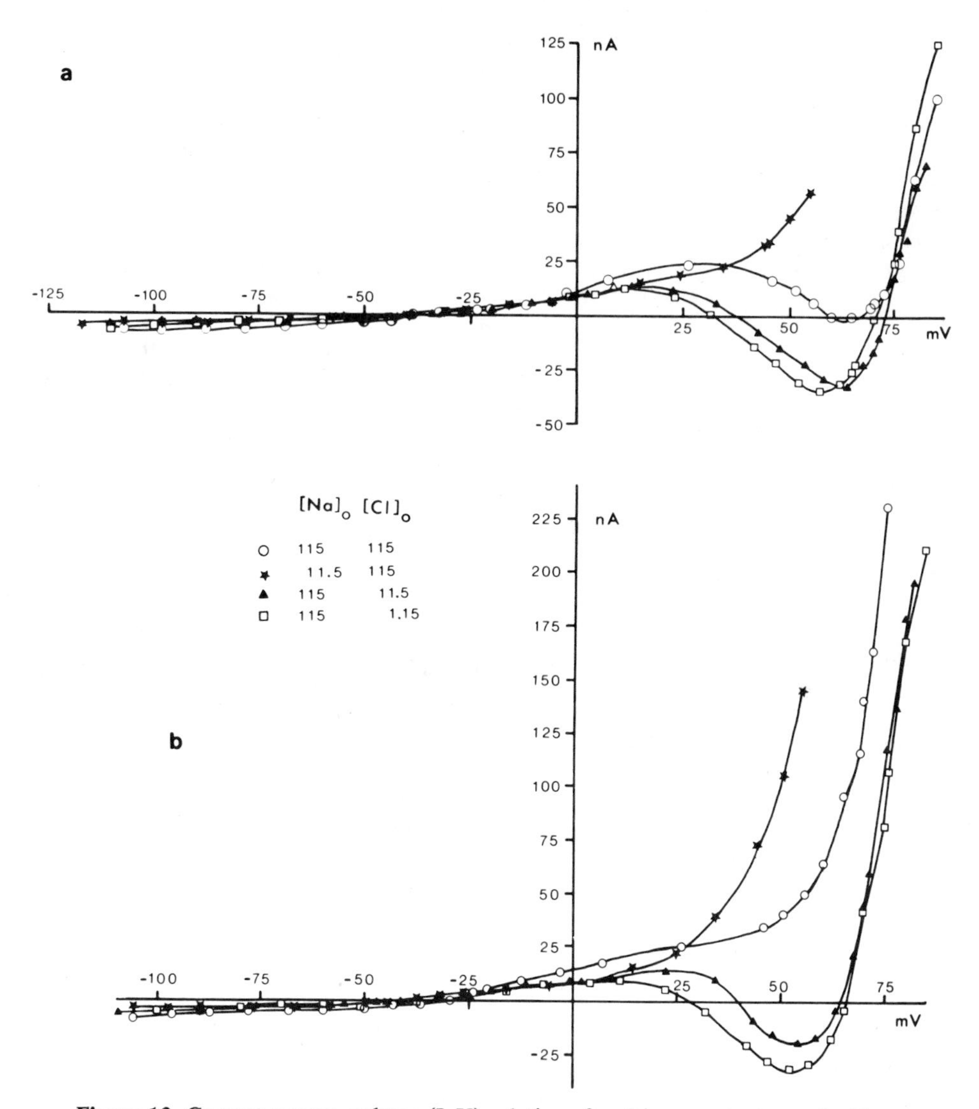

Figure 13. Current-versus-voltage (I-V) relations for the currents shown in Figure 13. **a)** Current was measured at 2s into the voltage-clamp pulse. **b)** Current was measured at the end of each 10s long voltage- clamp pulse. Solutions were: Normal Ringer's (○); low-Na Ringer's (✶); low-Cl Ringer's, 90% replacement (▲), 99% replacement (□) (from Schlichter 1989).

this outward current is carried by Cl^- influx. Over a small range of voltages (about +46 to 80 mV) an inwardly directed current is superimposed on the outward current. It will be shown that this current is carried by Na^+ influx through a novel type of voltage-dependent Na^+ channel; that the Na^+ current produces the upstroke of the action potential while the K^+ and Cl^- currents repolarize the membrane; and that the high K^+ conductance prevents oocytes at this stage from firing spontaneous trains of action potentials. The corresponding I-V relations show why the action potentials are brief and not spontaneous. Around the resting potential (zero-current potential) there is no net inward current either early in the voltage clamp pulse (2s, Figure 13a) or later (10s, Figure 13b). There is a negative-slope region at 2s, but not at 10s, hence, one would expect an action potential to be evoked by large depolarizing current pulses but lasting less than 10s.

Low Na^+ solution. When 90% of the Na^+ was replaced with choline no action potential could be evoked (e.g. see Figure 8b). The currents under voltage clamp were increasingly outward (Figure 12b) compared with Ringer's (Figure 12a) and the small inwardly directed current was eliminated. This is evidence that Na^+ carries the inward current. The I-V relation shows no negative slope region; therefore, no action potential can be evoked.

Low Cl^- solution. When Cl^- was replaced with increasing concentrations of methanesulfonate the action potentials became longer (see Figure 8c,d). The total current– (Figure 12c and d) had much larger inward and smaller outward components, showing that at least part of the outward current is carried by Cl^-. The inward (Na^+ current) reversed at about +75 to +78 mV (see arrows). The I-V relations show a net inward current at sufficiently depolarized potentials and strong negative-slope regions both at 2s (part a) and at 10s (part b) explaining why the action potential can last longer than 10s.

In summarizing the effects of Na_o and Cl_o: (1) reducing Na_o^+ showed that the existence of an action potential, inward current and negative-slope region depends on a Na^+ channel; (2) the low threshold outward current is not dependent on Cl_o^- (it will be shown later to be a K^+ current); and (3) the normally rapid end to the action potential is mainly due to a high-threshold outward Cl^- current (Cl^- influx).

F. THE Cl^- CURRENT (I_{Cl})

1. Separating I_{Cl}

Because the high-threshold outward current was reduced by Cl_o^- substitution, I tested the effect of a drug that blocks a Cl^- current in *Torpedo* electroplax membranes. SITS (4-acetamido -4′- isothiocyanostilbene- 2, 2′-disulfonic acid) is a well-established inhibitor of Cl^-/HCO_3^- exchange in red blood cells and other tissues (Knauf and Rothstein 1971; Thomas 1976) which also reversibly blocks the voltage-dependent Cl^- channel in *Torpedo* electroplax membranes with a K_i about 100 μM (White and Miller 1979).

Figure 14 shows the effects of SITS on the high-threshold outward current. Note–this oocyte was more mature than the ones shown in Figures 8-13 and showed the characteristic loss of the low-threshold (K^+) current that occurs during maturation. In normal Ringer's (Figure 14a) there were essentially no time-dependent currents below +10 mV, then the inward (Na^+) current appeared, followed by the outward (Cl^-) current which increased in amplitude with increasing voltage. The corresponding I-V relation taken at the end of the voltage-clamp pulse (10s, Figure 14d) shows a strongly increasing outward current above about +30 mV and no negative slope region. At shorter times (not shown) there was a negative-slope region in the I-V relation and action potentials briefer than 10s could be easily evoked by a depolarizing current pulse.

Adding SITS (100 μM, Figure 14b; 200 μM, Figure 14c) greatly reduced the Cl^- current (I_{Cl}) in a dose-dependent manner. This treatment exposed the inward (Na^+) current which turned on at about 0 to 10 mV and increased with voltage, then decreased and reversed at +58 to 59 mV (see arrows). The Na^+ current developed very smoothly with time. Some outward Cl^- current is still evident at high voltages, especially in 100 μM SITS. The I-V curves in 100 and 200 μM SITS show net inward current from around +20 to 50 mV and strong negative-slope regions. These I-V relations correspond with action potentials that reach a very

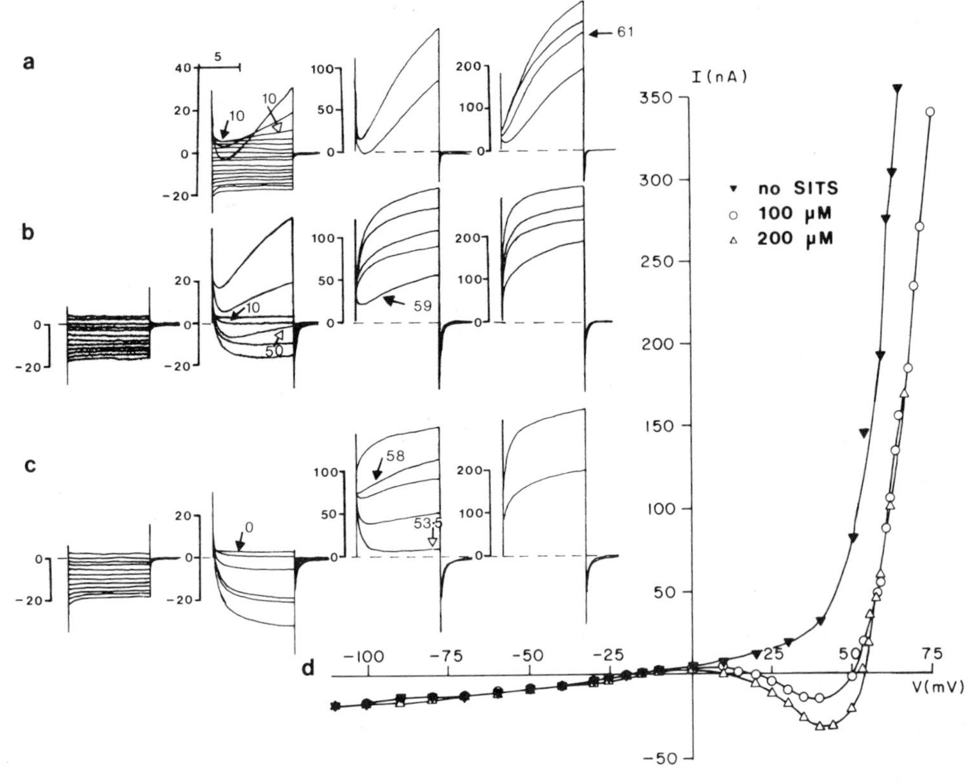

Figure 14. SITS blocks the Cl⁻ current (I_{Cl}). SITS (4-acetamido-4′-isothiocyanostilbene-2,2′-disulfonic acid) was added to normal Ringer's solution. This oocyte was older than those in Figures 8-13 and lacked the initial outward K^+ current. **a)** Normal Ringer's. Note I_{Cl} beginning at about +10 mV. **b)** 100 μM SITS reduced the outward current and exposed net inward (Na^+) currents. **c)** 200 μM SITS further reduced the outward Cl^- current and exposed the inward Na^+ current. Some I_{Cl} remained above about +50 mV. For complete elimination of I_{Cl} 400 to 500 μM SITS was used in subsequent experiments. **d)** I-V relations in Ringer's (▾) 100 μM SITS (○) or 200 μM SITS (△) (from Schlichter 1989).

positive value and last longer than 10s in the presence of SITS. Above 50 to 60 mV the outward current becomes very large and the I-V relation is nearly linear. This suggests that the Cl^- conductance is relatively constant above 50 mV, as will be shown below. The effect of SITS was fully reversible; washing for 10 to 15 min in normal Ringer's restored I_{Cl}.

To fully isolate the Cl^- current it would be necessary to block both the K^+ current (I_K) and the Na^+ current (I_{Na}). As I will show later: (1) I_K could be blocked by 20 mM tetraethylammonium (TEA) added to the bath; (2) I_{Na} could not be easily eliminated since tetrodotoxin (TTX) did not block it, even at 6 μM; and **(3)** substituting a non-permeant cation for Na_o (e.g. choline) eliminated the inward current but a large outward cation current passed through the Na^+ channel. Therefore, to separate I_{Cl}, I_K was blocked with TEA, then the remaining current (I_{Na}+ I_{Cl}) recorded, then 400 μM SITS was added to completely block I_{Cl}, and finally the remaining current (I_{Na}) was subtracted from I_{Na}+ I_{Cl}, on a point-by-point basis. These results are shown in Figure 15. The general features of I_{Cl} are: (1) the threshold voltage was between 0 and +10 mV; (2) I_{Cl} was always outward above threshold, indicating that the Nernst potential for Cl^- is negative (~-24 mV for mature oocytes); (3) I_{Cl} increased slowly with time and showed a definite lag phase at very positive potentials, suggesting complex activation kinetics; and (4) the tail current was very small at the resting (holding) potential of

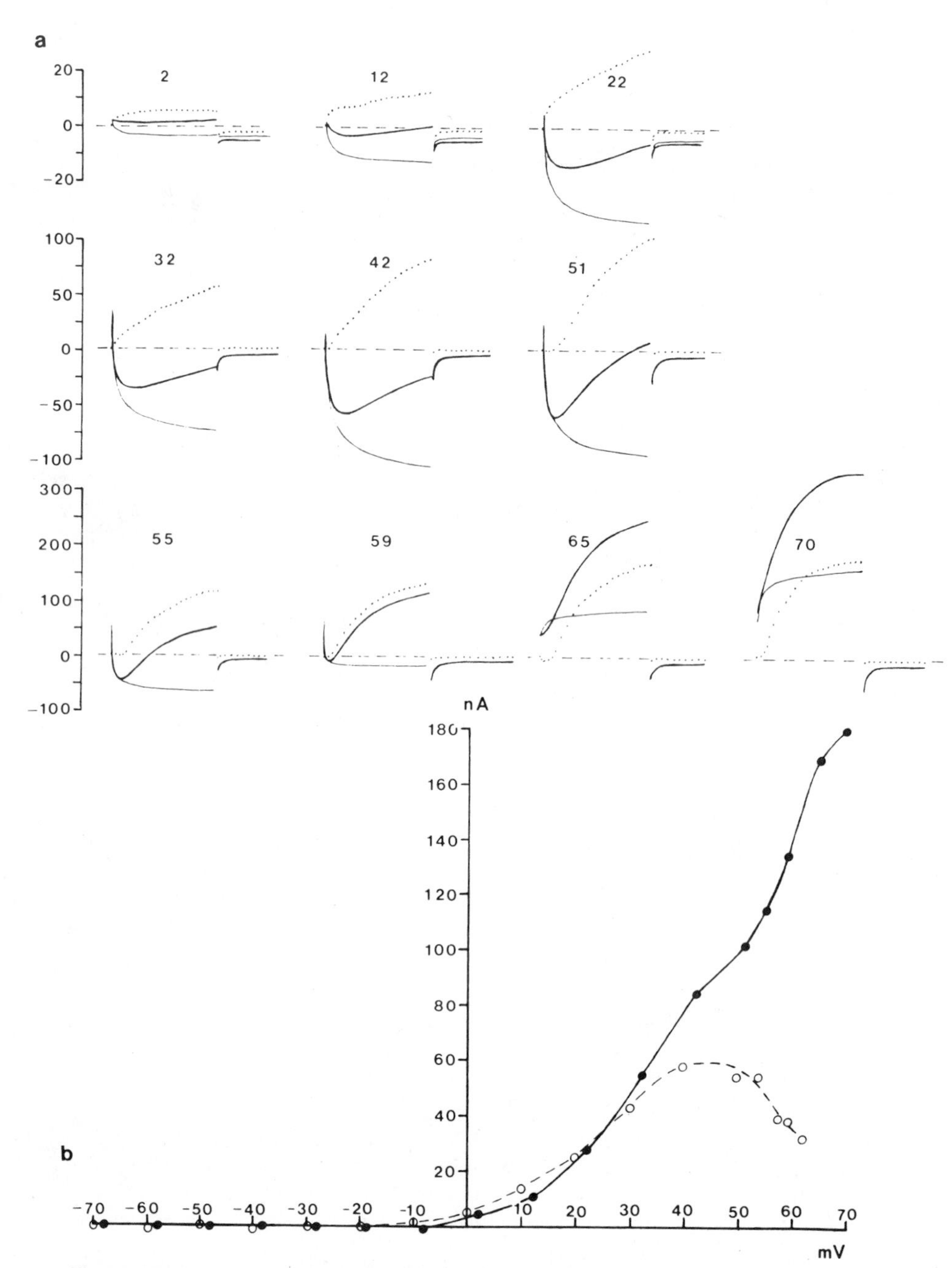

Figure 15. Cl^- current isolated by blocking I_K and subtracting I_{Na}. **a)** Currents were recorded during 10s long voltage-clamp pulses to the voltages indicated above each set of traces. All solutions contained 20 mM tetraethylammonium (TEA) to block the K^+ current as explained in Section G. The heavy curves are currents in normal Ringer's + TEA and represent $I_{Na} + I_{Cl} + I_{leak}$. The light curves are currents obtained after I_{Cl} was blocked with 400 μM SITS and represent $I_{Na} + I_{leak}$. The dotted curves were obtained from point-by-point subtractions of the light curves from the heavy curves. **b)** I-V relations constructed from the dotted curves in part a, measured at the end of each 10s long pulse (dots) and from a different oocyte that was 4.5h older (open circles) (from Schlichter 1989).

-20 to - 38 mV in different cells, suggesting the membrane potential is close to the Nernst potential for Cl^- when I_K is blocked with TEA. The I-V relation for I_{Cl} in this cell is shown in Figure 15b, solid curve. It is clearly an outwardly rectifying current that activates at about 0 mV and increases with voltage. The slope conductance seems to increase, then perhaps begins to level off at very positive potentials. As previously mentioned (section D2) the Cl^- current decreases with maturation, corresponding with an increase in the action potential duration. For comparison the I-V relation of I_{Cl} for an older oocyte is shown in Figure 15b, dotted curve. The slope conductance definitely decreases above +40 mV. The bell-shaped I-V relation looks like that of a Ca^{2+}-dependent outward current that decreases as the equilibrium potential for Ca^{2+} (E_{Ca}) is approached and Ca^{2+} influx decreases. Further evidence that this Cl^- current is Ca^{2+}-dependent is that in oocytes with small Cl^- currents and long action potentials, the Cl^- current was increased by raising external Ca^{2+} to five times normal (data not shown). The I-V relation for the Ca-stimulated outward current was dramatically bell-shaped, decreasing above about +40 mV. The increased I_{Cl} made the action potentials shorter in high Ca_o^{2+}.

The activation kinetics of I_{Cl} appeared to be complicated; that is, a delay or sigmoid onset appeared at quite positive potentials ($\geq$ +51 mV in Figure 15). Because of possible errors in the point-by-point subtraction of I_{Na} used to isolate I_{Cl}, I also measured I_{Cl} exactly at the reversal potential for Na^+(E_{Na}) so that I_{Na} was zero and I_{Cl} could be isolated without subtraction (data not shown). A definite delay in onset was obvious, lasting about 1.5s and if a 1.5s delay was allowed for, the rise in current was well described by a single exponential. A more complete kinetic description will be facilitated when a selective blocking agent for I_{Na} is found. Then, the rise in Cl^- current can be fit by non-linear least squares curve-fitting programs to determine the best model and extract the time constants as a function of voltage.

G. THE K^+ CURRENT (I_K) AND LEAKAGE CURRENTS

This section describes the voltage-dependent outward K^+ current which is observed when I_{Cl} is blocked with SITS and which is superimposed on the Na^+ current.

1. The K^+ Current

Ringer's. Figure 16a shows currents in normal Ringer's which contains 1.9 mM K^+. With depolarization the amplitude and rate of rise of the time-dependent outward current increased up to about + 20 mV. The amplitude of the tail currents also increased suggesting a voltage-dependent conductance increase over this potential range. At about + 20 mV the inward I_{Na} appeared and increased with voltage, obscuring the K^+ current and decreasing the tail current amplitude. The inward current reversed at +60 to 65 mV. These voltage-dependent changes in current are reflected in the I-V relation (Figure 16d) which shows an outward voltage-dependent current activating above the zero-current potential and a negative-slope region owing to the activation of the voltage-dependent Na^+ current. The net current reversed at about +60 mV and action potentials evoked by current injection reversed at about +60 mV in this solution. Currents below the resting potential will be explained in more detail later (section G2).

19 mM K_o - Figure 16b shows that high external K^+ decreased the amplitude of the outward currents and tail currents, owing to the smaller driving force on the K^+ ions (Nernst potential less negative). Thus, I_{Na} was less obscured by I_K and the inward current was larger (see I-V relation in Figure 16d). This is evidence that the outward voltage-dependent current is carried by K^+.

20 mM TEA - Tetraethylammonium (TEA) blocks some, but not all, K^+ currents in a wide variety of tissues (Thompson and Aldrich 1980). Therefore, I tested TEA on the K^+ currents in *Rana* oocytes. Figure 16c shows that TEA blocked both the outward "on"-current (during the pulse) and the outward tail currents. The remaining time-dependent current was I_{Na} and a leakage current which are described later. I_{Na} then showed voltage-dependent activation above about 0 mV and a clear reversal at +55 mV. Under current clamp (zero current) the membrane potential went spontaneously to +55 mV and remained there as long as no current

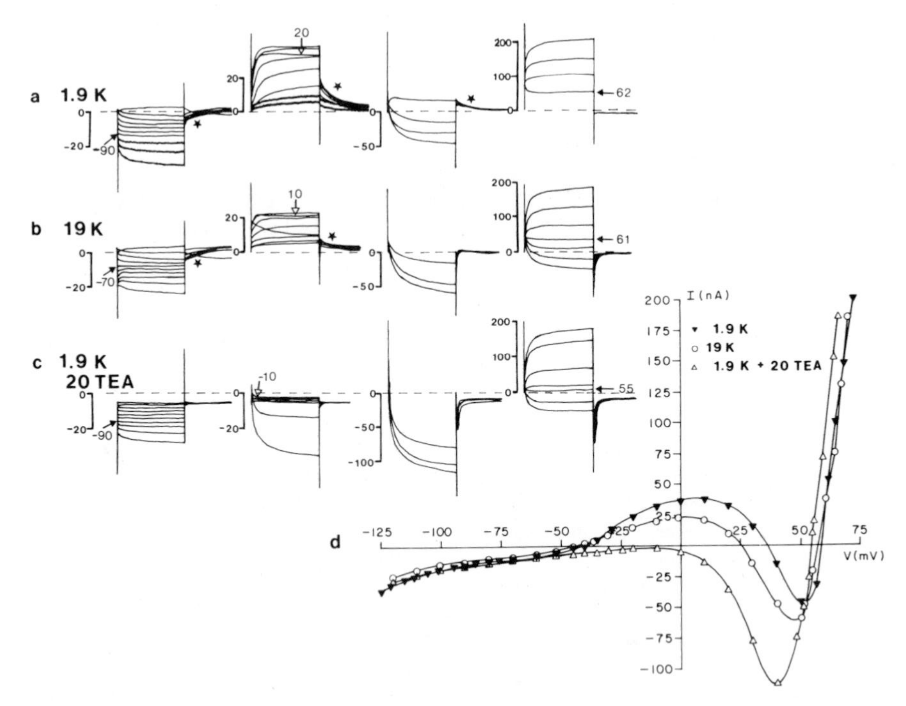

Figure 16. Effect of external K and tetraethylammonium (TEA) on the K^+ current in a young oocyte. All solutions contained 500 μM SITS to block I_{Cl}. Holding potential -40 mV in all solutions. **a)** Normal Ringer's solution. **b)** 19 mM K_o made with additional KNO_3. **c)** Normal Ringer's with 20 mM tetraethylammonium (TEA). **d)** I-V relations measured at the end of each 10s long voltage-clamp pulse.

Voltages are indicated for the appearance of the inward current (open arrows) and approximate reversal potentials for the net current (+62, +61, +55 mV). Leakage currents were subtracted manually before plotting the I-V relations in part d. Note the long tail currents (stars) (from Schlichter 1989).

was injected. This is expected since there were no outward currents to repolarize the membrane. The I-V relation for I_{Na} (open triangles) has a shape similar to Na^+ currents in nerve and muscle but translated along the voltage axis to more positive potentials.

2. Leakage Currents

In Figure 16 time-dependent currents were seen at hyperpolarized potentials and these leakage currents will be explored before further analyzing I_K. Figure 17 shows currents below the threshold for activating I_{Na} ; voltages are shown at the right. Compared with normal Ringer's (part a) the time-dependent currents and tail currents were somewhat reduced by raising external K (part b) showing they are carried in part by K^+. This is further demonstrated by the complete block of the time-dependent currents at -90 mV and above by 20 mM TEA (part c). However, there remained large inward currents both during and after the pulse which were independent of time except at very negative potentials (see -110 mV). This remaining inward current was greatly reduced by replacing external Na^+ with choline (part d), suggesting it is carried mainly by Na^+. The Na^+-dependent leak (I_N) is separated from the remaining leak (I_l) in the I-V relations in part e. Both leakage currents have a fairly linear I-V relation but I_l reverses at a negative potential and I_N reverses at a very positive potential (probably close to the Nernst potential for Na^+).

This separation of currents allows us to see the I-V relation for I_K in isolation (Figure 17f). When I_K was blocked with TEA (dots) the remaining leakage currents could be subtracted

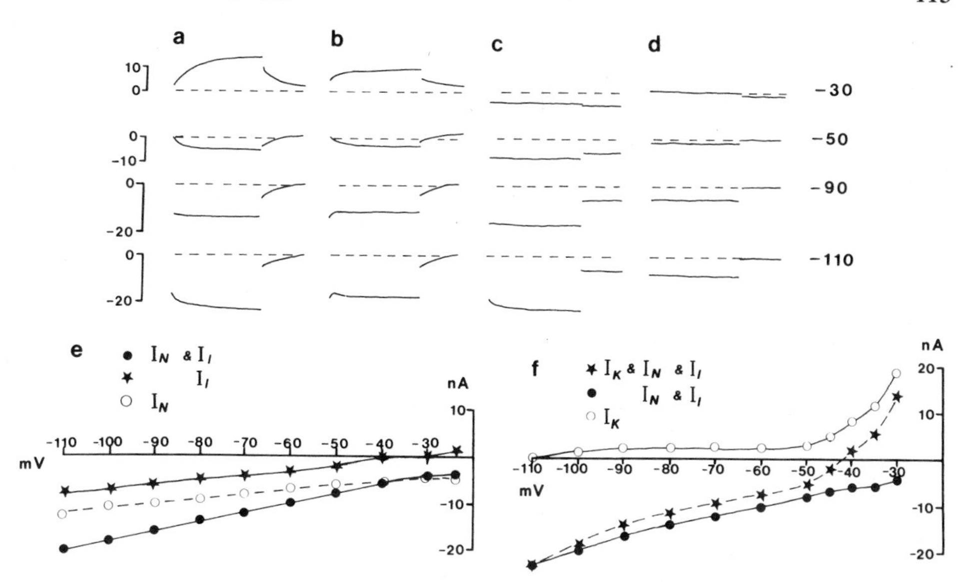

Figure 17. Separating I_K from the leakage currents. All solutions contain 400 μM SITS to block I_{Cl}. **a)** Ringer's. **b)** 19 mM K_o^+. **c)** Ringer's with 20 mM TEA, 1.9 mM K^+. **d)** Na-free choline solution, 1.9mM K^+. **e)** I-V relation measured at 10s in TEA solution (dots), called $I_N + I_l$ which represents the Na^+-dependent leak and remaining leak, respectively. Currents in Na-free solution (stars) represent I_l and the difference (open circles) represents I_N, the Na^+-dependent leak. **f)** I-V relation in Ringer's (stars) represents $I_K + I_N + I_l$. Currents in TEA solution (dots) include $I_N + I_l$ and the difference (open circles) represents I_K (from Schlichter 1989).

from the total current (stars) to yield I_K alone (open circles). I_K went to zero around -110 mV which is about the predicted Nernst potential for K^+. Some I_K is activated even at -100 mV but this current strongly activated about -50 mV and showed outward rectification.

3. Features of I_K

To further characterize the K^+ current, two methods of separating I_K from the other currents were used: **(1)** the voltage range was restricted to below the activation potential for I_{Na} (about 0 mV); **(2)** the current in the presence of SITS was recorded ($I_{Na} + I_K$) then I_K was blocked with TEA leaving I_{Na}, then these currents were subtracted point-by-point as had been done for I_{Cl} in section F1.

Figure 17f showed K^+ currents at hyperpolarized potentials, separated from the two leakage currents. Figure 18 shows point-by-point subtraction of ($I_{Na} + I_N + I_l$) from ($I_K + I_{Na} + I_N + I_l$) to yield I_K in isolation at each voltage. At the holding potential of -40 mV the net current was zero. Above this voltage the isolated K^+ current (dotted curve) activated strongly, increasing in amplitude and rate of rise with voltage, as is typical for a voltage-dependent outwardly rectifying K^+ current. The tail currents were outward and increased with voltage to about +40 mV then decreased, suggesting a decrease in K^+ conductance at very positive potentials.

The Na^+ current (light curves) appeared to activate at -10 mV and increase with voltage to about +40 mV, then decrease. The tail currents for Na^+ were always inward at the holding potential of -40 mV since the Nernst potential for Na^+ (E_{Na}) was very positive.

Figure 19 shows I-V and conductance-versus-voltage (g-V) curves for the K^+ current and the Na^+ current in a different cell. When all I_{Cl} was blocked with SITS and the leakage currents subtracted, the remaining current was I_{Na}+ I_K (crosses). Blocking I_K with TEA left only I_{Na} (dots) which shows voltage-dependent activation at about 0 mV and the typical shaped

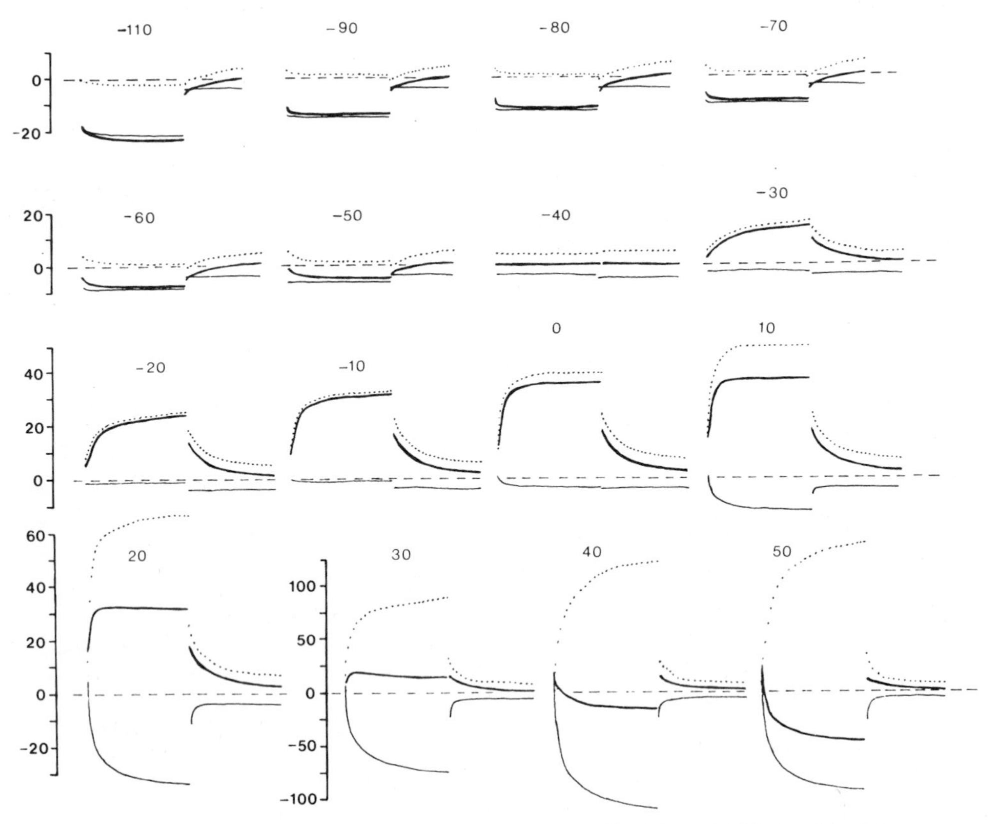

Figure 18. Isolating the K^+ current. All solutions contained 400 μM SITS to block I_{Cl}. Currents recorded in Ringer's (heavy curves) represent $I_{Na} + I_K + I_N + I_l$. Then 20 mM TEA was added to block I_K and the light, solid curves represent $I_{Na} + I_N + I_l$. The point-by-point difference (dotted curves) represents I_K. The cell was held at - 40 mV in all solutions.

I-V relation for a Na^+ current. I_{Na} reversed at +55 mV (E_{Na}) and this value was used to calculate conductance at each voltage (dashed line). The g-V curve for the Na^+ current had the typical sigmoid shape of most voltage-dependent Na^+ currents but was shifted along the voltage axis to more positive potentials.

The I-V relation for the K^+ current was then obtained by subtracting I_{Na} from $I_{Na} + I_K$ and showed a dramatically bell-shaped I-V relation with strong activation above -40 mV and decreasing current above +40 mV. The Nernst potential for K^+ was calculated as about -130 mV based on an E_K of -70 mV in ten-times normal external K concentration. The g-V curve for I_K was calculated using I_K = -130 mV and showed the same bell-shaped profile (dotted curve). Such bell-shaped g-V curves are very typical of Ca-activated K^+ currents in a wide variety of tissues. In these cases, as voltage is made more positive, voltage-dependent Ca channels open and stimulate more K^+ channels to open. Then as voltage approaches E_{Ca} the Ca^{2+} current decreases, hence fewer K channels open. I suspect the K^+ current in immature *Rana* oocytes is Ca^{2+}-dependent, because of the bell-shaped g-V curve and because mature *Rana* oocytes have Ca^{2+}-dependent K^+ channels that are opened upon fertilization or artificial activation (Jaffe and Schlichter 1985; Jaffe *et al.* 1985, and section I3).

H. THE Na^+ CURRENT (I_{Na})

1. Separating I_{Na}

Figure 19 showed the I-V relation and g-V relation for the Na^+ current in the presence of SITS (to block I_{Cl}) and TEA (to block I_K). Figure 20 shows the isolated Na^+ current in normal Ringer's (part a) and in the presence of 6 μM tetrodotoxin (TTX, part b), a specific toxin blocking most types of voltage-dependent Na^+ channels with K_i's in the nanomolar to micromolar range. However, in the frog oocyte even very high concentrations of TTX had little or no effect on the Na^+ current. The I-V relations with and without TTX virtually superimpose (part c). Figure 20d shows that the action potential in normal Ringer's containing SITS and TEA remains close to E_{Na} for many minutes (more than 6 min, upper trace), which shows that I_{Na} does not inactivate during this time. This conclusion is borne out by long voltage-clamp pulses which show I_{Na} developing smoothly with time and showing no sign of

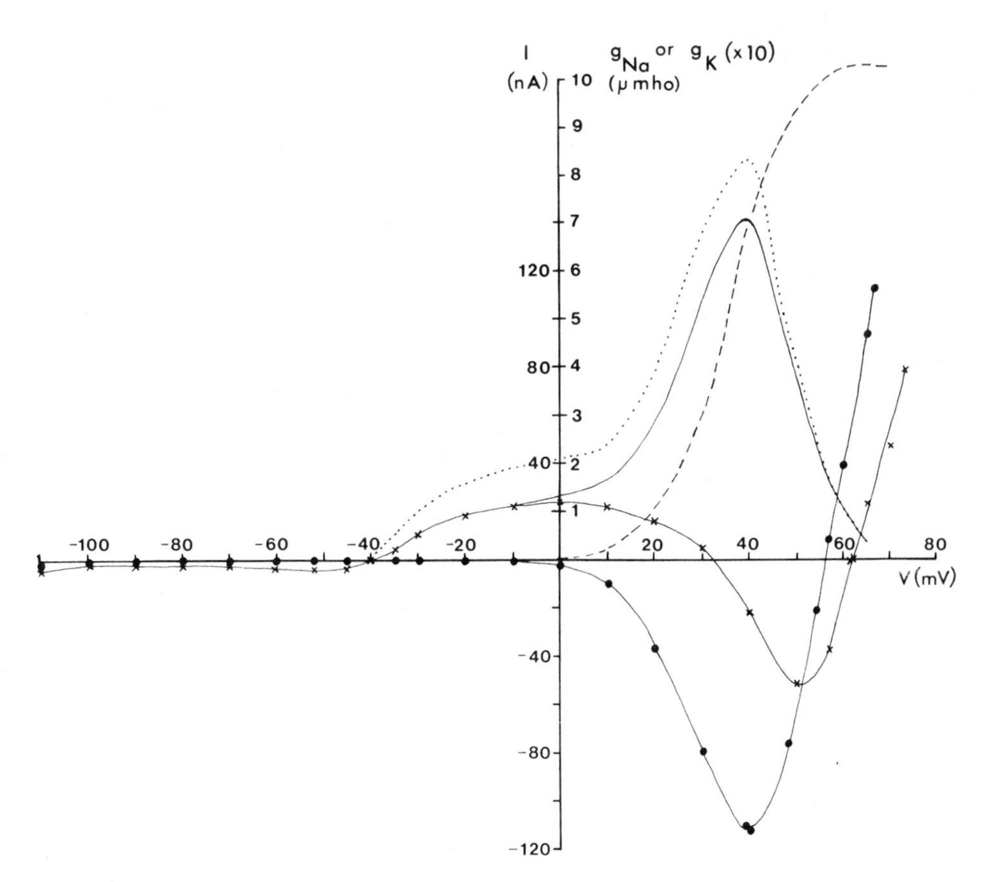

Figure 19. I-V and conductance-voltage relations for I_{Na} and I_K in a different egg from that in Figure 18. The leakage currents were subtracted from currents in Ringer's containing 400 μM SITS to yield I_{Na} + I_K (crosses). Then 20 mM TEA was added to eliminate I_K, leaving only I_{Na} (dots). Then the Na^+ conductance (g_{Na}, dashed curve) was calculated from the current at each voltage using a Nernst potential (E_{Na}) of +55 mV and the relation; $g_{Na} = I_{Na} / (V - E_{Na})$. Next the I-V relation for I_{Na} was subtracted from that of I_{Na} + I_K to yield the I-V for I_K (solid, bell-shaped curve). Finally the conductance-versus-voltage relation I_K (dotted, bell-shaped curve) was calculated from $g_K = I_K/(V-E_K)$ using -130 mV for the apparent Nernst potential for K^+ in this oocyte.

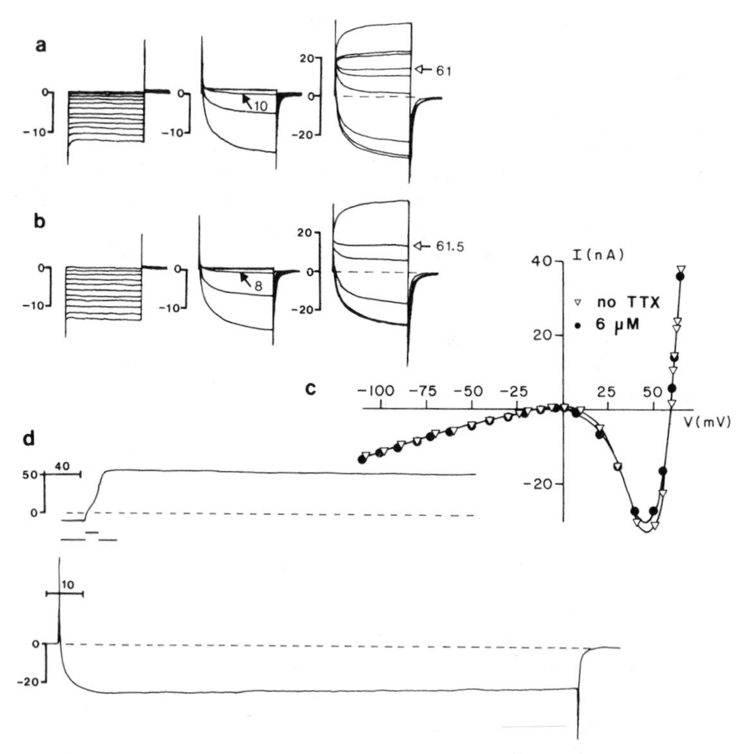

Figure 20. Isolated Na^+ current (I_{Na}). I_{Na} is insensitive to TTX and does not inactivate. Ringer's solutions contained 500 μM SITS and 20 mM TEA. **a)** Ringer's solution. **b)** Ringer's + 6 μM tetrodotoxin, TTX (Sankyo, no citrate). Voltages are indicated for the threshold of I_{Na} (solid arrows) and the reversal potential (E_{Na}, open arrows). **c)** I-V relations measured at the end of each 10s long pulse in Ringer's (∇)and TTX (•). **d)** Under current clamp a prolonged action potential was evoked by a depolarizing pulse (upper trace). Under voltage clamp to +38 mV a prolonged inward current (I_{Na}) was elicited which did not inactivate even during this 2.3 min long pulse (from Schlichter 1989).

inactivation (over 2 min, lower trace). The very slight decrease in current over a couple of minutes might reflect accumulation of Na^+ in the cytoplasm during Na^+ influx.

2. Effects of External Ca^{2+}

Because the Na^+ channel in the *Rana* oocyte differs fundamentally from other voltage-dependent Na^+ channels (e.g. no inactivation) it was important to ensure that it is truly selective for Na^+ and not, for example a Na^+/Ca^{2+} channel such as that in starfish eggs (Hagiwara *et al.* 1975). Figure 21 shows that external Ca^{2+} cannot substitute for Na^+ in carrying the inward current. I_{Cl} was blocked with SITS and I_K was blocked with TEA, thus isolating I_{Na} and any leakage currents. In normal Ringer's (part a) I_{Na} activated at about -8 mV and increased with voltage, then decreased and reversed at about +50 mV. Correspondingly, the I-V relation (part e, stars) shows a negative-slope region and reversal of the total current (leak not subtracted) at about 47 mV. Under current clamp (part d, left) the action potential was very long since there were no outward currents to repolarize it.

When all external Na^+ was removed and 55 mM Ca^{2+} added (part b) all inward current was eliminated. There remained an outward current that had the same time course as the Na^+ current suggesting Na^+ efflux through the Na^+ channel due to a reversed driving force. This current increased strongly with voltage above about +30 mV (part e, circles). The shift in activation potential was likely due to a positive shift in surface potential owing to the high

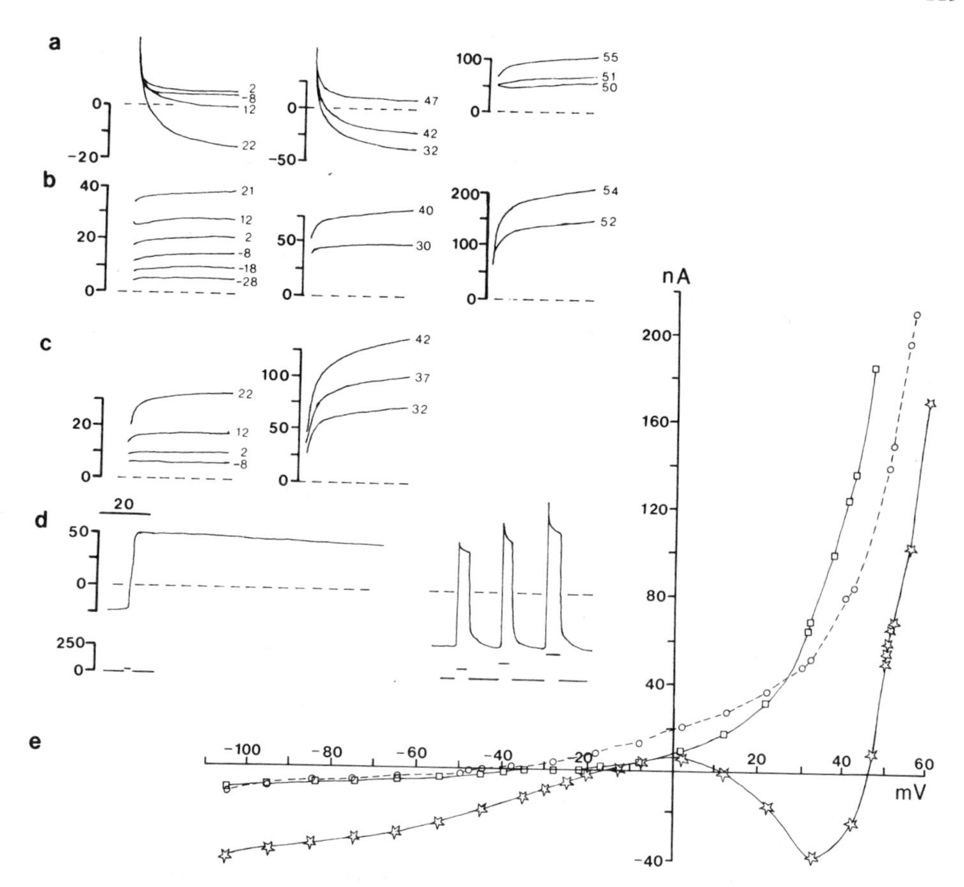

Figure 21. Can Ca^{2+} carry the inward current through Na channels? All solutions contained 400 μM SITS to block I_{Cl} and 20 mM TEA to block I_K. **(a) - (c)** A few representative current traces during voltage- clamp steps to the voltages indicated to the right of each trace. Holding potential was always the resting (zero-current) potential. **a)** Normal Ringer's; Na 115, Ca 1.1 mM. **b)** Ca-Ringer's; Na 0, Ca 55 mM. **c)** Na-free Ringer's; Na O, choline 115, Ca 1.1 mM. **d)** Left panel, action potential in normal Ringer's evoked by a current pulse. Right panel, in Ca 55 Ringer's, no action potentials could be evoked. **e)** I-V relations in normal Ringer's (☆), Ca 55 Ringer's (○), Na-free choline Ringer's (□). Currents were measured at the end of each 10s long voltage-clamp pulse.

divalent ion concentration. Note also that the leakage current was larger in high Ca^{2+} solution than in normal Ringer's. Consistent with the lack of inward current and no negative-slope region in the I-V relation, action potentials could not be evoked by current injection (part d, right).

As further evidence that Na^+ alone carries the time-dependent inward and outward currents, part c shows that replacing all external Na^+ with choline completely eliminated the inward currents. Remaining was a time-and voltage-dependent outward current which activated at about the same potential as the Na^+ current in Ringer's (about 0 mV) and looked kinetically the same. Hence the outward currents in choline are probably Na^+ efflux through the Na^+ channel under a reversed driving force. This efflux produces a strong outward rectification in the I-V relation (part e, squares) and, of course, no action potentials could be elicited (not shown).

Time Course of I_{Na} : In all records of isolated Na^+ current the inward or outward (above E_{Na}) currents appeared to develop smoothly with time but the kinetics appeared rather insensitive to voltage. This can be seen in Figures 16c, 20a,b which do not show a marked change in rate of activation with voltage, such as is typical of the classical fast-inactivating Na^+ channels in most excitable cells. When a single exponential was fitted to the activation of I_{Na} at different voltages the time constant was about 2.3 to 3s from +10 to +50 mV (data not shown); in other words rather insensitive to voltage in this range. The currents were not well described by a single exponential however and a faster component was present. Before a more complete kinetic description is undertaken, faster clamping speeds than I obtained with my equipment would be necessary.

3. Selectivity of the Na^+ Channel

Data in Figure 21 showed that Ca^{2+} does not permeate the Na^+ channel appreciably, and no inward current was observed. I then tested whether other monovalent cations could permeate and whether more Ca_o^{2+} could produce inward current. Figure 22 shows currents recorded in Ringer's solution (Na^+) and in Ringer's in which the 111 mM NaCl was replaced by equimolar amounts of LiCl, KCl, RbCl, CsCl, and $CaCl_2$. All solutions contained SITS to block I_{Cl} and TEA to block I_K and 4 mM NaOH to adjust pH. As usual, in normal Ringer's (Na^+) the Na^+ current had a threshold around 0 mV and was inward up to the reversal potential (E_{Na}) of about +61 mV. These currents corresponded with a very long action potential (Figure 23a, Na^+) and a typical I-V relation (Figure 23b, crosses).

In Li^+ solution (Figure 22), much smaller inward currents occurred between about +20 and +57 mV. Accordingly, the action potential was brief, but it did outlast the current pulse (Figure 23a, Li). The I-V relation showed a very slight negative-slope region (Figure 23b, open circles). Li^+ permeation is interesting in that it appeared to both permeate and block the Na^+ channel and this will be described further below.

None of the other monovalent cations (K^+, Rb^+, Cs^+) permeated the Na^+ channel. Above the threshold for opening Na^+ channels (about 0 mV) the currents were outward, probably carried by Na^+ efflux just as in choline-substituted solutions (Figure 21c). In these solutions E_{Na} was about -23 mV owing to the 4 mM NaOH used to adjust the pH of each solution. Consistent with the lack of inward current none of these ions (K^+, Rb^+, Cs^+; Figure 23a) could support action potentials and all the I-V relations (Figure 23b) were strongly outwardly rectifying. Further evidence that the outward current through the Na^+ channel is carried by Na^+ is that high K_o^+ yielded the largest outward currents (Figures 22 and 23). In contrast, if K^+ efflux carries outward current through the Na^+ channel, it should be greatly reduced by raising external K^+.

When external Ca^{2+} was raised to 111 mM (increased osmotic strength) by substitution for 111 mM Na^+, there was apparently a fast-developing, small inward Ca^{2+} current (Figures 22, Ca). This inward current did not look kinetically like the inward Na^+ current but very rapidly approached a steady-state during each pulse. The current reversed at about +45 mV, and the I-V relation did not rectify as strongly (Figures 23b, dots). Nevertheless, no action potentials could be elicited by current injection (Figures 23a, Ca). The interpretation of the currents in high Ca^{2+} solution is more difficult because high divalent cation concentrations shift the external surface potential to more positive values, thus g-V curves are also shifted to the right, to more positive values. Even so, the inward current in Ca- solution (Figures 22) was present even at -10 mV, suggesting it represents a tiny inward Ca^{2+} current (in normal Ringers, low Ca_o) that normally activates at much more negative potentials. I will speculate that this is a different current, a small Ca^{2+} current that may contribute to activation of I_K and I_{Cl} in the young oocyte. This current would be well worth further investigation but would require a cleaner dissection from I_{Na}. To do this, a specific blocker of I_{Na} must first be found.

4. Li^+ Blocks the Na^+ Channel

Figures 22 and 23 showed that Li^+ can permeate the Na^+ channel. But Figure 24 shows that Li^+ entry partially blocks the Na^+ channel (different oocyte from Figure 22). In Na^+ solution (normal Ringer's) I_{Na} had a typical I-V relation with a reversal potential of +61.5 mV. However, when LiCl replaced all NaCl in the solution the inward currents were much

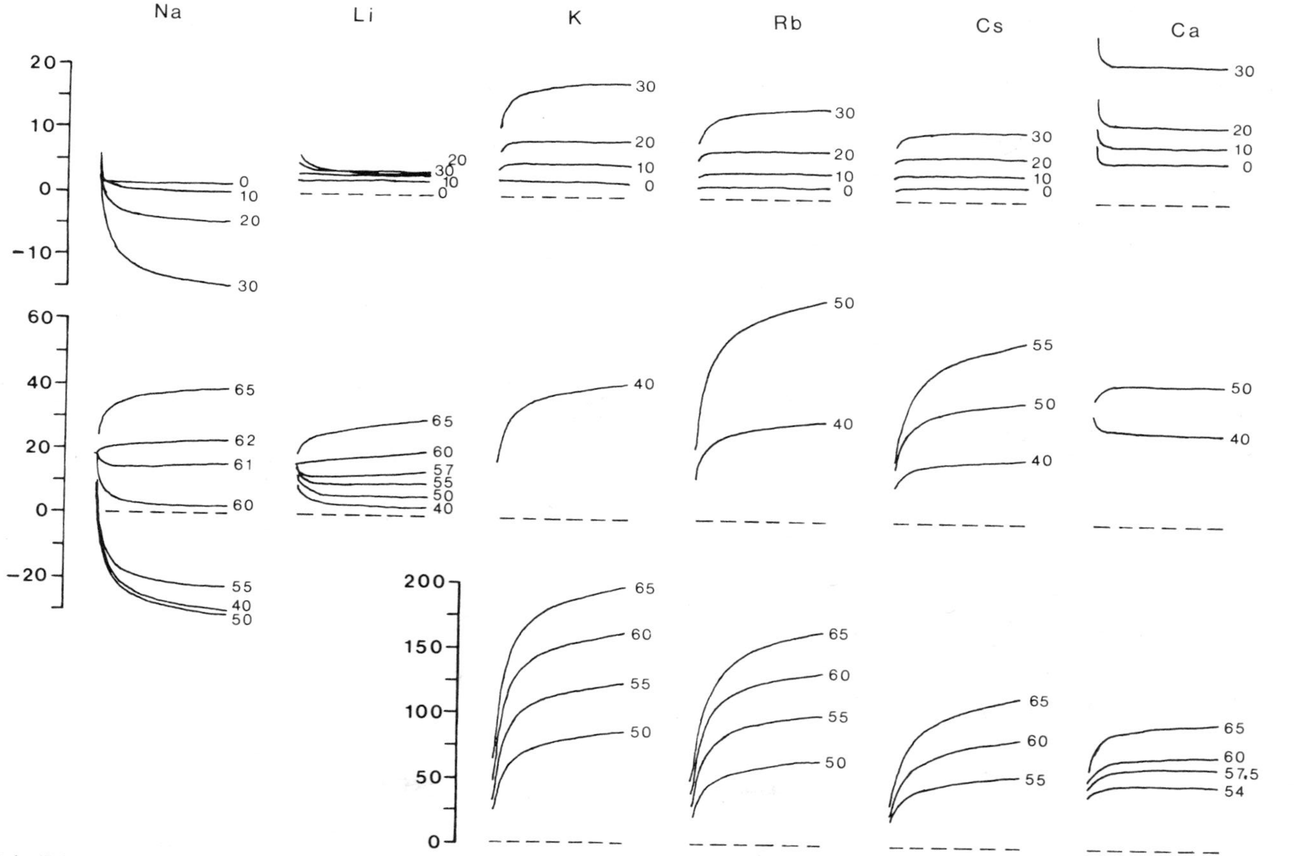

Figure 22. Selectivity of the Na channel. Current records during 10s long pulses to the voltages indicated at the right of each trace. All solutions contained 400 μM SITS and 20 mM TEA. Normal Ringer's (Na) contained (in mM); NaCl 111, NaOH 4, KCl 1.9, $CaCl_2$ 1.1, $MgSO_4$ 0.8, pH 7.8. For ion substitutions the NaCl was replaced with 111 mM LiCl, KCl, RbCl, CsCl, or $CaCl_2$. The osmolarity of the Ca solution was therefore higher than the other solutions. Note that inward-going currents were only observed in Na, Li and Ca solutions.

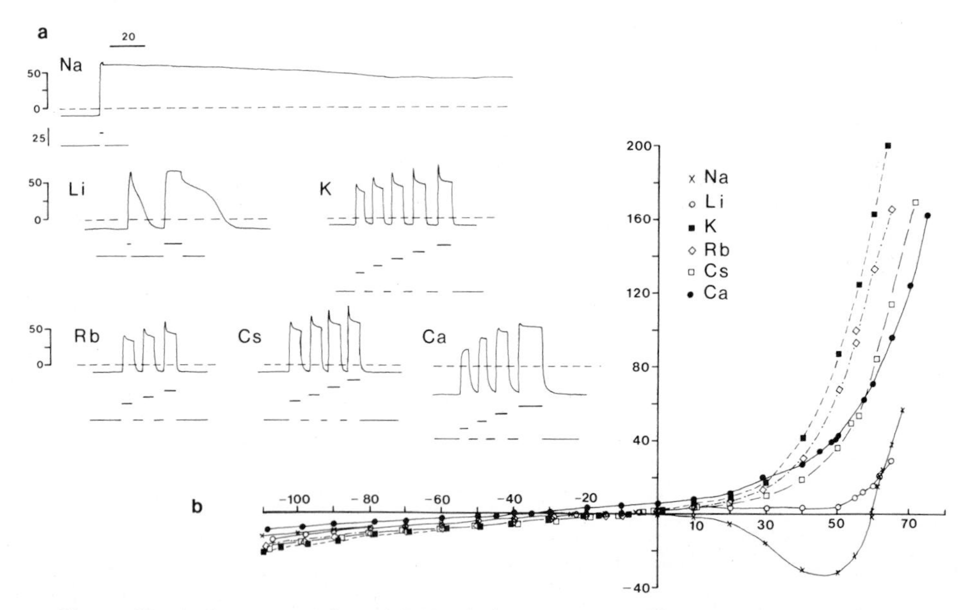

Figure 23. Action potentials and I-V relations corresponding with the currents in Figure 22 (same oocyte and solutions). **a)** Voltage responses to injected current pulses. Only Na^+ and Li^+ produce responses that outlast the current pulse. The time scale bar is 20s, V_m is in mV and current is in nA. **b)** I-V relations measured at the end of each 10s long voltage-clamp pulse Figure 22.

smaller and the reversal potential was +57.5 mV. Note that I-V relations in Figure 24 are not leak-subtracted, hence the I-V relations appear to reverse at less positive values. The ratio of Li^+ permeability to Na^+ permeability can be calculated from the reversal potential as follows:

$$E_{rev} = \frac{RT}{F} \ln \frac{P_{Li}\, Li_o}{P_{Na}\, Na_i} \qquad \textbf{(Eq. 10)}$$

This calculation yielded $P_{Li} / P_{Na} = 0.85$ which was then used to calculate predicted Li^+ currents at each voltage; *i.e.*

$$I_{Li} = g_{Li} (V - E_{rev}) \qquad \textbf{(Eq. 11)}$$

Where $g_{Li} = 0.85\ g_{Na^+}$ as observed and $E_{rev} = 57.5$ mV for Li solution. According to this equation the I-V relation for Li should have been the dashed line in Figure 24. Clearly the calculated Li^+ current predicts more than three times as much maximum inward Li^+ current than observed. This is evidence that Li^+ not only enters but partly blocks the Na^+ channel.

5. Effects of Internal pH

The data in this section are preliminary, but provocative. I hope by including them to stimulate further experiments by other egg electrophysiologists. I will not show figures but will simply describe the findings.

The reader will have noticed that methanesulfonate was used as a Cl^- substitute in all previous experiments. This anion is considered a very good "bystander" anion since it does not affect internal pH (Sharp and Thomas 1981) and neither blocks nor permeates most Cl^- transport pathways, including Cl^- channels. Methanesulfonate, in *Rana* oocytes, appeared to be impermeant, thus it reduced the Cl^- current but did not appear to affect I_{Na} or I_K. In constrast the anion, propionate, had dramatic effects on I_{Na}. When 90% of the Cl_o was replaced

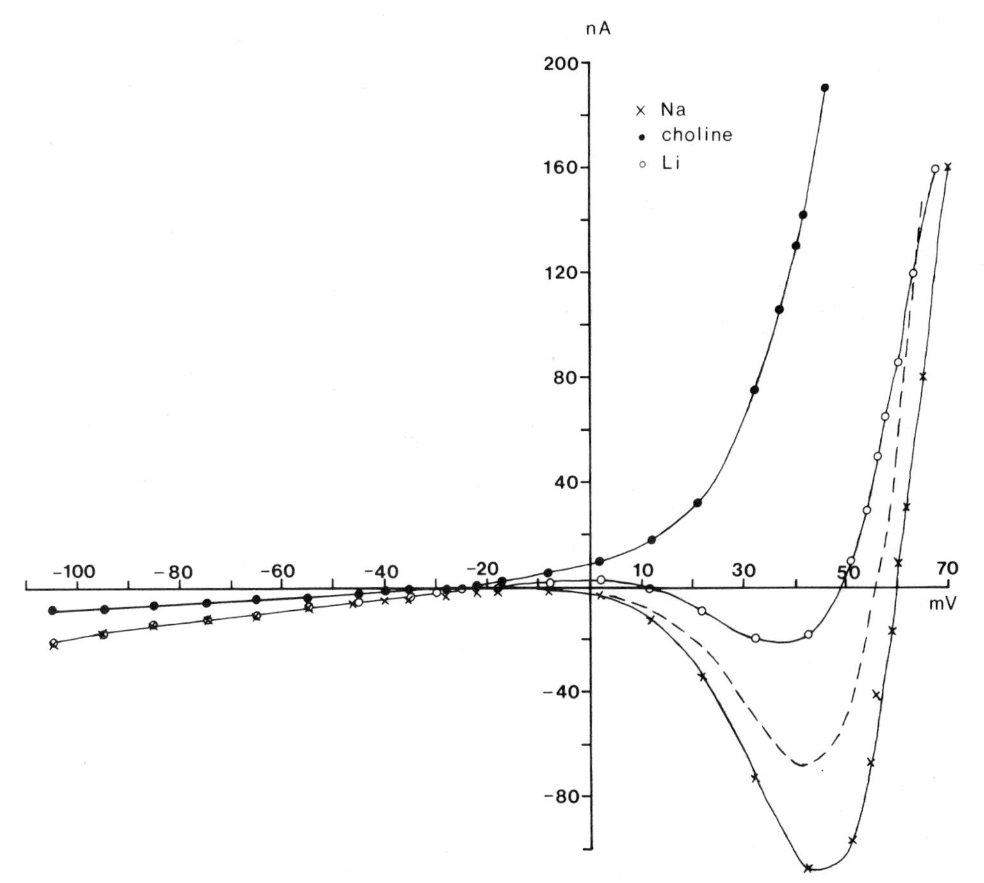

Figure 24. Li^+ partly blocks the Na channel. I-V relations for an oocyte with large Na^+ and Li^+ currents. All solutions contained 400 μM SITS and 20 mM TEA to block I_{Cl} and I_K. Solutions contained 111 mM NaCl (x), or choline Cl (•) or LiCl (○). Dashed curve is the I-V relation expected from the observed change in reversal potential, using a calculated P_{Li}/P_{Na}= 0.85 (see text).

with propionate the Na^+ current became much larger–maximum inward current was sometimes increased ten-fold compared with methanesulfonate-substituted media. The first possibility that comes to mind is that undissociated propionic acid (pK_a 4.87) entered the cell, dissociated and reduced the internal pH. For 100 mM Na^+ propionate at pH 7.8 in the bath, the undissociated acid would be about 0.1 mM. Acid could enter and dissociate to form propionate anions until a Nernst equilibrium was established. At a resting potential of -30 mV, for example, 20-30 mM H^+ ions might be released in this process. Then, depending on the buffering capacity of the cytoplasm, internal pH might be lowered 0.5 to 1 pH unit. It would be best, of course, to measure pH_i with an ion-selective microelectrode to establish the resultant pH_i value.

To test the hypothesis that a lowered pH_i value enhances I_{Na}, I bubbled 100% CO_2 into the medium to lower pH_i rapidly. On the contrary, inward Na^+ current was completely blocked. There remained a small outward current above E_{Na}, but it was much reduced compared with normal Ringer's. This block of I_{Na} is not too surprising since several types of ion channel are blocked by internal H^+ (e.g. Woodhull 1973). However, it leaves unexplained the difference between propionate and CO_2, both of which are expected to lower pH_i. One intriguing possibility is that the Na^+ channel is under very strict pH_i regulation and that the increase in Na^+ conductance observed during maturation is due to an increase in pH_i. It may be that a slight pH_i decrease enhances g_{Na} and a further lowering of pH_i suppresses g_{Na}. It would be interesting to simultaneously affect and monitor pH_i and g_{Na} to test these hypotheses.

I. ELECTRICAL RESPONSES TO FERTILIZATION AND ACTIVATION

Much of this work has been published (Schlichter and Elinson 1981; Jaffe and Schlichter 1985; Jaffe *et al.* 1985); therefore, I will only briefly describe our results. This section will concentrate on technical tips for making recordings during fertilization and activation and on the possible relationships between the ion channels underlying the action potentials and fertilization potentials.

1. Technical Tips

For fertilizing immature *Rana* oocytes the technique developed by Elinson (1977) to obtain immature, jelly-coated oocytes was extremely useful. A high success rate of sperm entry during electrical recordings was obtained with such oocytes. Briefly, a primer dose of macerated pituitaries is injected into the female frog 8h before injection of a full dose of pituitaries plus progesterone. Jelly-coated oocytes could be squeezed from the ovisac as early as 12-13h later. See Schlichter and Elinson (1981) for details and data on fertilization and activation potentials during maturation.

For electrical recordings jelly-coated oocytes were placed in a petri dish then flooded with a concentrated sperm suspension (10^6 to 10^7 sperm/ml). The recording electrode and ground agar bridge were quickly placed in contact with the sperm suspension and the potentials zeroed. One to 2 min were allowed for the sperm to penetrate into the jelly before adding 10% Ringer's to the dish. This was necessary because sperm penetrate swollen jelly less readily. An oocyte was then impaled with one or two microelectrodes and fertilization usually occurred 5 to 15 min after insemination. During the pre-fertilization period there was usually time for the membrane potential to stabilize and for delivery of a sufficient number of voltage-clamp pulses to construct an I-V relation for the resting oocyte.

For voltage-clamp recordings and capacitance measurements the two microelectrodes were similar to those described for voltage clamping in Section C2 (see also Jaffe and Schlichter 1985). Membrane capacitance was measured by applying an alternating current (AC) of 10^{-7} A at 400 Hz. If the frequency is sufficiently high, essentially all of this current goes through the membrane capacitance instead of the membrane resistance. The AC voltage produced across the membrane is inversely proportional to membrane capacitance, which is directly related to membrane area. For *Rana* oocytes we found that below about 400 Hz the membrane resistance became an increasingly significant parallel pathway for current, and that above this frequency AC coupling between the current and voltage electrodes became significant. For details of this technique see the Chapter by R.T Kado in this book. For specifics of our experiments see Jaffe and Schlichter (1985).

2. Fertilization and Activation Potentials During Maturation

Each immature oocyte exhibited one or more transient depolarizations such as those in Figure 25a,b. By counting the number of sperm nuclei inside or sperm entry sites on the surface, we found an excellent correlation between the number of large depolarizations and the number of sperm entries. We believe that each penetrating sperm elicits a transient depolarization in the most immature oocytes (Schlichter and Elinson 1981). As the oocyte matures there are fewer sperm entries and fewer transient depolarizations and ultimately a full fertilization potential occurs; either after one or more transients as in Figure 25b or without transients as in Figure 25c.

Unlike fertilization, the electrical response of immature oocytes to artificial activation by pricking or by the Ca ionophore, A23187, was not a transient depolarization. Pricking produced a rapid step-like depolarization that looked like the result of damage and increased leakiness. A23187 produced a sustained depolarization that resembled the activation potential

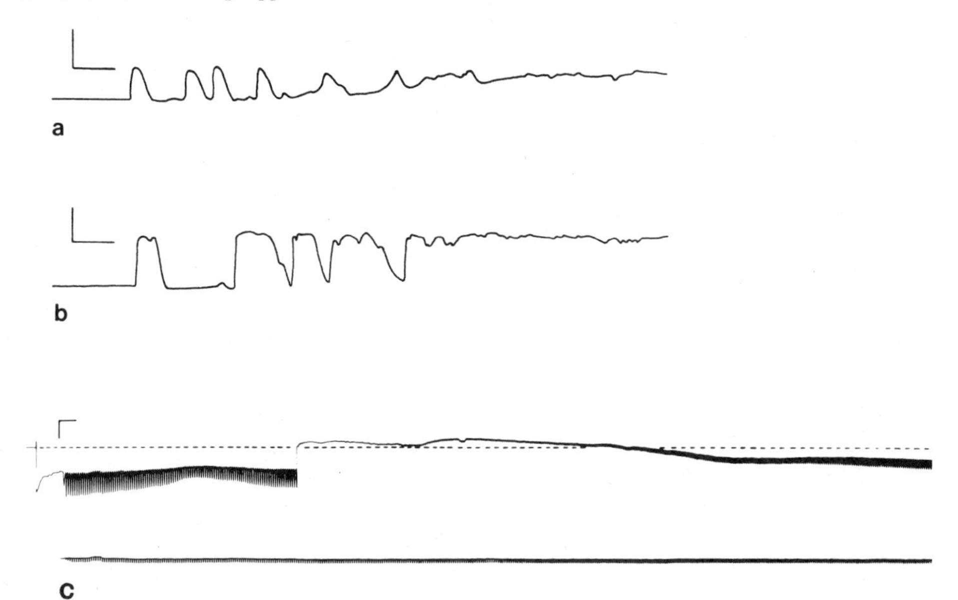

Figure 25. Fertilization potentials recorded in 10% Ringer's solution. **a)** Young (metaphase I) jelly-coated oocyte which had 4 large depolarizations and 4 sperm entries. Calibration bar 20 mV, 20s and the horizontal bar indicates 0 mV. **b)** Immature oocyte with 4 transient depolarizations and a full fertilization potential. There were five sperm entry sites. **c)** Fully mature oocyte showing impalement, some depolarization due to damage, a moderately high resistance (as measured by short current pulses) and a rapid fertilization potential with concomitant decrease in resistance. As the membrane repolarized the resistance increased.

of mature oocytes produced by A23187, by pricking or by sperm. Figure 26 a shows a typical activation potential produced by inserting a second microelectrode into a mature oocyte. Before activation V_m gradually hyperpolarized and R_m increased as the cell healed. Then the second electrode activated the egg and R_m decreased dramatically. After an extended plateau phase, V_m and R_m gradually increased.

3. Channels Underlying the Fertilization and Activation Potential of Mature Eggs

Before discussing the ionic basis of the fertilization and activation potentials throughout maturation, I will deal with the fully mature oocyte (egg).

Most of this work has been published in Jaffe and Schlichter (1985) and in the patch-clamp study by Jaffe *et al.* (1985). We found that the conductance increased transiently upon fertilization and usually decreased within a couple of minutes. This means that studies of the underlying channels must be done during a very short period. As a consequence there is not enough time to change ion solutions or add channel blockers or drugs during a single fertilization response. Comparisons must be made between oocytes bathed in different solutions.

Cross and Elinson (1980) first described the fertilization potential of *Rana pipiens* oocytes and Cross (1981) showed that the amplitude depended on external Cl^-, as predicted for a depolarization due to Cl^- efflux. A word on driving forces and directions of ion fluxes during fertilization may be helpful. Unlike the action potential and voltage-clamp studies in the previous sections, fertilization and activation studies were done in 10% Ringer's containing (in mM): NaCl 11.1, KCl 0.19, $CaCl_2$ 0.11, $MgSO_4$ 0.08, NaOH 0.4, HEPES 0.25, pH 7.8. This means the Nernst potentials for Na^+, K^+ and Cl^- are quite different from those in 100% Ringer's. Using ion-selective microelectrodes we found that the mature *Rana pipiens* oocyte had 121 mM K^+ and 44 mM Cl^- inside (Jaffe and Schlichter 1985). My action potential studies

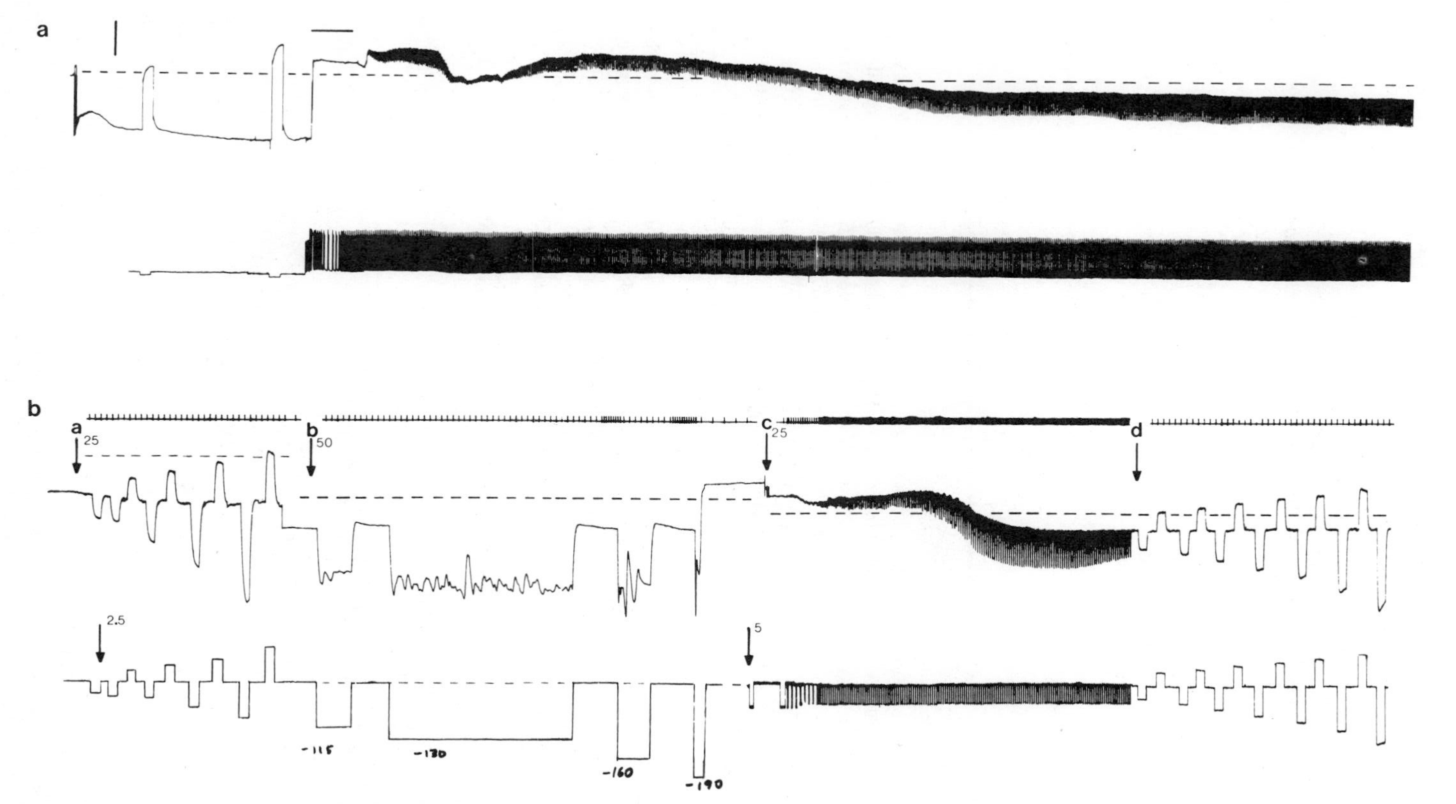

Figure 26. Activation potentials of mature *Rana* oocytes. **a)** Activation by impalement with a second microelectrode at arrow. Pulses of current were injected thereafter to monitor resistance. Scale bars 25 mV, 2 min; broken line indicates 0 mV. **b)** Activation by hyperpolarization. Egg was impaled with two microelectrodes and current pulses injected to construct an I-V relation (between a and b). At arrow marked b the egg was voltage clamped. Then, strongly hyperpolarizing pulses caused membrane instability and finally, activation. At arrow marked c the voltage clamp was switched to current clamp and current pulses were injected to monitor R_m and construct an I-V relation. Upper trace is voltage (mV) and lower trace is injected current (nA). Numbers beside each arrow are scale markings for voltage and current.

indicated that internal Na^+ is about 23 mM in the mature egg (Schlichter 1983 b). The corresponding Nernst potentials in 10% Ringer's are about: E_{Na} -17 mV; E_K -163 mV; E_{Cl} +34 mV. Therefore, any K^+ conductance will tend to hyperpolarize the egg, a Cl^- conductance will depolarize the cell toward +34 mV and a Na^+ conductance will drive V_m toward -17 mV.

4. Technical Tips

To artificially activate a mature egg during voltage clamping we found the easiest and most consistent method to be briefly clamping V_m to a very negative level e.g. -180 to -200 mV). Figure 26b shows such an activation. Pulses to -115, -130 and -160 mV produced membrane instability then a pulse to -190 mV activated the egg. V_m went positive abruptly and R_m decreased dramatically, as is typical of either fertilization or activation. The advantages of this method of activation are that a lot less damage (leakage) is inflicted than by pricking the egg and no mechanical movement occurs which might damage or dislodge one of the electrodes. Hyperpolarization-activation is a reliable and relatively gentle technique.

5. K^+ Conductance and Cl^- Conductance

We found that fertilization or activation involves two separate conductances; g_K and g_{Cl}. The fertilization current reversed between E_K and E_{Cl}, although much closer to E_{Cl}. The voltage-dependent Na^+ conductance present before fertilization (underlying the action potential) was much smaller than g_K and g_{Cl} and therefore contributed little to the fertilization potential. g_{Na} disappeared within minutes after fertilization. Presumably g_{Cl} is activated by a rise in internal Ca^{2+} since Cross (1981) found that Ca^{2+} injection produced a normal activation potential in the *Rana* egg. Tetraethylammonium (TEA), which blocked g_K in immature oocytes, also blocked the g_K elicited by fertilization/activation. TEA treatment revealed the Cl^- conductance activated by fertilization/activation. Like the g_{Cl} of immature oocytes it increased with voltage. Both g_K and g_{Cl} increased transiently after activation/fertilization and with a similar time course that produced the 1 to 2 min conductance increase previously mentioned.

We were interested in whether cortical granule exocytosis might insert new membrane containing these Cl^- and K^+ channels. Therefore we monitored the time course of exocytosis by continuously measuring membrane capacitance. By simultaneously monitoring the rise and fall of the conductance change we determined that exocytosis occurred after the channels opened maximally, hence the K^+ and Cl^- channels were already present in the plasma membrane before exocytosis.

Subsequently Jaffe and her collegues (Jaffe *et al.* 1985) found that Cl^- and K^+ channel opening propagates around the *Rana* egg from the animal to the vegetal region after fertilization/activation. These channels were very likely those described above; g_K was blocked by TEA and increased with increasing voltage, the same as g_K in the immature oocyte. The wave of K^+ and Cl^- channel opening might follow a wave of internal Ca^{2+} as observed in *Xenopus* eggs after fertilization (Busa and Nuccitelli 1985).

6. Relation Between Channels Underlying Action Potentials and Fertilization/Activation Potentials

As shown in the sections on ionic currents in the maturing oocyte, the youngest oocytes tested have a large K^+ conductance that is probably Ca^{2+}-dependent as well as activated by depolarization. This g_K disappears early during maturation, perhaps owing to a decrease in intracellular free Ca^{2+}. These young oocytes also possess a Cl^- conductance that may be Ca^{2+}-dependent and is also activated by depolarization. During maturation g_{Cl} also disappears, but later than g_K. Both conductances act to repolarize the membrane and end the action potential. Throughout maturation there is a voltage-dependent Na^+ conductance (g_{Na}, which may increase with maturation) and causes the upstroke of the action potential.

In 10% Ringer's solution opening the Cl^- channels would depolarize the oocyte ($E_{Cl} \sim +34$ mV) and opening Na^+ or K^+ channels would hyperpolarize the cell ($E_{Na} \sim -17$ mV; $E_K \sim -163$

mV). The electrical responses to fertilization of immature oocytes can be easily accounted for by these three channel types. I propose the following model: **(1)** In the youngest oocytes, sperm entry releases some Ca^{2+} internally and g_{Cl} is activated, producing a depolarization. But g_K is also activated such that the amplitude of the depolarization is small and V_m rapidly repolarizes. g_{Na} may aid in this repolarization. This is just the type of V_m change seen in the youngest oocytes (e.g. Figure 25a). **(2)** Slightly older oocytes have a lower g_K, therefore g_{Cl} produces a larger depolarization which also takes longer to repolarize. This is the pattern seen in "middle-aged" oocytes (e.g. Figure 25b). Possibly g_{Na} is the only repolarizing conductance at this stage. **(3)** Mature oocytes have lost all g_K and g_{Cl} and sperm entry reintroduces propagating g_K and g_{Cl}, probably owing to a propagating wave of Ca^{2+}. In order for the depolarization to last the many minutes (e.g. Figure 25c, 26) typical of fertilization potentials, g_{Cl} must dominate over g_K and g_{Na}.

Evidence that I_K is the same channel in immature and activating oocytes is: both are apparently Ca^{2+} dependent, activated by depolarization and blocked by external TEA. Evidence that I_{Cl} is the same channel is that both are apparently Ca^{2+} dependent and activated by depolarization. It would be very interesting to determine if the K^+ and Cl^- channels represent the same population in immature and activating oocytes. The difficulty will be the very short duration of channel opening during activation.

One prediction would be that blocking g_K with TEA would prolong the transient depolarization of immature oocytes. If a specific blocker of I_{Na} were found, even the youngest oocyte should then produce a long fertilization potential in the presence of TEA and the g_{Na} blocker. One would predict such oocytes to become monospermic regardless of their stage of maturity.

J. SPECULATIONS AND SUGGESTIONS FOR FURTHER WORK

1. Changes in Channel Activity During Maturation and Fertilization.

The following pattern of changes in channel expression (or ability to be opened) was characteristic of maturation.

a. The youngest oocytes tested (metaphase I) had a large g_K and g_{Cl} and a moderate g_{Na}, all of which were voltage dependent. In addition there was a Na^+ leakage conductance even at quite hyperpolarized potentials. The probable consequences of the interplay of these several conductances are: V_m stays negative and is probably determined by P_K / P_{Na}; V_m does not reach threshold for regenerative opening of the voltage-gated Na^+ channels, hence action potentials do not spontaneously fire; the membrane resistance is relatively low compared with mature oocytes; Na^+ probably enters through tonically activated channels and K^+ probably leaves, thereby changing Na_i^+ and K_i^+ unless Na^+/K^+ pump activity is present and compensating; and even though Cl^- channels are present they may not open at the resting potential, hence Cl^- fluxes may be small.

b. Slightly later during maturation g_K decreases as does the Na^+ leakage conductance. Probable consequences are: P_K and P_{Na} should decrease at rest; V_m may become less negative; spontaneous action potentials will begin if g_K is not large enough to "clamp" V_m negative; and P_K/P_{Na} will dramatically decrease, as has been observed (Ziegler and Morrill 1977). The probable interplay of conductances during these action potentials is: V_m depolarizes gradually driven by Na^+ influx through leakage channels; then voltage-dependent Na^+ channels open regeneratively, firing an action potential; during this depolarization voltage-gated K^+ and Cl^- channels open and V_m repolarizes; Cl^- and Na^+ channels close rapidly and V_m undershoots due to a large g_K; K^+ channels close more slowly and V_m slowly depolarizes again to the threshold for firing another action potential.

This sequence of channel opening and closing qualitatively accounts for rhythmic firing and is reasonable, based on the time- and voltage-dependence of I_{Na}, I_K and I_{Cl} as far as I studied them. However, to completely reconstruct such repetitive firing one would need to write a computer program that takes into account the voltage-and time-dependence of the K, Cl^- and Na^+ channels and the type and amount of leakage conductances. With such an approach Connor and Stevens (1971) were able to reconstruct the firing pattern of a molluscan nerve cell.

c. The next change during maturation was a gradual decrease in g_{Cl}. Action potentials continued to fire spontaneously but several changes took place: because g_K had decreased, the after- hyperpolarization was much less negative; the time between spikes decreased since Cl^- channels closed quickly (the previous long after-hyperpolarization was probably slow closing of K channels); each action potential lasted longer since only g_{Cl} was present to repolarize the membrane and this produced a gradual repolarization phase (plateau) which preceded the rapid repolarization brought on by voltage-dependent closing of Na^+ channels; and the spikes got longer as g_{Cl} decreased further.

There is considerable circumstantial evidence that both g_K and g_{Cl} are also Ca^{2+} dependent: both exhibited bell-shaped I-V relations; both currents increased in response to raising external Ca^{2+}; and Ca^{2+}-dependent Cl^- currents have been found in immature *Xenopus* ooyctes (Robinson 1979; Barish 1983). The simplest explanation for g_K and g_{Cl} decreasing with maturation would be a gradual decrease in free intracellular Ca^{2+} and it would be worth comparing g_K and g_{Cl} with Ca^{2+} throughout maturation. Alternatively one could test for the continued presence of K^+ and Cl^- channels by raising Ca_o^{2+} or Ca_i^{2+} (with ionomycin or A23187); or attempt to reduce g_K and g_{Cl} by preincubation in a Ca^{2+}-free medium or the Ca^{2+} buffer BAPTA-AM.

d. With the loss of g_K and dramatic reduction of g_{Cl}, at one stage V_m remained positive for many minutes because g_{Na} was larger than g_K, g_{Cl} and g_{leak}. With g_{Na} dominating, V_m remained near E_{Na}, but only because the Na^+ channels do not inactivate.

e. Finally, g_{Na} decreased such that V_m returned negative. The leakage conductances also decreased markedly, hence the total R_m increased considerably. It is at this stage (metaphase II) that the oocyte is ready to be ovulated into pond water and fertilized. There are several reasons why it would be advantageous to have a high R_m (few channels open) at this stage: pond water is of very low osmolarity and ions would diffuse out of the cell rapidly if channels were open; a "tight" membrane would reduce water uptake and osmotic swelling of the oocyte; and a high R_m would result in a faster change in membrane potential when sperm caused ion channels to open.

f. In response to fertilization the membrane rapidly depolarizes, due to opening of Cl^- channels. The amplitude is moderated by the opening of K channels, and to a lesser extent by Na^+ channels. (Recall that E_{Cl} is positive and E_K and E_{Na} negative in 10% Ringer's or pond water.) I suspect these are the same K^+ and Cl^- channels that contribute to the action potential of immature oocytes. To prove this point will be difficult since the fertilization-induced opening is very brief. Therefore, detailed studies of the voltage and time dependence and pharmacology of the channels will be difficult unless a method of opening the K and Cl^- channels without activating the egg is found. The fertilization-induced channels are probably Ca^{2+} dependent, since the fertilization potential can be induced by Ca^{2+} injection (Cross 1981) and by A23187 (Schlichter and Elinson 1981). The Cl^- and K^+ channel opening probably propagates around the oocyte from the site of sperm entry following a wave of Ca^{2+} (c.f. Jaffe *et al.* 1985).

2. Other Areas for Further Study

The possibility that all the changes in activity of K^+ and Cl^- channels simply reflect internal free Ca^{2+} has been addressed. There are probably other changes in the oocyte that can regulate channel expression; *i.e.*, the changes in g_{Na} and leakage conductances and loss of channels after fertilization. Because of the indications that g_{Na} is susceptible to changes in pH_i and external CO_2 it would be interesting to directly study g_{Na} (and g_K, g_{Cl}) as a function of pH_i. Since internal Ca^{2+} and pH are two physiologically relevant modulators of oocyte maturation and fertilization, and very commonly affect ion-channel function, they would perhaps be the first agents to test as modulators of the channels.

Other second messengers and factors that are known to be involved in maturation and fertilization are also interesting possible modulators of channel activity; e.g. cyclic nucleotides, inositol phosphates (especially $InsP_3$), GTP-binding proteins, and even Maturation Promoting

Factor (MPF). With the exception of MPF, all of these substances regulate ion channels in some types of cells and are therefore reasonable candidates for controlling ion-channel activity in the oocyte.

Regardless of what roles the ion channels play in maturation and fertilization, they are inherently interesting. For example, the Na^+ channel is distinctly different from most other voltage-gated Na^+ channels in lacking inactivation and lacking sensitivity to TTX. From the point of view of molecular biological studies of structure-versus-function of ion channels, it would be interesting to know if two distinct regions are missing from the DNA encoding the protein; *i.e.*, one region corresponding with the inactivation "gate" and one with the TTX-binding site. Could this Na^+ channel be a precursor of epithelial Na^+ channels in the mature frog? Could the Na^+ channel change its selectivity and become a Ca^{2+} channel (L or N type) in the mature animal? Is the Na^+ channel (which is blocked by Co^{2+}) sensitive to organic Ca^{2+} channel blockers? Is the Cl^- channel the precursor of epithelial Cl^- channels which may be involved in fluid or mucus secretion? What type of Ca^{2+}-dependent K^+ channels are present in the oocyte? Are these channels all present merely as precursors to channels in later embryos or mature animals or are they unique to oocytes?

I hope that this description of electrophysiological properties of the *Rana pipiens* oocyte will stimulate further research on frog egg channels and their roles and regulation.

ACKNOWLEDGEMENTS

I have benefitted in this work from stimulating discussions with many people, especially Drs. R. P. Elinson, J. Dainty, L. A. Jaffe and R. T. Kado. My original work presented here was supported by grants from the Natural Sciences and Engineering Research Council (Canada), the N.I.H., and the Medical Research Council (Canada).

RECOMMENDED READING

1. Membrane Transport

Andreoli, T. E., J. F. Hoffman, and D. D. Fanestil. 1980. Membrane Physiology. Plenum Medical, New York.

Stein, W. D. 1986. Transport and Diffusion Across Cell Membranes. Academic Press, New York.

2. Electrophysiology Techniques

Sakmann, B. and E. Neher. 1983. Single-Channel Recording. Plenum Press, New York.

Smith, T. G., H. Lecar, S. J. Redman and P. W. Gage. 1985. Voltage and Patch Clamping with Microelectrodes. American Physiological Society, Maryland.

Thomas, R. C. 1978. Ion-Sensitive Intracellular Microelectrodes. Academic Press, New York.

3. Ion Channels and Action Potentials

Hille, B. 1984. Ionic Channels of Excitable Membranes. Sinauer, Massachusetts.

Junge, D. 1981. Nerve and Muscle Excitation. Sinauer, Massachusetts.

REFERENCES

Barish, M. E. 1983. A transient calcium-dependent chloride current in the immature *Xenopus oocyte. J. Physiol.* 342:309-325.

Bellé, R., R. Ozon and J. Stinnakre. 1977. Free calcium in full grown *Xenopus laevis* oocyte following treatment with ionophore A23187 or progesterone. *Mol. Cell. Endocrinol.* 8:65-72.

Busa, W.B. and R. Nuccitelli. 1985. An elevated free cytosolic Ca^{2+} wave follows fertilization in eggs of the frog *Xenopus laevis*. *J. Cell Biol.* 100:1325-1329.

Connor, J. A. and C. F. Stevens. 1971. Inward and delayed outward membrane currents in isolated neural somata under voltage clamp. *J. Physiol.* 213:1-19.

Cross, N. L. 1981. Initiation of the activation potential by an increase in intracellular calcium in eggs of the frog, *Rana pipiens*. *Dev. Biol.* 85:380-384.

Cross, N. L. and R. P. Elinson. 1980. A fast block to polyspermy in frogs mediated by changes in the membrane potential. *Dev. Biol.* 75:187-198.

Dascal, N., T.P. Snutch, H. Lubbert, N. Davidson, and H. A. Lester. 1986. Expression and modulation of voltage-gated calcium channels after RNA injection in *Xenopus* oocytes. *Science* 231:1147-1150.

Ecker, R. E. and L. D. Smith. 1971. Influence of exogenous ions on the events of maturation in *Rana pipiens* oocytes. *J. Cell. Physiol.* 77:61-70.

Elinson, R. P. 1977. Fertilization of immature frog eggs: cleavage and development following subsequent activation. *J. Embryol. Exp. Morphol.* 37:187-201.

Hagiwara, S., S. Ozawa, and O. Sand. 1975. Voltage clamp analysis of two inward current mechanisms in the egg cell membrane of a starfish. *J. Gen. Physiol.* 65:617-644.

Houle, J. G. and W. J. Wasserman. 1983. Intracellular pH plays a role in regulating protein synthesis in *Xenopus oocytes*. *Dev. Biol.* 97:302-312.

Ito, S. 1972. Effects of media of different ionic composition on the activation potential of anuran egg cells. *Dev. Growth & Differ.* 14:217-227.

Jacobson, S. L. and G. A. R. Mealing. 1980. A method for producing very low resistance micropipettes for intracellular measurements. *Electroencephalogr. Clin. Neurophysiol.* 48:106-108.

Jaffe, L. A. and L. C. Schlichter. 1985. Fertilization-induced ionic conductances in eggs of the frog, *Rana pipiens*. *J. Physiol.* 358:299-319.

Jaffe, L. A., R. T. Kado, and L. Muncy. 1985. Propagating potassium and chloride conductances during activation and fertilization of the egg of the frog, *Rana pipiens*. *J. Physiol.* 368:227-242.

Kado, R. T., K. Marcher, and R. Ozon. 1979. Mise en evidence d'une depolarisation de longue duree dans l'ovocyte de *Xenopus laevis*. *C. R. Hebd. Seances Acad. Sci. Ser. D. Sci. Nat.* 288:1187-1189.

Knauf, P. A. and A. Rothstein. 1971. Chemical modification of membranes. I. Effects of sulfhydryl and amino reactive reagents on anion and cation permeability of the human red blood cell. *J. Gen. Physiol.* 58:190-210.

Maeno, T. 1959. Electrical characteristics and activation potential of *Bufo* eggs. *J. Gen. Physiol.* 43:139-157.

Morrill, G. A., A. B. Kostellow, and J. B. Murphy. 1971. Sequential forms of ATPase activity correlated with changes in cation binding and membrane potential from meiosis to first cleavage in *R. pipiens*. *Exp. Cell Res.* 66:289-298.

Morrill, G. A., J. Rosenthal, and D. E. Watson. 1966. Membrane permeability changes in amphibian eggs at ovulation. *J. Cell. Physiol.* 67:375-382.

Morrill, G. A. and D. E. Watson. 1966. Transmembrane electropotential changes in amphibian eggs at ovulation, activation and first cleavage. *J. Cell. Physiol.* 67:85-92.

Morrill, G. A., D. Ziegler, and V. S. Zabrenetzky. 1977. An analysis of transport, exchange, and binding of sodium and potassium in isolated amphibian follicles and denuded oocytes. *J. Cell Sci.* 26:311-322.

Robinson, K. R. 1979. Electrical currents through full-grown and maturing *Xenopus* oocytes. *Proc. Natl. Acad. Sci. USA* 76:837-841.

Schlichter, L. C. 1983a. Spontaneous action potentials produced by Na and Cl channels in maturing *Rana pipiens* oocytes. *Dev. Biol.* 98:47-59.

Schlichter, L. C. 1983b. A role for action potentials in maturing *Rana pipiens* oocytes. *Dev. Biol.* 98:60-69.

Schlichter, L. C. 1989. Ionic currents underlying the action potential of *Rana pipiens* oocytes. *Dev. Biol.* (in press).

Schlichter, L. C. and R. P. Elinson. 1981. Electrical responses of immature and mature *Rana pipiens* oocytes to sperm and other activating stimuli. *Dev. Biol.* 83:33-41.

Sharp, A. P. and R. C. Thomas. 1981. The effects of chloride substitution on intracellular pH in crab muscle. *J. Physiol.* 312:71-80.

Simoncini, L., M. L. Block, and W. J. Moody. 1988. Lineage-specific development of calcium currents during embryogenesis. *Science* 242:1572-1575.

Thomas. R. C. 1976. Ionic mechanism of the H^+ pump in a snail neurone. *Nature (Lond.)* 262:54-55.

Thompson, S. H. and R. W. Aldrich. 1980. Membrane potassium channels, p. 49-85. *In: The Cell Surface and Neuronal Function*, Vol. 6. C. W. Cotman, G. Poste and G. L. Nicolson (Eds.). Elsevier North-Holland, Amsterdam.

Wasserman, W. J.,L. H. Pinto, C. M. O'Connor, and L. D. Smith. 1980. Progesterone induces a rapid increase in $[Ca^{2+}]_{in}$ of *Xenopus laevis* oocytes. *Proc. Natl. Acad. Sci. USA* 77:1534-1536.

White, M. M. and C. Miller. 1979. A voltage-gated anion channel from the electric organ of *Torpedo californica. J. Biol. Chem.* 254:10161-10166.

Woodhull, A. 1973. Ionic blockage of sodium channels in nerve. *J. Gen. Physiol.* 61:687-708.

Ziegler, D. and G. A. Morrill. 1977. Regulation of the amphibian oocyte plasma membrane ion permeability by cytoplasmic factors during the first meiotic division. *Dev. Biol.* 60:318-325.

ELECTRICAL CAPACITANCE AND MEMBRANE AREA

Raymond T. Kado

Centre National de la Recherche Scientifique
Laboratoire de Neurobiologie Cellulaire et Moleculaire
Gif-sur-YVETTE, 91198 FRANCE

> "There seems, however, to be room in the fish for a battery of a sufficient size; for Mr. Hunter* has shown, that each of the prismatical columns of which the electrical organ is composed, is divided in a great number of partitions of fine membranes......and if the glass is five times as thin, which is perhaps not thinner than the membranes which form the partitions, it will contain five times as much electricity, or near fourteen times as much as my battery."

From: An account of some attempts to imitate the effects of the torpedo by electricity: By the Hon. HENRY CAVENDISH, F.R.S. *Philosophical Transactions* for 1776, Vol. LXVI Part I pp. 196-225. Art. 437.

*John Hunter, F.R.S. 1773. Anatomical observations on the Torpedo. Phil. Trans. LXIII. p.485. See Art. 614.

INTRODUCTION

Almost 200 years of scientific inquiry were necessary to understand that the electric skate produced its powerful electrical discharge by summing the membrane depolarizations of the cells in each prism in series and having all the prisms in parallel. Cavendishs' "battery" consisted of 40 Leyden jars, the only means for storing electricity at that time, made of especially thin glass and should have been capable of holding an appreciable charge. If we could know the capacitance of his jars, we would know how much charge was stored by the simple relation:

$$Q = C \times V \qquad \textbf{Eq.1}$$

where Q in coulombs is the charge stored in a capacitance of C Farads by the voltage V. If one volt of electrical force can store one coulomb of charge on a pair of conductors, the capacitance between the conductors is one FARAD by definition. Unfortunately, the laws of current and voltage had not yet been invented at the time Cavendish made his suggestion. The laws as we know them today were to come of such experiments as this one of Cavendish and those of Galvani and Volta (see also De Felice 1981). The membrane of the cell was already providing the intellectual link to understanding this strange force which we now know underlies the structure of all matter.

Capacitance: Its Principles

As the early workers in electricity came to understand that the electrical force can move charges through conductors, they defined the force in terms of Volts and the flow of charges as a current in Amperes. With the invention of better means to measure current, laws such as the law of Ohm were established and these provided the means of quantitatively describing the behavior of electrical circuits. Conductors and insulators were identified and classified as current carriers and non current carriers. However one electrical phenomenon posed a problem. If a conductor were cut somewhere in its path and the ends connected to metal plates facing each other and separated by a small space, a current could be shown to be present in the conductor when a voltage was applied to the ends. This was contradictory to the understanding that a space was an insulator through which, by definition, no current could pass. It took a while before Maxwell called this current a displacement current, to distinguish it from the currents normally found in a conductor carrying charges and to call two conductors separated by an insulator, a capacitor.

What is the Displacement Current?

Consider the circuit shown in Figure 1A. The two plates and the conductors connected to them all contain the same number of free charges which are uniformly distributed everywhere. These charges will be made to move if a voltage were applied as if the circuit consisted of one continuous conductor. In fact, at the instant the switch SW is closed in Figure 1A, the current is just equal to the voltage divided by the resistance in the circuit. In a resistor, the current is established as rapidly as the applied voltage as far as we know, taking place at the speed of light. The arrival of charges on one plate causes exactly the same number of charges to leave the other. It is the rate of movement of these charges which is the displacement current.

$$i_C = dQ/dt$$

and from Equation 1: **Eq. 2**

$$i_C = C \times dV/dt$$

As charges continue to move into one side and out of the other of the capacitor, their numbers will become more and more unequal. The difference in quantity of charge will result in a voltage across the capacitor of the same polarity as the voltage source applied. At some time the two voltages should become equal and reduce charge movement to zero, this never happens. The reason is simple, as the difference in voltage between the source and the capacitor becomes smaller, the force for moving charge also becomes smaller thus reducing

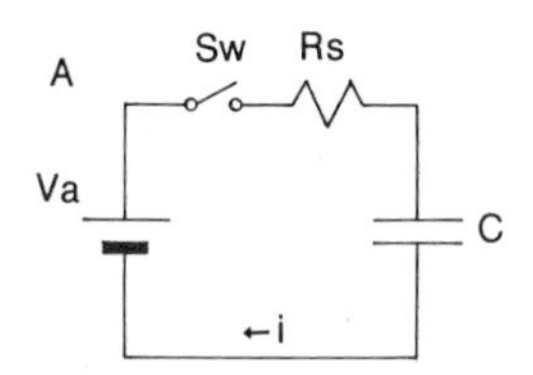

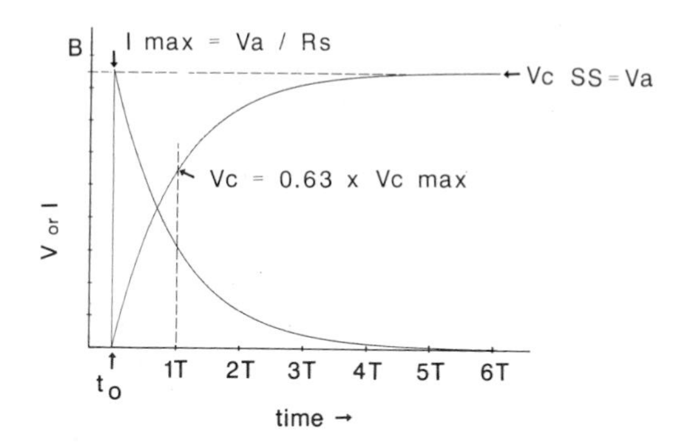

Figure 1. The basic R-C circuit. **A.** A battery (V_a), a switch (Sw), a resistor (R_s) and a capacitor (C) are shown connected in series. **B.** When Sw is closed, a current I_{max} is instantly established as shown at t_o. This current decrements as explained in the text and the voltage on the capacitance, V_C increases towards $V_{Css} = V_a$. Note that the scaling in this figure is such that I_{max} and V_{Css} fall on the same amplitude, this is just a result of programming convenience. 1T to 6T indicate the time in the number of time constants since t_o.

the rate of charging the capacitor. It would take an infinitely long time for all charge movement to cease. Just as in any other process where the quantity of accumulated product limits the rate of its production, the displacement current (i_C) diminishes as an exponential from its instantaneous maximum value, I_m (Figure 1B) as:

$$i_C = I_m \times \exp(-t/\tau) \quad \textbf{Eq. 3}$$

t, is the time in seconds from the instant of connecting the voltage source and τ is a constant called the time constant.

The Time Constant

The time constant in an R-C circuit is determined by the capacitance and resistance in the circuit as:

$$\tau = RC \quad \textbf{Eq. 4}$$

with R in Ohms and C in Farads. The unit for τ is seconds from (volts/coul/sec) (coul/volt.)

In the capacitor, we see that the current is established very rapidly, and falls off with time. The voltage across the capacitor increases proportionately with the accumulated charge and mirrors the fall in current. The voltage across the capacitor at any instant of time t can be found by:

$$v_C = V_m (1-\exp(-t/\tau)) \quad \textbf{Eq. 5}$$

Where V_m is the maximum voltage which will appear on the capacitor after a long enough time. V_m for all practical purposes is the same as V_a. Of course it is also possible to know what the voltage will be at any time if we know the rate of charge movement (the current) and the capacitance.

Charge storage in a capacitor involves a real displacement of charge as well as the distortion of the orbital electrons in the atoms of the insulator or dielectric between the plates. Because of these properties, strange but explicable things happen if the dielectric were to be changed. A detailed explanation of these effects will not be given here but it should be kept in mind that the capacitance depends on the properties of the dielectric as well as the size and separation of the conducting surfaces. Because of these properties any two conductors having different charges will have a measurable capacitance between them. An example all electrophysiologists know is the capacitance between the recording electrode and other wires in the laboratory carrying power line current. The displacement current in this case, drops a voltage across the resistance of the recording microelectrode and this voltage is seen by the voltage follower (VF) as hum.

Capacitance and Alternating Current (AC)

At time zero (t_O, switch closure) the current in Figure 1B is maximal and the voltage across the capacitor (v_C) is zero. By the time i_C approaches zero, v_C has become maximal. This relation between the current in a capacitor and the voltage across it always holds because of its physical properties. This property leads to an interesting effect when the applied voltage is constantly changing such as in the case of a sinusoidal voltage which is constantly changing in amplitude and periodically changes direction. With a sinusoidal applied voltage to the capacitor in Figure 1A and $R_s = 0$, i_C and v_C will have the relations shown in Figure 2A. Note that i_C is maximal at the times that v_C is zero and vice-versa, they are 90 degrees out of phase. This phase relation is shown vectorially in Figure 2B. The current is said to lead the voltage because the entire vector diagram is rotating counter clockwise around the origin at an angular speed, ω, given by:

$$\omega = 2\pi \text{ x Frequency} \qquad \textbf{Eq. 6}$$

At the instant shown in Figure 2B, v_C is zero and i_C is already maximal, i_C "leads" v_C. The magnitude of v_C or i_C at any instant in time can be found from:

$$v_{C(t)} = v_{max} \sin(\omega t)$$
$$i_{C(t)} = i_{max} \mathrm{Cos}(\omega t) \qquad \textbf{Eq. 7}$$

This holds if there is no other time reference for the sine wave so that the sine wave is its own reference. For sinusoidally alternating currents the properties of the capacitor can be described as a capacitive reactance X_C defined as:

$$X_C = 1/\omega C \qquad \textbf{Eq. 8}$$

Note that X_C varies inversely with both frequency and capacitance and can also be found from Ohm's law, dividing the AC voltage of the capacitor by its AC current. X_C is equivalent to resistance for calculating AC currents and voltages in the capacitor:

$$i_C = v_C / X_C \qquad \textbf{Eq. 9}$$

where both i_C and v_C are taken in the same units, for example the rms value (also known as the effective value, is the one most commonly used as with the 110V line voltage).

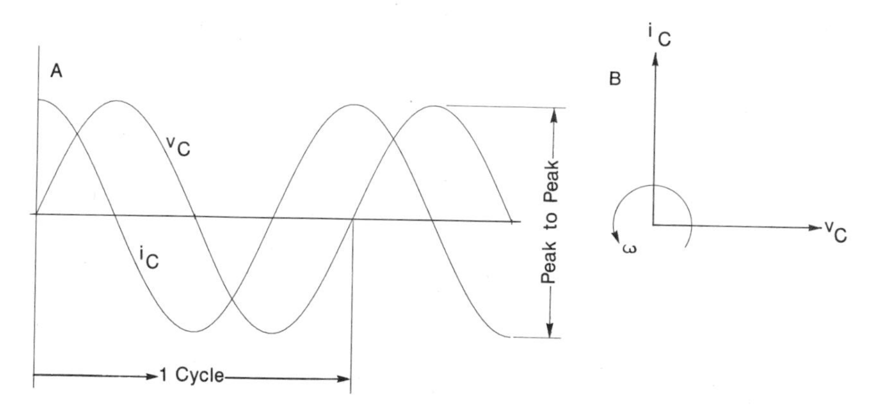

Figure 2. Sinewave voltage and current in a capacitance. **A.** Capacitive current, i_C, is shown leading the voltage V_C in time. The amplitude is indicated as peak-to-peak (pp) with both i_C and V_C scaled equally. The RMS Value is found by:

$$V_{rms} = 0.707 \times V_{pp}/2$$

which says that the 120V line voltage is really 340 V_{pp}. **B.** The vectorial representation of the two sinewaves in A are shown. The vector diagram is actually a "snapshot" taken at the time that V_C was zero and i_C maximal. The diagram rotates at the angular speed, ω as given in Equation 6.

Impedance

In a circuit having both resistance and capacitance as current pathways, the current source must provide a leading current in the capacitance and an in-phase current in the resistance with respect to its voltage. This is the case for Figure 3A. The generator must supply the sum of two orthogonal currents. If the series resistance R_s in Figure 3A is very small or zero, the voltage of the generator will be the same as that across R_m and C_m and the current it supplies will be leading the voltage by an angle determined by the magnitude of the currents in R_m and C_m. This is clearly a case of conductances in parallel (even though X_C is an AC conductance) and the rules for parallel conductances applies. A slight further complication arises because these conductances are also orthogonal so they will add according to Pythagoras' rule. The result of this addition is the impedance Z_m.

$$Z_m = (R_m^{-2} + X_C^{-2})^{-\frac{1}{2}} \qquad \textbf{Eq. 10}$$

Note that this equation calculates Z_m only, it does not take into consideration any other resistances in the circuit. In the case of R and C in series, Z is calculated simply as the vector sum. The over-all impedance of any circuit can be found by dividing the source voltage (v_a) by the supplied current (i_a). The unit for impedance is the ohm.

The phase relations of the voltages and currents in parallel paths are most conveniently illustrated by vector diagrams as in Figure 3B. It is a bit complicated because we are not accustomed to thinking in terms of vector diagrams, let alone those which spin. Diagrams such as these can become very complicated especially if there are many pathways for the current.In spite of the complication in the rest of the circuit, i_C and v_C always retain their orthogonality. The source then must produce a current component which is "imaginary" or orthogonal with its voltage it is producing. This current does not dissipate power so it does not "cost anything" unlike the current component which is in-phase with the voltage. This is a tricky point in AC current analyses with reactive elements. Study carefully what is shown in Figure 3.

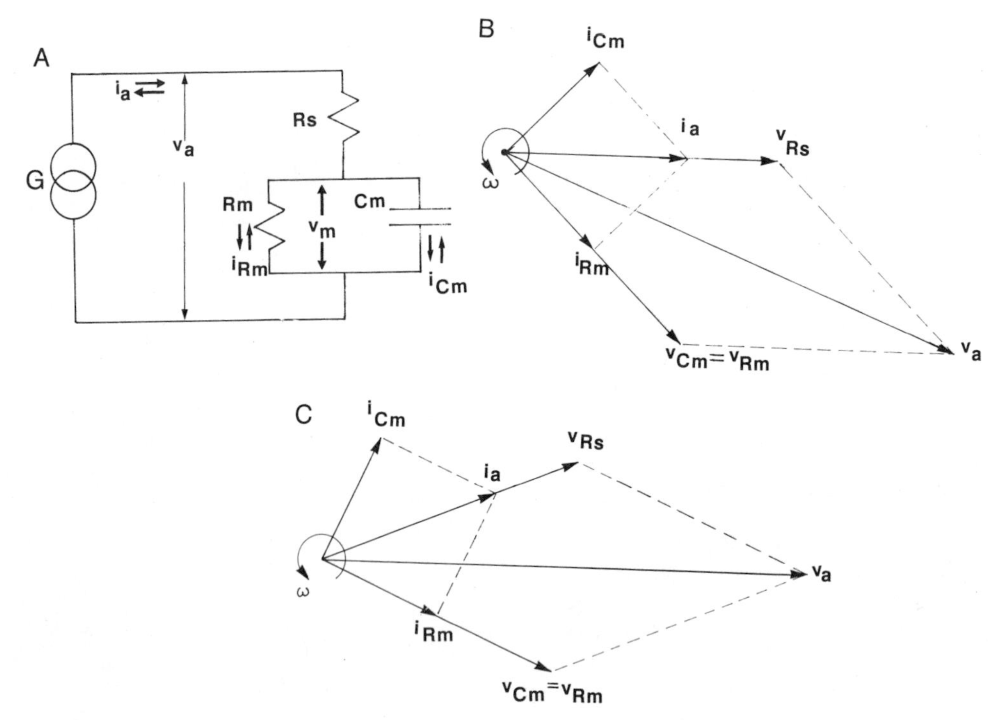

Figure 3. Sinewave currents and voltages for a parallel R-C circuit. The circuit in A is supplied from a generator (G) which could be a constant voltage, low internal resistance unlimited current output, or constant current, high internal resistance, unlimited voltage output source. B and C are the resulting vector diagrams for the phase relationships for all the indicated currents and voltages in the case of constant current source (B) or constant voltage (C). Note: 1. that i_a in B and v_a in C are in the zero phase position, the reference phase. 2. that while i_{Cm} and v_{Cm} are orthogonal with each other, they are no longer at 90 degrees to the reference. 3. that i_{Rm} is in-phase with v_{Cm} which is the same as v_{Rm}. 4. that if C were rotated clockwise (phase lagged) a little, it can be superposed on B.

DC CAPACITANCE MEASUREMENT METHODS

The principles sketched above are the ones which are used to measure the capacity of cell membranes. It is already clear that the capacitance should vary directly with the membrane surface area. Changing membrane area is the same as changing the size of the two conducting surfaces. If the membrane increases in thickness, its capacitance will decrease since the two conductors will be separated by a greater space and vice-versa if the membrane thins. Changes in thickness of the membrane have not yet been reported in cells but thinning of lipid bilayers has been monitored by the increase in capacitance (Niles *et al.* 1985). One way in which the measured capacitance may indicate a decreased dielectric constant is if the measuring current has to traverse two closely apposed membranes. Another question is whether the dielectric (the lipid bilayer of the cell membrane) might not change its properties as proteins are implanted in it or as water and ions enter channels. While these possibilities cannot be eliminated at this time, we tend to believe that they do not significantly change the dielectric properties of the membrane. Since capacitance is due to unequal charge accumulations, changes in the surface charges on the membrane may also affect the measured capacitance. With these conditions and assumptions in mind, we will examine how some capacitance measurements have been done.

Before microelectrodes, membrane capacitance determinations were made on large numbers of eggs in suspension using AC methods (see Cole 1968). Many careful studies showed that as an approximation, the unit cell membrane capacitance is about 1 microfarad per square centimeter. This figure has been more or less confirmed, more or less because it is very difficult to know the exact extent of the cell membrane. Today, microelectrodes can be placed in the cell or its membrane could be sealed onto a patch electrode, allowing direct measurement across the membrane. Here we will be concerned only with measurements made using some kind of electrical connection to the inside of the cell.

The Time Constant Method

The membrane capacitance is determined from measurements made on the recorded membrane voltage change produced by injecting a constant current into the cell. It can also be called the "current-clamp" method. There are three conditions which must be met to obtain valid results. The first is that the membrane potential shift produced by the injected current, does not activate voltage-dependent conductances. The second is that the current pulse producing the voltage change must be sufficiently long to ensure that the membrane capacity has been fully charged. The third, is that all parts of the cell membrane are equally accessible for the injected current.

We make use of the fact that the time constant of the membrane, τ, is just due to R_m and C_m (Equation 4) and that a membrane current, after a long time, will result in a steady state voltage, V_{mss}, equal to R_m x I_a. At the time, t, equaling one time constant, V_m will be just 0.63 V_{mss} (from Equation 5). The procedure is to first determine V_m a long time after I_m started, and has attained steady state (V_{mss}). Multiply V_{mss} by 0.63 and find this value on the V_m trace. The time corresponding to this value of V_m is just equal to one time constant, τ. Now find R_m by dividing V_{mss} by the injected current. Dividing the time, τ, in seconds by the calculated R_m in Ohms gives C_m in Farads. You will find that dividing Ohms by Farads will give the units of time if you use the physical units for these parameters. An example of the voltage responses of the membrane potential to injected constant current pulses and how to calculate C_m are shown in Figure 4. One check for the validity of the V_{mss} and τ is that V_{mss} should occur only after about 6 time constants or 6τ seconds after the current started (see Figure 1).

A confusing point here may be raised by the fact that the membrane is a parallel circuit as in Figure 3A, yet the time constant is calculated in exactly the same way as for the series case shown in Figure 1A. There is no contradiction in fact. In Figure 1A, a voltage source of

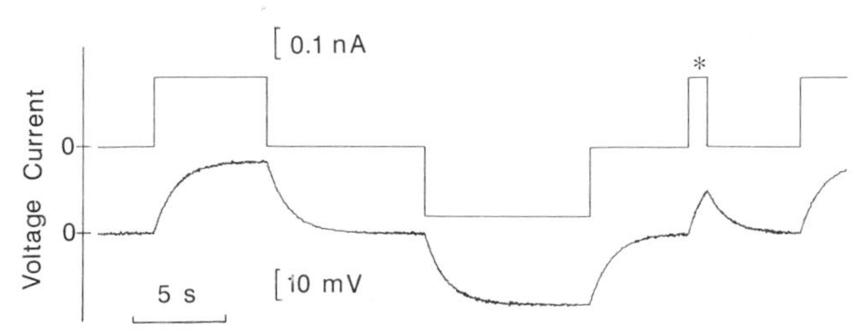

Figure 4. Parallel R-C voltage responses to injected constant currents. These traces are the current and voltage recorded with a digital oscilloscope and plotted with an X-Y plotter. The current was injected with a VF (Biodyne* AM-2) and recorded with a second VF both electrodes were 36 MΩ and the R-C time constant was 1 second. Note: the first pulse is not quite long enough to allow an accurate estimate of R_m and the third is much too short. See text and Figure 1 for finding the time constant and capacitance.

*Biodyne is now known as General Bioengineering, Mission Viejo, CA.

low internal resistance supplies the voltage and current, the sum of the voltages across the resistor and capacitor will equal the source voltage and the current will be the same in both. In the case of Figure 3A, a current source, of very high internal resistance supplies a constant current which is equal to the sum of the currents in the resistor and capacitor and the voltage is the same on both. Both circuits have R and C connected in parallel. In the series case the capacitor is paralleled by the low resistance of the voltage source plus the series resistor. In the case of the membrane, R_m and C_m are paralleled by the very high internal resistance of the constant current source. The time constant for both is given by R x C.

Most animal eggs are by far the most suitable types of cells for measuring membrane capacitance by this technique. Long nerve or muscle fibers, neurons with complex morphology and cells with electrical coupling to many others such as at certain embryonic stages, are poor candidates for this method. The explanation for this limitation is that cytoplasmic resistances tend to isolate the membrane of remote cell regions from the site of current injection. Since the membrane voltage is usually measured very near or at the site of current injection, it no longer behaves as a single exponential. If a problem of this kind is suspected, plotting the change in voltage on a log scale against time will immediately reveal whether or not the measurement will be valid. A pure single exponential V_m change will yield a straight line on a log plot.

Surprisingly, the microvilli of the *Xenopus* oocyte which are about 6 x 0.1 microns in size, appear to use all of their surface membrane to carry the capacitance measuring current. Given their length-to-diameter ratio, it might have been expected that they would behave as if some part of their ends were electrically isolated from the measuring current but their voltage responses plotted in log gave straight lines (Kado *et al.*1981). Starfish oocytes which also have microvilli which disappear during maturation were also shown to decrease in membrane capacitance by Miyazaki *et al.* (1975). Moody and Bosma (1985) found in starfish oocytes that while membrane capacitance decreased by about 48% during maturation, surface area was decreased by about 40% by their morphometric method. This discrepancy allowed these authors to conclude that the potassium conductance which decreased in the same proportion as the capacitance was located in the membrane which disappeared but the calcium conductance which was slightly increased, might be in the approximately 8% of remaining membrane. How much of the microvilli membrane participates in the measured capacitance of the oocyte, especially if they are branched, can only be guessed at this time.

Voltage-dependent membrane conductances activated during the voltage shift will alter the time course of the membrane voltage and a false time constant will be read from the record. Any voltage shift record which does not smoothly approach the maximal value asymptotically is suspect. Even if it does, check by passing the same current in the opposite direction. The same potential shift of the opposite polarity should be obtained. If not, the membrane is rectifying. In any case, use only enough current to obtain a readable record, a few millivolts will suffice if there is not too much noise.

Measurement of Capacitive Currents

In voltage clamp, an amplifier is used to impose fixed membrane voltages against membrane conductance changes. The conductance at any time for a given voltage is found by dividing the clamp current at that time by the voltage. The control voltage must therefore be changed to different levels to find how the conductance depends on the membrane potential. Changing the membrane voltage requires a clamp current proportional to dV/dt (Equation 2) to charge the membrane capacitance, the well known capacitive transients of voltage clamp. This transient current can be very large because the voltage changes are usually very fast. Since the resistance of the current-passing electrode is electrically in parallel with the membrane resistance, it effectively shunts the membrane and shortens its time constant. Depending on the current electrode resistance, the membrane voltage can be changed much more rapidly than in current clamp where the series resistance (electrode plus amplifier) is very much larger than the resistance of the membrane. Note that this shunting effect does not change the membrane conductance pathway since it is present only in the current supply path. The speed can be further increased by having a high voltage output clamp amplifier to pass more current through the electrode resistance.

If the quantity of charge needed to produce a particular voltage change were known, it is possible to know how much capacitance was present as defined by Equation 1. For the case of a constant current, it suffices to multiply the current (coulombs/sec) by the time the current passed to find the quantity of charge displaced. If the current has a complex time course, one easy way to determine the total charge is to photograph it from an oscilloscope or record it on paper. Cut the current trace out of the paper along with a square equal to a unit charge, then weight the two. The weight of the complex current cut-out divided by the weight of the unit charge gives the quantity of charge which was displaced by the current. Elegantly simple but labor intensive. Electronically, the charging current can be integrated with a capacitor, the resulting voltage on the capacitor is proportional to the total current. This was used by Adrian and Almers (1974) to measure capacitance in voltage-clamped frog muscle fibers.

Today, measurements of currents and voltages are easily and conveniently made and stored digitally. Discrete samples of an analog signal are taken and converted electronically into numerical values or "digitized". Sequential sampling then permits an analog signal to be stored as a series of numbers proportional to the signal amplitude at the time each sample was taken. No matter how complex the waveform might be, it can be digitized at appropriate intervals to preserve the waveform in numerical form. If the current were digitized in this way, the quantity of charge moved can be easily found by summing the charge in each of the samples. The charge in each sample is found by simply multiplying the magnitude of the current sample by its duration.

Measurement of the capacitive charge from digitized current data has been used by Peres and Bernardini (1985) to measure membrane capacitance in *Xenopus* eggs during activation. These authors experienced two problems which complicated their measurements. The first was that the current in the capacitance of the membrane is a function of the rate of change of the voltage as well as the capacitance of the membrane (see Equation 2). The second was that the membrane capacitance is in parallel with a membrane conductance which becomes very large during activation. The voltage-clamp current during activation therefore had two components, i_C and i_{Rm}. The first problem was solved by fixing the rate of change of the control potential shift slow enough that the peak capacitive current was held to a reasonable range. The second was accounted for by assuming that the voltage-dependent part of the activation conductance increased linearly during the period of measurement. With this assumption a linearly increasing current due to the conductance could be subtracted from the measured current, thus leaving only the capacitive current. That these solutions worked was shown by the fact that the capacitive currents for the "on" and "off" of the test voltage pulse were nearly identical.

Peres and Bernardini were able to show that the capacitance of the *Xenopus* egg did not change with membrane voltage over the range from -30 to +60 mV. However with activation, C_m increased on the average of 60.4% and this increase was blocked by exposure of the egg to CO_2 which blocked exocytosis of the cortical granules. A similar blockage of cortical granule exocytosis has been reported for *Xenopus* eggs in the presence of ammonia and procaine by Charbonneau and Webb (1987). These authors showed the blockage with electron micrographs but did not measure the capacitance.

Both the time constant and capacitive transient methods suffer from the fact that they are in essence sampling methods. They can only yield discrete measurements of an ongoing process. While it seems logical to just apply shorter pulses at a faster rate, the time constant of many eggs, especially the larger ones, are too long to allow the measurement of capacitance with pulses which are too short, even in voltage clamp where the time constant is decreased. The alternative method is to use a triangular voltage waveform in voltage clamp to obtain a squarewave current.

The Triangular Wave Method

A voltage changing at a constant rate first in one direction then at the same rate in the opposite direction is described as being triangular. If a triangular voltage is applied to a circuit with a capacitance, the current in the capacitor will be a square wave. This follows from Equation 2 where dV/dt will be constant over some time interval as illustrated in Figure 5 and

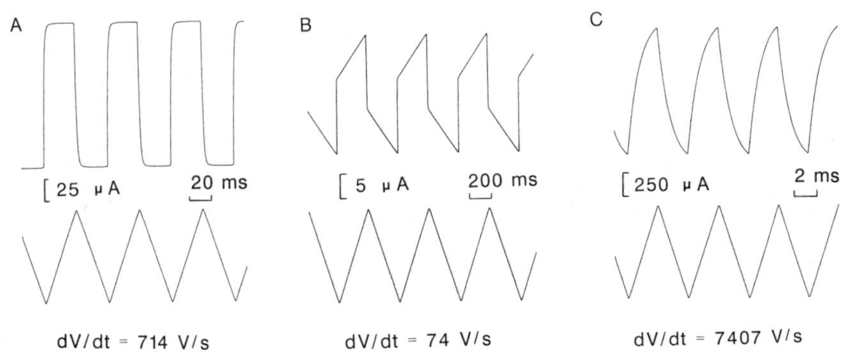

Figure 5. Capacitive currents due to triangular voltages. Upper traces are the capacitive currents for a 100 nf capacitance shunted by a 2 MΩ R_m and the lower traces are the imposed voltages. The currents resulting from three different dV/dt as indicated under each voltage trace shows that a square wave of current is obtained within a certain range of dV/dt **(A)**. A smaller dV/dt results in a smaller i_C and the current due to R_m becomes evident **(B)**. However, a valid i_C is still obtained if only the fast part of the current is measured. In the case where dV/dt is too great (frequency too high), i_C begins to follow dV/dt **(C)**. There will be a time constant unless all resistances in series with the membrane capacitance were zero Ohms.

then suddenly change direction. The instant dV/dt changes direction, the current must change to begin charging the capacitance at the rate of voltage change dictated by dV/dt. This change gives rise to the switching in direction of the current generating a square wave. The current will go positive with positive dV/dt, and negative with negative dV/dt even if the triangular wave is all positive or negative (that is, superposed on a constant voltage).

A squarewave current is obtained only if the frequency of the triangular voltage wave is slow relative to the voltage clamp time constant of the membrane that is, series resistance is not too large. The rising and falling phases of the squarewave cannot be any faster than the time constant of the voltage-clamped membrane because the change in voltage must be present long enough for the capacitance to charge. At high frequencies, the current will just follow dV/dt as a triangular wave and for peak voltages of sufficient magnitude the squarewave will be distorted. If the membrane conductance is high, the clamp current will be a squarewave with a superimposed triangle as shown in Figure 5.

This method for obtaining a square wave current was used in the late 50's and 60's for adjusting the "negative capacitance" compensation of the voltage follower (VF). A triangular voltage was coupled to the input of the voltage follower through a small capacitor (about 1 pF) producing a square current in the electrode. This current, in principle, produced a squarewave voltage drop across the resistance of the microelectrode. The "negative capacitance" compensation was then adjusted until the output of the voltage follower was as square as possible. With this method since the current could be known, it was also possible to measure the electrode resistance.

The voltage clamped triangular wave was used by Moody and Bosma (1985) to follow membrane area changes during maturation of the tunicate egg. By choosing a 50 Hz triangular wave of 10mV peak amplitude, which gave a dV/dt of 1V/s, the resulting squarewave current in nanoamps could be read directly as nanofarads of capacitance in the positive or negative direction. If the peak-to-peak current is read, it must be divided by two to obtain the capacitance. Perhaps it would be more convenient to use a dV/dt of 0.5 V/s, the peak-to-peak current would then correspond to the capacitance. In any case, if the cell is being studied under voltage clamp, applying a small triangular control voltage would be a simple and direct way to make periodic capacitance measurements. A large capacitance current would be desirable if small changes in capacitance are to be measured accurately but they might mask the conductance current. For example, at 1 V/s, 100 nF membrane capacitance will give 100 nA of current, large enough to periodically mask conductance currents of only a few nanoamperes.

The triangular wave method allows membrane capacitance changes to be followed at high rates. At 50 Hz, measurements of capacitance will be made every 20 ms which is certainly fast enough to follow most expected membrane changes. However, the large currents needed to resolve small changes in capacitance may make it difficult to resolve small changes in the conductance currents. For simultaneous measurement of the membrane conductance and capacitance this method may not be ideal.

ALTERNATING CURRENT METHODS

There are several ways of using AC currents to measure membrane capacitance. From the relatively straightforward to some very sophisticated phase detection techniques as well as statistical techniques. These all depend on the sinusoidal voltage and current phase relationship and the fact that X_C varies inversely with frequency. Alternating currents can be used in the usual current- or voltage-clamp modes but their implementation is more complicated. In experiments where simultaneous membrane electrical and capacitance measurement is necessary, the AC methods are probably the best way to go in spite of the complications. The major advantage is that most membrane conductance changes take place in the range of frequencies very much slower than the AC current frequencies used to make the capacitance measurements. For example the time constant method described above to find Rm is essentially a zero frequency measurement. Making measurements using AC currents provides a second probe which is independent of the DC since with appropriate filtering the two can be made completely independent of each other.

The R-C Low Pass Filter Methods

The fact that X_C varies inversely with frequency permits the most straightforward AC membrane capacitance measurement method. For a membrane such as in Figure 3A, as the frequency of the injected constant current, i_a, increases, X_C decreases (see Equation 8) and the impedance Z_m will also decrease (see Equation 10). The decreased Z_m results in a decreased V_m ($V_m = i_a \times Z_m$). As frequency continues to increase, V_m becomes ever smaller therefore the circuit shown in Figure 3A is a low-pass filter. Note that at zero frequency, X_{Cm} is infinitely large and Z_m is just equal to R_m. A consequence of decreasing X_{Cm} is that almost all the injected current will be in the capacitance of the membrane. The AC membrane voltage will then be due almost entirely to X_{Cm}. The membrane resistance is essentially shorted for the AC current although for DC currents it will show its normal resistance. With a little algebra, Equation 10 can show that for $\omega^2\tau^2$ much greater than 1, $Z_m = X_C$.

The capacitance is found from:

$$X_{Cm} = V_m/i_a \text{ and } C_m = 1/\omega X_{Cm} \qquad \textbf{Eq. 11}$$

Where V_m is the AC voltage at the membrane (before amplification) and i_a is the injected AC current. Both must be in the same value, rms, average or peak-to-peak. C_m will be in Farads. Note also that V_m varies inversely with C_m.

In order that the calculated capacitance be the real membrane capacitance, X_C must be significantly smaller than R_m. This requirement introduces a serious problem in cells where the membrane resistance will be greatly reduced at some time during the process under study. During egg activation, the massive opening of ion channels leads to as much as 20-fold increases in membrane conductance. Since X_C must be equal to or less than $R_m/10$ during this time, it will be necessary to know in advance the lowest resistance to be expected, and approximately the capacitance, to choose the frequency to use.

The choice of frequency will ultimately be a compromise between several factors. The AC current will be limited by rectification in the tip of the current electrode, even at moderately high frequencies (to about 1 kHz). The AC signal will be very small, on the order of microvolts because of the X_C *vs.* R_m requirement but this also avoids activating membrane conductances. Shielding will be necessary because at high frequencies, AC currents will be capacitively coupled between the shanks of the voltage and current electrodes. This coupled current could produce a larger voltage across the tip resistance of the voltage electrode than that from the membrane. The current or voltage electrode must be shielded to ground. Since V_m will be on the order of microvolts, amplification is needed before further treatment of the signal is possible. Fortunately, high-gain AC amplifiers are not difficult to achieve and with a bandpass filter tuned to the same frequency as the current, a 1000 times amplification with little added noise can be obtained.

The low pass filter method can produce excellent results and was used by Jaffe *et al.* (1978) to measure the capacitance change in sea urchin eggs during fertilization as well as in frog eggs (Jaffe and Schlichter 1985, Kline *et al.* 1988) and partial exocytosis in sea urchin eggs (Kado 1978). The set-up used in the work of Kline *et al.*, assembled with commercially available components is shown in Figure 6. This method does not depend on the phase relations between i_a and V_m and can yield reliable C_m measurements as long as the R_m remains much larger than X_{Cm}.

Noise Analysis Method

Since the membrane acts as a low-pass filter, if a current consisting of many different frequencies such as a "white noise" is injected into the cell, it can be expected that the membrane voltage at each frequency will depend on the impedance of the membrane at that frequency. Voltages measured at discrete frequencies can be plotted against frequency to give the frequency response curve of the membrane. If the current at each of the frequencies were also measured, the impedance at each frequency can be found and C_m can be determined. Measurements of currents and voltage at discrete frequencies can be done by computer using a "fast Fourier transform" or FFT program on digitized data. The Fourier analysis yields measurements for all frequencies from nearly zero to an upper limit set by the user. The intricacies of applying the FFT method, as well as its underlying principles are described by DeFelice in his book (1981). This method has been used to measure membrane capacitance in many kinds of cells but to the best of my knowledge, never in eggs (cf. Clausen and Fernandez 1981).

Obviously the capacitance should be constant for all the frequencies. However, if the voltage amplitude exceeds a few mV, it is possible that membrane conductances with a delay of their own will be activated. At the lower frequencies (below about 10 Hz) these can act as a negative capacitance, i.e. current lags voltage. This has been shown to happen using the noise analysis method in *Chara* (Ross *et al.* 1985) and in Necturus gallbladder epithelium (Gogelein and Van Driessche 1981) and can probably be shown to occur also in eggs. At this time at least, it would appear that demonstrating the presence of a negative capacitance in the egg membrane will be of limited value in understanding how the egg does what it does.

Phase Detection Methods

Every electrophysiologist has seen a sine wave synchronized to stay in one place on the oscilloscope screen even as it changes amplitude. If a mask having a triangular slit in it were placed in front of the screen, the amount of light from the exposed segment of the trace will vary almost linearly with the amplitude of that part of the sinewave. This was used as a means of detecting one of the components of a signal having a normal and an orthogonal component (Kolin and Kado 1959a). The process is illustrated in Figure 7A. It soon became evident that one did not need an oscilloscope to do this, fairly reliable transistors were becoming available and so it was possible to substitute a transistor switch which synchronously sampled the sinewave to produce the phase detected signal as shown in Figure 7B (Kolin and Kado 1959b). The use of phase detection permitted sensing extremely small (nanovolt) signals because it looked at the incoming signal only when it should have been there (Kolin and Kado 1959b), the noise which is present all of the other times is eliminated. Using phase detection of the

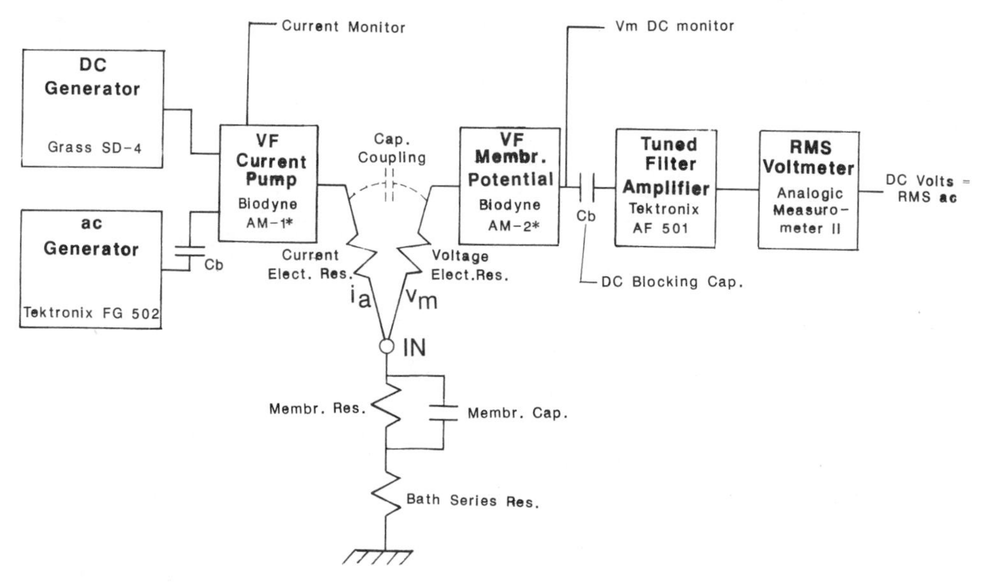

Figure 6. A set-up for AC measurement of membrane capacitance. This schematic diagram should be read from left to right. The generators supply the DC pulse and the AC control signals for the current pump (VF used as a constant current source) which can also supply polarizing currents. The combined currents are injected through one intracellular electrode (i_a) into the cell (IN) and exit through the membrane (R_m and C_m) returning to the injecting circuit through the bath resistance and ground. Note that the second electrode will sense both the voltage drop,of the membrane and the series resistance of the bath. The latter must be made very small since only the AC voltage across the membrane will be useful in the capacitance determination. A very large grounding Ag-Agcl wire near the cell will be necessary.
The voltage output of the second VF is used directly to record membrane potential and is coupled through a blocking capacitor (Cb) to the AC amplifying and measuring system. The blocking capacitor is used here, and at the output of the AC generator, to prevent DC and low frequency voltages from entering the high gain ae amplifier or the current pump. The AC Amp. is tuned to the same frequency as that of the AC generator by adjusting the tuning until a maximal output is obtained from the RMS voltmeter. This voltmeter gives a DC output equal to the true RMS value of its input AC signal which can then be read from its front panel digital read-out and/or recorded with a chart recorder. This output must be divided by the amplifier gain to find the AC membrane voltage (V_m). To measure the injected ae current (i_a) in RMS value, this voltmeter can be temporarily connected to the current monitor output of the current pump VF.

output of a vibrating electrode in a current field, nV voltages dropped by local cell membrane currents were measured by Jaffe and Nuccitelli (1974).

Since phase detection depends on sampling, the sampling pulse could be timed to occur at any part of a signal. If the signal was a composite of two sinewaves, each of the components could be detected. Two switches, 90 degrees out-of-phase with each other, could be used to detect the resistive and capacitive components simultaneously. An application of this approach is illustrated in Figure 7C (Kado and Adey 1965). Phase detection can now be done quite easily with commercially available instruments called "Lock-in amplifiers" which basically work on the principle described above, and was used to detect the capacitance changes following activation in fish eggs (Nuccitelli 1980). A lock-in amplifier simulated on a computer has also been used to detect the capacitance changes produced by endo- or exocytoic pits in bovine adrenal chromaffin cells (Neher and Marty 1982) and in rat mast cells (Joshi and Fernandez 1988).

It is a little tricky to be sure that the sampling is being done at the correct time. Both the above authors and Kolin and Kado (1959b) resorted to the same trick, introduce a small signal at the phase of interest, adjust the sampling phase to get a zero output (sampling at the zero crossing time of the desired signal), then move the sampling phase back exactly 90 degrees. The sampling now occurs at the peak of the desired signal. Commercial lock-in amplifiers have this switched 90 degree feature built-in.

While these studies reveal that the method allows resolving very small changes in capacitance (less than 0.1 pF), it cannot tell which way the membrane addition is going. One needs to know in advance that the process will be endo- or exocytotic or if indeed both processes may not be going on at the same time.

The most interesting and recent application of this technique is that by D. McCulloch and T. Chambers to detect the capacitance increase added to a small patch of the sea urchin egg membrane in which a sperm makes electrical continuity with the cytoplasm of the egg (D. McCulloch this meeting and McCulloch and Chambers 1986). In what is clearly an experimentally monumental task, they were able to get a sperm to swim down a patch electrode to the egg membrane at the end and fuse. The patch electrode was voltage-clamped with an AC voltage of a few mV which gave a current due to the resistance and the capacitance of the patched membrane and the capacitance of the electrode. This current was then cancelled in the output so that at the resting condition the output was zero. Suppression of this larger "baseline" current then permitted the small change in the AC current due to the sperm membrane to be greatly amplified. The current signal was phase detected with a lock-in amplifier adjusted to detect currents leading the voltage in the electrode. In this way they found changes in the current which appears to show when sperm fusion occurs.

Phase detection is a powerful tool in detecting very small signals and is widely used in many fields. Small signals, often buried in noise can be detected because they are looked for when they should be there. The fact that the AC frequencies used can be much higher than the frequency components of the intrinsic electrical properties of the cell membrane allows simultaneous monitoring of both parameters with a minimum of interference between them.

Figure 7. Phase detection. There are various ways of extracting a signal from an amplitude modulated sinewave. Phase detection is one of them but requires that the detecting is done with a fixed phase relation to the sinewave which is changing in magnitude. The simplest is shown in A where the synchronization of the detection with the sinewave is done by an oscilloscope. The masks allow only a segment of the bright trace to be coupled to the photocells which provide an output voltage proportional to amount of light. Two photocells connected in opposition allow phase reversal to be detected and output to the chart recorder (adapted from Kolin and Kado 1959a). This can be done electronically as shown in B. Electronic switching connects the output to the input only during the positive peaks of the input sinewave. Unless the frequency of the input signal is much higher than the response of the recording system the output should be integrated with a low-pass filter. A composite input sinewave may be sampled as shown in C to independently measure its cosine and sine components at their peaks.

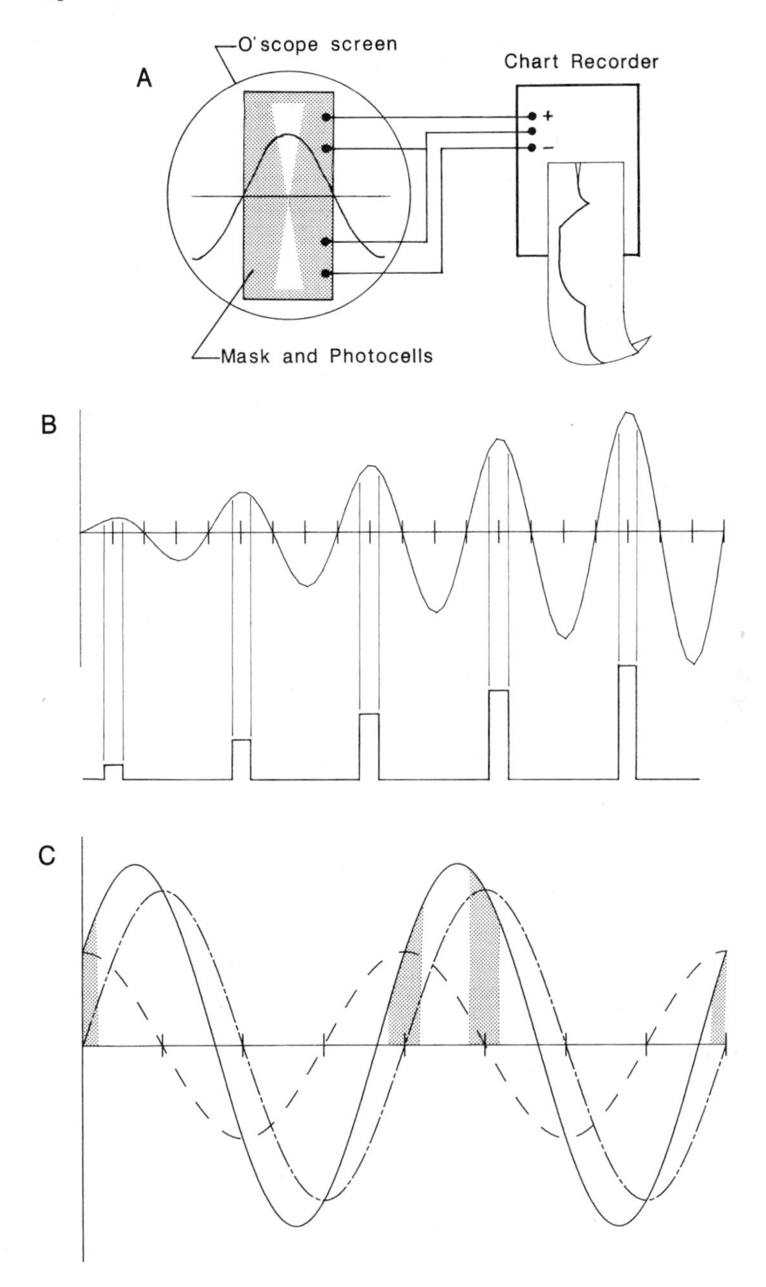

SOME CONCLUSIONS

From the above it can be concluded that membrane capacitance is what we measure electrically by manipulating the current in, or the voltage across the membrane. The behavior of the current or voltage can be interpreted as an electrical capacitance because it fits the definition. It is probable that the membrane capacitance is not a single identifiable element. It is more likely the result of the charges of the lipids, the membrane proteins, the surface glycoproteins, and the fact that these as well as other membrane elements maintain ionic gradients across the plasmalemma. Furthermore, given the sometimes complex morphologies of the cell membrane, the extent to which our present measurement methods allow us to know the cell surface area remains to be fully established. How accurately are our methods reporting the microscopic extensions of the membrane both inwardly and outwardly oriented?

In spite of its complex origin and complicating electrical considerations, the measured membrane capacitance seems to agree well with surface areas or their changes expected from morphological evidence. It is a valid and useful way to know when the membrane area changes. When measured simultaneously with other parameters, changes in surface area can be studied in relation with other on-going processes in the cell.

REFERENCES

Adey, W. R., R. T. Kado, and J. Didio. 1962. Impedance measurements in brain tissue of chronic animals using microvolt signals. *Exp. Neurol.* 5:47-66.

Adrian, R. H. and W. Almers. 1974. Membrane capacity measurements on frog skeletal muscle in media of low ion content. *J. Physiol. (Lond.)* 237:573-605.

Charbonneau, M. and D. J. Webb. 1987. Weak bases partially activate *Xenopus* eggs and permit changes in membrane conductance whilst inhibiting cortical granule exocytosis. *J. Cell Science.* 87:205-220.

Clausen, C. and J. M. Fernandez. 1981. A low cost method for rapid transfer function measurement with direct application to biological impedance analysis. *Pfluegers Arch. Eur. J. Physiol.* 390:290-295.

Cole, K. S. 1968. Membranes, ions and impulses. University of California Press, Berkeley.

De Felice, L. J. 1981. Introduction to membrane noise. Plenum Press, New York.

Gogelein, H. and W. Van Driessche. 1981. Capacitive and inductive low frequency impedances of Necturus gallbladder epithelium. *Pfluegers Arch. Eur. J. Physiol.* 389:105-113.

Jaffe, L. A., S. Hagiwara, and R. T. Kado. 1978. The time course of cortical vesicle fusion in sea urchin eggs observed as membrane capacitance changes. *Dev. Biol.* 67:243-248.

Jaffe, L. A. and L. C. Schlichter. 1985. Fertilization-induced ionic conductances in eggs of the frog, *Rana pipiens. J. Physiol.* 358:299-319.

Jaffe, L. F. and R. Nuccitelli. 1974. An ultrasensitive vibrating probe for measuring steady extracellular currents. *J. Cell Biol.* 63:614-628.

Joshi, C. and J. M. Fernandez. 1988. Capacitance measurements. *Biophys. J.* 53:885-892.

Kado, R. T. and W. R. Adey. 1965. Method for the measurement of impedance changes in brain tissue. 6th Intl. Conf. on Medical Electronics and Biological Engineering, Tokyo.

Kado, R. T. 1978. The time course of cortical granule fusion in the fertilized and non-fertilized sea urchin egg. *Biol. Cell.* 32:141-148.

Kado, R. T., K. Marcher, and R. Ozon. 1981. Electrical membrane properties of the *Xenopus laevis* oocyte during progesterone-induced meiotic maturation. *Dev. Biol.* 84:471-476.

Kline, D., L. Simmoncini, G. Mandel, R. Maue, R. T. Kado, and L. A. Jaffe. 1988. Fertilization events induced by neurotransmitters after injection of mRNA in *Xenopus* eggs. *Science* 24:464-467.

Kolin, A. and R. T. Kado. 1959a. Simple photoelectric demodulator. *J. Sci. Instrum.* 37:107.

Kolin, A. and R. T. Kado. 1959b. Miniaturization of the electromagnetic blood flow meter and its use for the recording of circulatory responses of conscious animals to sensory stimuli. *Proc. Natl. Acad. Sci. USA.* 45:1312-1321.

McCulloch, D. and E. L. Chambers. 1986. When does the sperm fuse with the egg? *Abstr. 40th Annu. Meeting Soc. of Gen. Physiol.* 38a.

Moody, W. J. and M. M. Bosma. 1985. Hormone-induced loss of surface membrane during maturation of starfish oocytes: differential effect on potassium and calcium channels. *Dev. Biol.* 112:396-404.

Miyazaki, M., H. Ohmori, and S. Sasaki. 1975. Potassium rectifications of the starfish oocyte membrane and their changes during oocyte maturation. *J. Physiol. (Lond.)* 246:55-78.

Neher, E. and A. Marty. 1982. Discrete changes of cell membrane capacitance observed under conditions of enhanced secretion in bovine adrenal chromaffin cells. *Proc. Natl. Acad. Sci USA.* 79:6712-6716.

Niles, W. D., R. A. Levis, and F. S. Cohen. 1985. Planar bilayer membrane made from phospholipid monolayers form by a thinning process. *Biophys. J.* 53:327-335.

Nuccitelli, R. 1980. The electrical changes accompanying fertilization and cortical vesicle secretion in the medaka egg. *Dev. Biol.* 76:483-498.

Peres, A. and G. Bernardini. 1985. The effective membrane capacity of *Xenopus* eggs: its relations with membrane conductance and cortical granule exocytosis. *Pfluegers Arch. Eur. J. Physiol.* 404:266-272.

Ross, S. M, J. M. Ferrier, and J. Dainty. 1985. Frequency-dependent membrane impedance in *Chara coralm* estimated by Fourier analysis. *J. Membr. Biol.* 85:233-243.

RECEPTORS, G-PROTEINS, AND EGG ACTIVATION

Laurinda A. Jaffe

Department of Physiology
University of Connecticut Health Center
Farmington, CT 06032

At fertilization, an intricate series of signals passes back and forth between the sperm and egg. This paper concerns the signal from the sperm that causes the egg to activate and begin development. Evidence is presented in support of the hypothesis that this process involves a guanine nucleotide binding protein (G-protein) in the egg membrane, and perhaps a related receptor as well. In this respect, the process of sperm-egg interaction may have a close resemblance to the interaction of certain neurotransmitters and hormones with their target cells.

G-proteins are a class of membrane proteins that function in many cells to couple membrane receptors to membrane effector enzymes; they are named G-proteins because their function depends on the binding of guanine nucleotides. Several excellent reviews of G-proteins and related receptors and enzymes have been published recently (Stryer and Bourne 1986; Gilman 1987; Dohlman *et al.* 1987). The function of G-proteins in eggs and oocytes has been reviewed by Turner and Jaffe (1989), and evidence for a physiological role of G-proteins in sperm has been described by Endo *et al.* (1987).

Among the enzymes known to be activated by a G-protein, phosphatidylinositol-4,5 bisphosphate phosphodiesterase (PIP_2PDE, or phospholipase C) is of particular interest to the problem of egg activation. The substrate of this enzyme, the lipid PIP_2, as well as its products, inositol trisphosphate ($InsP_3$) and diacylglycerol (DAG), rise within 15s after fertilization in sea urchin eggs (Turner *et al.* 1984; Ciapa and Whitaker 1986).

The rise in $InsP_3$ causes Ca^{2+} release, leading to cortical vesicle exocytosis, ion channel opening, and other responses of egg activation (Whitaker and Irvine 1984; Busa *et al.* 1985; Swann and Whitaker 1986; Turner *et al.* 1986; Miyazaki 1988; Kline 1988). The rise in DAG, which stimulates protein kinase C, may be important in controlling such events as the rise in pH and the increase in protein synthesis (Swann and Whitaker 1985; Lau *et al.* 1986; Shen and Burgart 1986; but see Shen, this volume).

The model we propose is that the sperm activates a receptor in the egg plasma membrane, which in turn activates a G-protein, leading to activation of PIP_2 PDE and/or other enzymes of the phosphatidylinositol pathway. Although receptors are well known to function in the process of sperm-egg binding (Ruiz-Bravo and Lennarz 1986; Wassarman 1987), no sperm receptor in the egg plasma membrane has as yet been identified. G-proteins, however, are known to be present. Using cholera toxin (CTX) and pertussis toxin (PTX), which ADP-ribosylate certain G-proteins and cause labelling if radioactive NAD is present, two

G-proteins have been identified in sea urchin eggs (Oinuma *et al.* 1986; Turner *et al.* 1987). The CTX substrate has a molecular weight of 47 kD, and the PTX substrate has a molecular weight of 39-40 kD, in the same range as for α subunits of other G-proteins.

Three experimental approaches have been used to investigate the role of these G-proteins in the process of egg activation. These are: 1) injection of guanine nucleotide analogs, 2) injection of cholera and pertussis toxins, and 3) expression of mRNA for G-protein related receptors.

EXPERIMENTS WITH GUANINE NUCLEOTIDE ANALOGS

The first evidence that a G-protein might be involved in the events of fertilization came from injection into sea urchin eggs of metabolically stable guanine nucleotide analogs, guanosine 5′-O-(3-thiotriphosphate) (GTP-γ-S) and guanosine 5′-O-(2-thiodiphosphate) (GDP-β-S). G-proteins are active when GTP is bound, and inactive when GDP is bound. Receptor stimulation activates G-proteins by causing GDP to come off the protein and GTP to bind. The G-protein then stays activated until its inherent GTPase activity hydrolyzes GTP to GDP. GTP-γ-S, a hydrolysis resistant analog of GTP, can activate G-proteins in the absence of receptor stimulation, because of a slow but significant exchange of GDP for GTP even at rest. GTP-γ-S can thus become bound to G-proteins, and because it is hydrolysis resistant, the G-protein remains in an active state. GDP-β-S can inactivate G-proteins, because it competes with GTP for binding; when a GDP-like molecule is bound, the G-protein remains in an inactive state.

GTP-γ-S (30 μM) was found to initiate cortical vesicle exocytosis in *Lytechinus variegatus* eggs, although not in eggs preinjected with EGTA (Turner *et al.* 1986). In *Lytechinus pictus* eggs, GTP-γ-S (100 μM) caused a rise in intracellular calcium, as detected with fura-2 (Swann *et al.* 1987). GDP-β-S (3 mM) was found to inhibit cortical vesicle exocytosis in response to sperm, but not in response to $InsP_3$ injection (Turner *et al.* 1986). GDP-β-S (5 mM) did not inhibit sperm-egg fusion (Swann *et al.* 1987). These results, along with the results of GTP-γ-S injection, support the interpretation that a G-protein is involved in the activation of the egg at fertilization. Sperm-egg fusion appears to proceed by a pathway that does not depend on a G-protein.

Subsequent experiments with frogs and hamsters have given similar results. GTP-γ-S (100 μM) caused an activation potential and cortical vesicle exocytosis in frog eggs (Kline and Jaffe 1987). GTP-γ-S (> 12 μM) caused calcium release, and an electrical response like that occurring at fertilization in hamster eggs, and GDP-β-S (8 mM) inhibited this electrical response to sperm but not to $InsP_3$ injection (Miyazaki 1988, and this volume).

In sea urchin eggs, responses to GTP-γ-S appear to be somewhat more complex than originally supposed. While some batches of *Lytechinus variegatus* eggs underwent exocytosis in response to GTP-γ-S injection, others did not. In some tests of *L. variegatus* eggs, exocytosis occurred at 20-100 μM GTP-γ-S, but not at 300 μM-1 mM GTP-γ-S (P. R. Turner and L. A. Jaffe, unpublished results).

One possible explanation of these variable results is the observation that at higher concentrations of GTP-γ-S, the cortical cytoplasm of the egg becomes blebbed and amoeboid. This response might represent an exaggerated form of the polymerization of actin that occurs during normal fertilization (Hamaguchi and Mabuchi 1988). It has been reported that at the site of fertilization cone formation, where actin polymerization at fertilization is most massive, cortical granule exocytosis does not occur (Epel and Patton 1985). Thus in a situation where massive actin polymerization occurs all around the egg, cortical granule exocytosis might be generally inhibited. Perhaps the exocytotic response to GTP-γ-S depends on whether or not the effect of GTP-γ-S on the cortical cytoplasm predominates.

Variation in the response to GTP-γ-S is not surprising, since GTP-γ-S is likely to activate multiple G-proteins, some of which may have opposing physiological effects. Furthermore, activation of G-proteins by GTP-γ-S depends critically on intracellular concentrations of GTP and GDP (Northup *et al.* 1982). In sea urchin eggs, [GTP] has been measured to be about 300 μM (Yanagisawa and Isono 1966), but could conceivably vary between batches of eggs.

The activation of sea urchin eggs by a hydrolysis resistant form of GTP was confirmed by use of a different hydrolysis-resistant analog, 5-guanylyl imidodiphosphate (GppNHp). GppNHp has been shown to activate G-proteins in other cellular systems, although higher concentrations are required compared with GTP-γ-S (Northup *et al.* 1982). GppNHp (2-5 mM) was found to stimulate cortical vesicle exocytosis in eggs of both sea urchin *(Lytechinus pictus)* and starfish (*Patiria miniata).* In addition, GppNHp (4-5 mM) caused a pH increase in the sea urchin, like that occurring at fertilization. Measurements of pH using the fluorescent indicator BCECF-dextran (Bright *et al.* 1987) showed a pH shift from 6.99 ± 0.03 to 7.41 ± 0.17 (SD,n=5) after GppNHp injection, compared with a shift from 7.04 ± 0.06 to 7.45 ± 0.03 (n=5) after insemination (L.A. Jaffe and R.T. Kado, unpublished results).

GTP-γ-S and GppNHp, while hydrolysis-resistant in terms of formation of GDP, can be hydrolyzed to form cGMP (Brandwein *et al.* 1982). Since injection of 100 μM cGMP has been reported to cause cortical vesicle exocytosis in sea urchin eggs, it has been suggested that the action of GTP-γ-S might be due to formation of cGMP (Swann *et al.* 1987). However, since GTP is a better substrate for guanylate cyclase than is GTP-γ-S (Brandwein *et al.* 1982), and since injection of GTP (100 μM) does not cause exocytosis (Swann *et al.* 1987), this hypothesis seems unlikely.

EXPERIMENTS WITH TOXINS

Our next experiments to investigate the role of G-proteins in egg activation involved the use of cholera toxin (CTX) and pertussis toxin (PTX) which ADP-ribosylate G-proteins and as a result modify their function. CTX modification causes activation of G-proteins, while PTX modification causes inactivation; only certain G-proteins are sensitive to toxin modifications. Neither toxin had a significant effect when applied externally to sea urchin eggs. However, microinjection of CTX (30 μg/ml) or its catalytically active A-subunit caused cortical vesicle exocytosis (Turner *et al.* 1987). Preinjection of EGTA blocked the CTX-induced exocytosis, indicating that CTX acts by a Ca^{2+}-dependent pathway. The pathway is cAMP-independent, since injection of cAMP or a hydrolysis-resistant analog, adenosine 3′:5′-cyclic monophosphothioate cyclic Sp- isomer (cAMP-S), did not cause exocytosis.

PTX injection did not cause or inhibit exocytosis in sea urchin eggs. We therefore wondered what function the PTX substrate present in sea urchin eggs (see above) might serve. One possibility was a role in meiotic maturation. This hypothesis was examined using starfish oocytes, in which meiosis is reinitiated in response to the surface-acting hormone 1-methyladenine (1-MA). Injection of PTX (2-5 μg/ml) inhibited the reinitiation of meiosis in response to 1-MA (Shilling *et al.* 1989). This inhibition could, however, be bypassed by subsequent injection of maturation promoting factor (MPF), indicating that PTX blocks an early step in the 1-MA pathway, consistent with action on a G-protein. PTX-injected, MPF-matured starfish eggs underwent normal cortical vesicle exocytosis and cleavage when subsequently inseminated, indicating that the PTX-sensitive G-protein that appears to be involved in mediating meiotic reinitiation differs from the G-protein believed to function in mediating exocytosis at fertilization.

EXPERIMENTS WITH MESSENGER RNA

As another approach to investigating the role of G-proteins in fertilization, as well as to begin to investigate the possibility of a receptor-mediated activation of the G-protein, we introduced into eggs certain neurotransmitter receptors that are known to act by way of G-proteins (Kline *et al.* 1988). The receptors that activate G-proteins belong to a structurally homologous family (Dohlman *et al.* 1987; Julius *et al.* 1988). We chose to express serotonin and muscarinic M1 acetylcholine receptors, both known to act by way of G-proteins to stimulate $InsP_3$ production. The receptors were introduced by injection of mRNA into *Xenopus* oocytes (rat brain poly A^+ mRNA to obtain serotonin receptors, and a specific mRNA derived from a DNA clone to obtain muscarinic M1 receptors). The oocytes were matured to eggs by addition of progesterone, and then agonists for the corresponding receptors were applied. Serotonin or acetylcholine caused at least four fertilization-like responses in mRNA-injected eggs: an activation potential, cortical vesicle exocytosis, endocytosis, and cortical contraction. Non-injected eggs did not show these responses. These results indicated that exogenously-introduced receptors could act, by way of egg G-proteins, to stimulate egg activation. They also suggested the possibility that an analogous "sperm receptor" might be present in the egg membrane.

The use of mRNA injection may provide a way to investigate more directly the possible role of a plasma membrane sperm receptor. We are currently studying the interactions between sperm and egg species that do not cross-fertilize, with the idea that expression of mRNA prepared from oocytes of one species into eggs of another species might introduce receptivity to the foreign sperm. A positive result would provide the opportunity to analyze the nature of the receptor by molecular biological techniques. Starfish cross-fertilization may provide a useful model for this study. We are encouraged in this possibility by evidence that serotonin receptors can be expressed in *Patiria miniata* oocytes following injection of rat brain mRNA; after application of 1-MA to mature the oocytes, serotonin caused cortical vesicle exocytosis, as in *Xenopus* (F. Shilling, G. Mandel and L. A. Jaffe, unpublished).

ACKNOWLEDGEMENTS

The work and ideas described here are the result of collaborations with many colleagues -- Paul Turner, Doug Kline, Fraser Shilling, Gail Mandel and Ray Kado in particular. Support was provided by NIH grant HD14939.

REFERENCES

Brandwein, H. J., J. A. Lewick, S. A. Waldman, and F. Murad. 1982. Effect of GTP analogues on purified soluble guanylate cyclase. *J. Biol. Chem.* 257:1309-1311.

Bright, G. R., G. W. Fisher, J. Rogowska, and D. L. Taylor. 1987. Fluorescence ratio imaging microscopy: temporal and spatial measurements of cytoplasmic pH. *J. Cell Biol.* 104:1019-1033.

Busa, W. B., J. E. Ferguson, S. K. Joseph, J. R. Williamson, and R. Nuccitelli. 1985. Activation of frog (*Xenopus laevis*) eggs by inositol trisphosphate. I. Characterization of Ca^{2+} release from intracellular stores. *J. Cell Biol.* 101:677-682.

Ciapa, B. and M. Whitaker. 1986. Two phases of inositol polyphosphate and diacylglycerol production at fertilization. *FEBS Lett.* 195:347-351.

Dohlman, H. G., M. G. Caron, and R. J. Lefkowitz. 1987. A family of receptors coupled to guanine nucleotide regulatory proteins. *Biochemistry* 26:2657-2668.

Endo, Y., M. A. Lee, and G. S. Kopf. 1987. Evidence for the role of a guanine nucleotide-binding regulatory protein in the zona pellucida-induced mouse sperm acrosome reaction. *Dev. Biol.* 119:210-216.

Epel, D. and C. Patton. 1985. Cortical granules of sea urchin eggs do not undergo exocytosis at the site of sperm-egg fusion. *Dev. Growth & Differ.* 27:361-369.

Gilman, A. G. 1987. G proteins: transducers of receptor-generated signals. *Annu. Rev. Biochem.* 56:615-649.

Hamaguchi, Y. and I. Mabuchi. 1988. Accumulation of fluorescently labelled actin in the cortical layer in sea urchin eggs after fertilization. *Cell Motil. Cytoskeleton* 9:153-163.

Julius, D., A. B. MacDermott, R. Axel, and T. M. Jessell. 1988. Molecular characterization of a functional cDNA encoding the serotonin 1c receptor. *Science* 241:558-564.

Kline, D. 1988. Calcium-dependent events at fertilization of the frog egg: injection of a calcium buffer blocks ion channel opening, exocytosis, and formation of pronuclei. *Dev. Biol.* 126:346-361.

Kline, D. and L. A. Jaffe. 1987. The fertilization potential of the *Xenopus* egg is blocked by injection of a calcium buffer and is mimicked by injection of a GTP analog. *Biophys. J.* 51:398a.

Kline, D., L. Simoncini, G. Mandel, R. A. Maue, R. T. Kado, and L. A. Jaffe. 1988. Fertilization events induced by neurotransmitters after injection of mRNA in *Xenopus* eggs. *Science* 241:464-467.

Lau, A. F., T. C. Rayson, and T. Humphreys. 1986. Tumor promoters and diacylglycerol activate the Na^+/H^+ antiporter of sea urchin eggs. *Exp. Cell Res.* 166:23-30.

Miyazaki, S. 1988. Inositol 1,4,5-trisphosphate-induced calcium release and guanine nucleotide-binding protein-mediated periodic calcium rises in golden hamster eggs. *J. Cell Biol.* 106:345-353.

Northup, J. K., M. D. Smigel, and A. G. Gilman. 1982. The guanine nucleotide activating site of the regulatory component of adenylate cyclase. *J. Biol. Chem.* 257:11416-11423.

Oinuma, M., T. Katada, H. Yokosawa, and M. Ui. 1986. Guanine nucleotide-binding protein in sea urchin eggs serving as the specific substrate of islet-activating protein, pertussis toxin. *FEBS Lett.* 207:28-34.

Ruiz-Bravo, N. and W. J. Lennarz. 1986. Isolation and characterization of proteolytic fragments of the sea urchin sperm receptor that retain species specificity. *Dev. Biol.* 118:202-208.

Shen, S. S. and L. J. Burgart. 1986. 1,2-diacylglycerols mimic phorbol 12-myristate 13-acetate activation of the sea urchin egg. *J. Cell. Physiol.* 127:330-340.

Shilling, F., K. Chiba, M. Hoshi, T. Kishimoto, and L. A. Jaffe. 1989. Pertussis toxin inhibits 1-methyladenine-induced maturation in starfish oocytes. *Dev. Biol.* (In press).

Stryer, L. and H. R. Bourne. 1986. G-proteins: a family of signal transducers. *Annu. Rev. Cell Biol.* 2:391-419.

Swann, K., B. Ciapa, and M. Whitaker. 1987. Cellular messengers and sea urchin egg activation. p. 45-69. *In: Molecular Biology of Invertebrate Development.* J. D. O'Connor (Ed.). Alan R. Liss, New York.

Swann, K. and M. Whitaker. 1985. Stimulation of the Na/H exchanger of sea urchin eggs by phorbol ester. *Nature (Lond.)* 314:274-277.

Swann, K. and M. Whitaker. 1986. The part played by inositol trisphosphate and calcium in the propagation of the fertilization wave in sea urchin eggs. *J. Cell Biol.* 103:2333-2342.

Turner, P. R. and L. A. Jaffe. 1989. G-proteins and the regulation of oocyte maturation and fertilization. p. 297-318. *In: The Cell Biology of Fertilization.* H. Schatten and G. Schatten (Eds.). Academic Press, Orlando.

Turner, P. R., L. A. Jaffe, and A. Fein. 1986. Regulation of cortical vesicle exocytosis in sea urchin eggs by inositol 1,4,5-trisphosphate and GTP-binding protein. *J. Cell Biol.* 102:70-76.

Turner, P. R., L. A. Jaffe, and P. Primakoff. 1987. A cholera toxin-sensitive G-protein stimulates exocytosis in sea urchin eggs. *Dev. Biol.* 120:577-583.

Turner, P. R., M. P. Sheetz, and L. A. Jaffe. 1984. Fertilization increases the polyphosphoinositide content of sea urchin eggs. *Nature (Lond.)* 310:414-415.

Wassarman, P. M. 1987. Early events in mammalian fertilization. *Annu. Rev. Cell Biol.* 3:109-142.

Whitaker, M. and R. F. Irvine. 1984. Inositol 1,4,5-trisphosphate microinjection activates sea urchin eggs. *Nature (Lond.)* 312:636-639.

Yanagisawa, T. and N. Isono. 1966. Acid-soluble nucleotides in the sea urchin egg. I. Ion-exchange chromatographic separation and characterization. *Embryologia* 9:170-183.

WHAT HAPPENS DURING THE LATENT PERIOD AT FERTILIZATION

Michael Whitaker, Karl Swann and Ian Crossley
Department of Physiology
University College London
Gower Street, London WC1E 6BT
UNITED KINGDOM

ABSTRACT

A period of several seconds to half a minute can elapse between the initial interaction of the fertilizing sperm with the egg and the initiation of the calcium wave that causes global egg activation. It has been called the latent period. We discuss experiments that have shed light on what is happening during the latent period. We suggest that the latent period reflects the transfer of an activating messenger from the sperm to the egg through a labile fusion pore. The latent period ends when the messenger triggers the increase in cytoplasmic calcium that consolidates the labile fusion state and then sweeps across the egg, activating it.

INTRODUCTION

How does a sperm activate an egg? That is, what is the crucial interaction between egg and sperm that re-initiates the cell division cycle at fertilization?

It has been known for some time that the global signal that stimulates the transition from quiescence to proliferation at fertilization is a large, rapid and transient increase in intracellular free calcium concentration (Steinhardt and Epel 1974; Ridgeway *et al.* 1977; Steinhardt *et al.* 1977; Whitaker and Steinhardt 1982; Jaffe 1983; Busa and Nucitelli 1985). In many eggs, the global calcium signal is in the form of a wave of calcium increase that sweeps across the egg from the point of sperm-egg interaction (Gilkey *et al.* 1978; Eisen *et al.* 1984; Busa and Nuccitelli 1985; Swann and Whitaker 1986; Miyazaki *et al.* 1986; Kubota *et al.* 1987; Yoshimoto *et al.* 1987). In sea urchin eggs, the calcium wave propagates through positive feedback: elevated intracellular calcium concentrations (Ca_i) lead to the production of inositol trisphosphate ($InsP_3$), a cellular messenger that induces further calcium release from intracellular stores (Berridge and Irvine 1984; Whitaker and Irvine 1984; Whitaker and Aitchison 1985; Swann and Whitaker 1986; Ciapa and Whitaker 1986). The calcium wave is a property of the egg, not the sperm. It can be induced by microinjection of very small quantities of $InsP_3$ (Whitaker and Irvine 1984; Swann and Whitaker 1986) and, indeed, $InsP_3$ is a very efficient parthenogenetic activator that stimulates initiation of chromatin condensation cycles (I. Crossley and M. Whitaker, unpublished). $InsP_3$-injected eggs do not undergo mitosis or cell division, however, presumably because they lack the competent centrosome normally donated by the sperm (Brandriff *et al.* 1975). The sperm's other obvious gift at fertilization is the DNA and protein of the sperm nucleus. Neither the nucleus nor the centrosome is essential for re-initiation of the cell cycle. Haploid parthenogenetic embryos can be reared by first inducing the calcium wave in an unfertilized egg and then inducing the formation of centrosomes (Brandriff *et al.* 1975). The re-initiation of the cell cycle by the sperm at fertilization is due solely to the triggering of the activating calcium wave by the sperm. This paper is concerned with how the triggering of the calcium wave comes about.

PHOSPHOINOSITIDES AND G-PROTEINS

The calcium-releasing messenger, $InsP_3$, is formed by the hydrolysis of the plasma membrane phospholipid, phosphatidyl-inositol bisphosphate ($PtdInsP_2$). The phosphoinositide messenger system is ubiquitous (Berridge 1987). In cells that respond to hormones with the hydrolysis of $InsP_3$,it is thought that receptor occupancy is coupled to $PtdInsP_2$ hydrolysis by GTP-binding proteins that stimulate phosphoinositidase C (Cockcroft and Gomperts 1985). The analogy is with the G-proteins that couple receptors to adenylate cyclase (Rodbell 1980). The idea that the egg, too, possesses a sperm receptor coupled via a G-protein to phosphoinositidase C (PIC) is attractive. It is discussed in detail in a paper by Laurinda Jaffe in this volume (p. 151). There is good indirect evidence that the sea urchin egg PIC can be stimulated by a G-protein. The guanine nucleotide thiophosphate analogue GTP-γS causes constitutive activation of G-proteins and activates eggs when microinjected (Turner *et al.* 1986) by causing a transient increase in Ca_i^{2+} identical to the Ca_i^{2+} transient at fertilization (Swann *et al.* 1987). Cholera toxin, another G-protein activator, also activates eggs when microinjected and substrates for both cholera and pertussis toxin (putative G-proteins) are found associated with the egg plasma membrane (Turner *et al.* 1987). There is also evidence that the sperm activates the egg via a G-protein: microinjection of the G-protein inhibitor GDP-βS blocks egg activation as judged by cortical granule exocytosis (Turner *et al.* 1986). One tentative answer to the question of how the sperm initiates the calcium wave at fertilization has been that binding of the sperm to a receptor causes $InsP_3$ production because the receptor activates a G-protein that in turn stimulates PIC.

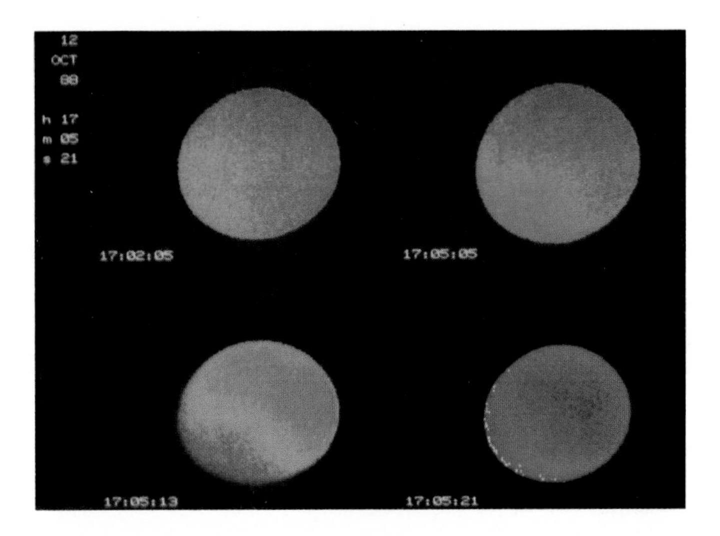

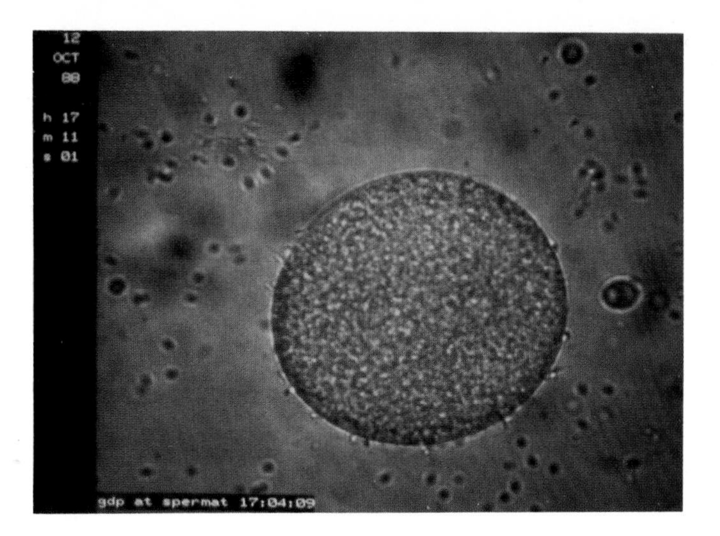

Figure 1. The sperm-induced calcium wave in an egg microinjected with 3 mM GDP-βS measured using the calcium-sensitive dye fura-2 and a fluorescence imaging system. TOP: The increase in Ca_i^{2+} was first noted in the bottom left quadrant of the egg at 17:05:05. The calcium wave is halfway across the egg at 17:05:13 (8s later) and Ca_i^{2+} is uniformly high in the egg at 17:05:21. Blue indicates the lowest Ca_i^{2+} and purple the highest. Resting Ca_i^{2+} is around 100 nM, peak Ca_i^{2+} around 2 μM. BOTTOM: A brightfield image of the same egg after the passage of the calcium wave. The absence of a fertilization membrane indicates that the exocytosis of the egg's cortical granules has been inhibited by the GDP-βS injection. *Lytechinus pictus*. 16°C. 50 μM fura-2 injected. For methods see Swann and Whitaker (1986) and Williams *et al.* (1985).

GDP-βS BLOCKS EXOCYTOSIS, NOT THE CALCIUM WAVE

Egg activation is most easily judged by looking for the the fertilization envelope that comes up off the egg surface at fertilization. The fertilization envelope forms as a result of the exocytosis of secretory granules that fuse with the plasma membrane when Ca_i^{2+} increases (Whitaker and Steinhardt 1982). GDP-βS microinjection blocks egg activation judged by this criterion (Turner *et al.* 1986). We wanted to see whether GDP-βS also blocked the initiation of the calcium wave. To visualize the calcium wave we used the calcium-sensitive fluorescent dye, fura-2, and an imaging system. The results of one of three experiments in which we fertilized an egg after microinjecting GDP-βS are shown in Figure 1. GDP-βS clearly prevents exocytosis, but the calcium wave still sweeps across the egg with a time course similar to controls. The conclusion is that the G-protein inhibitor GDP-βS does not, in fact, prevent the sperm from initiating the calcium wave. These observations raise doubts that the link between the sperm and the calcium wave involves the egg's G-proteins. The effect of GDP-βS seems to be simply to inhibit exocytosis.

AN ACTIVATING MESSENGER PROVIDED BY THE SPERM

An alternative, though equally tentative, answer to the question of the the initiation of the calcium wave is that some activating substance is transferred from sperm to egg. Leaving aside the identity of this putative messenger (it might be calcium itself [Jaffe 1980], $InsP_3$ [Whitaker and Irvine 1984], cGMP [Swann *et al.* 1987] or a nicotinamide nucleotide derivative [Clapper *et al.* 1987], for example), the utility of this idea depends on how quickly after the initial sperm-egg interaction a messenger could diffuse from sperm to egg. This can only happen once cytoplasmic continuity has been established between sperm and egg. When does this occur? Is it before or after the initiation of the calcium wave? We can best answer these questions by considering what happens during what has been called the latent period of egg activation.

THE LATENT PERIOD

Several seconds elapse between the first interaction of sperm and egg and the elevation of the fertilization envelope. This gap between sperm-egg contact (when sperm cease to be susceptible to the spermicide KCl) and activation has been called the latent period (Allen and Griffin 1958). Electrophysiology has provided the best indication of what is going on during the latent period. Figure 2 shows an experiment in which the electrical changes at fertilization are correlated with the calcium wave measured with fura2. The first electrical sign of sperm-egg interaction is a depolarization due to sperm-gated channels (Dale *et al.* 1978) that gives rise to an action potential (Chambers and deArmendi 1979). The action potential is followed by the activation potential. The activation potential is due to the opening of calcium-activated cation channels as Ca_i^{2+} rises (Chambers and deArmendi 1979; David *et al.* 1988). In the experiment shown in Figure 2, a period of 12s elapses between the onset of the sperm-gated current and the onset of the calcium wave. This 12s period corresponds to the latent period defined by Allen and Griffin, but here it is measured in a single egg.

A striking characteristic of the latent period measured in an egg population is that it has a large statistical variance (Allen and Griffin 1958). The meaning of the large variance is best illustrated by the elegant experiments of Shen and Steinhardt (1984) in single eggs. Shen and Steinhardt used the fact that sperm-egg interaction is prevented at positive egg membrane potentials (Jaffe 1976) to measure the time that elapsed from initial sperm-egg interaction to the initiation of the calcium wave. They very briefly hyperpolarized the egg to permit sperm-egg interaction and measured the onset of the calcium wave by measuring the time of onset of the calcium-induced activation current. Their results are shown in Figure 3. The data indicate that the time to initiation of the calcium wave after initial sperm-egg interaction is very variable and can be longer than 30s. In their experiments, the mean time to onset of the

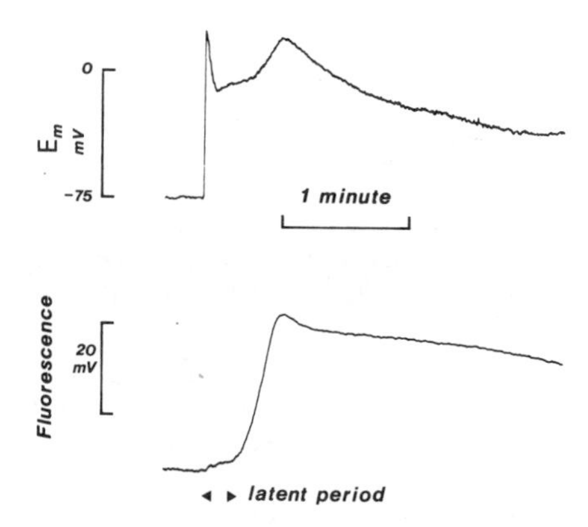

Figure 2. Simultaneous recording of changes in Ca_i^{2+} and membrane potential at fertilization. Fura-2 fluorescence at 380 nm is shown on an inverted scale since 380 nm fluorescence decreases as Ca_i^{2+} increases. The action potential marks the beginning of the latent period. The latent period ends with the marked increase in Ca_i^{2+} that represents the initiation of the calcium wave. This is coincident with the second phase of depolarization due to the activation current. The small increase in Ca_i^{2+} at the beginning of the latent period is due to calcium entry during the action potential, since it can be seen in unfertilized eggs each time the egg is depolarized and is absent when eggs are fertilized under conditions in which the action potential has been inactivated. This is one of four experiments in each of which the onset of the calcium wave cas cemented with the second phase of depolariation due to the activation cellent. The latent period measured in this way ranged from 12s to 22s. *Lytechinus pictus*. 16°C. 50 μM fura-2 injected. For methods see Swann and Whitaker (1986) and David *et al.* (1988).

calcium wave was 17s. The other interesting aspect of these data is that initiation of the calcium wave occurs with a substantial absolute latency. These experiments show that at 16-18°C, there is an absolute latent period of 7s, followed by a variable latent period that can be as long as 40s. The calcium wave is a regenerative process (Swann and Whitaker 1986). The variable latency of the calcium wave is a characteristic of regenerative phenomena (Calvin 1975). What seems to be happening during the latent period is the accumulation of a triggering substance.

WHEN DO SPERM AND EGG FUSE?

Preliminary results by McCulloh and Chambers (1986) simultaneously measuring capacitance and fertilization potential indicate that sperm-egg fusion is the first detectable interaction between egg and sperm. The capacitance increase that accompanies sperm-egg fusion is coincident with the sperm-gated membrane current, indicating that the current is due to channels in the sperm membrane that reveal themselves as part of the egg membrane conductance when fusion establishes electrical continuity between sperm and egg cytoplasm. By this criterion, sperm-egg fusion begins at the very onset of the latent period.

Other methods have also been used to define the time of sperm-egg fusion. Continuity between sperm and egg using electron microscopy (Longo *et al.* 1986) or dye transfer (Hinckley *et al.* 1986) is established 5-10s after the onset of the sperm-gated current (that

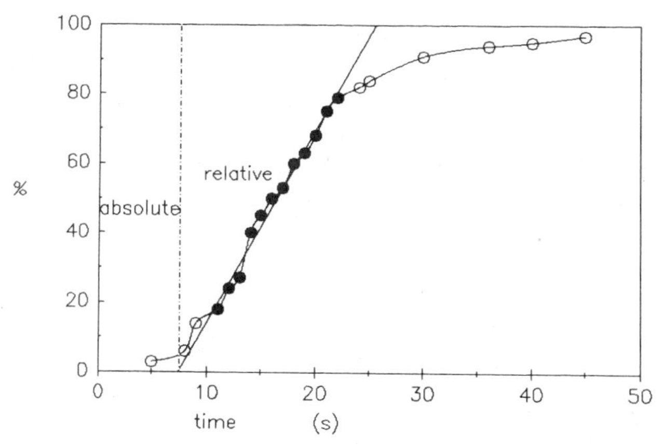

Figure 3. Proportion of eggs activating with time after initiation of sperm-egg interaction. The data are taken from Shen and Steinhardt (1984). At zero time in each separate experiment, sperm was allowed to interact with the egg by a brief window of permissive hyperpolarization from a positive holding potential. The time of activation (the end of the latent period) was taken to be the time at which the Ca_i^{2+}-induced activation current appeared. The latent period can be divided into two epochs: the absolute latent period is the epoch in which no egg activation occurs; the relative latent period is the epoch in which there appears to be a roughly constant probability of egg activation. *Lytechinus pictus* 16-18°C.

is, the onset of the latent period). There is a discrepancy here that needs explaining. In fact, the discrepancy has been instrumental in revealing more about what happens during the latent period. We return to it below.

REVERSIBLE SPERM-EGG FUSION

Reversible Sperm-Egg Fusion in Neomycin-Injected Eggs

Eggs injected with the aminoglycoside antibiotic, neomycin, do not initiate a calcium wave when fertilized. Nor do sperm fuse by the criterion of dye transfer from egg to sperm (Swann *et al.* 1987). However, such eggs have electrical responses to sperm. Figure 4 shows the currents induced in voltage-clamped, neomycin-injected eggs after insemination. The sperm-induced currents have a sharp onset and generally terminate equally abruptly. After the termination of such a current, the current-voltage relation of the membrane is identical to that of an unfertilized egg. McCulloh and Chambers (1986) have shown that sperm-gated currents of this magnitude are due to the insertion of sperm channels into the egg membrane when sperm and egg fuse. The current steps in neomycin-injected eggs are therefore a manifestation of a reversible sperm-egg fusion.

Reversible Sperm-Egg Fusion in Voltage-Clamped Eggs

Sperm-gated currents indicative of sperm-egg fusion are also seen in eggs voltage-clamped at negative holding potentials. Figure 5 shows an egg in which two sperm fused transiently without initiating the calcium wave. A third episode of sperm-egg fusion led to activation of the calcium wave. Transient sperm-egg fusion without initiation of the calcium wave is common in eggs held at negative membrane potentials. At a holding potential

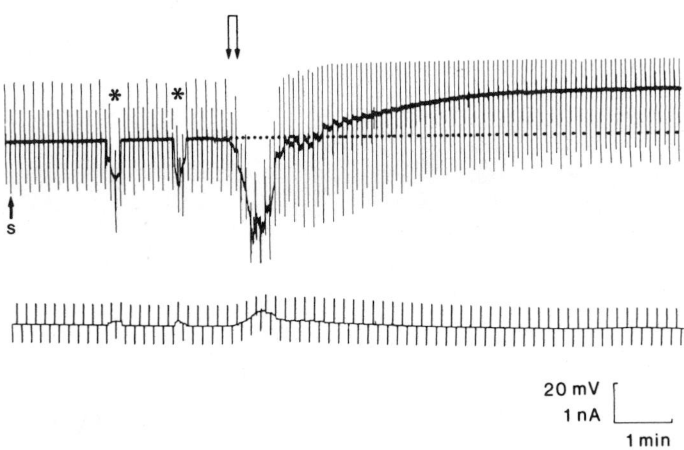

Figure 4. Unsuccessful sperm-egg interactions in an egg voltage-clamped at -50 mV. Upper trace, current. Lower trace, voltage. The egg was alternately stepped to -20 mV and -80 mV every 5s. Sperm were added at S. The two starred current episodes were associated with sperm attachment. At the end of each, the current-voltage relation of the egg membrane reverted to that of the unfertilized egg and the egg remained unactivated. A third episode of sperm-related current (double arrow) resulted in egg activation – judged from the appearance of the larger activation current and the raising of a fertilization envelope. *Lytechinus pictus*. 16°. See David *et al*. (1988).

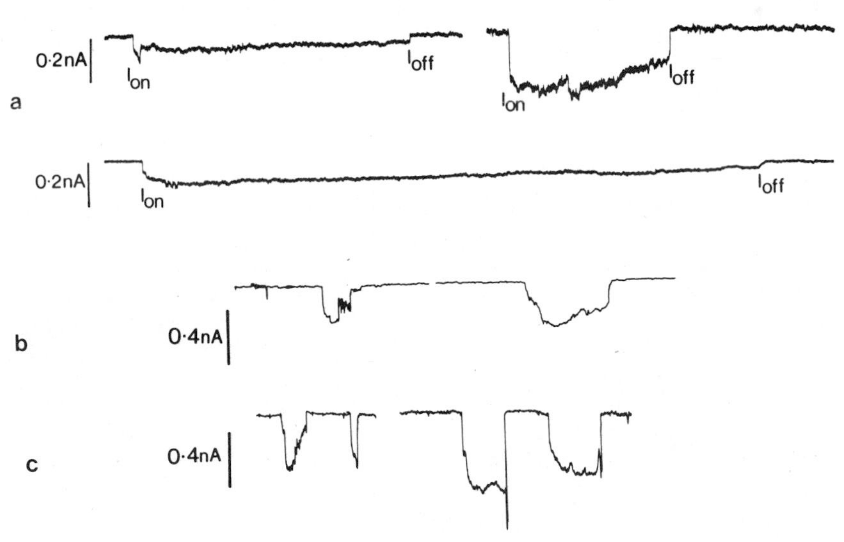

Figure 5. Episodic current steps in voltage-clamped eggs injected with 10 mM final concentration of neomycin. Examples at **(a)** -20 mV; **(b)** -50 mV; **(c)** -70 mV. The effects of holding potential on the length and duration of the current episodes are shown in Table 1. *Lytechinus pictus* 16°C. For methods see Swann and Whitaker (1986) and David *et al*. (1988).

of -70 mV, 30% of sperm-egg interactions do not result in activation and 50% lead to activation without sperm incorporation (Lynn *et al.* 1988). In eggs that did not incorporate sperm in these experiments, sperm-egg fusion reversed before the eggs activated, indicating very graphically that the processes underlying egg activation during the latent period can continue under some circumstances even in the absence of continued sperm-egg interaction.

The Spermicide Uranyl Nitrate Reverses Sperm-Egg Fusion

Fertilization can be prevented by treating insemination mixtures with spermicides such as KCl, lauryl sulphate and uranyl nitrate. In egg populations, the beginning of the latent period is defined as the time at which the sperm ceases to be susceptible to KCl or lauryl sulphate. Uranyl nitrate can prevent fertilization at a later stage than KCl or lauryl sulphate (Baker and Presley 1969). It acts 5-10s after KCl, at the end of the latent period (Presley and Baker 1970; Shen and Steinhardt 1984). Since sperm-egg fusion measured electrically occurs at the beginning of the latent period, uranyl nitrate must prevent fertilization by reversing sperm-egg fusion.

TWO STAGES OF SPERM-EGG FUSION AT FERTILIZATION

These various findings suggest that there are two experimentally-defined stages of sperm-egg fusion at fertilization. The first phase of fusion reverses spontaneously at negative membrane potentials and can also be reversed by the spermicide, uranyl nitrate. It is also reversed by fixation for electron microscopy or dye transfer experiments, since sperm-egg fusion in the electron microscope or dye transfer in fixed eggs can only be detected 5-10s into the latent period (Hinckley *et al.* 1986; Longo *et al.* 1986). EM-detectable fusion defines the second phase of sperm-egg fusion. In the second stage, fusion is no longer reversible and dye transfer from egg to sperm occurs easily. When during the latent period does the transition from reversible to permanent fusion occur?

All the data relevant to the timing of the fusion transition are collected in Figure 6. The most important point to consider is the relation between the absolute and relative latent periods and the EM-fusion and dye transfer data. Looking at Shen and Steinhardt's (1984) data, we construe the latent period to consist of two epochs. The first epoch, the absolute latent period, is the shortest time that can elapse between sperm-egg interaction (that is, electrically-detectable fusion) and the initiation of the calcium transient. The second epoch, the relative latent period, is the time that elapses between the end of the absolute latent period and the initiation of the calcium transient in any particular egg. Because of the stochastic nature of calcium-wave initiation, the question of when the calcium wave is initiated can have two equally valid answers. One answer is at 5s (Figure 6), the time at which initiation is first detected. The second answer is at 11s, the time after electrically-detected sperm-egg fusion at which, on average, 50% of eggs in a population will have initiated the calcium transient. This point is best illustrated by considering EM-and dye transfer-detectable fusion on the one hand and cessation of sperm motility on the other (Figure 6). The dye transfer and EM experiments ask when fusion is *first* detected using these methods. The answer seems to be: at the same time that the initiation of the calcium transient is *first* detected. Observations on the cessation of sperm motility (Hulser and Schatten 1982; Shen and Steinhardt 1984; Lynn and Chambers 1984) ask when cessation of sperm motility occurs *on average*. The answer is: when, on average, half the eggs in the population have initiated a calcium transient. It appears, therefore, that both EM-detectable fusion and cessation of sperm motility occur at or around the time at which the calcium transient is initiated, suggesting that it is increase in Ca_i may be responsible for both these events.

So, when during the latent period does the transition from reversible to permanent fusion occur? Figure 6 shows that susceptibility to uranyl nitrate, appearance of fusion in the EM and dye transfer all *first* occur at the end of the absolute latent period. This suggests that the consolidation of the initial transient fusion state *first* occurs at this point, 5s into the latent period, the earliest time at which a calcium transient can be initiated.

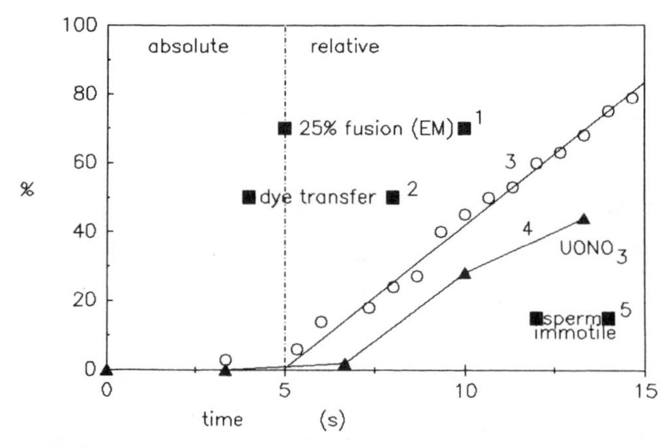

Figure 6. Sequence of events during the latent period. EM-detectable fusion and dye transfer from sperm to egg are first seen at the end of the absolute latent period. The cumulative curve of egg activation (3) from Figure 3 and the time course of loss of susceptibility to the spermicide uranyl nitrate (4) are shown. They are corrected for the 3°C difference in temperature between these and the EM and dye transfer experiments, using a value of 15 as the Q_{10} of the latent period (Presley and Baker 1970; Allen and Griffin 1958). With this correction, the times of cessation of sperm motility coincide. The mean time of cessation of sperm motility (5) approximately coincides with the mean time to 50% activation. (1) Longo *et al.* 1986. (2) Hinckley *et al.* 1986. (3,4) Shen and Steinhardt 1984. (5) Hulser and Schatten 1982; Shen and Steinhardt 1984; Lynn and Chambers 1984.

THE TRANSIENT FUSION STATE

Neomycin Prevents the Consolidation of the Initial Fusion State

Dye transfer experiments show that sperm never fuse permanently with neomycin-injected eggs (Swann *et al.* 1987). This behavior implies that neomycin prevents the consolidation of the fusion pore between sperm and egg, perhaps by preventing initiation of the calcium wave. We have used neomycin-injected eggs to investigate some of the properties of the transient fusion state.

Transient Fusion is Membrane Potential-Dependent

The mean duration of transient fusion recorded by measuring current steps in voltage-clamped, neomycin-injected eggs changes with holding potential. At negative holding potentials, fusion episodes are shorter than at more positive holding potentials (Table 1). We cannot at present suggest why this should be, but this behavior does provide an explanation of why normal eggs held at -70 mV show transient fusion events and do not always initiate a calcium wave. We discuss this below.

Transient Fusion Does Not Lead to Dye Transfer between Egg and Sperm

At positive membrane potentials, an episode of transient fusion between egg and sperm may last more than 200s in neomycin-injected eggs (Table 1). However, despite prolonged

Table 1. Sperm-associated Membrane Currents in Voltage-clamped Neomycin-injected Eggs (*Lytechinus pictus*)

Holding potential (mV)	Current step duration (s)	I_{max} (nA)
-20	134 ± 31 (9)	0.05 ± 0.005 (9)
-50	30 ± 8 (15)	0.21 ± 0.021 (19)
-70	19 ± 3 (16)	0.42 ± 0.36 (21)

mean and SEM are shown, with n in parentheses

fusion of sperm and egg, no dye transfer takes place (Swann *et al.* 1987). This suggests that the fusion pore between egg and sperm in neomycin-treated eggs is too narrow to permit the rapid diffusion of dye. In fact, even sperm that have been incorporated fully into eggs take up dye rather slowly (Hinckley *et al.* 1986), so we cannot make a good maximal eastimate of pore size from this observation. We can, however, place a lower estimate on the size of the transient fusion pore by considering its conductance. The conductance inserted into the egg membrane by the sperm in neomycin-injected eggs is around 5 nS. The pore conductance cannot be smaller than this. The acrosomal rocess is about 500 nm in length (Longo *et al.* 1986). A pore of 60 nm diameter and 500 nm long would have a conductance of 5 nS, given a cytoplasmic resistivity of 150 ohm-cm (Cole and Curtis 1938). The smallest pore diameter during sperm-egg fusion detected by EM is 100-200 nm (Longo *et al.* 1986). It seems reasonable to conclude that this is the approximate size of the transient fusion pore.

TRANSIENT FUSION AND THE INITIATION OF THE CALCIUM WAVE

An Activator Diffuses from Sperm to Egg

Three rather different sets of experimental observations can be combined to link transient fusion to the initiation of the calcium wave:

1. The length of transient fusion episodes is potential-dependent and decreases at negative membrane potentials (Table 1).
2. Initiation of the calcium wave is stochastic. The latent period can vary considerably in length (Shen and Steinhardt 1984 and Figure 3).
3. The incidence of transient fusion events that do not lead to egg activation increases when eggs are held at negative membrane potentials (Lynn *et al.* 1988).

These findings suggest that a diffusible activator may pass from sperm to egg during the phase of transient fusion. On this hypothesis, the decreased likelihood of egg activation at negative membrane potentials (Lynn *et al.* 1988) is due to the shorter episodes of transient fusion at these membrane potentials; shorter fusion episodes do not permit the transfer of sufficient activator and the calcium wave is not initiated. These three sets of observations are also *quantitatively* consistent with the idea of a diffusible messenger: the mean length of fusion episodes at -70 mV is 19s (Table 1); Shen and Steinhardt's (1984) data predict that association of sperm and egg for 19s will lead to initiation of the calcium wave in 65% of eggs; Lynn *et al.* (1988) find that activation occurs in 70% of cases at -70 mV. The good quantitative agreement is a strong indication that the likelihood of initiation of the calcium wave at negative potentials is determined by the lifetime of individual transient fusion events: events shorter than the absolute latent period will not lead to activation, while events longer than this will lead to activation with a frequency predicted by the Shen and Steinhardt curve shown in Figure 3.

Transient Fusion and Membrane Potential

It is not clear why the length of the transient fusion events should be sensitive to membrane potential, but this behavior has also been observed in oocytes (McCulloh *et al.* 1987). In oocytes it is clear that negative membrane potentials discourage the formation of fertilization cones, so one must assume that the connection between egg and sperm during the early fusion state is more tenuous at negative membrane potentials. Perhaps the sperm's own motion is responsible for detaching it and putting an end to the reversible fusion state. A more tenuous connection at more negative membrane potentials would make motion-induced detachment more likely. This is a limp argument, but no better one is available.

Rate of Diffusion of a Putative Activator

How quickly could an activator diffuse through the transient fusion pore from the sperm into the egg? The calculated steady-state concentration distribution of an activator is shown in Figure 7, for a pore of 100 nm radius and 500 nm in length (see above). At the very mouth of the fusion pore, the steady state concentration is 10% of the concentration in the sperm: there is a steep concentration gradient within the pore. The steady-state concentration distribution within the egg will to a first approximation have radial symmetry. Half a micron from the mouth, the concentration is 2% of the sperm concentration and one micron away, it is less than 1%. The steady-state concentration profile is independent of the diffusion constant, but the time to steady state is not. It increases with distance from the pore. The inset shows that the time to steady state for a molecule of around 500 dalton is less than a second for distances up to 1 μM from the mouth of the pore. For a molecule of 20,000 dalton, the time to steady state at 0.5 μM is 1s and at 1 μM, 5s. The calculations indicate than if an egg is activated by a diffusible messenger from the sperm, then the messenger must have its effects within 1 μM of the transient fusion pore. The time to steady-state at 1 μM from the pore is small compared to the length of the absolute latent period for a 500 dalton activatorbut comparable for a 20,000 dalton activator. This suggests that either (i) a low molecular weight activator initiates the calcium wave not directly, but by activating an intermediate enzyme or (ii) a higher molecular weight activator initiates the calcium wave once it reaches a critical concentration some distance from the fusion pore mouth. Diffusion of an activating substance through the transient fusion pore is evidently a feasible mechanism for initiation of the calcium wave, provided that the activator exerts its effects close to the fusion pore, but diffusion itself can account for only a small fraction of the time that elapses during the latent period.

CONSOLIDATION OF THE TRANSIENT FUSION STATE BY CALCIUM

An analysis of the sequence of events during the latent period has led us to the idea that the transition from the transient fusion state to permanent sperm-egg fusion is caused by the local increase in cytoplasmic Ca_i that occurs at the site of sperm-egg interaction when the calcium wave is initiated. At first sight this may seem an unlikely notion because eggs microinjected with the calcium chelator EGTA incorporate sperm quite adequately (Swann *et al.* 1987). There may be inadequate local buffering of Ca_i in EGTA-injected eggs, however. Small increases in Ca_i are detectable after insemination in EGTA-injected eggs (Swann and Whitaker, unpublished observations) and cortical granule exocytosis (which is a calcium-dependent process) may also occur at the site of sperm entry in eggs injected with EGTA to a final concentration of approximately 5 mM (Hamaguchi and Mabuchi 1988), though it is possible that the cortical granules are merely displaced into the cytoplasm (Epel and Patton 1985). Sperm incorporation involves the formation of filamentous actin in the fertilization cone and is inhibited by cytochalasin B, an actin filament antagonist (Longo 1978; Byrd and Perry 1980; Cline *et al.* 1983). Eggs treated with cytochalasin B activate, but fail

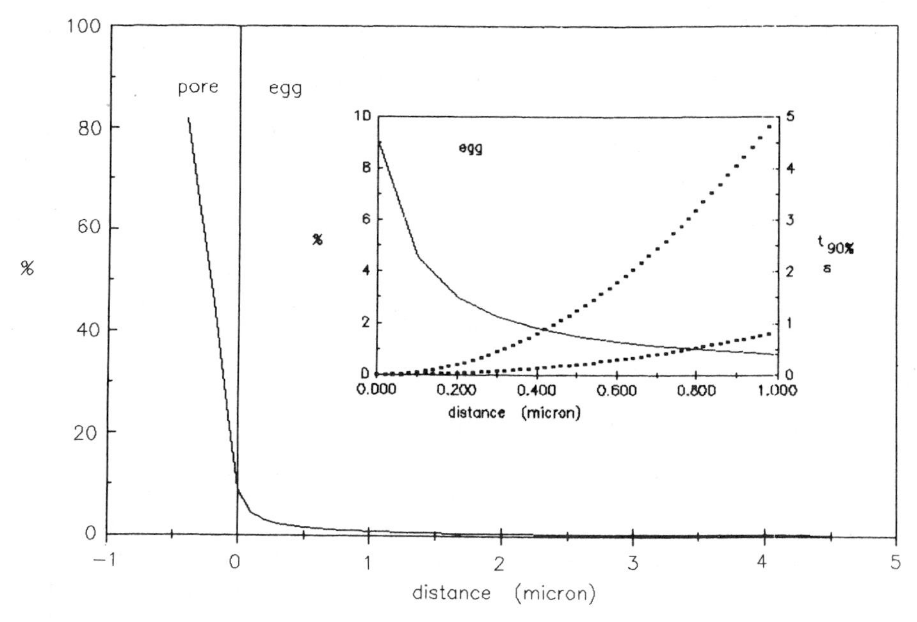

Figure 7. Diffusion of an activator from sperm to egg. The steady-state concentration in the fusion pore and egg as a percentage of the concentration in the sperm is shown. The calculations are for a pore of 100 nm radius and 500 nm long. The calculation shows that the diffusible activator is present at significant concentration in a hemisphere of 1 μM radius centered on the pore mouth. **INSET**: The concentration profile and time to steady state in this 1 μM hemisphere are indicated. Continuous line: steady state concentration profile. Dotted lines: time to 90% steady-state concentration. $t_{90\%}$ is inversely proportional to the diffusion constant. The lower dotted line is for D = 3. 10^{-7} cm^2s^{-1} (a molecule of approximately 500 dalton [Mannhertz 1968]) and the upper dotted line for D = 5. 10^{-8} cm^2s^{-1} (a molecule of approximately 20,000 dalton [Blinks *et al.* 1976]). The flux through the pore, Q, is: $DA(C_s-C_p)/x$, where D is the diffusion constant, A the pore area, C_s the concentration of activator in the sperm, C_p the concentration at the mouth of the pore and x the pore length. The concentration profile in the egg for a constant flux, Q, at a distance r from the pore is given by: $C_r = Q\,(\mathrm{erfc}[r/2Dt])/2Dr$ (Crank 1975). The steady-state concentration profile is therefore independent of D, provided that the pore radius is small compared to r. $t_{90\%}$ is given by: $r^2/0{\cdot}04$ D. The calculated fluxes do not lead to a significant decrease in C_s with the above diffusion constants at the time shown.

to incorporate sperm (Byrd and Perry 1980). Actin polymerization at the egg cortex is triggered by a local increase in Ca_i^{2+} (Hamaguchi and Mabuchi 1988). The simplest interpretation of these data is that the local increase in Ca_i^{2+} triggered by the sperm results in the formation of actin filaments that first anchor the sperm and consolidate fusion and then draw it into the egg.

WHAT HAPPENS DURING THE LATENT PERIOD

The sequence of events during the latent period appears to be as follows (Figure 8). The first interaction between sperm and egg is the formation of a narrow fusion neck (I and II). An activator diffuses through the neck (II) and builds up in the cytoplasm. When a critical activator concentration is attained in the egg cytoplasm, or, more likely (see above), when the activator has triggered a separate enzymatic process, the calcium wave is initiated (III). At a time almost co-incident with (III), the fusion pore is consolidated by the growth of actin filaments due to the local increase in Ca_i and the sperm enters the egg (IV).

Figure 8 also illustrates what happens in eggs held at negative membrane potentials or injected with neomycin. If (Figure 8, III and IV, right) the labile fusion state (II) reverts before a critical activator concentration is achieved, for example, at negative membrane potentials or after treatment with uranyl nitrate, then the egg does not activate but is still receptive to subsequently-arriving sperm. If (Figure 8, III and IV, left) the labile fusion state reverts before the calcium wave initiates but after having delivered an adequate amount of activator, the egg activates but the sperm is not incorporated. This is what happens at negative membrane potentials when the sperm current abruptly reverses before egg activation (Lynn *et al.* 1988). Finally, in eggs treated with cytochalasin B the actin filaments of state IV-center do not form in response to calcium and the sperm is swept off the egg plasma membrane by the rising fertilization envelope (Longo 1978; Byrd and Perry 1980).

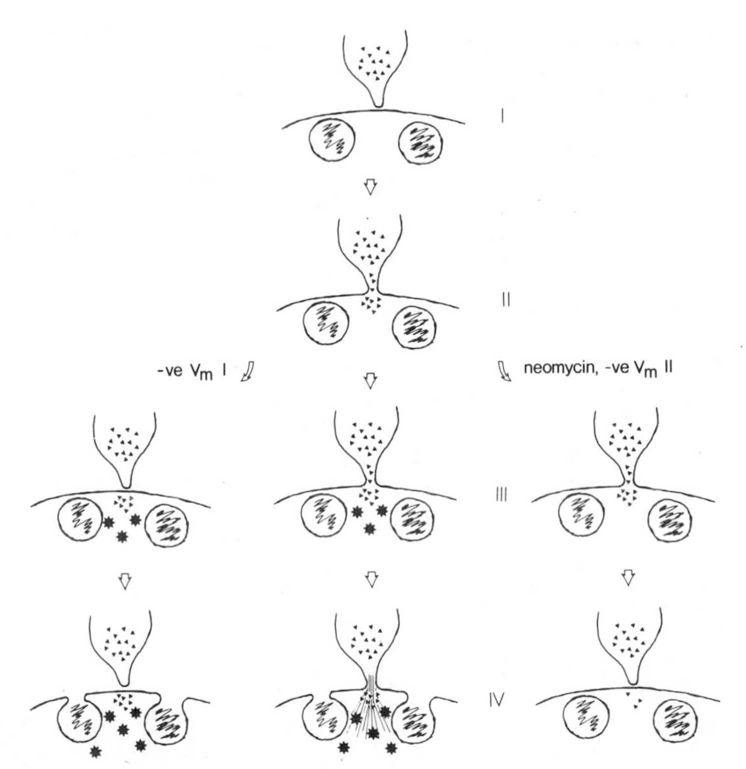

Figure 8. A schematic representation of the stages of sperm-egg interaction during the latent period. The normal progress of sperm egg fusion is shown in the centre of the diagram. The sperm attaches to the egg **(I)**. A diffusing messenger (▲) passes from sperm to egg through a labile fusion pore **(II)**. The messenger initiates the calcium (*) wave **(III)**. The local increase in Ca_i^{2+} consolidates the fusion pore **(IV)**. Pathological examples of sperm-egg interaction are shown to left and right. **LEFT:** The transient fusion state reverts before the calcium wave is initiated. This leads to activation without sperm incorporation and can occur when the egg membrane potential is held negative. **RIGHT:** No calcium wave is initiated and the transient fusion state reverts. This occurs in eggs injected with neomycin and can also occur in eggs held at negative membrane potentials when the fusion state reverts before the calcium wave is initiated.

HOW DOES THE SPERM ACTIVATE THE EGG?

The obvious unanswered question is the identity of the putative activator. Our guess is cGMP, because cGMP reaches high levels in activated sperm (Kopf *et al.* 1979) and will induce a calcium transient when microinjected into eggs (Swann *et al.* 1987), but cGMP is by no means the only candidate (Whitaker and Irvine 1984; Clapper *et al.* 1987) among low molecular weight activators; a higher molecular weight compound that activates eggs when microinjected has also been isolated from sperm (Dale *et al.* 1985; G. Ehrenstein, personal communication)

ACKNOWLEDGEMENTS

Some of these experiments were begun during a visit to Dr. E. L. Chambers at the University of Miami. We thank Dr. Chambers, Dr. J. W Lynn and Dr. D. H. McCulloh for showing us and discussing with us their unpublished data. This work was supported by funds from the Wellcome Trust, the Browne Fund of the Royal Society and the Science and Engineering Research Council. IC is a Wellcome Trust Scholar.

REFERENCES

Allen, R. D. and J. L. Griffin. 1958. The time sequence of early events in the fertilisation of sea urchin eggs. 1. The latent period and the cortical reaction. *Exp. Cell Res.* 15:163-173.

Baker, P. F. and R. Presley. 1969. Kinetic evidence for an intermediate stage in the fertilization of the sea urchin egg. *Nature (Lond.)* 221:488-490.

Berridge, M. J. 1987. Inositol trisphosphate and diacylglycerol: two interacting second messengers. *Annu. Rev. Biochem.* 56:159-163.

Berridge, M. J. and R. F. Irvine. 1984. Inositol trisphosphate, a novel second messenger in signal transduction. *Nature (Lond.)* 312:315-318.

Blinks, J. R., F. G. Prendergast, and D. G. Allen. 1976. Photo-proteins as biological calcium indicators. *Pharmacol. Rev.* 28:1-93.

Brandriff, B, R. I. Hinegardner, and R. A. Steinhardt. 1975. Development and lifecycle of the parthenogenetically activated sea urchin embryo. *J. Exp. Zool.* 192:13-24.

Busa, W. B. and R. Nuccitelli. 1985. An elevated cytosolic calcium wave follows fertilization in the eggs of the frog *Xenopus laevis*. *J. Cell Biol.* 100:1325-1329.

Byrd, W. and G. Perry. 1980. Cytochalasin B blocks sperm incorporation but allows activation of the sea urchin egg. *Exp. Cell Res.* 126:333-342.

Calvin, W. H. 1975. Generation of spike trains in CNS neurones. *Brain Res.* 84:1-22.

Chambers, E. L. and J. de Armendi. 1979. Membrane potential, action potential and activation potential of the eggs of the sea urchin *Lytechinus variegatus*. *Exp. Cell Res.* 122:203-218.

Ciapa, B. and M. J. Whitaker. 1986. Two phases of inositol polyphosphate and diacylglycelol production at fertiliation. *FEBS Lett.* 195:137-140.

Clapper, D. L., T. F. Walseth, P. J. Dargie, and H.-C. Lee. 1987. Pyridine nucleotide metabolites stimulate calcium release from sea urchin egg microsomes desensitized to inositol trisphosphate. *J. Biol. Chem.* 262:9561-9568.

Cline C. A., H. Schatten, R. Balczon, and G. Schatten. 1983. Actin-mediated surface motility during sea urchin fertilization. *Cell Motil.* 3:513-524.

Cockcroft, S. and B. D. Gomperts. 1985. Role of guanine nucleotides in the activation of phosphoinositide phospho-diesterase. *Nature (Lond.)* 314:534-536.

Cole, K. S. and H. J. Curtis. 1938. Transverse electrical impedance of the squid giant axon. *J. Gen. Physiol.* 22:757-765.

Crank, J. 1975. The Mathematics of Diffusion. Oxford University Press, London.

Dale, B., L. J. DeFelice, and V. Taglietti. 1978. Membrane noise and conductance increase during single spermatozoon-egg interactions. *Nature (Lond.)* 275:217-219.

Dale, B, L. J. DeFelice, and G. Ehrenstein. 1985. Injection of a soluble sperm fraction into sea-urchin eggs triggers the cortical reaction. *Experientia* 41:1068-1070.

David, C., J. Halliwell, and M. J. Whitaker. 1988. Some properties of the membrane currents underlying the fertilization potential in sea urchin eggs. *J. Physiol. (Lond.)* 402:139-154.

Eisen, A., D. P. Kiehart, S. J. Wieland, and G. T. Reynolds. 1984. Temporal sequence and spatial distribution of early events of fertilization in single sea urchin eggs. *J. Cell Biol.* 99:1647-1654.

Epel, D. and C. Patton. 1985. Cortical granules of sea urchin eggs do not undergo exocytosis at the site of sperm-egg fusion. *Dev. Growth & Differ.* 27:361-369.

Gilkey, J. C., L. F. Jaffe, E. B. Ridgway, and G. T. Reynolds. 1978. A free calcium wave traverses the activating egg of the medaka, *Oryzias latipes. J. Cell Biol.* 76:448-466.

Hamaguchi, Y. and I. Mabuchi. 1988. Accumulation of fluorescently labeled actin in the cortical layer in sea urchin eggs after fertilization. Cell Motil. Cytoskeleton. 9:153-163.

Hinckley, R. E., B. D. Wright, and J. W. Lynn. 1986. Rapid visual detection of sperm-egg fusion using the DNA-specific fluorochrome Hoechst 33342. *Dev. Biol.* 118:148-154.

Hulser, D. and G. Schatten. 1982. Bioelectric responses at fertilization: separation of the events associated with insemination from those due the the cortical reaction in the sea urchin *Lytechinus variegatus. Gamete Res.* 5:363-377.

Jaffe, L. A. 1976. Fast block to polyspermy in sea urchin eggs is electrically mediated. *Nature (Lond.)* 261:68-71.

Jaffe, L. F. 1980. Calcium explosions as triggers of development. *Annu. NY Acad. Sci.* 339:86-101.

Jaffe, L. F. 1983. Sources of calcium in egg activation: a review and hypothesis. *Dev. Biol.* 99:256-276.

Kopf, G. S., D. J. Tubb, and D. L. Garbers. 1979. Activation of sperm respiration by a low molecular weight egg factor and by 8-bromoguanosine 3′,5′-monophosphate. *J. Biol. Chem.* 254:8554-8560.

Kubota, H. Y., Y. Yoshimoto, Y. Yoneda, and Y. Hiramoto. 1987. Free calcium wave upon activation in *Xenopus* eggs. *Dev. Biol.* 119:126-136.

Longo, F. J. 1978. Effects of cytochalasin B on sperm-egg interactions. *Dev. Biol.* 67:249-265.

Longo, F. J., J. W. Lynn, D. H. McCulloh, and E. L. Chambers. 1986. Correlative ultrastructural and electrophysiological studies of sperm-egg interactions of the sea urchin, *Lytechnus variegatus. Dev. Biol.* 118:155-166.

Lynn, J. W. and E. L. Chambers. 1984. Voltage clamp studies of fertilization in sea urchin eggs. I. Effect of clamped membrane potential on sperm entry, activation and development. *Dev. Biol.* DD 102:98-109.

Lynn, J. W., D. H. McCulloh, and E. L. Chambers. 1988. Voltage clamp studies of fertilization in sea urchin eggs. II. Current patterns in relation to sperm entry, nonentry and activation. *Dev. Biol.* 128:305-323.

Mannhertz, H. G. 1968. ATP-Spaltung und ATP-Diffusion in oscillierenden extrahiertenMuskelfasern. *Pfluegers Arch. Gesamte Physiol.* 303:230-248.

McCulloh, D. H. and E. L. Chambers. 1986. When does the sperm fuse with the egg? *J. Gen. Physiol.* 88:38-39a.

McCulloh, D. H., J. W. Lynn, and E. L. Chambers. 1987. Membrane depolarization facilitates sperm entry, large fertilization cone formation and prolonged current responses in sea urchin oocytes. *Dev. Biol.* 124:177-190

Miyazaki, S., N. Hashimoto, Y. Yoshimoto, T. Kishimoto, Y. Igusa, and Y. Hiramoto. 1986. Temporal and spatial dynamics of the periodic increase in intracellular calcium at fertilization of golden hamster eggs. *Dev. Biol.* 118:259-267.

Presley, R. and P. F. Baker. 1970. Kinetics of fertilization in the sea urchin: a comparison of methods. *J. Exp. Biol.* 52:455-468.

Ridgway, E. B., J. C. Gilkey, and L. F. Jaffe. 1977. Free calcium increases explosively in activating medaka eggs. *Proc. Natl. Acad. Sci. USA. 74:623-627.*

Rodbell, M. 1980. The role of hormone receptors and GTP regulatory proteins in membrane transduction. *Nature (Lond.)* 284:17-20.

Shen, S. S. and R. A. Steinhardt. 1984. Time and voltage windows for reversing the electrical block to fertilization. *Proc. Natl Acad. Sci. USA.* 81:1436-1439.

Steinhardt, R. A., R. S. Zucker, and G. Schatten. 1977. Intracellular calcium release at fertilization in the sea urchin egg. *Dev. Biol.* 58:185-196.

Swann, K. and M. J. Whitaker. 1986. The part played by inositol trisphosphate and calcium in the propagation of the fertilization wave in sea urchin eggs. *J. Cell Biol.* 103:2333-2342.

Swann, K., B. Ciapa, and M. J. Whitaker. 1987. Cell messengers and sea urchin egg activation. p. 45-69. *In: Molecular Biology of Invertebrate Development*. D. O'Connor (Ed). Alan R. Liss, New York.

Turner, P. R., L. A. Jaffe, and A. Fein. 1986. Regulation of cortical granule exocytosis by inositol 1,4,5-trisphosphate and GTP binding protein. *J. Cell Biol.* 102:70-76.

Turner, P. R., L. A. Jaffe, and P. Primakoff. 1987. A cholera-toxin sensitive G-protein stimulates exocytosis in sea urchin eggs. *Dev. Biol.* 120:577-583.

Whitaker, M. J. and R. A. Steinhardt. 1982. Ionic regulation of egg activation. *Q. Rev. Biophys.* 15:593-666.

Whitaker, M. J. and R. F. Irvine. 1984. Microinjection of inositol torsposphate activates sea urchin eggs. *Nature (Lond.)* 312:636-638.

Whitaker, M. J. and J. Aitchison. 1985. Calcium-dependent phosphoinositide hydrolysis is associated with exocytosis *in vitro FEBS Lett.* 182:119-124.

Williams, D. A., K. E. Fogarty, R. Y. Tsien, and F. S. Fay. 1985. Calcium gradients in single smooth muscle cells revealed by the digital imaging microscope using Fura-2. *Nature (Lond.)* 318:558-561.

Yoshimoto, Y., T. Iwamatsu, K. Hirano, and Y. Hiramoto. 1987. The wave pattern of free calcium released upon fertilization in medaka and sand dollar eggs. *Dev. Growth & Differ.* 28:583-596.

PROTEIN KINASE C AND REGULATION OF THE Na^+-H^+ ANTIPORTER ACTIVITY DURING FERTILIZATION OF THE SEA URCHIN EGG

Sheldon S. Shen
Department of Zoology
Iowa State University
Ames, IA 50011

INTRODUCTION

The sea urchin egg is poised to initiate or accelerate a number of metabolic activities, which are quite different in kind. These changes in activities during fertilization have been causally linked to changes in intracellular ion activities (Whitaker and Steinhardt 1985), in particular, the transient increase in internal Ca^{2+} activity (Steinhardt and Epel 1974; Chambers *et al.* 1974; Steinhardt *et al.* 1977) and the rise in cytoplasmic pH (Epel *et al.* 1974; Johnson *et al.* 1976; Shen and Steinhardt 1978). The importance of these ionic changes is readily evident by their dramatic effects when imposed on eggs. Successful embryogenesis has been reported with parthenogenetic activation of sea urchin eggs with agents that raise internal Ca^{2+} activity and cytoplasmic pH (Brandriff *et al.* 1975). The activating capabilities of these parthenogenetic treatments are well documented for a wide variety of marine invertebrates (Shen 1983). A reason for continuing interests in the dual ionic signal associated with fertilization is the observations of similar dual ionic signal being a characteristic of many mitogenic stimulation of mammalian cells in culture (Hesketh *et al.* 1985, 1988; Moolenaar *et al.* 1986). There are many observations suggesting intracellular Ca^{2+} transient acts as a primary trigger for cell growth (Whitfield *et al.* 1980; Campbell 1983; Hesketh *et al.* 1985). More recently, increase in intracellular pH through activation of Na^+-H^+ antiporter during growth stimulation has been demonstrated (Moolenaar *et al.* 1986; Soltoff and Cantley 1988).

Similar second messenger pathways coupling stimulus and intracellular ionic responses have been demonstrated in both sea urchin fertilization and mitogenic stimulation. While a number of differences exist between cell systems and details remain to be worked out for a variety of growth factors, the generation of a bifurcating signal pathway by inositol lipid metabolism appears to be a central component in the control mechanisms of a wide variety of mammalian cellular responses (Rozengurt 1986; Berridge 1987; Zachary *et al.* 1987). Evidences for a similar pathway coupling sperm-egg binding and metabolic derepression in sea urchin fertilization have been reported. An early event during fertilization appears to be the hydrolysis of phosphatidylinositol 4,5-bisphosphate (PIP_2) by G-binding protein- regulated phospholipase C (PLC). A transient increase in inositol trisphosphate (IP_3) and 1,2-diacylglycerol (DAG), the two metabolic products of PIP_2 are detected within 20s of insemination (Turner *et al.* 1984; Ciapa and Whitaker 1986). PLC activity is associated with the egg plasma membrane in isolated egg corticies preparations (Whitaker and Aitchison 1985). Injections of nonhydrolyzable GTP or GDP analogs into unfertilized eggs can activate the egg or block the fertilization response, respectively (Turner *et al.* 1986). Further evidence of G-binding protein involvement during fertilization is stimulation of cortical granule exocytosis by injection of preactivated cholera toxin (Turner *et al.* 1987). Microinjection of IP_3 triggers intracellular Ca^{2+} release (Whitaker and Irvine 1984; Turner *et al.* 1986; Slack *et al.* 1986) and exogenous DAG or phorbol diesters (PMA), which can activate protein kinase C (PKC), will activate the Na^+-H^+ antiporter and protein synthesis (Swann and Whitaker 1985; Shen and Burgart 1986; Lau *et al.* 1986).

In many cell types, a large number of studies suggest that regulation of Na^+-H^+ antiporter activity is mediated by activation of PKC by DAG, which is generated by an increase in inositol lipid turnover. However, several studies have suggested multiple pathways for regulating Na^+-H^+ antiporter activity (Owen and Villereal 1982; Vara and Rozengurt 1985; Rosoff and Terres 1986; Letterio *et al.* 1986; Chambard *et al.* 1987; Ober and Pardee 1987; Huang *et al.* 1987). An alternative regulatory pathway may be Ca^{2+}-calmodulin (CaM)-dependent. In some cell lines, CaM antagonists have been reported to block mitogen- induced activation of Na^+-H^+ antiport (Rosoff and Terres 1986; Ober and Pardee 1987). Ca^{2+}-CaM-dependent pathways play important roles during the early events associated with sea urchin fertilization, including cortical granule exocytosis (Steinhardt and Alderton 1982; Stapleton *et al.* 1985) and NAD kinase activity (Epel *et al.* 1981).

A requirement for PKC as part of a signal transduction mechanism is inferred from the actions of synthetic diacylglycerols or tumor-promoting phorbol esters (Nishizuka 1986). PKC was first identified in rat brain and is now considered to be ubiquitous in tissues and organs (Kikkawa and Nishizuka 1986). However its identity and characterization in non-mammalian tissue have been sparse and mostly with adult tissues (Kuo *et al.* 1980; Blumberg *et al.* 1981; Rosenthal *et al.* 1987; Laurent *et al.* 1988). Differing effects of PKC activators on embryonic cells have been reported. DAG and PMA have been reported to stimulate maturation of rat oocyte (Aberdam and Dekel 1985), polychaete oocyte (Eckberg and Carroll 1987) and clam oocyte (Dubé *et al.* 1987; Eckberg *et al.* 1987; Dubé 1988), as well as inhibit maturation of mouse oocyte (Urner and Schorderet-Slatkine 1984; Bornslaeger *et al.* 1986) and starfish oocyte (Kishimoto *et al.* 1985). Injection of PKC purified from rat brain into *Xenopus* oocytes did not affect progesterone-induced maturation but accelerated insulin-induced maturation (Stith and Maller 1987). In view of possible involvement of kinases other than PKC in the action of DAG or PMA (Blenis *et al.* 1984; Morin *et al.* 1987), we have studied the kinetic characteristics and requirements of PKC activity in a partially purified cytosolic fraction of unfertilized sea urchin eggs. Furthermore, we have examined with specific protein kinase inhibitors the significance of PKC activity for regulation of Na^+-H^+ antiport during fertilization. These studies show the presence of PKC in sea urchin eggs, which differs immunologically, but show similar Ca^{2+}, phospholipid and diacylglyceride dependences of mammalian PKC preparations (Shen and Ricke 1989). While the antiporter activity may be regulated by PKC activity, we have found that regulation of the antiporter during fertilization appears to be Ca^{2+}-CaM-dependent.

MATERIALS and METHODS

Materials and Solutions

Lytechinus pictus were purchased from Pacific Biomarine Laboratories, Inc. (Venice, CA) and Marinus, Inc. (Long Beach, CA) and maintained in Instant Ocean culture systems with biweekly feedings of the *Macrocystis* algae. Eggs were obtained by injection of 0.5 M KCl into the coelomic cavity. The jelly coats were removed from the eggs by passage of the eggs through fine mesh silk and then washing the eggs twice in artificial sea water (ASW) of the following composition (millimolar): NaCl, 470; KCl, 10; $CaCl_2$, 11; $MgSO_4$, 29; $MgCl_2$, 27; $NaHCO_3$, 2.5; pH 8 with 0.1 M NaOH or 0.1 M HCl. We do not dejelly eggs by a brief acid (pH 4-5) wash because this treatment may raise intracellular pH (pH_i; Shen 1982) and cause partial activation of synthetic activities (Grainger *et al.* 1979). The dejellied eggs are maintained at 16 to 18°C and constantly stirred at 60 rpm. All experiments used eggs within 4h of shedding. Eggs used for PKC studies were also washed in Ca^{2+} and Mg^{2+}-free ASW of the following composition (millimolar): NaCl, 510; KCl, 10; EGTA, 10; $NaHCO_3$, 2.5; pH 8 with 0.1 M NaOH or 0.1 M HCl with 5 μg/ml aprotinin. The eggs were then washed and resuspended in intracellular buffer of the following composition (millimolar): K acetate, 220; glycine, 500; NaCl, 40; $MgCl_2$, 59; $CaCl_2$, 4.3; EGTA, 10; pH 6.9 with 20 mM Tris with 5 μg/ml aprotinin and 1 mg/ml soybean trypsin inhibitor. All chemicals, protease inhibitors, phospholipids, diacylglycerols and phorbol esters were purchased from Sigma Chemicals (St. Louis, MO), except for [Γ-^{32}P]-ATP from New England Nuclear (Boston, MA), sn- 1,2-dioctanoylglycerol from Avanti Polar Lipids (Birmingham, AL), protein kinase inhibitor K252a from Kamiya Biomedical Co. (Thousand Oaks, CA), inhibitors H-7, HA-1004, and W-7 from Seikagaku America, Inc. (St, Petersburg, FL), CaM inhibitor calmidazolium (R24571) from Janssen Pharmaceutica Inc. (Piscataway, NJ) and Sigma. The kinase and CaM inhibitors were prepared and used as recommended by the manufacturers. The phorbol diester, diacylglycerol, retinoic acid, K252a and R24571 were prepared as a stock solution in 100% dimethyl sulfoxide (DMSO, Fisher Scientific, Chicago, IL) and therefore an equivalent volume of this solvent was used as control (less than 0.5% of assay or bath volume). At this concentration of DMSO, no physiological effect was observed. Diethylaminoethyl cellulose (DE-52) was obtained from Whatman (Hillsboro, OR).

Manufacture of Microelectrodes

Conventional microelectrodes for recording membrane potentials were pulled from 1.5-mm o.d. borosilicate tubing with filament (Glass Co. of America, Inc., Bargaintown, NJ and W-P Instruments, Inc., New Haven, CT) and filled directly with 3 M KCl. The tip resistances of these electrodes in ASW ranged from 20 to 25 MΩ. Ion-sensitive microelectrodes were constructed by preheating the pulled conventional microelectrodes for 2h at 200°C in a drying oven, exposed to ~0.5 ml of hexamethyldisilazane (Sigma Chemical Co.) vapor under cover of a beaker for 20 to 30 min, and baked for 30 min more after the vapor was evacuated. The electrodes were allowed to cool in a desiccator and could be stored for a few days without deterioration. The silaned electrodes were backfilled with a proton-sensitive liquid exchanger (#95291, Fluka Chemical Corp., Ronkonkoma, NY), described by Ammann *et al.* (1981), and then backfilled with filtered solution of 0.1 M NaCl and 0.1 M citrate buffer, pH 6. To minimize the exchanger column length and contact exchanger with reference solution, a second micropipette with a long tip taper was pulled from 1.2 mm o.d. omega-dot tubing, broken below the shoulder, and inserted into the back of the ion-sensitive electrode (Orme 1969). The tip of the insert was positioned within 100 to 200 μm of the tip of the outer electrode. This effectively reduced the length of exchanger, lowering electrode resistance and greatly improving the electrode performance. The electrodes were calibrated before and after use in intracellular buffer with 100 mM phosphate buffer in place of Tris. The slopes of the electrodes used in this report varied from 54 to 61 mV/pH unit from pH 6 to 8. Response times were within 15s and the electrodes were sufficently sensitive to record 0.02 unit change in pH.

Electrophysiological Recording

The eggs were held on poly(lysine)-coated plastic petri dishes (Falcon 1008) and maintained at 18°C on a Wild dissection scope stage. The membrane potential (E_m) was recorded as the potential difference between the 3 M KCl-filled electrode and a 3 M KCl-agar bridge connected by an Ag-AgCl wire to ground with a Biodyne AM-4 preamplifier (Biodyne Electronics, Santa Monica, CA). The ion-sensitive microelectrode was connected by an Ag-AgCl pellet to an FD-223 electrometer (W-P Instruments, Inc.). Impalement of eggs with conventional microelectrodes was made by bringing the electrode against the egg membrane, then briefly overtuning the negative capacitance compensation. This method of impalement does not work with ion-sensitive microelectrodes, whose entry into the egg requires a sharp mechanical tapping of the manipulator or stage. The criteria for successful penetration of sea urchin eggs by both electrodes have been detailed elsewhere (Shen 1982). To determine that both electrodes are implanted in the egg with minimal damage, current pulses are passed through the KCl electrode periodically and the corresponding voltage deflections are monitored by the ion-sensitive microelectrode. Excessive membrane damage or vesiculation at the electrode tips results in a loss of electrical coupling. The outputs of the two amplifiers were displayed on a Tektronix 5111 storage oscilloscope (Tektronix, Inc., Beaverton, OR) and stored on magnetic tape with an instrumentation cassette recorder (A.R. Vetter Co., Rebersburg, PA). Pen recordings were made with a differential chart recorder (Soltec Corp., San Fernando, CA). The intracellular ion-sensitive electrode potential was electronically corrected for the simultaneously measured E_m. Due to the high tip resistance (10^{10} to 10^{11} Ω) of ion-sensitive microelectrodes, slight movements of the microelectrodes induced sudden voltage deflections. We did not use a low pass RC filter, in order to minimize the duration of coupling voltage pulses.

Acid Release

Acid production by sea urchin eggs during activation was followed by using a pH stat system consisting of a Radiometer (Copenhagen) pH meter (PHM 82), titrator (TTT 80), and autoburette (ABU 12). Three ml aliquots of 0.5 % ovicrits were maintained in suspension with a magnetic "flea". Stirring rate was carefully monitored to minimize egg breakage, which is reflected by gradual acidification of the bath. The pH of the egg suspension was allowed to stabilize at 8.00. After activation of the egg suspension, the decrease in the ASW pH was titrated with the addition of 1 mM NaOH. A chart recorder (Radiometer SBR 3) plotted the delivery of base as a function of time. The largest pH deviation during clamping to 8.00 was 0.03 pH unit. Acid release was expressed as mM by: molar acid release from eggs = (volume of base added x molarity of base)/egg volume determined by ovicrit.

Separation of PKC

A 50% egg suspension in intracellular buffer was homogenized by 5 strokes of a Dounce homogenizer. The homogenate was centrifuged at 100,000 x g for 60 min. The supernatant of this centrifugation is referred to as the soluble or cytosolic fraction. The pellet was resuspended in intracellular buffer with 0.5% Triton X-100 and stirred for 1h. This mixture was then centrifuged at 100,000 x g for 60 min with Amberlite XAD-2 to remove the Triton X-100 (Holloway 1973). The supernatant of this centrifugation is referred to as the particulate or membrane fraction. Both soluble and particulate fractions were dialyzed in 20 mM Tris (pH 7 at 4°C), 2 mM EGTA, 2 mM EDTA, 5 mM dithiothriotol and 0.1 mM phenylmethylsulfonyl fluoride. Routinely 10 to 20 ml of soluble or particulate fraction was loaded on 12 x 1.5 cm columns of DE-52, which had been equilibrated with the dialysis buffer. After washing with 150 to 300 ml of the same buffer, the activity was eluted with 50 to 100 ml of linear gradient of NaCl in dialysis buffer. Fractions of approximately 6 ml were collected and assay of the eluates was measured with conductivity measurements. All operations were performed at 4°C.

Protein Kinase Assay

Protein kinase C was assayed at 20°C with calf thymus H1 histone as a phosphate acceptor. Assays (250 μl) contained 20 mM Tris (pH 7), 50 μg H1 histone, 5.5 mM Mg acetate, 10 μM ATP, 1μCi [Γ-^{32}P]-ATP (1000-3000 Ci/mmole) and 0.5 mM EGTA or 0.5 mM free Ca^{2+} with 20 μg/ml of phosphatidylserine and 2 μg/ml of diolein. Reactions were initiated by addition of 25 μl of enzyme preparation, incubated for 5 min and terminated by addition of 1 ml of ice cold 25% (w/v) trichloroacetic acid (TCA) and 50 μl of 6 mg/ml bovine serum albumin (BSA). Precipitated protein was collected on GA-6 (Whatman) filters, washed with ice-cold 5% TCA and 1% $Na_4P_2O_7$, and counted in a scintillation counter (model 1217, LKB, Stockholm). The assay was linear from 1 to 10 min of incubation and 0.5 to 5 μg of protein. Phospholipids and neutral lipids were dissolved in chloroform, mixed appropriately, dried under nitrogen or in vacuum and resuspended in 20 mM Tris (pH 7) by sonication for about 1 min with a sonicator (model 300, Artek System, Farmingdale, NY). The sonicated samples were prepared fresh. Protein concentration was determined according to Bradford (1976) using BSA as standard. Data expressed in the text are presented as mean ± standard deviation (number of experiments).

RESULTS

Purification and Characterization of PKC

In homogenates of sea urchin eggs, a Ca^{2+} and phosphatidylserine/diglyceride (PS/D) dependent protein kinase activity was measurable and this activity was defined as PKC activity (Nishizuka 1984). Separation of the homogenate into soluble and particulate fractions showed the bulk of kinase activity to be associated with the cytosolic fraction (Table 1). Nearly 30% of the kinase activity measured by the phosphorylation of H1 histones was defined as PKC and greater than 90% of the total PKC activity in the homogenates was in the cytosolic fraction (Table 1).

DEAE cellulose chromatography using a linear gradient of NaCl resolved PKC activity separate from the Ca^{2+} and PS/D independent kinase activity (Figure 1). In elutions of four separate cytosol preparations, a mean of 57% of the total PKC activity eluted between 10 - 26 mM salt. This is in contrast to the higher salt concentration of near 90 to 100 mM necessary to elute mammalian PKC from DEAE cellulose (Kikkawa *et al.* 1986; Woodgett and Hunter 1987). Needless to say in our initial attempts which followed protocols for mammalian preparations, (these included a preliminary wash of buffer containing 20 mM salt) very little PKC activity was recovered during subsequent elution. This single step in elution resulted in nearly a 150-fold increase in enzyme specific activity, although the possible presence of endogenous inhibitors or activators of PKC and other phospholipid-stimulated histone kinases cannot be precluded. The majority of protein kinases, which are not regulated by Ca^{2+} and/or

Table 1. Subcellular Distribution of Protein Kinase C*

Fraction	EGTA	0.5 mM Ca^{2+} + PS/D
Soluble	1.01 ± .17	1.44 ± .24
Particulate	0.12 ± .04	0.15 ± .04

* PKC activity in sea urchin egg homogenates was separated into soluble and particulate fractions as described in Materials and Methods. Protein kinase activity was measured either in the presence of 2 mM EGTA without PS/D or in the presence of 0.5 mM free Ca^{2+} concentration and PS/D at 20 μg/ml and 2 μg/ml respectively. Activity is expressed as nanomoleof phosphate incorporated into histones per min per ml of packed eggs. The date are mean ± SE of three different batches of eggs.

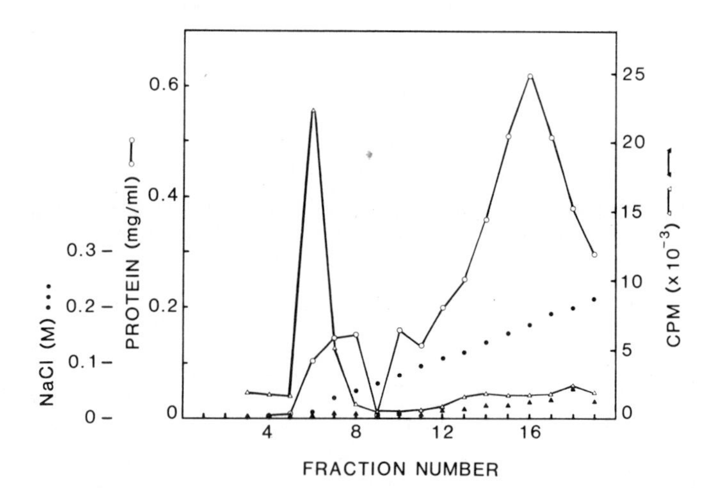

Figure 1. DEAE cellulose chromatography elution profile of PKC. PKC activity was measured either in the presence of EGTA (▲) or in the presence of 0.5 mM free Ca^{2+} concentration, 20 μg/ml PS and 2 μg/ml diolein (Δ). (○), protein concentration; (•), NaCl concentration.

PS/D, were eluted at concentrations greater than 0.2 M NaCl. Similar to the reports for mammalian DE-52 eluates, the enzyme activity was completely inactivated by freezing at -20°C and thawing. The PKC activity in DE-52 fractions stored at 4°C was very stable, losing less than 10% of their activity per month.

The activity of PKC was strongly dependent upon the addition of Ca^{2+} at 10 μg/ml of PS (Figure 2). As seen in a typical assay of PKC activity with varying Ca^{2+} activity, the addition

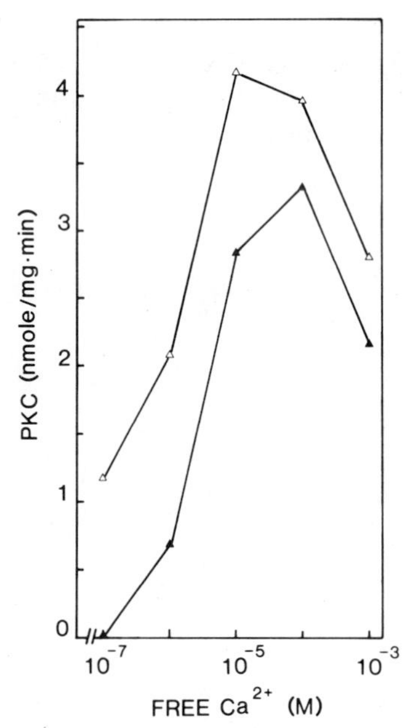

Figure 2. Effect of Ca^{2+} on PKC. The activity of PKC was dependent upon free Ca^{2+} concentration in the presence of 10 μg/ml of PS (▲) or in the presence of 10 μg/ml of PS and 10 μg/ml of diolein (Δ).

of 10 μg/ml of diolein enhanced the enzyme activity at all concentrations of free Ca^{2+} tested. More significantly, the enzyme activity was increased 11-fold in this example, by the addition of diolein at 10^{-7} M Ca^{2+}, which is near the intracellular level of free Ca^{2+} in unfertilized eggs (Poenie *et al.* 1985). In this case, near maximal enzyme stimulation was observed near 0.1 mM Ca^{2+}. In all cases, maximal enzyme stimulation was observed at 0.1 to 1 mM Ca^{2+} and 0.5 mM free Ca^{2+} concentration was routinely used in all activity assays of the effects of phospholipids and diacylglycerols. A wide variety of phospholipids and diacylglycerols were tested as costimulators of PKC activity (Table 2). At 50 μg/ml of phospholipid concentration without the presence of diolein, only phosphatidylserine significantly stimulated the enzyme. Activation by phosphatidylinositol was only 9% of the activity stimulated by PS, while phosphatidylcholine and phosphatidylethanolamine were even less stimulatory. The activity of PKC was stimulated nearly 200-fold with the addition of PS to 20 μg/ml and nearly 250-fold with the addition of PS to 60 to 80 μg/ml (Figure 3). In this assay, the addition of 2 μg/ml of the neutral lipid diolein enhanced by nearly 20% the activation of PKC at maximal PS stimulation. This effect of diolein was dose dependent. In the presence of 10 μg/ml of PS, the addition of diolein to 40 μg/ml stimulated the enzyme activity 2-fold (Figure 4).

Specificity of the isomer of DAG has been suggested from intact cell preparations (Shen and Burgart 1986); however, under these assay conditions, DAG with differing fatty acid moieties and of different isomers were equally effective in stimulating PKC activity (Table 2). The naturally occurring DAG will have the 1,2- sn stereochemistry and this enantiomer activates the enzyme. The 1,3-DAG has been reported to activate mammalian enzyme preparation, due to possible racemization to the 1,2 isomer (Boni and Rando 1985). The difference in stereospecific DAG requirement for stimulating PKC activity between intact

Table 2. Lipid Specificity for Activation of Protein Kinase C*

	nanomole / mg x min	% of standard[1] activation
PHOSPHOLIPID		
L-α-phosphatidyl-L-serine	2.57 ± .34	100
L-α-phosphatidylinositol (from soybean)	.23 ± .03	9
L-α-phosphatidylcholine (dioleoyl)	.08 ± .01	3
L-α-phosphatidylcholine (lecithin)	.08 ± .01	3
L-α-phosphatidylethanolamine (dioleoyl)	.13 ± .03	5
DIACYLGLYCEROL		
Diolein (C18:1, [cis]-9; 85% 1,3; 15% 1,2)	3.09 ± .32	100
1,3-Di-[(cis)-9-octadecenoyl]-rac-glycerol	3.63 ± .3	118
1,2-Dihexanoyl-sn-glycerol	3.05 ± .37	99
1,2-Dioctanoyl-sn-glycerol	3.69 ± .31	119
1,3-Dioctanoyl-rac-glycerol	4.17 ± .25	135
Phorbol 12-myristate 13-acetate (0.2 μg/ml)	2.83 ± .26	92
Phorbol 12-myristate 13-acetate-4-O-methyl ether (2 μg/ml)	2.20 ± .16	71

* PKC activity was tested as described in Materials and Methods. The phospholipid specificity was tested at 50 μg/ml without diolein. Diacylglycerol specificity was tested at 10 μg/ml in the presence of 10 μg/ml of L-α-phosphatidyl-L-serine. The phorbol esters were also tested with 10 μg/ml of phosphatidylserine present. In all experiments the free Ca^{2+} concentration was 0.5 mM. Data are the mean ± SE of three experiments.

[1] Standard activation was considered to be the PKC activity in the presence of PS for phospholipid experiments and PS with diolein for diacylglycerol replacements.

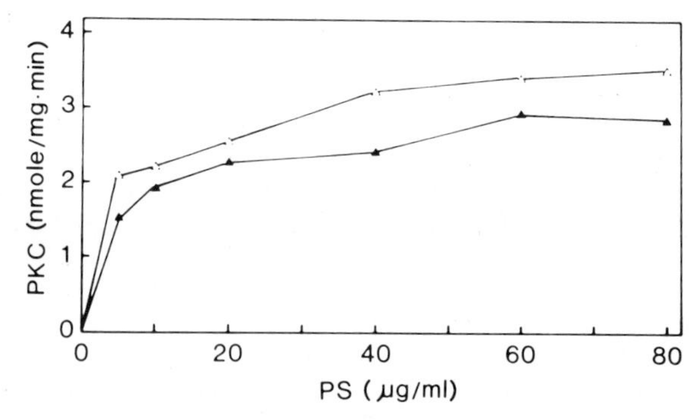

Figure 3. Effect of PS on PKC. The activity of PKC was dependent upon PS in the presence of 0.5 mM free Ca^{2+} concentration (▲) or in the presence of 0.5 mM free Ca^{2+} concentration and 2 μg/ml of diolein (△).

eggs and the cytosol fraction is unclear. Tumor promoting phorbol esters have been shown to activate unfertilized eggs. At a concentration similar to that used in intact cell preparations, phorbol 12-myristate 13- acetate (PMA) stimulated PKC activity. The non-biologically active moiety, PMA 4-0-methyl ether was less active even at a 10- fold greater concentration (Table 2).

Further attempts to purify PKC were unsuccessful. We used a variety of different strategies, which have been described for purification of mammalian preparations. These procedures using affinity chromatography included threonine-Sepharose (Kikkawa *et al.* 1986), phenyl-Sepharose (Woodgett and Hunter 1987) and polyacrylamide-immobilized phosphatidylserine (Uchida and Filburn 1984). While parallel purification of rat brain preparations proceeded uneventfully, in all cases of sea urchin preparations, PKC activity was not recovered. The cause of this difficulty remains uncertain. Evidence of polypeptide difference between sea urchin PKC and mammalian preparations was a lack of immuno cross reaction. The commercially available mouse monoclonal antibody MC5 (RPN. 536, Amersham Corp., Arlington Hts., IL), which recognizes mammalian α and ß species of PKC (Nishizuka 1988), was tested against sea urchin homogenates and DE-52 fractions using standard immunoblotting protocols recommended by the manufacturer. The epitope recognized by MC5 has been determined to be a 14 amino acid sequence lying in the hinge domain between the regulatory and catalytic domains. While a positive cross reaction was observed with the 80kD band of the rat brain preparation, no specific cross reaction was ever observed with sea urchin egg preparations.

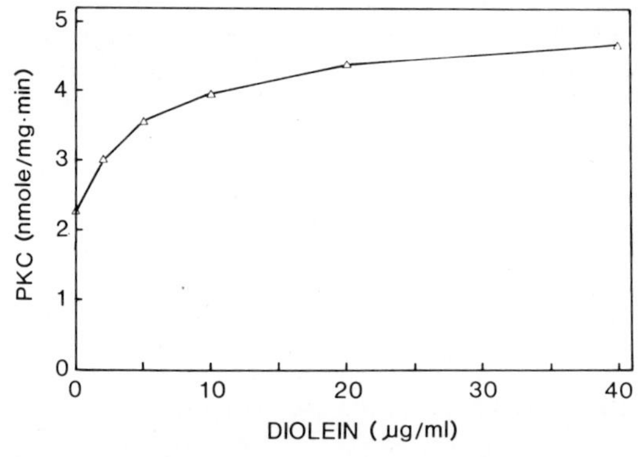

Figure 4. Effect of diolein on PKC. The activity of PKC was enhanced by the addition of diolein in the presence of 0.5 mM free Ca^{2+} concentration and 10 μg/ml of PS.

Protein Kinase Inhibitors

One of the primary purpose for trying to purify PKC from sea urchin eggs was to generate antibodies and specifically block the enzyme to evaluate its role during fertilization. Given the difficulties encountered with purifying the enzyme by affinity chromatography techniques and the absence of cross reaction with the mammalian monoclonal antibody, we turned to the recently developed protein kinase inhibitors of microbial origin and isoquinolinesulfonamide derivatives. Their structures and inhibition constants (K_i) are shown in Figure 5. K252a is of microbial origin and a very potent inhibitor of protein kinases (Kase *et al.* 1986). The inhibitions were of the competetive type with respect to ATP, which suggest direct interaction with the catalytic domain of the protein kinase. Maximal inhibition of 90% of PKC activity occurred with K252a greater than 50 nM. Since K252a is not water soluble, a fine precipitate is seen to occur with addition of stock into the bath and crystals of K252a can be seen in the dish. To be certain of penetration of the inhibitor across the intact egg membrane, 0.5 μM was routinely used. Eggs fertilized in the presence of 0.5 μM K252a have fertilization envelope (FE) formation, which appeared similar to untreated controls. Thus suggesting occurrence of the Ca^{2+} transient and Ca^{2+}-CaM-dependent events. However, these eggs became developmentally arrested prior to pronuclei fusion. Addition of K252a to unfertilized eggs had no significant effect on pH_i or cytoplasmic alkalinization during fertilization (Figure 6). In this example, the addition of K252a caused a drop in pH_i from 7.03

K-252a

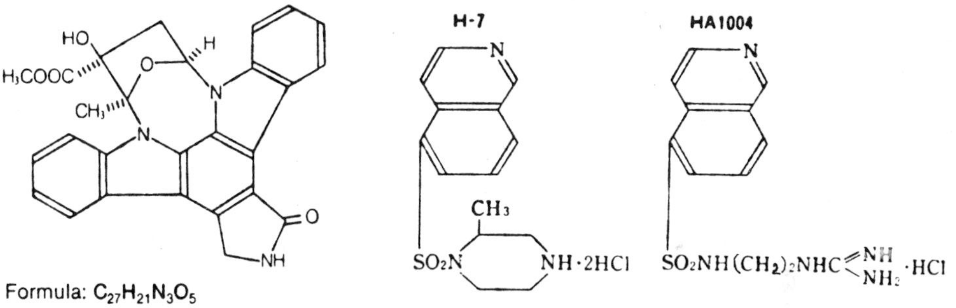

Formula: $C_{27}H_{21}N_3O_5$
Molecular weight: 467

H-7

SO_2N NH·2HCl

HA1004

$SO_2NH(CH_2)_2NHC(=NH)NH_2$·HCl

	Inhibition Constant, K_i (μM)		
Enzyme	K252a	H-7[3]	HA-1004[3]
Protein Kinase C	0.025[1]	6.0	40
cAMP-Dependent Protein Kinase	0.018[1]	3.0	1.3
cGMP-Dependent Protein Kinase	0.020[1]	5.8	2.3
Myosin Light Chain Kinase	0.020[2]	97	150

[1]Kase *et al.* 1987.
[2]Nakanishi *et al.* 1988.
[3]Hidaka *et al.* 1984.

Figure 5. Structures and K_i of Protein Kinase Inhibitors.

to 7.00 in the unfertilized egg and the cytoplasm alkalinized to 7.31 by 6 min post-fertilization. Normal FE elevation was observed by 100 s. In 6 successful recordings, pH_i rose from 6.95 ± .08 to 7.22 ± .16 during fertilization in K252a. In contrast, K252a blocked 1,2-dioctanoylglycerol (DiC_8)-induced rise in pH_i (Figure 7). Due to the formation of micelles, the actual effective concentration of DiC_8 is uncertain. However, incorporation of DAG into the egg membrane is often accompanied by membrane depolarization and subsequently an increase in membrane resistance. In 7 recordings, pH_i in eggs were 6.89 ± .08 and 6.91 ± .04 before and after the addition of DiC_8. Surprisingly, the presence of K252a did not block PMA-induced rise in pH_i (Figure 8). In this example, addition of K252a caused a rise in pH from 6.85 to a new stable value of 6.94 over 12 min. The addition of 250 nM PMA induced a rise in pH to a new stable value of 7.39 by 28 min. In a total of 4 experiments, the cytoplasm of eggs in K252a alkalinized from 6.92 ± .02 to 7.29 ± .14 with the addition of 250 nM PMA. Inhibition of DiC_8-induced rise in pH by K252a was specific, since subsequent fertilization or PMA treatment of these eggs was accompanied by cytoplasmic alkalinization (Figure 9). In these examples, pH rose from 6.85 to 7.09 and 6.82 to 7.03 for fertilization and PMA treatment respectively. In 6 cases, pH_i rose from 6.91 ± .04 to 7.09 ± .06 with fertilization of eggs previously treated with DiC_8 in the presence of K252a.

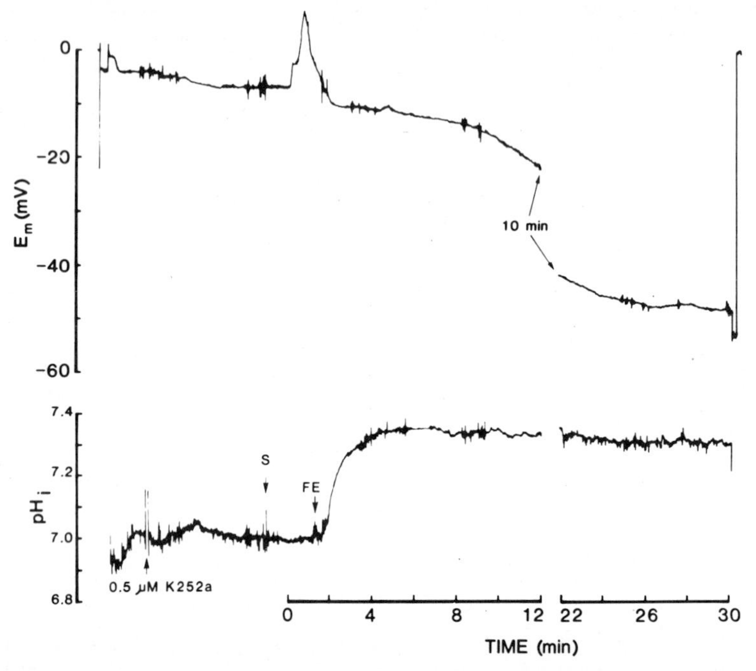

Figure 6. Changes in E_m and pH_i during fertilization in the presence of K252a. The unfertilized egg was first impaled with the conventional microelectrode and then a H^+-sensitive microelectrode. After pH_i stablized near 7.03, 0.5 μM K252a was added to the bath. By 7 min, E_m and pH_i stabilized at -8 mV and 7.0, respectively, sperm (S) was added and after a min reached the egg and fertilization ensued. Time zero is set with the onset of the fertilization potential. K252a in all cases had no effect on fertilization envelope (FE) elevation, which was noted near 90s. Subsequently, the cytoplasm alkalinized to 7.31 and the membrane hyperpolarized, all features being similar to normal fertilization responses. At about 30 min post-fertilization, the electrodes were removed. The ion-sensitive electrodes used in all experiments were calibrated before and after the experiments. 10 min of the continuous recording during fertilization was deleted in this figure. Due to the high tip resistance of the ion-sensitive electrodes, slight movements induced sudden voltage deflections. Most of the voltage shifts occurred during viewing through the dissection scope and especially during the addition and mixing of drugs into the bath.

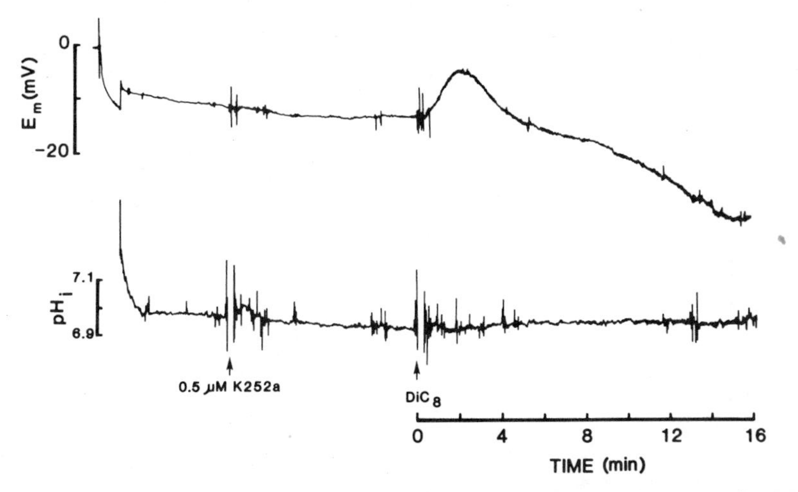

Figure 7. Changes in E_m and pH_i during DiC_8 activation in the presence of K252a. The unfertilized egg in ASW was first penetrated with the conventional microelectrode and then the H^+-sensitive microelectrode. K252a was added and subsequently DiC_8. Since the DAG formed micelles in the recording chamber, the actual concentration of DiC_8 is uncertain; however, incorporation of DAG is often accompanied by membrane depolarization and hyperpolarization, as well as, changes in membrane resistances. In this and all other cases, no change in pH_i is observed with DiC_8 activation of K252a-treated eggs.

Another group of well characterized protein kinase inhibitors are isoquinolinesulfonamide derivatives, including H-7 and HA-1004 (Figure 5; Kawamoto and Hidaka 1984; Hidaka *et al.* 1984). The inhibitions are reversible and competetive with respect to ATP. Similar to K252a, H-7 is a potent inhibitor of PKC. At 100 µM or higher concentration of H-7, greater than 95% of PKC activity in sea urchin preparations was inhibited. Near normal FE elevation

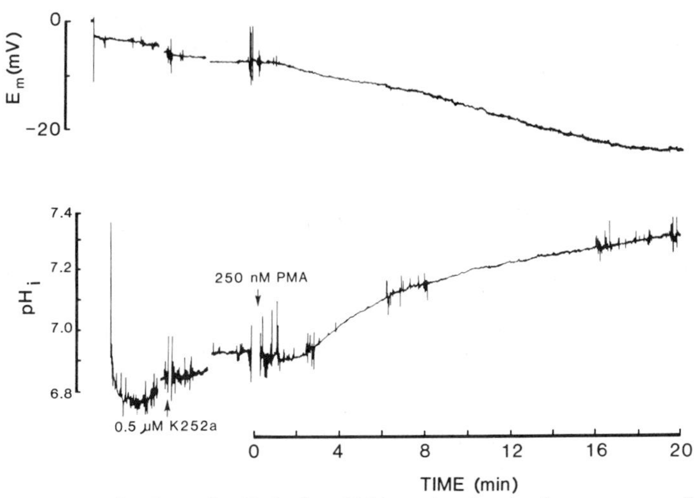

Figure 8. Changes in E_m and pH_i during PMA activation in the presence of K252a. In this example, 250 nM PMA was added to unfertilized eggs after treatment with K252a. In contrast to DAG activated eggs, PMA activated eggs were accompanied by rise in pH; however, no fertilization envelope elevation was observed. 3 min and 10 min are deleted from the figure during stabilization of pH after impalement and after addition of K252a, respectively.

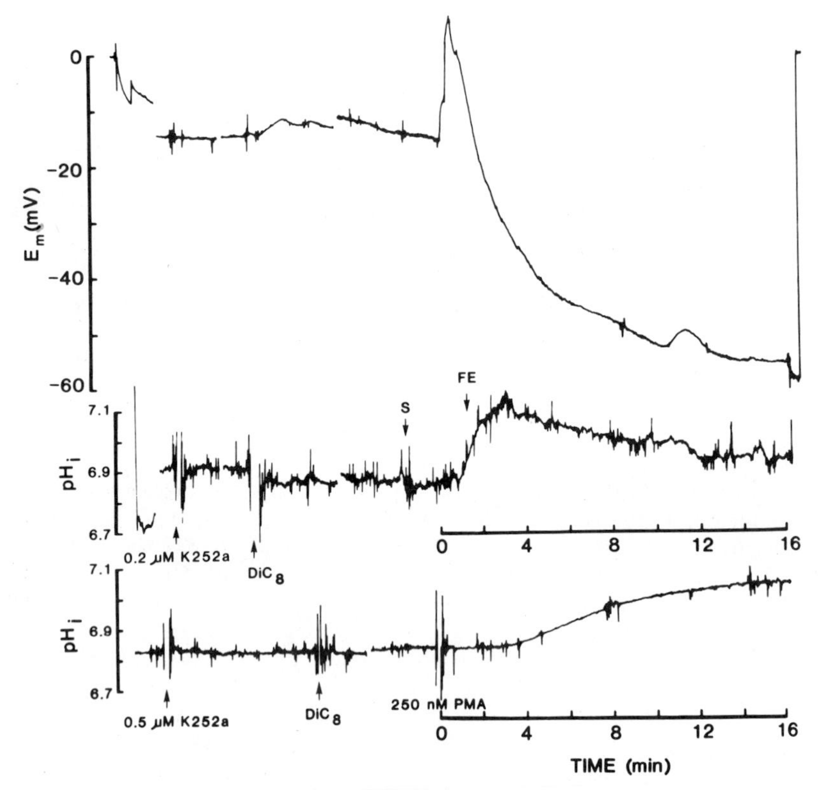

Figure 9. K252a inhibition of DiC_8 activation is specific. The top two traces show the changes in E_m and pH_i during K252a addition, DiC_8 addition and fertilization in a single egg. As in all cases, no changes in pH_i is observed during DiC_8 treatment in the presence of K252a; however, following addition of sperm, fertilization ensued and changes in E_m and pH_i, which normally occur during fertilization were observed. 6 min, 3 min and 10 min of the record during recovery from impalement, K252a and DiC_8 are deleted from the figure. Similarly, addition of PMA to K252a-treated and DiC_8-activated eggs resulted in a rise in pH. The accompanying E_m record is not shown and 6 min of the record during DiC_8 exposure is deleted.

was always observed during fertilization of eggs in 100 µM H-7 and in general the embryos were developmentally arrested prior to cell division. Even at this high concentration of H-7, there always appeared to be 2 to 5% of the embryos dividing and an occasional blastula in the dish next morning. At 10 µM of H-7, PKC activity was inhibited nearly 67%, but near control values of embryos divided and developed to blastula stage. The addition of 100 µM H-7 had little effect on pH_i in unfertilized eggs or cytoplasmic alkalinization during fertilization (Figure 10). In this case, pH_i rose from 6.91 to 7.23 during fertilization and in 3 experiments, pH rose from 6.93 ± .05 to 7.26 ± .02. In contrast to the effect of K252a, the presence of H-7 blocked PMA-induced rise in pH (Figure 11). In this most extreme example, pH_i rose slightly from 6.96 to 7.05 and in 3 experiments, pH_i changed from 6.9 ± .07 to 6.96 ± .09 with 250 nM PMA treatment in the presence of H-7. The inhibitory action of H-7 was specfic to PMA, fertilization of these eggs was accompanied by pH rise to 7.13 ± .02 (Figure 12). Since K252a and H-7 at the concentrations used in these experiments may block cyclic-nucleotide-dependent kinases, this possibility was tested by using 200 µM HA-1004. This agent is a potent inhibitor of c-AMP and c-GMP-dependent protein kinases and a weak inhibitor of PKC (Hidaka *et al.* 1984). At 200 µM HA-1004, eggs were observed to undergo normal develoment to blastula stage. Subsequent development became increasingly abnormal. The

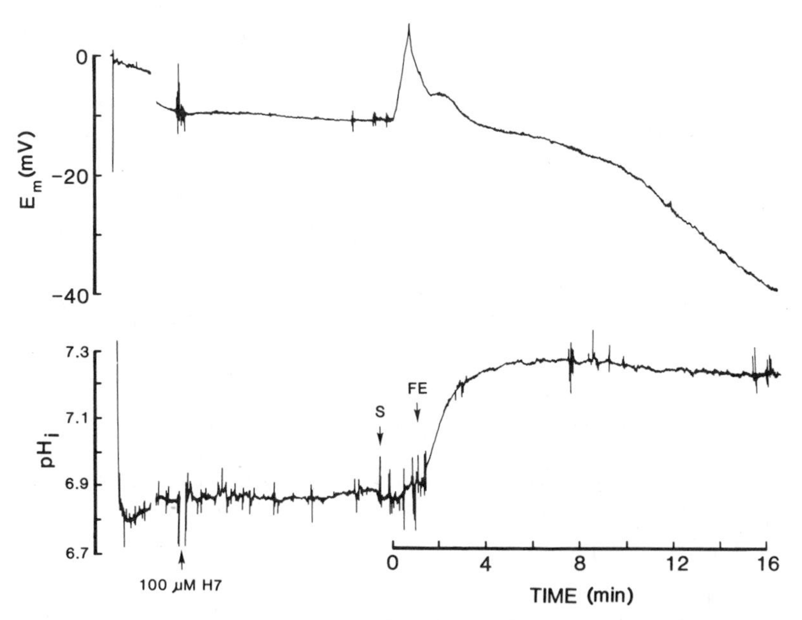

Figure 10. Changes in E_m and pH_i during fertilization in the presence of H-7. Similar to eggs treated with K252a, H-7 did not block the normal changes in E_m and pH_i during fertilization. 4 min of the record is deleted during stabilization of pH_i after impalements.

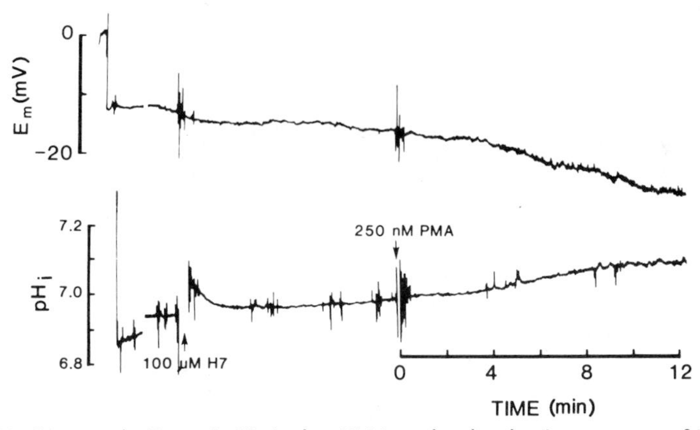

Figure 11. Changes in E_m and pH_i during PMA activation in the presence of H-7. In contrast to K252a, the presence of H-7 blocked the rise in pH during PMA activation. In this example, PMA treatment resulted in 0.09 rise in pH. In the other two recordings under similar conditions, pH_i rose only 0.06 and 0.04. These values are considerably less than the 0.3 pH unit rise seen with PMA activation of unfertilized eggs. 4 min of the record during recovery from impalement was deleted.

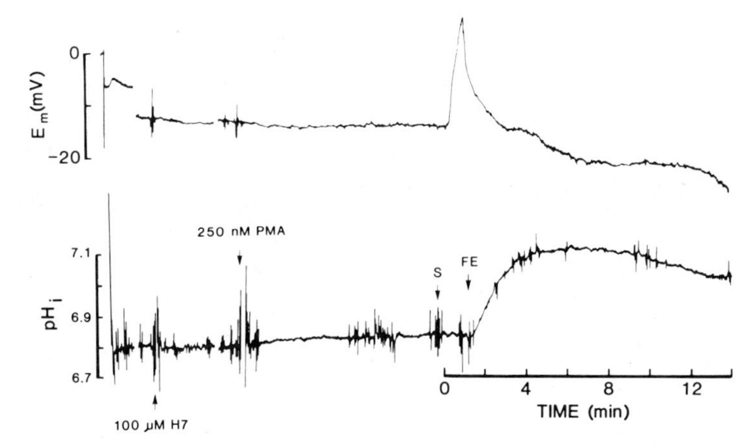

Figure 12. Changes in E_m and pH_i during fertilization of an egg, which had been treated with H-7 and PMA-activated. Similar to Figure 11, H-7 blocked PMA-induced rise in pH and the effect is specific to PMA action. Subsequent fertilization of the egg resulted in fertilization potential, FE elevation, cytoplasmic alkalinization and membrane hyperpolarization. A 7 min span of recording is deleted from the figure during recovery from impalements.

presence of HA-1004 had little effect on PMA-induced rise in pH (Figure 13). In this case, the cytoplasm alkalinized from 7.0 to 7.31 by 16 min after the addition of 250 nM PMA and in another experiment PMA induced a rise in pH from 6.99 to 7.33. These observations suggest the inhibitory actions of H-7 and K252a were mediated through the inhibition of PKC. Furthermore, these results suggest that regulation of Na^+-H^+ antiporter during fertilization may occur by a PKC-independent pathway.

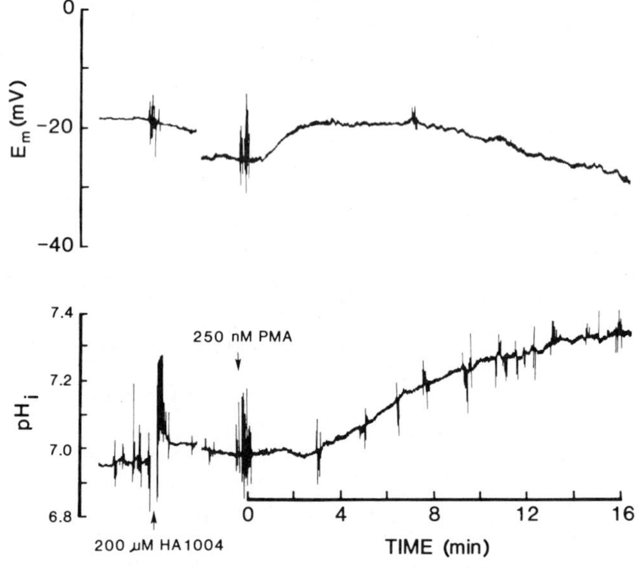

Figure 13. Changes in E_m and pH_i during PMA activation in presence of HA-1004. Since H-7 inhibits cyclic nucleotide-dependent protein kinases and PKC, the effect of HA-1004, which is weakly inhibitory of PKC activity, was tested. HA-1004 does not block early development during fertilization or PMA-induced changes in pH_i, thus the inhibitory actions of H-7 during egg activation is most likely mediated by inhibition of PKC activity.

Calmodulin Inhibitors

In order to characterize the PKC-independent pathway for regulating the Na^+-H^+ antiporter, two agents, which have been reported to inhibit a variety of enzyme reactions and biological processes that are Ca^{2+}-CaM-dependent (Van Belle 1981; Hidaka *et al.* 1979) were tested for their effects on changes in pH_i in sea urchin eggs. Their structures and reported K_i upon various kinases are shown in Figure 14. Calmidazolium (R24571), a derivative of miconazole, was found to have variable effects during fertilization. At 10 μM R24571, FE formation was observed to vary between 17% to 54% with different batches of eggs. Eggs with FE were observed to divide with varying degrees of abnormal development to blastula stage. In order to test if the action of R24571 was directly upon cortical granule exocytosis, cortical granule lawns were prepared (Vacquier 1975; Whitaker and Baker 1983) and assayed for the effect of 10 μM R24571 upon exocytosis, which was triggered by Ca^{2+} addition. In 3 trials, no differences from control values were observed. Electrophysiological recordings showed no effect of R24571 upon cytoplasmic alkalinization (Figure 15) when fertilization was detected by membrane depolarization. In this case pH_i rose from 6.91 to 7.38 and in another recording, pH_i rose from 6.89 to 7.27. The presence of 10 μM R24571 had no effect on PMA- induced alkalinization of the cytoplasm. In 3 experiments, the addition of 250 nM PMA induced a rise in pH from 6.95 ± .09 to 7.3 ± .14 in 15 to 20 min.

A naphthalenesulfonamide derivative, W-7, has been reported to be a potent inhibitor of Ca^{2+}-CaM-dependent enzyme reactions (Hidaka *et al.* 1979; 1981) and Ca^{2+}-PS-dependent kinase (Tanaka *et al.* 1982; Schatzman *et al.* 1983). At 200 μM W-7, FE formation was observed on less than 5% of the eggs and no cell division was observed. At lower concentrations of W-7, increasing percentages of eggs undergo FE formation and cell division, such that at 50 μM W-7, nearly 80% of the eggs elevated FE and subsequently divided. However, these embryos were not observed to develop past the 4-cell stage. In the presence of 10 μM W-7, near normal numbers (>95%) of the eggs elevated FE and divided. These embryos did not form blastulas but remained as disorganized clumps of cells. The addition of 200 μM W-7 caused a rise in pH_i in unfertilized eggs. In 14 completed recordings, the addition of W-7 caused pH_i to rise 0.15 ± .06 pH unit. After a new stable cytoplasmic pH had been

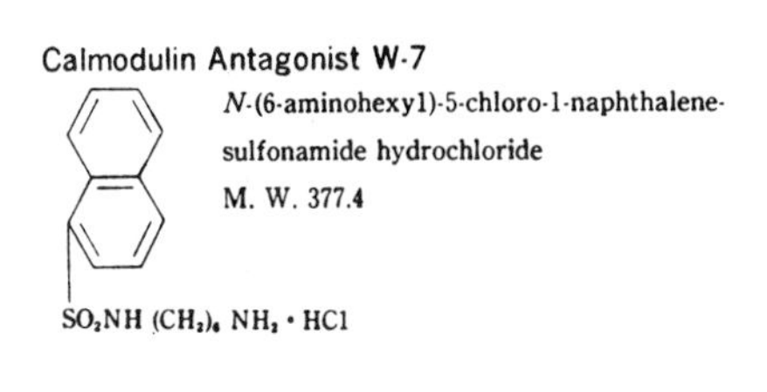

Structure: R 24 571

M.W.: 688

	Inhibition Constant, K_i (μM)	
Enzyme	R24571[1]	W-7
Phosphodiesterase	0.1	7.5[2]
Myosin Light Chain Kinase	2.1	12[3]
Protein Kinase C	14.5	110[3]

[1]Mazzei *et al.* 1984.
[2]Hidaka *et al.* 1979.
[3]Tanaka *et al.* 1982.

Figure 14. Structure of calmodulin antagonists, W-7 and R24571, and their respective K_i's for phosphodiesterase, myosin light chain kinase and protein kinase C activities.

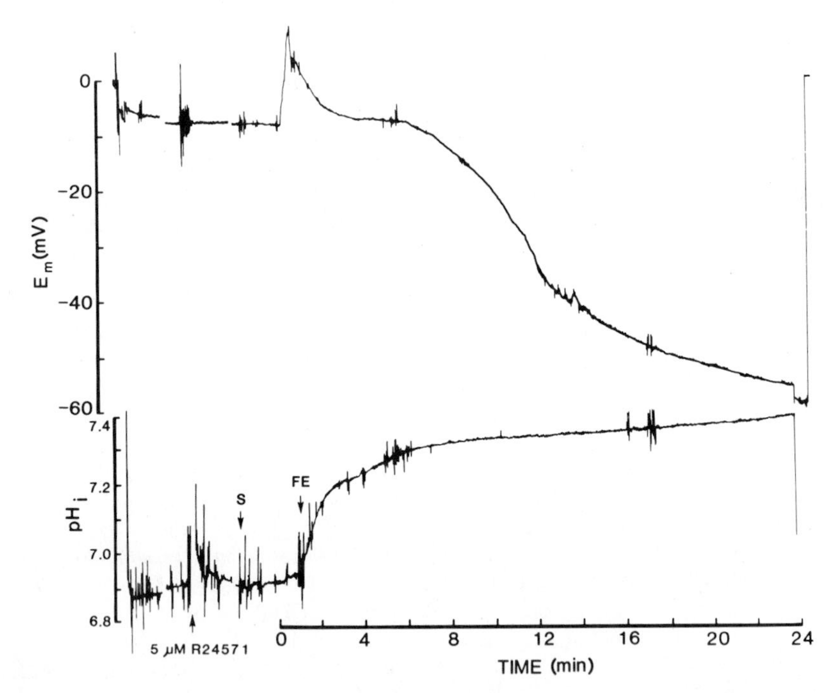

Figure 15. Changes in E_m and pH_i during fertilization in the presence of R24571. Similar records were observed at 10 μM R24571. In some cases, eggs did not fertilize in the presence of R24571, which may be due to its inhibition of sperm activation. Nonetheless, when fertilization potential was observed, FE elevation, cytoplasmic alkalinization and membrane hyperpolarization were observed.

attained, fertilization of these eggs was not accompanied by cytoplasmic alkalinization (Figure 16). In this example, the addition of W-7 elevated pH_i from 6.9 to 7.05 during the 10 min incubation. Fertilization was marked by depolarization of the membrane, but even 10 min after fertilization, pH_i remained at 7.07. In many recordings, onset of a slow alkalinization was seen by 15 min post-fertilization. From 6 experiments, the addition of W-7 triggered a rise

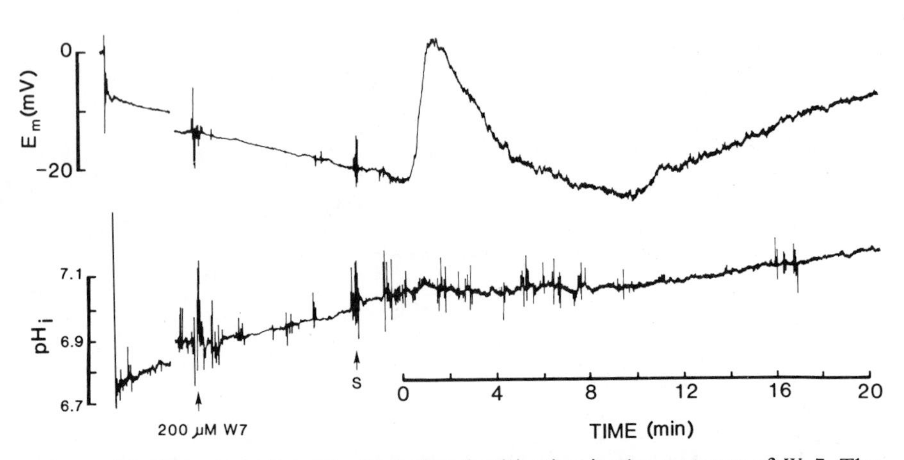

Figure 16. Changes in E_m and pH_i during fertilization in the presence of W-7. The addition of 200 μM W-7 caused an alkalinization in this case from 6.9 to 7.05. Subsequent to the addition of sperm, a fertilization potential was observed; however, no rise in pH_i or FE elevation was observed in this or any other similar experiments. 5 min of the record during recovery from impalements is deleted.

in pH_i from 6.82 ± .06 to 6.97 ± .1 and at 10 min post-fertilization the pH_i was 7.03 ± .07. Presence of 200 μM W-7 did not block the pH rise induced by PMA or DiC_8 treatment of the egg (Figure 17). In this example with 250 nM PMA treatment, pH_i rose from 6.86 to 7.03 with the addition of W-7 and after the addition of PMA, the cytoplasm alkalinized to 7.32. For 6 successful experiments, pH_i were for the unfertilized egg, 6.84 ± .04, rising to 6.99 ± .09 after addition of W-7, and rising further to 7.26 ± .1 with exposure to 250 nM PMA. A more gradual rise in pH was seen with DiC_8 treatment (Figure 17). In this case, DiC_8 addition caused a rise in pH_i from 7.02 to 7.27, and in another experiment, pH_i rose from 7.06 to 7.28. Both alkalinization required nearly 20 min to attain new stable values. The inhibitory actions of W-7 on cytoplasmic alkalinization during fertilization was specific. Exposure of fertilized eggs to 250 nM PMA triggered a rapid rise in pH (Figure 18). In this case the pH rose from 7.16

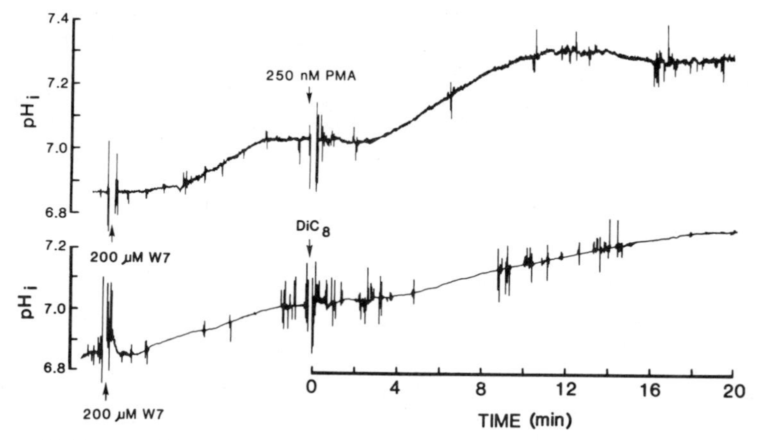

Figure 17. Changes in pH_i during PMA or DiC_8 activation of unfertilized eggs in the presence of W-7. In nearly all cases, the addition of W-7 caused a rise in pH_i. The cause of this change in pH is uncertain. After stabilization of pH in W-7, PMA or DiC_8 addition resulted in further rise in pH, which is similar to those observed earlier with untreated eggs. The inhibitory actions of W-7 at 200 μM in sea urchin eggs do not include blockage of PKC activation by PMA or DiC_8.

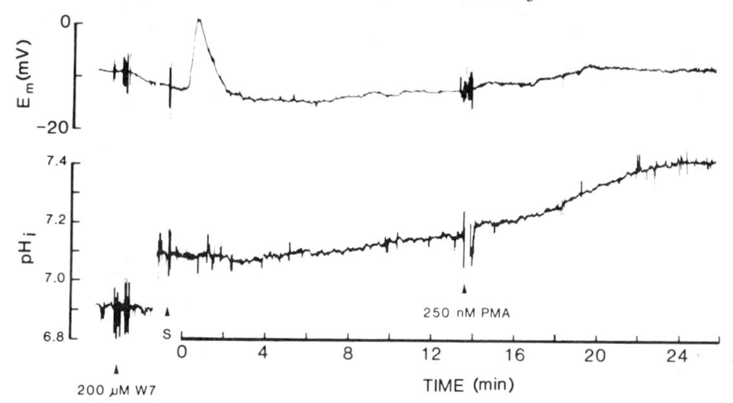

Figure 18. Inhibition by W-7 of fertilization-induced rise in pH is specific. PMA treatment of eggs fertilized in the presence of W-7 resulted in a rise in pH. In this case, W-7 addition caused a rise of 0.18 pH unit (8 min is deleted from the figure during this alkalinization). Fertilization is accompanied by membrane depolarization and a gradual alkalinization 7 min post- fertilization. The cause of this slow rise in pH is uncertain, visual inspection of eggs fertilized in W-7 suggested gradual lysis, which might be accompanied by a slow alkalinization. Nonetheless, the addition of PMA resulted in a near normal appearing rise in pH from 7.16 to 7.42.

to 7.42. In 4 recordings, the addition of PMA caused a rise in pH from 7.04 ± .07 of fertilized eggs to 7.29 ± .08. A summary of the effects of the inhibitors on pH_i and the effects of fertilization, PMA and DiC_8 treatments is compiled in Table 3.

Acid Release

Further corroboration of the effects of the inhibitors on changes in pH_i during fertilization or PMA activation was measurements of acid release from egg suspensions. As seen in Figure 19, acid release was detected during fertilization and PMA treatment similar to values reported previously (Shen and Burgart 1986). The addition of 100 μM H-7 caused no significant acid release during the 10 min incubation period and had no effect on the acid release normally seen during fertilization. However, the acid release normally seen during PMA treatment was blocked in the presence of H-7 but subsequent fertilization of these eggs was accompanied by near normal acid release (Figure 19A). With another batch of eggs, acid release was observed during fertilization and PMA activation (Figure 19B). In these studies, acid release during incubation in 200 μM W-7 occurred and is not shown. The addition of 250 nM PMA to W-7 treated eggs resulted in near normal acid release. In contrast, fertilization of W-7 treated eggs showed no significant release of acid and PMA treatment of these eggs was accompanied by acid release. Eggs fertilized in the presence of W-7 appeared to be more fragile with a slightly sloped baseline, which suggests gradual egg lysis.

Table 3. The Effect of Protein Kinase and Calmodulin Inhibitors On pH_i in Unfertilized Eggs and Changes in pH_i During Fertilization, PMA or DAG Activation.

	Changes in Intracellular pH (pH_i)			
Inhibitor	Unfertilized Egg	Fertilization	250 nM PMA	DiC_8
None	—	0.43 ± .03[1] (n=15)	0.29 ± .07[2] (n=9)	0.22 ± .06[2] (n=5)
K252a (0.5 μM)	0.01 ± .01[3] (n=17)	0.27 ± .09 (n=6)	0.37 ± 0.1 (n=4)	0.03 ± .03[4] (n=7)
H-7 (100 μM)	0.03 ± .02 (n=7)	0.33 ± .04 (n=3)	0.06 ± .02[5] (n=3)	—
HA-1004 (200 μM)	0.04 (n=2)	[6]	0.33 (n=2)	—
W-7 (200 μM)	0.15 ± .06 (n=14)	0.06 ± .03[7] (n=6)	0.27 ± .05 (n=6)	0.23 (n=2)
R24571 (10 μM)	0.01 ± .01 (n=5)	0.43 (n=2)	0.35 ± .05 (n=3)	—

[1] Steinhardt *et al.* (1978)
[2] Shen and Burgart (1986)
[3] Mean ± standard deviation (Number of experiments).
[4] Fertilization of these eggs resulted in a rise of 0.17 ± .06 (n=6) pH unit.
[5] Fertilization of these eggs resulted in a rise of 0.18 ± .07 (n=3) pH unit.
[6] Eggs fertilized in the presence of HA-1004 undergo cleavages.
[7] PMA treatment of these eggs resulted in a rise of 0.25 ± .07 (n=4) pH unit.

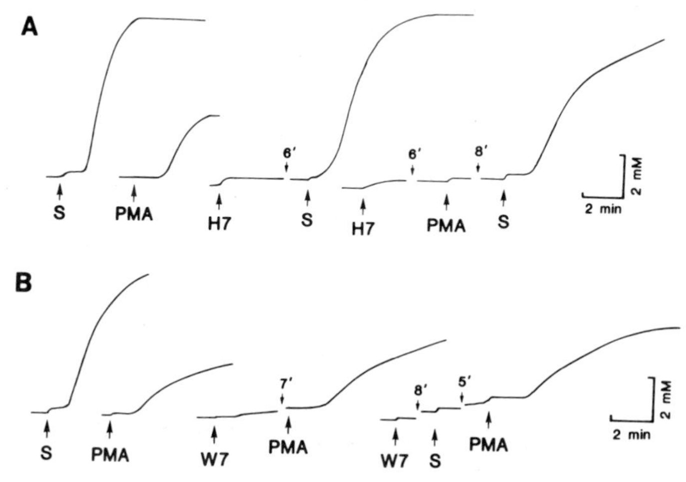

Figure 19. Acid release during fertilization or PMA activation. **A.** In the presence of 100 μM H-7. The first two traces are acid release during fertilization (S) and 250 nM PMA activation, respectively. Acid release by PMA is less than that of fertilization because PMA does not trigger cortical granule exocytosis, which contributes to the acid release measured during fertilization. The addition of H-7 had no effect on fertilization-induced acid release. 6 min of the record is deleted. In contrast, H-7 blocked 250nM PMA-induced acid release, but fertilization of this batch of eggs resulted in near normal acid release. 6 min and 8 min of the record during H-7 and PMA incubations, respectively, were deleted. **B.** In the presence of 200 μM W-7. The first two traces are again of fertilization and 250nM PMA-induced acid release from another batch of eggs. The presence of W-7 did not block PMA-induced acid release. Acid release is seen during W-7 incubation and 7 min was deleted. The addition of sperm did not cause acid release, but 250 nM PMA treatment of these eggs was accompanied by near normal acid release. 8 min and 5 min during W-7 incubation and fertilization, respectively, were deleted.

DISCUSSION

Protein Kinase C from Sea Urchin Eggs

In recent years, inositol phospholipid hydrolysis acting as a bifurcating second messenger pathway, which regulates a wide variety of physiological processes, has been established (Sekar and Hokin 1986; Berridge 1987). Since the principal cellular target of DAG is PKC (Nishizuka 1984; 1986), occurrence of PKC in a wide variety of organisms has been demonstrated. While the presence of PKC is implicated by the actions of DAG or PMA throughout development, most of the direct studies of PKC have been limited to adult tissues. We have established that sea urchin eggs contain a Ca^{2+}-PS-dependent protein kinase, which is similar, but not identical to that described in mammalian tissues (Shen and Ricke 1989). While PKC activity was detectable in homogenates of unfertilized sea urchin eggs, its activity was greatly enhanced following fractionation by DE-52 chromatography, suggesting the presence of PKC inhibitor(s) and/or phosphatases (Ingebritsen and Cohen 1983; McDonald *et al.* 1987). Similar to a recent report that established the presence of PKC in *Xenopus* oocytes (Laurent *et al.* 1988), sea urchin PKC was eluted from DEAE- cellulose at a lower NaCl concentration than those reported for mammalian preparations. Sea urchin PKC required Ca^{2+} and phsopholipid, particularly PS, for its activation. Similar to mammalian PKC (Kishimoto *et al.* 1980; Kaibuchi *et al.* 1981), the addition of DAG decreased the Ca^{2+} requirement for PKC activation and thus activating the enzyme without an increase in Ca^{2+}

at levels similar to those detected in unfertilized eggs (Poenie *et al.* 1985). The resulting picture for enzyme activation is of a synergistic action of an increase in Ca^{2+} and formation of DAG in the presence of PS.

Given the broad range of PKC effects, it is not surprising that PKC is now known to be a large family of proteins, with at least 7 subspecies (Nishizuka 1988). The relationship of the sea urchin PKC to these subspecies is unknown. Using the commercially available antibody to PKC, which cross reacts with types II and III of PKC from mammalian tissues, we failed to find specific cross reaction with the sea urchin PKC. In many mammalian cell types, stimulation of the enzyme is associated with its translocation from cytoplasm to the plasma membrane (Wolf *et al.* 1985; Vaartjes *et al.* 1986). Differing localization of PKC has been reported during oocyte maturation. PKC activity was demonstrated in both cytosolic and membrane fractions of immature *Xenopus* oocytes, with greater than 90% of the activity in cytosol (Laurent *et al.* 1988), which is similar to our findings in sea urchin eggs (Shen and Ricke 1989). By Metaphase II, the oocytes no longer showed membrane-associated PKC activity; however, enzyme activity in the cytosolic fraction did not increase. In *Spisula* oocytes, PKC has been reported to shift from membrane-associated in the oocytes to cytosol post-fertilization (Eckberg *et al.* 1987). In preliminary experiments, we did not observe any evidence of translocation during the first 10 min of fertilization. However, PKC activity is found in the particulate fraction by pluteus stage (Blumberg *et al.* 1981).

Action of Protein Kinase and Calmodulin Inhibitors

The involvement of PKC in cellular processes is often implied from PMA and/or DAG induction of events that mimic the cellular process. Several studies with PKC agonists suggest that they may have other differing biochemical effects aside from activation of PKC (Morin *et al.* 1987) or they may have differing effects on PKC substrate specificity (Ramsdell *et al.* 1986). An alternative approach to discern the contribution of PKC is to block its activity with inhibitors. In this study, I have used members of two different classes of PKC antagonists. While both inhibitors were effective in blocking PKC activity *in vitro*, they differed in their actions in the intact egg. Both K252a and H-7 blocked the action of DiC_8, while K252a was ineffective in blocking PMA- induced rise in pH. K252a and H-7 have both been reported to compete with ATP for the catalytic site of PKC (Hidaka *et al.* 1984; Kase *et al.* 1987; Loomis and Bell 1988). These PKC antagonists have been reported to have different effects on events during PMA-induced neutrophil activation. K252a inhibited superoxide anion production by neutrophil during PMA stimulation (Smith *et al.* 1988), while 100 μM H-7 was ineffective (Wright and Hoffman 1986). However, using electropermeabilized neutrophils and H-7 at 200 μM, inhibition of PMA-induced oxygen consumption has been reported (Grinstein and Furuya 1988). Since both K252a and H-7 blocked DAG action and early development of fertilized eggs, their differential effect on PMA action in our studies cannot be a problem of cell permeability. Also, HA-1004 did not block PMA-induced pH rise, the difference between K252a and H-7 is most likely unrelated to their different effects on cyclic nucleotide-dependent kinases.

A third reported class of PKC inhibitor is retinoids (Cope *et al.* 1986). Retinol has been reported to block PMA-induced maturation events in *Chaetopterus* and *Spisula* oocytes (Eckberg and Carroll 1987; Eckberg *et al.* 1987). At 100 μM retinoic acid, I did not observe inhibition of PKC activity *in vitro* or altered cytoplasmic alkalinization during fertilization, although cleavage was arrested. Regardless of the bases of the differential actions of the reported PKC antagonists, they blocked PKC activity and the action of PKC agonists, but failed to block cytoplasmic alkalinization during fertilization. Thus, activation of the Na^+- H^+ antiporter can occur by PKC-independent pathway.

In contrast to PKC antagonists, the Ca^{2+}-CaM antagonist, W-7, was found to block fertilization-associated rise in pH. While W-7 has been reported to also block PKC activity (Schatzman *et al.* 1983; Tanaka *et al.* 1982), the reported K_i is 5 to 10-fold greater than those for Ca^{2+}-CaM-dependent kinases. Furthermore, W-7 at 200 μM was not observed to block PKC agonist induction of rise in pH. W-7 has been reported to block PMA-induced maturation of invertebrate oocytes (Eckberg and Carroll 1987; Eckberg *et al.* 1987). However, a variety of CaM antagonists have been reported to block maturation in invertebrate oocytes

(Dorée *et al.* 1982; Carroll and Eckberg 1986), which suggests oocyte maturation may be dependent upon both Ca^{2+}-CaM-dependent and PKC-dependent pathways.

Similar to PKC antagonists, different effects of CaM inhibitors during sea urchin fertilization were observed. Calmidazolium (R24571) has been reported to inhibit germinal vesicle breakdown in *Spisula* oocytes (Carroll and Eckberg 1986). We have found R24571 to have variable effects on intact eggs and no effect on cortical granule exocytosis *in vitro*. In contrast, other CaM inhibitors, including W-7, have been reported to block the cortical reaction (Picard and Dorée 1982) and acid release (Komukai *et al.* 1985). Kinetic analysis of W-7 and R24571 indicate that the primary effect of these agents is mediated through competitive inhibition of the enzyme activation by interaction with cofactors (Tanaka *et al.* 1982; Mazzei *et al.* 1984). Using CaM antibodies, Steinhardt and Alderton (1982) concluded that CaM in sea urchin eggs may be cryptic, thus one possible cause of the differential effect of R24571 and W-7 is their ability to interact with CaM.

W-7 has been found to block the acrosome reaction of sea urchin spermatozoa and thus the fertilization reaction (Sano 1982). In contrast, aster formation has been reported in eggs fertilized in the presence of W-7 and inhibition of FE elevation was unaffected by calcium-ionophore A23187 treatment of these eggs (Picard and Dorée 1982). Further, normal Ca^{2+} fluxes were measured during fertilization of eggs treated with W-7 (Komukai *et al.* 1985). The addition of W-7 following fertilization blocked acid release normally associated with sea urchin fertilization (Komukai *et al.* 1985). The W-7 sensitive period for inhibition of acid release was between 30 and 120s, which is near the lag time for activation of Na^+-H^+ antiport (Shen and Steinhardt 1979). In the experiments reported here, membrane depolarization was detected in all cases when W-7-treated eggs were fertilized. Thus, it appears that cytoplasmic alkalinization during fertilization is Ca^{2+}-CaM-dependent.

Regulation of Na^+-H^+ Antiporter Activity

Several mechanisms for regulating the Na^+-H^+ antiporter activity have been demonstrated. The stimulation of Na^+-H^+ antiport by many stimuli is mediated by PKC (Rozengurt 1986; Soltoff and Cantley 1988). A variety of stimuli; however, must activate Na^+-H^+ antiport, independently of PKC activity. Members of mitogens, which are known to bind receptors with tyrosine kinase activity can stimulate Na^+-H^+ antiport and reinitiate DNA synthesis without activating inositol phospholipid hydrolysis (Hunter and Cooper 1985; Vara and Rozengurt 1985; Besterman *et al.* 1986; L'Allemain and Pouysségur 1986; Berk *et al.* 1987; Chambard *et al.* 1987).

Since increase in cytoplasmic Ca^{2+} and pH are coupled during early mitogenic events in many cell types, Ca^{2+} mobilization has been implicated in the control of Na^+-H^+ antiporter activity (Owen and Villereal 1982). Na^+ influx in human fibroblasts is stimulated by the divalent cation ionophore, A23187 and serum, but not by phorbol esters (Muldoon *et al.* 1987). Serum and A23187-stimulated Na^+ influx is inhibited in a dose dependent manner by calmodulin (CaM) antagonists (Owen and Villereal 1982). In other cell types, both PKC-dependent and Ca^{2+}-CaM-dependent regulation of Na^+-H^+ antiport have been inferred by differential inhibitor sensitivities. Agents that act as CaM antagonist blocked epithelial growth factor-stimulated Na^+-H^+ antiport in Chinese hamster embryo fibroblasts (Ober and Pardee 1987) and T3-T cell receptor-stimulated Na^+-H^+ exchange during T cell activation (Rosoff and Terres 1986). In both cell types, a separate PKC- dependent system regulating Na^+-H^+ antiport remained intact, which could be blocked by H-7. The action of Ca^{2+} may also be mediated through arachidonic acid release, which appears to play a role in thrombin activation of Na^+-H^+ antiport in vascular smooth muscle (Huang *et al.* 1987; Mitsushai and Ives 1988).

The work I have presented here suggests the presence of both PKC-dependent and -independent regulation of Na^+-H^+ antiporter in sea urchin eggs. The data does not suggest the relative roles of the two pathways during fertilization. The effect of W-7 on inositol phospholipid hydrolysis in sea urchin eggs is uncertain. W-7 probably does not block inositol phospholipid hydrolysis in Chinese hamster embryo fibroblasts. In these cells, thrombin stimulates inositol phospholipid hydrolysis (Raben and Cunningham 1985) and W-7 does not block thrombin-induced activation of Na^+-H^+ antiport, which is inhibited by H-7 (Ober and Pardee 1987). Since FE elevation is absent when eggs are fertilized in W-7, it is uncertain if

Ca^{2+} release has occurred. However, W-7 has been reported to have no effect on Ca^{2+} in sea urchin eggs during fertilization (Komukai *et al.* 1985). Thus, W-7 inhibition of fertilization events is most likely a result of its action as a CaM antagonist. Thus despite evidence for generation of DAG during fertilization (Ciapa and Whitaker 1986) and induction of Na^+-H^+ antiport by exogenous DAG (Shen and Burgart 1986), activation of Na^+-H^+ antiporter activity during fertilization occurs by a Ca^{2+}-CaM-dependent pathway, independent of PKC activity. Why sea urchin eggs have both PKC-dependent and - independent regulatory mechanisms is another issue that awaits further study.

ACKNOWLEDGEMENTS

I thank Drs. M. M. Winkler and T. S. Ingebritsen for helpful discussions during this study and A. L. Sui and L. A. Ricke for technical assistance. This research was supported by National Science Foundation DCB 86- 02263.

REFERENCES

Aberdam, E. and N. Dekel. 1985. Activators of protein kinase C stimulate meiotic maturation of rat oocytes. *Biochem. Biophys. Res. Commun.* 132:570-574.

Ammann, D., F. Lanter, R. A. Steiner, P. Schulthess, Y. Shijo, and W. Simon. 1981. Neutral carrier based hydrogen selective microelectrodes for extra- and intracellular studies. *Anal. Chem.* 53:2267-2269.

Berk, B. C., M. S. Aronow, T. A. Brock, E. Cragoe, Jr., M. A. Gimbrone, Jr., and R. W. Alexander. 1987. Angiotensin II-stimulated Na^+/H^+ exchange in cultured vascular smooth muscle cells. Evidence for protein kinase C-dependent and independent pathways. *J. Biol. Chem.* 262:5057-5064.

Berridge, M. J. 1987. Inositol trisphosphate and diacylglycerol: two interacting second messengers. *Annu. Rev. Biochem.* 56:159-193.

Berridge, M. J. and R. F. Irvine. 1984. Inositol trisphosphate, a novel second messenger in cellular signal transduction. *Nature (Lond.)* 312:315-321.

Besterman, J. M., S. P. Watson, and P. Cuatrecasas. 1986. Lack of association of epidermal growth factor-, insulin-, and serum-induced mitogenesis with stimulation of phosphoinositide degradation in BALB/c 3T3 fibroblasts. *J. Biol. Chem.* 261:723-727.

Blenis, J., J. G. Spivack, and R. L. Erikson. 1984. Phorbol esters, serum, and Rous sarcoma virus transforming gene product induce similar phosphorylations of ribosomal protein S6. *Proc. Natl. Acad. Sci. USA.* 81:6408-6412.

Blumberg, P. M., K. B. Delclos, and S. Jaken. 1981. Tissues and species specificity for phorbol ester receptors. p. 201-227. *In: Organ and Species Specificity in Chemical Carcinogenesis.* R. Langenbach, S. Nesnow and J. M. Rice (Eds.). Plenum Press, New York.

Boni, L. T. and R. R. Rando. 1985. The nature of protein kinase C activation by physically defined phospholipid vesicles and diacylglycerols. *J. Biol. Chem.* 260:10819-10825.

Bornslaeger, E. A., W. T. Poueymirou, P. Mattei, and R. M. Schultz. 1986. Effects of protein kinase C activators on germinal vesicle breakdown and polar body emmission of mouse oocytes. *Exp. Cell Res.* 165:507-517.

Bradford, M. M. 1976. A rapid and sensitive method for quantitation of microgram quantities of protein utilizing the principle of protein-dye binding. *Anal. Biochem.* 72:248-254.

Brandriff, B., R. T. Hinegardner, and R. A. Steinhardt. 1975. Development and life cycle of the parthenogenetically activated sea urchin embryo. *J. Exp. Zool.* 192:13-24.

Campbell, A. K. 1983. Intracellular Calcium. Its Universal Roles as Regulator. J. Wiley & Sons Ltd., New York.

Carroll, A. G. and W. R. Eckberg. 1986. Inhibition of germinal vesicle breakdown and activation of cytoplasmic contractility in *Spisula* oocytes by calmodulin antagonists. *Biol. Bull.* 170:43-50.

Chambard, J. C., S. Paris, G. L'Allemain, and J. Pouyssegur. 1987 Two growth factor signalling pathways in fibroblasts distinguished by pertussis toxin. *Nature (Lond.)* 326:800-803.

Chambers, E. L., B. C. Pressman, and B. Rose. 1974. The activation of sea urchin eggs by the divalent ionophores A23187 and X-537A. *Biochem. Biophys. Res. Commun.* 60:126-132.

Ciapa, B. and M. Whitaker. 1986. Two phases of inositol polyphosphate and diacylglycerol production at fertilisation. *FEBS Lett.* 195:347-351.

Cope, F.O., B. D. Howard, and R. K. Boutwell. 1986. The *in vitro* characterization of the inhibition of mouse brain protein kinase-C by retinoids and their receptors. *Experentia* 42:1023-1027.

Dorée, M., A. Picard, J. C. Cavadore, C. Le Peuch, and J. G. Demaille. 1982. Calmodulin antagonists and hormonal control of meiosis in starfish oocytes. *Exp. Cell Res.* 139:135-144.

Dubé, F. 1988. The relationships between early ionic events, the pattern of protein synthesis, and oocyte activation in the surf clam, *Spisula solidissima*. *Dev. Biol.* 126:233-241.

Dubé, F., R. Golsteyn, and L. Dufresne. 1987. Protein kinase C and meiotic maturation of surf clam oocytes. *Biochem. Biophys. Res. Commun.* 142:1072-1076.

Eckberg, W. R. and A. G. Carroll. 1987. Evidence for involvement of protein kinase C in germinal vesicle breakdown in *Chaetopterus*. *Dev. Growth & Differ.* 29:489-496.

Eckberg, W. R., E. Z. Szuts, and A. G. Carroll. 1987. Protein kinase C activity, protein phosphorylation and germinal vesicle breakdown in *Spisula* oocytes. *Dev. Biol.* 124:57-64.

Epel, D., R. Steinhardt, T. Humphreys, and D. Mazia. 1974. An analysis of the partial derepression of sea urchin eggs by ammonia; the existence of independent pathways. *Dev. Biol.* 40:245-255.

Epel, D., C. Patton, R. W. Wallace, and W. Y. Cheung. 1981. Calmodulin activates NAD kinase of sea urchin eggs. An early event of fertilization. *Cell* 23:543-549.

Grainger, J. L., M. M. Winkler, S. S. Shen, and R. A. Steinhardt. 1979. Intracellular pH controls protein synthesis rate in the sea urchin egg and early embryo. *Dev. Biol.* 68:396-406.

Grinstein, S. and W. Furuya. 1988. Receptor-mediated activation of electropermeabilized neutrophils. Evidence for a Ca^{2+}- and protein kinase C-independent signalling pathway. *J. Biol. Chem.* 263:1779-1783.

Hesketh, T. R., J. P. Moore, J. D. H. Morris, M. V. Taylor, J. Rogers, G. A. Smith, and J. C. Metcalfe. 1985. A common sequence of calcium and pH signals in the mitogenic stimulation of eukaryotic cells. *Nature (Lond.)* 313:481-484.

Hesketh, T. R., J. D. Morris, J. P. Moore, and J. C. Metcalfe. 1988. Ca^{2+} and pH responses to sequential additions of mitogens in single 3T3 fibroblasts: correlations with DNA synthesis. *J. Biol. Chem.* 263:11879-11886.

Hidaka, H., T. Yamaki, T. Totsuka, and M. Asano. 1979. Selective inhibitors of Ca^{2+}-binding modulators of phosphodiesterase produce vascular relaxation and inhibit actin-myosin interaction. *Mol. Pharmacol.* 15:49-59.

Hidaka, H., M. Asano, and T. Tanaka. 1981. Activity-structure relationship of calmodulin antagonists. *Mol. Pharmacol.* 20:571-578.

Hidaka, H., M. Inagaki, S. Kawamoto, and Y. Sasaki. 1984. Isoquinolinesulfonamides, novel and potent inhibitors of cyclic nucleotide dependent protein kinase and protein kinase C. *Biochemistry* 23:5036-5041.

Holloway, P. W. 1973. A simple procedure for removal of Triton X-100 from protein samples. *Anal. Biochem.* 53:304-308.

Huang, C. L., M. G. Cogan, E. J. Cragoe, Jr., and H. E. Ives. 1987. Thrombin activates Na^+/H^+ exchanger in vascular smooth muscle cells. Evidence for a kinase C-independent pathway which is Ca^{2+}- dependent and pertussis toxin-sensitive. *J. Biol. Chem.* 262:14134-14140.

Hunter, T. and J. A. Cooper. 1985. Protein-tyrosine kinases. *Annu. Rev. Biochem.* 54:897-930.

Ingebritsen, T. S. and P. Cohen. 1983. Protein phosphatases: Properties and role in cellular regulation. *Science* 221:331-338.

Johnson, J. D., D. Epel, and M. Paul. 1976 Intracellular pH and activation of sea urchin eggs after fertilisation. *Nature (Lond.)* 262:661-664.

Kaibuchi, K., Y. Takai, and Y. Nishizuka. 1981. Cooperative roles of various membrane phospholipids in the activation of calcium-activated, phospholipid-dependent protein kinase. *J. Biol. Chem.* 256:7146-7149.

Kase, H., K. Iwahashi, and Y. Matsuda. 1986. K-252a, potent inhibitor of protein kinase C from microbial origin. *J. Antibiot. (Tokyo)* 39:1059-1965.

Kase, H., K. Iwahashi, S. Nakanishi, Y. Matsuda, K. Wamada, M. Takahashi, C. Murakata, A. Sato, and M. Kaneko. 1987. K252 compounds, novel and potent inhibitors of protein kinase C and cyclic nucleotide-dependent protein kinases. *Biochem. Biophys. Res. Commun.* 142:436-440.

Kawamoto, S. and H. Hidaka. 1984. 1-(5-Isoquinolinesulfonyl)-2-methylpiperazine (H-7) is a selective inhibitor of protein kinase C in rabbit platlets. *Biochem. Biophys. Res. Commun.* 125:258-264.

Kikkawa, U. and Y. Nishizuka. 1986. The role of protein kinase C in transmembrane signalling. *Annu. Rev. Cell Biol.* 2:149-178.

Kikkawa, U., M. Go, J. Koumoto, and Y. Nishizuka. 1986. Rapid purification of protein kinase C by high performance liquid chromatography. *Biochem. Biophys. Res. Commun.* 135:636-643.

Kishimoto, A., Y. Takai, T. Mori, U. Kikkawa, and Y. Nishizuka. 1980. Activation of calcium and phospholipid-dependent protein kinase by diacylglycerol, its possible relation to phosphatidylinositol turnover. *J. Biol. Chem.* 255:2273-2276.

Kishimoto, T., M. Yoshikuni, H. Ikadai, and H. Kanatani. 1985. Inhibition of starfish oocyte maturation by tumor-promoting phorbol esters. *Dev. Growth & Differ.* 27:233-242.

Komukai, M., A. Fujiwara, Y. Fujino, and I. Yasumasu. 1985. The effects of several ion channel blockers and calmodulin antagonists on fertilization-induced acid release and $^{45}Ca^{2+}$ uptake in sea urchin eggs. *Exp. Cell Res.* 159:463-472.

Kuo, J. F., R. G. G. Anderson, B. C. Wise, L. Mackerlova, I. Salomonsson, N. L. Brackett, N. Katoh, M. Shoji, and R. W. Wrenn. 1980. Calcium-dependent protein kinase: widespread occurrence in various tissues and phyla of the animal kingdom and comparison of effects of phospholipid, calmodulin, and trifluoperazine. *Proc. Natl. Acad. Sci. USA.* 77:7039-7043.

L'Allemain, G. and J. Pouysségur. 1986. EGF and insulin action in fibroblasts. Evidence that phosphoinsitide hydrolysis is not an essential mitogenic signalling pathway. *FEBS Lett.* 197:344-348.

Lau, A. F., T. C. Rayon, and T. Humphreys. 1986. Tumor promoters and diacylglycerol activate the Na+/H+ antiporter of sea urchin eggs. *Exp. Cell Res.* 166:23-30.

Laurent, A., M. Basset, M. Dorée, and C. J. Le Peuch. 1988. Involvement of a calcium-phospholipid-dependent protein kinase in the maturation of *Xenopus laevis* oocytes. *FEBS Lett.* 226:324-330.

Letterio, J. J., S. R. Coughlin, and L. T. Williams. 1986. Pertussis toxin-sensitive pathway in the stimulation of c-myc expression and DNA synthesis by bombesin. *Science* 234:1117-1119.

Loomis, C. R. and R. M. Bell. 1988. Sangivamycin, a nucleoside analogue, is a potent inhibitor of protein kinase C. *J. Biol. Chem.* 263:1682-1692.

Mazzei, G. J., R. C. Schatzman, S. Turner, W. R. Vogler, and J. F. Kuo. 1984. Phospholipid-sensitive Ca^{2+}-dependent protein kinase inhibition by R-24571, a calmodulin antagonist. *Biochem. Pharmacol.* 33:125-130.

McDonald, J. R., U. Gröschel-Stewart, and M. P. Walsh. 1987. Properties and distribution of the protein inhibitor (M_r 17000) of protein kinase C. *Biochem. J.* 242:695-705.

Mitsuhashi, T. and H. E. Ives. 1988. Intracellular Ca^{2+} requirement for activation of the Na^+/H^+ exchanger in vascular smooth muscle cells. *J. Biol. Chem.* 263:8790-8795.

Moolenaar, W. H., L. H. K. Defize, P. T. Van Deer Saag, and S. W. De Laat. 1986. The generation of ionic signals by growth factors. *Curr. Top. Membr. Transp.* 26:137-156.

Morin, M. J., D. Kreutter, H. Rasmussen, and A. C. Sartorelli. 1987. Disparate effects of activators of protein kinase C on HL-60 promyelocytic leukemia cell differentiation. *J. Biol. Chem.* 262:11758-11763.

Muldoon, L. L., G. A. Jamieson, Jr., A. C. Kao, H. C. Palfrey, and M. L. Villereal. 1987. Mitogen stimulation of Na^+-H^+ exchange: differential involvement of protein kinase C. *Am. J. Physiol.* 253:C219-C229.

Nakanishi, S., K. Yamada, H. Kase, S. Nakamura, and Y. Nonumura. 1988. K- 252a, a novel microbial product, inhibits smooth muscle myosin light chain kinase. *J. Biol. Chem.* 263:6215-6219.

Nishizuka, Y. 1984. The role of protein kinase C in cell surface signal transduction and tumor promotion. *Nature (Lond.)* 308:693-698.

Nishizuka, Y. 1986. Studies and perspectives of protein kinase C. *Science* 233:305-312.

Nishizuka, Y. 1988. The molecular heterogeneity of protein kinase C and its implications for cellular regulation. *Nature (Lond.)* 334:661-665.

Ober, S. S. and A. B. Pardee. 1987. Both protein kinase C and calcium mediate activation of the Na^+/H^+ antiporter in Chinese hamster embryo fibroblasts. *J. Cell. Physiol.* 132:311-317.

Orme, F. W. 1969. Liquid ion-exchanger microelectrodes. p.376-403. *In: Glass Microelectrodes.* M. Lavallee, O. F. Schanne, and N. C. Hebert (Eds.). John Wiley and Sons, New York.

Owen, N. E. and M. L. Villereal. 1982. Evidence for a role of calmodulin in serum stimulation of Na^+ influx in human fibroblasts. *Proc. Natl. Acad. Sci. USA.* 79:3537-3541.

Picard, A. and M. Dorée. 1982. Intracellular microinjection of anticalmodulin drugs does not inhibit the cortical reaction induced by fertilization, ionophore A23187 or injection of calcium buffers in sea urchin eggs. *Dev. Growth & Differ.* 24:155-162.

Poenie, M., J. Alderton, R. Y. Tsien, and R. A. Steinhardt. 1985. Changes of free calcium levels with stages of the cell division cycle. *Nature (Lond.)* 315:147-149.

Raben, D. M. and D. D. Cunningham. 1985. Effects of EGF and thrombin on inositol-containing phospholipids of cultured fibroblasts: stimulation of phosphatidylinositol synthesis by thrombin but not EGF. *J. Cell. Physiol.* 125:582-590.

Ramsdell, J. S., G. R. Pettit, and A. H. Tashjian, Jr. 1986. Three activators of protein kinase C, bryostatins, dioleins, and phorbol esters, show differing specificities of action on GH_4 pituitary cells. *J. Biol. Chem.* 261:17073-17080.

Rosenthal, A., L. Rhee, R. Yadegari, R. Paro, A. Ullrich, and D. V. Goeddel. 1987. Structure and nucleotide sequence of a *Drosophila melanogaster* protein kinase C gene. *EMBO J.* 6:433-441.

Rosoff, P. M. and G. Terres. 1986. Cyclosporine A inhibits Ca^{2+}-dependent stimulation of the Na^+/H^+ antiport in human T cells. *J. Cell Biol.* 103:457-463.

Rozengurt, E. 1986. Early signals in the mitogenic response. *Science* 234:161-166.

Sano, K. 1982. Inhibition of the acrosome reaction of sea urchin spermatozoa by a calmodulin antagonist, N-(6-aminohexyl)-5-chloro-1-naphthalenesulfonamide (W-7). *J. Exp. Zool.* 226:471-473.

Schatzman, R. C., R. L. Raynor, and J. F. Kuo. 1983. N-(6-aminohexyl)-5-chloro-1-naphthalenesulfonamide (W-7), a calmodulin antagonist, also inhibits phospholipid-sensitive calcium-dependent protein kinase. *Biochim. Biophys. Acta* 755:144-147.

Sekar, M. C. and L. E. Hokin. 1986. The role of phosphoinositides in signal transduction. *J. Membr. Biol.* 89:193-210.

Shen, S. S. 1982. The effects of external ions on pH_i in sea urchin eggs. p.269-282. *In: Intracellular pH: Its Measurement, Regulation and Utilization in Cellular Functions.* R. Nuccitelli and D. W. Deamer (Eds.). Alan R. Liss, New York.

Shen, S. S. 1983. Membrane properties and intracellular ion activities of marine invertebrate eggs and their changes during activation. p. 213-267. *In: Mechanism and Control of Animal Fertilization.* J. F. Hartmann (Ed.). Academic Press, New York.

Shen, S. S. and L. J. Burgart. 1986. 1,2-Diacylglycerols mimic phorbol 12-myristate 13-acetate activation of the sea urchin egg. *J. Cell. Physiol.* 127:330-340.

Shen, S. S. and L. A. Ricke. 1989. Protein kinase C from sea urchin eggs. *Comp. Biochem. Physiol.* (in press).

Shen, S. S. and R. A. Steinhardt. 1978. Direct measurement of intracellular pH during metabolic derepression of the sea urchin egg. *Nature (Lond.)* 272:253-254.

Shen, S. S. and R. A. Steinhardt. 1979. Intracellular pH and the sodium requirement at fertilisation. *Nature (Lond.)* 282:87-89.

Slack, B. E., J. E. Bell, and D. J. Benos. 1986. Inositol-1,4,5-trisphosphate injection mimics fertilization potentials in sea urchin eggs. *Am. J. Physiol.* 250:C340-C347.

Smith, R. J., J. M. Justen, and L. M. Sam. 1988. Effects of protein kinase C inhibitor, K252a, on human polymorphonuclear neutrophil responsiveness. *Biochem. Biophys. Res. Commun.* 152:1497-1503.

Soltoff, S. P. and L. C. Cantley. 1988. Mitogens and ion fluxes. *Annu. Rev. Physiol.* 50:207-223.

Stapleton, C. L., L. L. Mills, and D. E. Chandler. 1985. Cortical granule exocytosis in sea urchin eggs is inhibited by drugs that alter intracellular calcium stores. *J. Exp. Zool.* 234:289-299.

Steinhardt, R. A. and J. M. Alderton. 1982. Calmodulin confers calcium sensitivity on secretory exocytosis. *Nature (Lond.)* 295:154-155.

Steinhardt, R. A. and D. Epel. 1974. Activation of sea-urchin eggs by a calcium ionophore. *Proc. Natl. Acad. Sci. USA* 71:1915-1919.

Steinhardt, R. A., R. S. Zucker, and G. Schatten. 1977. Intracellular calcium release at fertilization in the sea urchin egg. *Dev. Biol.* 58:185-196.

Steinhardt, R. A., S. S. Shen, and R. S. Zucker. 1978. Direct evidence for ionic messengers in the two phases of metabolic derepression at fertilization in the sea urchin egg. *ICN-UCLA Symp. Mol. Cell. Bio.* 12:415-424.

Stith, B. and J. L. Maller. 1987. Induction of meiotic maturation in *Xenopus* oocytes by 12-0-tetradecanoylphorbol 13-acetate. *Exp. Cell Res.* 169:514-523.

Swann, K. and M. Whitaker. 1985. Stimulation of the Na^+/H^+ exchanger of sea urchin eggs by phorbol ester. *Nature (Lond.)* 314:274-277.

Tanaka, T., T. Ohmura, T. Yamakado, and H. Hidaka. 1982. Two types of calcium-dependent protein phosphorylations modulated by calmodulin antagonists. *Mol. Pharmacol.* 22:408-412.

Turner, P. R., M. P. Sheetz, and L. A. Jaffe. 1984. Fertilization increases the polyphosphoinositide content of sea urchin eggs. *Nature (Lond.)* 310:414-415.

Turner, P. R., L. A. Jaffe, and A. Fein. 1986. Regulation of cortical vesicle exocytosis in sea urchin eggs by inositol 1,4,5-trisphosphate and GTP-binding protein. *J. Cell Biol.* 102:70-76.

Turner, P. R., L. A. Jaffe, and P. Primakoff. 1987. A cholera toxin-sensitive G-protein stimulates exocytosis in sea urchin eggs. *Dev. Biol.* 120:577-583.

Uchida, T. and C. R. Filburn. 1984. Affinity chromatography of protein kinase C-phorbol ester receptor on polyacrylamide-immobilized phosphatidylserine. *J. Biol. Chem.* 259:12311-12314.

Urner, F. and S. Schorderet-Slatkine. 1984. Inhibition of denuded mouse oocyte meiotic maturation by tumor-promoting phorbol esters and its reversal by retinoids. *Exp. Cell Res.* 154:600-605.

Vaartjes, W. J., C. G. M. de Haas, and S. G. van den Bergh. 1986. Phorbol esters, but not epidermal growth factor or insulin, rapidly decrease soluble protein kinase C activity in rat hepatocytes. *Biochem. Biophys. Res. Commun.* 138:1328-1333.

Vacquier, V. D. 1975. The isolation of intact cortical granules from sea urchin eggs: calcium ions trigger granule discharge. *Dev. Biol.* 43:62-74.

Van Belle, H. 1981. R24571: A potent inhibitor of calmodulin activated enzymes. *Cell Calcium* 2:483-494.

Vara, F. and E. Rozengurt. 1985. Stimulation of Na^+-H^+ antiport activity by epidermal growth factor and insulin occurs without activation of protein kinase C. *Biochem. Biophys. Res. Commun.* 130:646-653.

Whitaker, M. and M. Aitchison. 1983. Calcium-dependent polyphosphoinositide hydrolysis is associated with exocytosis *in vitro*. *FEBS Lett.* 182:119-124.

Whitaker, M. and P. F. Baker. 1983. Calcium-dependent exocytosis in an *in vitro* secretory granule plasma membrane preparation from sea urchin eggs and the effect of some inhibitors of cytoskeletal function. *Proc. Roy. Soc. Lond. B Biol. Sci.* 218:397-413.

Whitaker, M. and R. F. Irvine. 1984. Inositol 1,4,5-trisphosphate microinjection activates sea urchin eggs. *Nature (Lond.)* 312:636-639.

Whitaker, M. and R. A. Steinhardt. 1985. Ionic signaling in the sea urchin egg at fertiliation. p. 167-221. *In: Biology of Fertilization,* Vol. 3. C. B. Metz and A. Monroy (Eds.) Academic Press, New York.

Whitfield, J. F., A. L. Boynton, J. P. MacManus, R. H. Rixon, M. Sikorska, B. Tsang, and P. R. Walker. 1980. The roles of calcium and cyclic AMP in cell proliferation. *Annu. NY Acad. Sci.* 339:216-240.

Wolf, M., H. Le Vine, III, W. Stratford May, Jr., P. Cuatrecasas, and N. Sahyoun. 1985. A model for intracellular translocation of protein kinase C involving synergism between Ca^{2+} and phorbol esters. *Nature (Lond.)* 317:546-549.

Woodgett, J. R. and T. Hunter. 1987. Isolation and characterization of two distinct forms of protein kinase C. *J. Biol. Chem.* 262:4836-4843.

Wright, C. D. and M. D. Hoffman. 1986. The protein kinase C inhibitors H-7 and H-9 fail to inhibit human neutrophil activation. *Biochem. Biophys. Res. Commun.* 135:749-755.

Zachary, I., P. J. Woll and E. Rozengurt. 1987. A role for neuropeptides in the control of cell proliferation. *Dev. Biol.* 124:295-308.

HOW DO SPERM ACTIVATE EGGS IN *URECHIS* (AS WELL AS IN POLYCHAETES AND MOLLUSCS)?

Meredith Gould and José Luis Stephano
Escuela Superior de Ciencias
Universidad Autónoma de Baja California
A.P. 2921, Ensenada, B.C.N., MEXICO

In this chapter we summarize data showing that sperm acrosomal protein activates *Urechis* eggs, causing a fertilization potential and morphological changes like those produced by sperm. We then review the evidence in *Urechis* that Ca^{2+} uptake and acid release at fertilization cause rises in intracellular Ca^{2+} and pH necessary for activation, and propose that both ion movements occur primarily through voltage-gated channels that open during the fertilization potential. Available data from related species (polychaetes and molluscs) relating to this model are also reviewed.

WHAT EVENT INITIATES THE SERIES OF REACTIONS THAT LEAD TO EGG ACTIVATION?

Hypotheses

Several models can be proposed for how sperm initiate egg activation (Figure 1). One possibility is that molecules on the sperm surface interact with components of the egg surface to initiate activation (Figure 1a). Such molecules could be within the acrosome and exposed during the acrosome reaction, or could already be exposed on the external surface of sperm in species that do not undergo an acrosome reaction at fertilization. Another possibility is that fusion of the sperm lipid bilayer with the egg lipid bilayer produces a conformational change in the latter that activates molecules to initiate egg activation (Figure 1b). A third possibility is that "activator" molecules inside the sperm pass into the egg cytoplasm through the cytoplasmic bridge formed at sperm-egg fusion (Figure 1c). Data from *Urechis* favor the hypothesis (1a) that molecules exposed on the sperm surface interact with the egg to set in motion the chain of events leading to egg activation, since protein isolated from sperm activates eggs, and sperm can activate eggs when fusion (required for models 1b and c) is inhibited. So far in other species, there is indirect evidence favoring model 1a, and fusion (models 1b and c) has neither been ruled out nor shown to be required (see below and other chapters in this volume).

Acrosomal Protein Activates *Urechis* Eggs

When *Urechis* sperm undergo the acrosome reaction, a ring of electron dense material is exposed on the apical surface of the sperm. This material binds the sperm tightly to the egg surface coat and appears to contact egg microvilli that extend into the outer border of the coat (Tyler 1965; Gould-Somero and Holland 1975a; Gould *et al* 1986). Dense material which may be from the acrosomal granule also coats acrosomal processes (Figure 2). The latter point is being investigated with antibody to acrosomal protein.

The major component of *Urechis* sperm acrosomal granules is highly basic protein (50% lysine + arginine) that migrates as a single band in acetic acid-urea and CTAB (cetyltrimethyl-ammonium bromide) polyacrylamide gel electrophoresis (Gould *et al.* 1986). The apparent molecular mass is 25-30,000 daltons in CTAB gels and no detectable carbohydrate is present (Gould *et al.* 1986). When dispersed in seawater at concentrations as low as 30 μg/ml, AP (acrosomal protein) will activate 100% of eggs, causing the same morphological changes as sperm (rounding out, surface coat elevation, germinal vesicle breakdown and the formation of both polar bodies) as well as the establishment of the permanent polyspermy block (Gould *et al.* 1986). AP also opens Na^+ channels in the egg membrane, producing an activation potential with the same ion dependence, duration and general form as the fertilization potential induced by sperm (Gould and Stephano 1987) (Figure 3).

Eggs activated with AP do not cleave, presumably because they lack the sperm centrioles normally used (Conklin 1904). However, we have found that peptide fractions from AP will cause substantial numbers of eggs to cleave and even form larvae (unpublished results; see Figure 5). Thus, by every parameter looked at, AP mimics the effect of intact sperm on eggs.

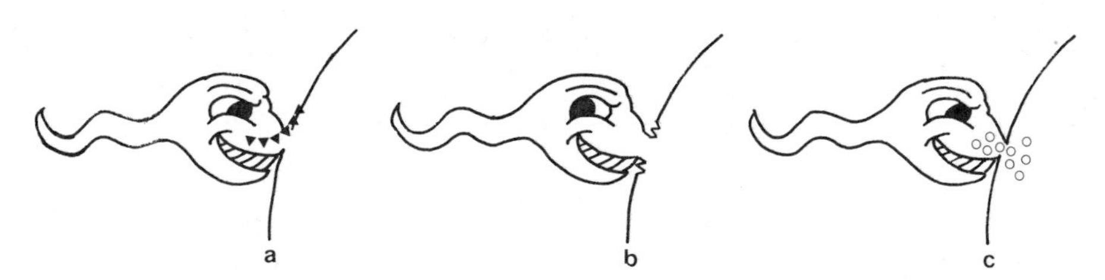

Figure 1. Models for the initiation of egg activation by sperm. In **a**, molecules exposed on the sperm surface interact with molecules on or in the egg plasma membrane. In **b**, the fusion of sperm and egg lipid bilayers produces a conformational change in the latter that activates egg signal molecules. In **c**, "activator" molecules inside the sperm pass into the egg through the cytoplasmic bridge formed at sperm-egg fusion.

AP will also cause activation potentials and activate eggs when applied locally to the egg surface from pipets with 20-30 μm tip diameters (Gould and Stephano 1987). In fact, in two of these experiments, just touching the tip of the pipet to the egg surface was sufficient to initiate activation (Gould and Stephano 1987). These pipets contained AP at concentrations of 4.4 and 19 μg/μl, respectively. In comparison, it is interesting to consider how sperm might deliver AP to the egg. From electron micrographs such as the one in Figure 2, the volume of the "doughnut" of AP on the egg surface was calculated to be about 6 x 10^{-10} μl. With 0.5 pg AP per sperm (Gould *et al.* 1986), there would be 850 μg AP/μl in a patch about 1.5 μm in diameter on the egg surface, a higher concentration and smaller patch than in the experimental applications. Nevertheless, it is impressive that the responses to AP are so similar to those produced by sperm.

AP also mimics *Urechis* sperm in a cross-species application. When sea urchin eggs are inseminated with high concentrations of *Urechis* sperm, some percentage of them are induced to undergo partial or complete fertilization envelope elevation (Jaffe *et al.* 1982). The same response is elicited by high concentrations of *Urechis* AP (Gould *et al.* 1986).

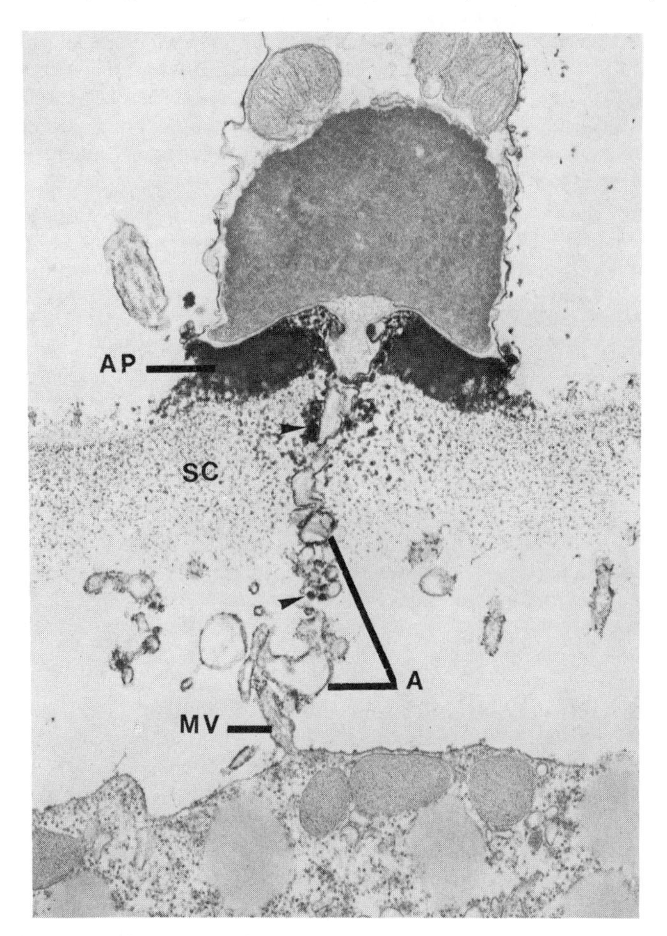

Figure 2. Electron micrograph of an acrosome-reacted *Urechis* sperm bound to the surface coat (SC) of an egg by the ring of acrosomal protein (AP). Electron dense material (arrows) is also associated with the surface of the acrosomal process (A) which has extended through the surface coat to come into contact with an egg microvillus (MV). Although not visible in this section, microvilli also extend into the outer region of the surface coat where it is bound by AP (see text). This is almost certainly a supernumerary sperm, since fixation was at one minute after insemination. X 30,000. Reprinted by permission from Gould *et al.* 1986).

Activation of *Urechis* Eggs Without Fusion

The fact that isolated AP activates eggs indicates that sperm fusion is not necessary. However, we considered the possibility that sperm membrane lipids that normally insert into the egg plasma membrane during fusion (see Figure 1b) might be present in the isolated AP. AP viewed by electron microscopy contains no membranes or other organelles (Gould *et al.* 1986), but the possibility remained that sperm lipids were present in AP *in situ*, or became associated with it during the detergent extraction used in the isolation procedure. When AP was exhaustively extracted to remove both covalently and non-covalently bound lipid, it retained the same biological activity (Gould and Stephano 1987). Thus it is the protein in AP that activates eggs.

The hypothesis that sperm-egg fusion is not necessary for egg activation was also tested by clamping egg membrane potentials to positive values to inhibit fusion (the electrical polyspermy block; Gould-Somero *et al.* 1979), then asking if sperm were still able to induce fertilization potentials, the earliest known response of eggs to fertilizing sperm (reviewed by Hagiwara and Jaffe 1979). When egg membrane potentials were clamped to about +50 mV, sperm still initiated fertilization potentials (Gould and Stephano 1987). In a previous study, sperm also initiated fertilization potentials without fusion in 3/6 eggs clamped to about -5 mV (they fused with the other three; Gould-Somero *et al.* 1979). Isolated AP also initiated fertilization potentials in eggs clamped to +50 mV (Gould and Stephano 1987). These results argue that, although sperm must normally fuse with eggs very rapidly – before the AP causes the egg membrane potential to go positive – fusion is not necessary for egg activation.

Electrical responses in the absence of sperm-egg fusion also occur when previously fertilized *Urechis* eggs are reinseminated after the fertilization potential is over and the egg membrane potential has returned to a negative value. These eggs respond to sperm with short,

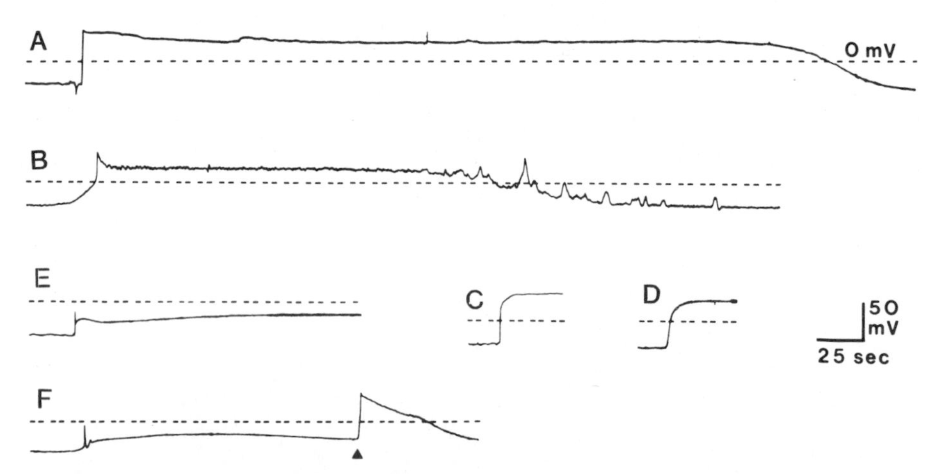

Figure 3. Electrical responses induced in *Urechis* eggs by sperm and acrosomal protein (AP). **A.** Fertilization potential induced by sperm. **B.** Electrical response to AP. Oscillations in the falling phase were occasionally observed in both sperm and AP potentials. **C** and **D.** Rising phase of the electrical responses to sperm and AP, respectively, in 1/10 Ca^{2+} seawater. In both, the initial peak produced by the Ca^{2+} action potential is absent, but the amplitude of the response is otherwise the same. **E.** Fertilization potential in 1/10 Na^+ seawater. Amplitudes were an average of 55 mV more negative than in normal seawater, showing the Na^+-dependence of the fertilization potential. **F.** AP potential in 1/10 Na^+ seawater. At the arrow, the egg was momentarily flooded with normal seawater: the potential immediately rose from -25 to +38 mV showing that the AP potential was also Na^+-dependent. Experimental details are in Gould and Stephano 1987. Reprinted by permission.

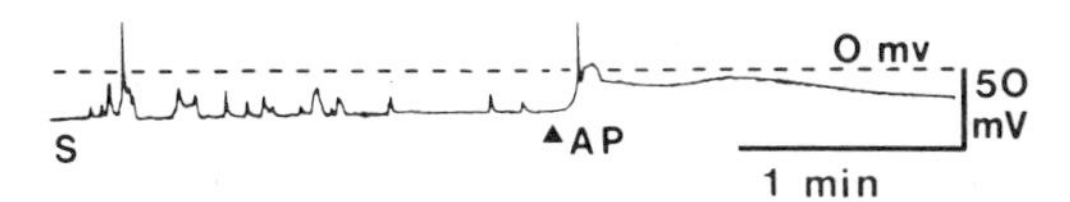

Figure 4. Electrical response to sperm(S) and to AP in an egg which had been activated 30 min earlier with AP. In this experiment AP was tested after eggs had been reinseminated; in other experiments AP gave the same responses when supernumerary sperm were not present.

low-amplitude depolarizations (Figure 4; also see Gould-Somero *et al.* 1979). The supernumerary sperm undergo the acrosome reaction, binding to the egg surface coat and extending their acrosomal processes, although fusion of egg and sperm lipid bilayers is never observed (Paul and Gould-Somero 1976; Gould *et al.* 1986; see Figure 2). When isolated AP is applied to previously fertilized eggs it causes electrical responses like those produced by supernumerary sperm (Figure 4). We propose it is the AP on sperm that causes these responses. Currently we are investigating the ionic basis for the supernumerary sperm- and AP-induced depolarizations, and how these depolarizations differ from the responses to sperm and AP in previously unfertilized eggs.

In summary, the following evidence supports the conclusion that acrosomal protein of sperm initiates egg activation. (a) The protein is in the right place (contacting the egg plasma membrane) at the right time (immediately upon sperm binding) (Tyler 1965; Gould *et al.* 1986). (b) When applied in suspension or locally to unfertilized eggs, AP at low concentrations activates 100% of them to undergo the same morphological and physiological changes that are induced by sperm (Gould *et al.* 1986; Gould and Stephano 1987). Peptide fractions from AP have been able to induce cleavage and larva formation (unpublished). (c) Sperm can initiate fertilization potentials in eggs when fusion of acrosomal processes with the egg plasma membrane is inhibited by clamping egg membrane potentials to positive values (Gould-Somero *et al.* 1979; Gould and Stephano 1987). AP also initiates fertilization potentials under the same conditions (Gould and Stephano 1987). (d) AP elicits fertilization potentials and activates eggs when any associated sperm lipids have been removed (Gould and Stephano 1987).

Do Isolated Sperm Components Activate Eggs of Other Species?

In *Urechis*, the acrosomal protein that activates eggs is also the "bindin" that adheres sperm to eggs (Gould *et al.* 1986). Bindin from sea urchin sperm does not activate sea urchin eggs (Vacquier and Moy 1977). However, since active material might have been denatured or lost during bindin isolation, it would be interesting to try other methods for isolating acrosomal components from sea urchin sperm that might activate eggs.

Isolation of "bindin" from oyster sperm was also reported (Brandriff *et al.* 1978). The preparation neither activated nor agglutinated living eggs, but agglutinated eggs that were previously fixed with formaldehyde (Brandriff *et al.* 1978). By a modification of their procedure, we isolated acrosomal granule contents from oyster sperm that strongly agglutinate living eggs and activate a small percentage of them (up to 20%) to form polar bodies (unpublished). However, when viewed by electron microscopy, our preparation looks full of holes (see also Brandriff *et al.* 1978), suggesting loss of material during isolation. We are modifying the isolation procedure in an attempt to improve biological activity. Nevertheless, results are encouraging that an acrosomal fraction that activates eggs may also be obtained from oyster sperm.

In *Mytilus*, it was reported that material released into the supernate when sperm were induced to undergo the acrosome reaction caused some eggs to form polar bodies (Tamaki and Osanai 1985). However, percentages of activation were usually low (<50) even with high (>1 mg protein/ml) concentrations (Tamaki and Osanai 1985). It would be interesting to determine whether a fraction with higher specific activity might be obtained from these sperm.

There have also been reports that crude extracts of testes or whole sperm could induce some activation events when added to eggs in seawater (Robertson 1912; Osanai 1976) or

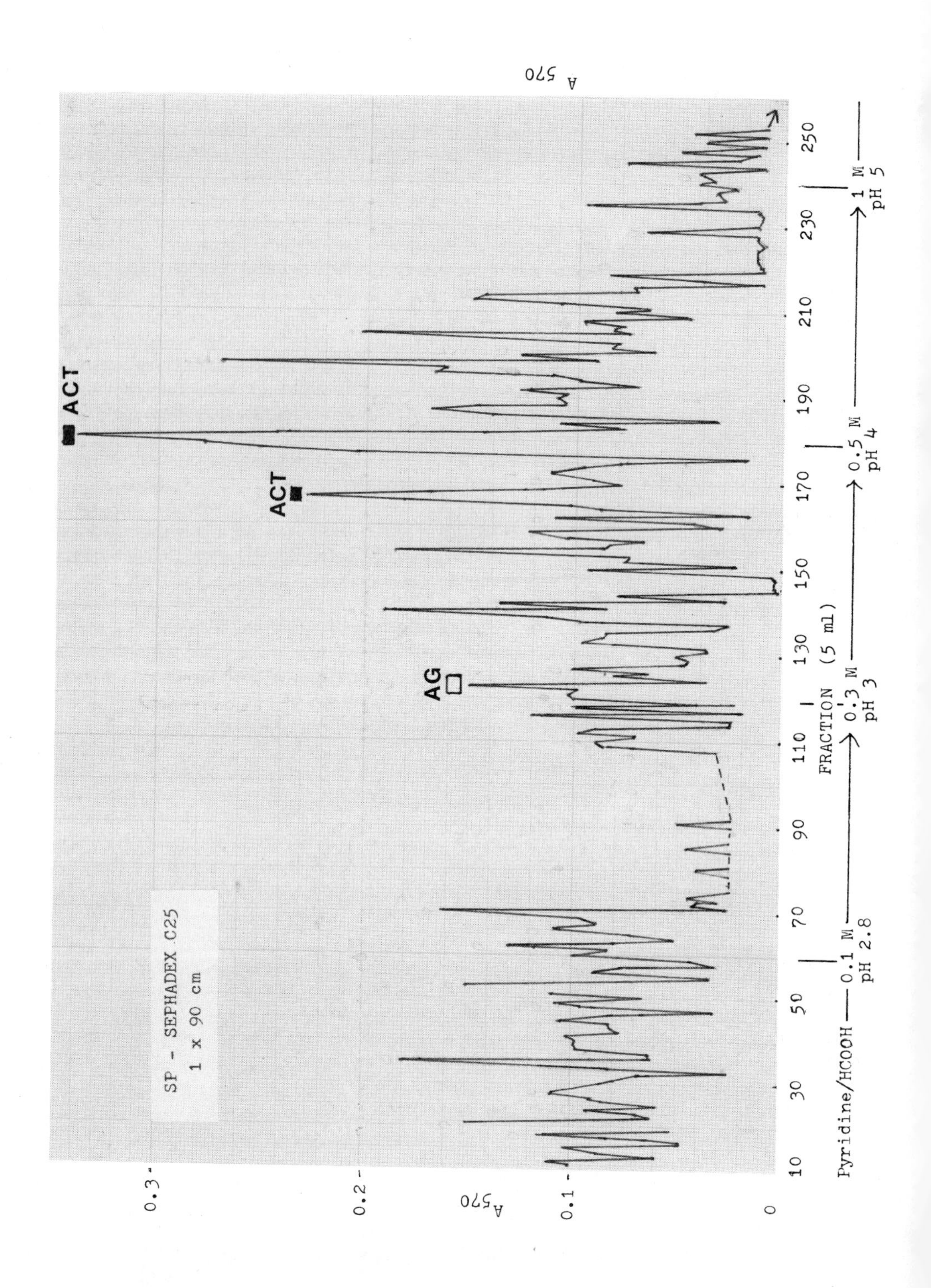
SP - SEPHADEX C25
1 x 90 cm
ACT
ACT
AG
A_{570}
0.3
0.2
0.1
0
10
30
50
70
90
110
130
150
170
190
210
230
250
FRACTION (5 ml)
Pyridine/HCOOH
0.1 M
pH 2.8
0.3 M
pH 3
0.5 M
pH 4
1 M
pH 5

injected into them (Ehrenstein *et al.* 1984; Dale *et al.* 1985; Dale 1988). However, responses were sometimes absent or partial, and the active ingredients and their source in the sperm are unknown. Thus it is premature to claim (Dale *et al.* 1985; Dale 1988) that such data support the model (Figure 1c) in which sperm activate eggs by injecting soluble components into egg cytoplasm.

Evidence for the hypothesis that sperm activate echinoderm, frog and perhaps mammalian eggs by G protein-linked mechanisms is described in chapters by Jaffe, Nuccitelli and Miyazaki in this volume. Since G proteins are controlled by membrane receptors in other systems (see review by Gilman 1987) such receptors may be present in the above eggs (model 1a), although there is so far no direct evidence for their existence. Below we discuss evidence that the initial events in egg activation of *Urechis* and related species may not require such second messenger systems.

Mechanism of Activation by AP in *Urechis*

With what components of the egg does AP interact to open Na^+ channels and cause egg activation? Many possibilities can be imagined. There could be a specific ligand-receptor interaction between AP and a site on the Na^+ channel or between AP and an "activator" molecule that opens the channels directly or *via* a second messenger. Alternatively, AP, which is a polycation and has hydrophobic properties (Gould *et al.* 1986) could act in a more non-specific manner. For example, it could change the surface charge on the membrane, or insert itself into the lipid bilayer of the egg causing a conformational change that would open the Na^+ channels. Perturbation of the lipid bilayer might also result in a Ca^{2+} leak that could open the channels. A local effect on Na^+ channel opening is expected, since sperm open only a localized patch of Na^+ channels to produce the fertilization potential (Gould-Somero 1981).

Since AP is relatively insoluble in seawater and is a polycation, binding non-specifically to the surfaces of eggs from a variety of species (Gould *et al.* 1986), we broke it into peptide fragments which we thought would be more suitable for identifying with which components of the egg surface AP interacts. Thermolysin digests (Gould *et al.* 1986) of AP have consistently produced peptide fractions that retain the ability to activate eggs (unpublished). Furthermore, egg activating ability has been recovered in fractions that do not agglutinate eggs and *vice versa* (Figure 5). We are purifying active peptides, will determine their sequences, and use them in the attempt to identify their "receptors".

A very interesting property of the peptide fractions is that they induce a high percentage of eggs to cleave and develop (unpublished; Figure 5), whereas eggs treated with the parent protein stop development after second polar body formation (Gould *et al.* 1986). How peptides induce parthenogenetic development is under investigation.

Figure 5. SP Sephadex column fractionation of a thermolysin digest of acrosomal protein. 100 mg AP was hydroysed with thermolysin (method of Suzuki and Ando 1972), and the supernate soluble in 0.09 M acetic acid (the small precipitate had no biological activity) was lyophilized, resuspended in 0.1 M pyridine adjusted to pH 2.8 with formic acid and applied to the column. The column was eluted with 60 ml of the same buffer, then with a gradient of increasing pH and ionic strength (Ohe *et al.* 1979) as indicated. Protein concentrations were measured with ninhydrin in 100 μl aliquots of the fractions. The fractions were lyophilized, resuspended in seawater and tested for biological activity. Fractions marked AG (121-125) agglutinated without activating eggs, and fractions marked ACT (166-168, 179-181; 177-178 not tested) activated eggs without visibly agglutinating them. Eggs incubated in some of these fractions underwent segmentation and developed further. For example, in a test of fraction 181, 1/50 of the fraction in 50 μl seawater activated 91% of eggs, with 60% of those activated forming ciliated blastulae. Controls included eggs with no additions (7% germinal vesicle breakdown, no blastulae) and eggs with sperm (96% blastula formation).

IONIC BASIS FOR ACTIVATION IN *URECHIS*, POLYCHAETES AND MOLLUSCS

Below we propose a simple working hypothesis for egg activation in *Urechis* and review available evidence in polychaetes and molluscs as well. The elements of the hypothesis are as follows.

The Fertilizing Sperm Opens a Localized Patch of Na^+ Channels at the Site of Interaction with the Egg

In *Urechis* eggs, the sperm opens a localized patch of Na^+ channels to produce the fertilization potential (Gould-Somero 1981). The resulting positive shift in membrane potential opens voltage-gated Ca^{2+} action potential channels (Jaffe *et al.* 1979). As described above, local application of sperm AP causes fertilization potentials in *Urechis* eggs, and we are currently investigating whether AP directly activates Na^+ channels or acts *via* a second messenger such as Ca^{2+}.

Very little is known about fertilization potentials in polychaete and mollusc eggs. Positive-going fertilization potentials have been recorded in *Spisula* (Finkel and Wolf 1980), *Chaetopterus* (Jaffe 1983), oyster (unpublished) and abalone (unpublished) eggs, but their ionic basis, and mode of opening (local only as in *Urechis*; Gould-Somero, 1981; or local followed by wave as in frog; Jaffe *et al.* 1985; Kline and Nuccitelli 1985) have not been determined. Voltage-dependent Ca^{2+} action potentials occur in *Chaetopterus* (Hagiwara and Miyazaki 1977) and *Dentalium* (Baud *et al.* 1987) eggs, but their presence in other polychaetes and molluscs has not been investigated.

There is No Wave of Exocytosis

Cortical granules are present in *Urechis* eggs, but most of them remain intact at fertilization (Gould-Somero and Holland 1975b). The perivitelline space does enlarge after fertilization, partly due to elevation of the surface coat and partly due to shrinkage of the egg (Tyler 1932; Holt 1934). However, surface coat elevation is a slow response that does not proceed as a wave from the point of sperm entry, but begins irregularly at different points around the egg surface and requires about 20 min for completion (Tyler 1932). The elevated surface coat has no obvious biochemical differences (Cross *et al.* 1985), continues to induce acrosome reactions and bind sperm (Paul and Gould-Somero 1976), and is no barrier to sperm penetration (Johnston and Paul 1977), indicating it is not modified by any egg secretions at fertilization.

To our knowledge, none of the mollusc eggs studied so far undergo cortical granule exocytosis at fertilization. Species examined include *Barnea* (Pasteels and deHarven 1962), *Crassostrea* (Alliegro and Wright 1983), *Dentalium* (Dufresne-Dubé *et al.* 1983a), *Mytilus* (Humphreys 1967) and *Spisula* (Rebhun 1962). Data on cortical reactions in polychaetes are scarcer. *Nereis* eggs undergo exocytosis at fertilization, but published accounts do not refer to any wave (Pasteels 1966; Fallon and Austin 1967). *Sabellaria* eggs undergo cortical granule exocytosis before fertilization (Peaucellier 1977b).

In contrast, waves of cortical granule exocytosis beginning at the site of sperm entry occur in echinoderms, fish and amphibians (reviewed by Jaffe 1983, 1985). Jaffe (1983, 1985) also reviews evidence that there are no "fertilization waves" among protostomes in general.

External Ca^{2+} Is Required for Activation and Enters the Egg Primarily Through Action Potential Channels Opened During the Fertilization Potential

In *Urechis* eggs, Ca^{2+} uptake from seawater begins rapidly after insemination (10% complete within 10s; Johnston and Paul 1977). If eggs are transferred to calcium-free seawater during the first 1.5 min after activation is initiated by exposing them to trypsin, they will not undergo surface coat elevation and germinal vesicle breakdown even though a fertilization potential has been initiated and considerable Ca^{2+} uptake has occurred (Johnston and Paul 1977; Jaffe *et al.* 1979). Activation by sperm also requires external Ca^{2+} (Figure 6). Evidence

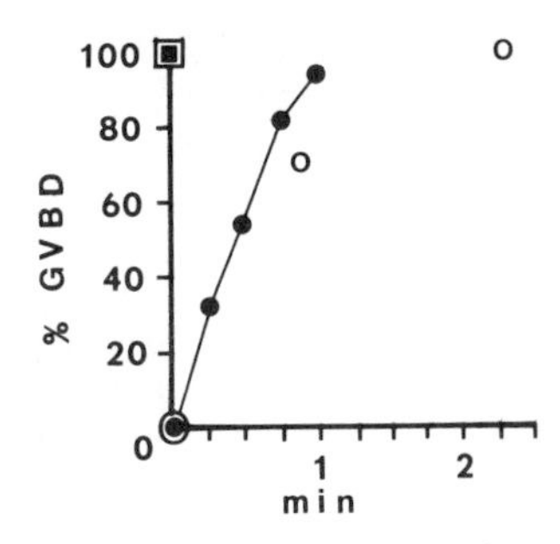

Figure 6. Dependence on external Ca^{2+} of *Urechis* egg activation by sperm. Eggs in normal seawater (10 mM Ca^{2+}) were inseminated and transferred at the times indicated to Ca^{2+}-free seawater containing 2 mM EGTA. Transfer was by resuspension following centrifugation in experiment 1 (open symbols), or by dilution into 10 volumes of OCa2EGTA seawater in experiment 2 (closed symbols). Control samples (squares) were transferred at the same times and in the same manner to normal seawater: 100% of these eggs activated (only time "0" samples are shown), showing that failure to activate in 0Ca seawater was not due to inadequate prior exposure to sperm.

that Ca^{2+} enters *Urechis* eggs primarily through the Ca^{2+} action potential channels is in Jaffe *et al.* (1979).

Barnea (Dubé and Guerrier 1982a) and *Spisula* (Dubé 1988) eggs also take up Ca^{2+} from seawater when they are fertilized or artificially activated, and transfer to Ca^{2+} free seawater during the first 4-5 min following insemination in *Spisula* inhibits activation despite the fact that sperm penetrate (Allen 1953). Transfer of *Barnea* eggs to Ca^{2+} free seawater during the first minute after insemination was also reported to block activation, although whether sperm had entered was not reported (Dubé and Guerrier 1982a). Transfer of *Macoma nasutis* (a bivalve) eggs to Ca^{2+} free seawater during the first minute after insemination also inhibits activation despite sperm entry (unpublished).

High K^+ seawater will artificially activate eggs of *Urechis* (Gould-Somero *et al.* 1979 and unpublished), *Spisula* (Allen 1953; Ii and Rebhun 1979), *Barnea* (Dubé and Guerrier 1982a,b), and *Chaetopterus* (Ikegami *et al.* 1976), provided extracellular Ca^{2+} is present. *Mytilus* (Dufresne-Dubé *et al.* 1983b) and *Urechis* (Gould-Somero *et al.* 1979) eggs are also activated by high K^+ seawater but Ca^{2+} requirements were not investigated. It has been proposed that high external K^+ activates these eggs by depolarizing the membrane potential, thus opening the voltage-gated Ca^{2+} action potential channels and causing Ca^{2+} uptake (see Jaffe 1983, 1985). By contrast, in sea urchin eggs, external Ca^{2+} is not required for the cortical reaction, and high K^+ seawater does not artificially activate eggs although it does cause Ca^{2+} uptake (Schmidt *et al.* 1982).

If, as originally proposed by Jaffe (1983, 1985), the main source of Ca^{2+} for activation is external in *Urechis* and related species, the mechanism involved could be very simple: the sperm opens fertilization potential channels and the resulting depolarization opens Ca^{2+} action potential channels. Ca^{2+} entering through these channels could cause a rise in intracellular Ca^{2+} activity without the need for any second messenger systems. Whether any second messenger systems are involved in either channel opening or a putative rise in intracellular Ca^{2+} has not yet been determined in *Urechis*, polychaetes or molluscs. Bloom *et al.* (1988) proposed that IP3 mediates activation in *Spisula* eggs based on their observations that PIP2 loses radiolabel within 30s of insemination and that IP3 injection can cause germinal vesicle breakdown. However, there were no observations with respect to fertilization potentials or Ca^{2+}, and the activation induced by IP3 was incomplete since the injected eggs failed to form polar bodies.

The above data indicate that extracellular Ca^{2+} is necessary for activation in *Urechis* and related species. Whether internal Ca^{2+} stores play a significant role in activation of these eggs remains to be determined. That internal Ca^{2+} stores exist is suggested by the observations that eggs of *Urechis* (Paul 1975), *Sabellaria* (Paucellier 1977a) and *Chaetopterus* (Eckberg and Carroll 1982) can be activated by the Ca^{2+} ionophore A23187 in the absence of external Ca^{2+}. Presumably a rise in intracellular Ca^{2+} activity is required for activation in eggs of *Urechis* and related species. However, there have so far been no studies in which intracellular pCa has been directly measured and controlled (e.g. by Ca^{2+} buffer injections). Only a preliminary report that Quin 2 fluorescence increases during artificial activation of *Barnea* eggs has been published (Brassard *et al.* 1988)

Finally, while there is evidence that Ca^{2+} uptake is necessary for activation in *Urechis* and related species, in *Urechis* it is clear that Ca^{2+} uptake is not sufficient. Acid release must also occur (see below). Ca^{2+} uptake may not be sufficient for activation in *Spisula* either, since Dubé (1988) observed that eggs in 21 mM K^+ seawater did not undergo germinal vesicle breakdown although they took up more Ca^{2+} than did fertilized controls in normal (10 mM K^+) seawater.

Where Acid Release is Required for Activation it May Occur Primarily Through Voltage-Gated Channels

Acid release is required for activation in *Urechis*. *Urechis* eggs release ca. 3 pM protons/egg during the first six minutes after insemination or artificial activation (Paul 1975). The amount of acid released decreases with external pH and none is released at pH 7 (Paul 1975; Holland *et al.* 1984). Egg activation is correlated with acid release and does not occur at pH 7 although sperm enter (Paul 1975; Holland *et al.* 1984). If eggs are fertilized at pH 8, then transferred within 1-2 min to seawater at pH 7, they do not activate although sperm enter, and fertilization potentials and Ca^{2+} uptake are normal (Tyler and Schultz 1932; Johnston and Paul 1977; Gould-Somero *et al.* 1979). The eggs round out, but the indentation reappears and further signs of activation (surface coat elevation and germinal vesicle breakdown) do not occur. If these eggs are reinseminated at pH 7, more sperm will enter and there will be a second fertilization potential accompanied by Ca^{2+} uptake, but no further activation (Tyler and Schultz 1932; Johnston and Paul 1977; Gould-Somero *et al.* 1979). If the eggs are returned to pH 8, they remain unactivated until inseminated again, then normal activation occurs although the eggs are polyspermic (Tyler and Schultz 1932). Thus activation is reversible in *Urechis* when eggs are transferred to either Ca^{2+} free (see above) or acid seawater during the first few minutes after insemination.

The correlation between acid release and egg activation strongly suggests that a rise in intracellular pH is required. This remains to be tested by direct measurements of intracellular pH at fertilization.

Fertilization acid release occurs and may be required for activation in some polychaetes and molluscs. *Spisula* and *Barnea* eggs release acid at fertilization or artificial activation (Allen 1953; Ii and Rebhun 1979; Dubé 1988; Dubé and Guerrier 1982b). In *Spisula* acid release is inhibited at pH <6.5 and activation is reversible despite sperm penetration if eggs are transferred to acid seawater during the first 4-5 min after insemination (Allen 1953). Acid release was also correlated with egg activation in studies using high K^+ as the activating agent (Ii and Rebhun 1979). *Barnea* eggs failed to activate when they were transferred one minute after insemination to seawater containing 100 mM Na acetate, pH 6.3 (Dubé and Guerrier 1982b). Thus, the data available are consistent with an acid release requirement for activation in *Spisula*, but the question has not received much attention in other species.

Sabellaria eggs can be fertilized and will develop whether they are inseminated at the germinal vesicle stage, or after they have undergone GVBD and arrested at meiotic metaphase I (Peaucellier 1977b, 1978). When they are fertilized at the germinal vesicle stage, they release acid, but no acid release occurs when fertilization is at metaphase I (Peaucellier 1978). Proteases caused GVBD in unfertilized eggs and this was accompanied by acid release (Peaucellier 1978). Thus, in *Sabellaria*, acid release accompanies GVBD, but not the resumption of meiosis after metaphase I. Effects of blocking acid release were not studied.

Not all Species Release Fertilization Acid. (Paul 1975) showed that *Mytilus* and *Acmaea* eggs do not release acid at fertilization. It was also stated without data that *Patella* eggs don't release fertilization acid either (Guerrier *et al.* 1986b). In these three species eggs undergo GVBD and proceed to meiotic metaphase I before fertilization, whereas *Urechis*, *Spisula* and *Barnea* eggs undergo GVBD after fertilization. Thus an interesting hypothesis emerges; eggs that undergo GVBD after fertilization release fertilization acid, whereas eggs that undergo GVBD during spawning may release "maturation acid", but will show no further acid release at fertilization. If acid release indicates a rise in intracellular pH, the implication is that a rise in pH_i accompanies GVBD, and may be required for this event, but that no pH_i change occurs when meiosis is resumed after metaphase I. In further support of this hypothesis is the observation that *Patella* oocytes with intact germinal vesicles were induced to undergo GVBD following injections of Hepes-KOH solutions buffered to pH 7.8 - 8.5, but not 7.5 (Guerrier *et al.* 1986b). Experiments in which intracellular pH is measured directly are required for a critical test of this hypothesis, and these remain to be performed.

Voltage-dependent acid release. The proton efflux at fertilization in *Urechis* is not inhibited by amiloride and is not stoichiometric with Na^+ uptake: Na^+ uptake and acid release can be varied independently by manipulating experimental conditions (Gould and Holland 1984). The amount of acid released appears to be a function of the electrochemical gradient for protons across the egg membrane since it varies with external pH and correlates with the amplitude of the fertilization potential (Holland *et al.* 1984; Gould and Holland 1984). In fact, unfertilized eggs will release acid and activate in sodium-free seawater when Na^+ is replaced by K^+, but not when it is replaced by choline (unpublished results). In high K^+ seawater the membrane potential goes positive, but in choline seawater it stays negative (Jaffe *et al.* 1979; Gould-Somero *et al.* 1979). Thus the available data strongly suggest that at least a substantial portion of acid release in *Urechis* is *via* voltage-gated proton channels similar to those described in molluscan neurons (Byerly *et al.* 1984; Meech and Thomas 1987) and salamander oocytes (Barish and Baud 1984).

Among polychaetes and molluscs where acid release has been described, there is no conclusive evidence regarding the mechanism. In Ii and Rebhun's (1979) study with *Spisula* no acid release occurred in sodium-free seawater, but Na^+ was replaced by choline, not K^+. In *Barnea*, activation by high (63 mM) K^+ seawater required external Na^+, but Na^+ was replaced by Tris (an impermeant substitute like choline) and acid release was not measured (Dubé and Guerrier 1982b). These data certainly are not sufficient to conclude that Na^+/H^+ exchange is occurring. More studies are needed to determine the mechanism of acid release in these eggs.

SUMMARY

The following model for the ionic basis of egg activation in *Urechis* is proposed. The sperm acrosomal protein opens a patch of Na^+ channels in the egg membrane. This drives the egg membrane potential positive, opening voltage-gated Ca^{2+} and H^+ channels. While the fertilization potential remains positive, Ca^{2+} enters and protons exit, producing a rise in intracellular free Ca^{2+} and increase in intracellular pH. Changes in the concentrations of these two ions trigger the series of events leading to GVBD and further development. A similar model can be proposed for activation in related polychaete and mollusc species, recognizing the possibility that intracellular pH may not change in species that do not release fertilization acid.

ACKNOWLEDGMENTS

The preparation of this article was supported in part by grants from the Universidad Autónoma de Baja California, CONACyT and the Secretaría de Educación Pública.

REFERENCES

Allen, R. D. 1953. Fertilization and artificial activation in the egg of the surf-clam, *Spisula solidissima. Biol. Bull.* 105:213-239.

Alliegro, M. and D. Wright. 1983. Polyspermy inhibition in the oyster, *Crassostrea virginica. J. Exp. Zool.* 227:127-137.

Barish, M. and C. Baud. 1984. A voltage-gated hydrogen ion current in the oocyte membrane of the axolotl, *Ambystoma. J. Physiol.* 352:243-263.

Baud, C., M. Moreau, and P. Guerrier. 1987. Ionic mechanism of the action potential and its disappearance after fertilization in the *Dentalium* egg. *Dev. Biol.* 122:516-521.

Bloom, T., E. Szuts, and W. Eckberg. 1988. Insitol triphosphate, inositol phospholipid metabolism, and germinal vesicle breakdown in surf clam oocytes. *Dev. Biol.* 129:532-540.

Brandriff, B., G. Moy, and V. Vacquier. 1978. Isolation of sperm bindin from the oyster (*Crassostrea gigas*). *Gamete Res.* 1:89-99.

Brassard, M., H. Duclohier, M. Moreau, and P. Guerrier. 1988. Intracellular pH change does not appear as a prerequisite for triggering activation of *Barnea candida* (Mollusca, Pelecypoda) oocytes. *Gamete Res.* 20:43-52.

Byerly, L., R. Meech, and W. Moody, Jr. 1984. Rapidly activating hydrogen ion currents in perfused neurones of the snail, *Lymnaea stagnalis. J. Physiol.* 351:199-216.

Conklin, E. G. 1904. Experiments on the origin of the cleavage centrosomes. *Biol. Bull.* 7:221-226.

Cross, N. L., T. Slezynger, and L. Z. Holland. 1985. Isolation and partial characterization of *Urechis caupo* egg envelopes. *J. Cell Sci.* 74:193-205.

Dale, B. 1988. Primary and secondary messengers in the activation of ascidian eggs. *Exp. Cell Res.* 177:205-211.

Dale, B., L. J. DeFelice, and G. Ehrenstein. 1985. Injection of a soluble sperm fraction into sea urchin eggs triggers the cortical reaction. *Experientia* 41:1068-1070.

Dubé, F. 1988. The relationships between early ionic events, the pattern of protein synthesis, and oocyte activation in the surf clam, *Spisula solidissima. Dev. Biol.* 126:233-241.

Dubé, F. and P. Guerrier. 1982a. Activation of *Barnea candida* (Mollusca, Pelecypoda) oocytes by sperm or KCl but not by NH_4Cl requires a calcium influx. *Dev. Biol.* 92:408-417.

Dubé, F. and P. Guerrier. 1982b. Acid release during activation of *Barnea candida* (Mollusca, Pelecypoda) ooycytes. *Dev. Growth & Differ.* 24:163-171.

Dufresne-Dubé, L., B. Picheral, and P. Guerrier. 1983a. An ultrastructural analysis of *Dentalium vulgare* (Mollusca, Scaphopoda) gametes with special reference to early events at fertilization. *J. Ultrastruct. Res.* 83:242-257.

Dufresne-Dubé, L., F. Dubé, P. Guerrier, and P. Couillard. 1983b. Absence of a complete block to polyspermy after fertilization of *Mytilus galloprovincialis* (Mollusca, Pelecypoda) oocytes. *Dev. Biol.* 97:27-33.

Eckberg, W. and A. Carroll. 1982. Sequestered calcium triggers oocyte maturation in *Chaetopterus. Cell Differ.* 11:155-160.

Ehrenstein, G., B. Dale, and L. J. DeFelice. 1984. A soluble fraction of sperm triggers cortical granule exocytosis in sea urchin eggs. *Biophys. J.* 45:23a.

Fallon, J. F. and C. R. Austin. 1967. Fine structure of gametes of *Nereis limbata* (Annelida) before and after interaction. *J. Exp. Zool.* 166:225-242.

Finkel, T. and D. Wolf. 1980. Membrane potential, pH and the activation of surf clam oocytes. *Gamete Res.* 3:299-304.

Gilman, A. 1987. G Proteins: transducers of receptor-generated signals. *Annu. Rev. Biochem.* 56:615-649.

Gould, M. and L. Holland. 1984. Fertilization acid release in *Urechis* eggs II. The stoichiometry of Na^+ uptake and H^+ release. *Dev. Biol.* 104:329-335.

Gould, M., J. L. Stephano, and L. Holland. 1986. Isolation of protein from *Urechis* sperm acrosomal granules that binds sperm to eggs and initiates development. *Dev. Biol.* 117:306-318.

Gould, M. and J. L. Stephano. 1987. Electrical responses of eggs to acrosomal protein similar to those induced by sperm. *Science.* 235:1654-1656.

Gould-Somero, M. 1981. Localized gating of egg Na^+ channels by sperm. *Nature (Lond.)* 291:254-256.

Gould-Somero, M. and L. Z. Holland. 1975a. Oocyte differentiation in *Urechis caupo*. (Echiura): A fine structural study. *J. Morphol.* 147:475-506.

Gould-Somero, M. and L. Holland. 1975b. Fine structural investigation of the insemination response in *Urechis caupo. Dev. Biol.* 46:358-369.

Gould-Somero, M., L. A. Jaffe, and L. Holland. 1979. Electrically mediated fast polyspermy block in eggs of the marine worm, *Urechis caupo. J. Cell Biol.* 82:426-440.

Guerrier, P., C. Guierrier, I. Neant, and M. Moreau. 1986a. Germinal vesicle nucleoplasm and intracellular pH requirements for cytoplasmic maturity in oocytes of the prosobranch mollusk *Patella vulgata. Dev. Biol.* 116:92-99.

Guerrier, P., M. Brassart, C. David, and M. Moreau. 1986b. Sequential control of meiosis reinitiation by pH and Ca^{2+} in oocytes of the prosobranch mollusk *Patella vulgata. Dev. Biol.* 114:315-324.

Hagiwara S. and L. A. Jaffe. 1979. Electrical properties of egg cell membranes. *Annu. Rev. Biophys. Bioeng.* 8:385-416.

Hagiwara, S. and S. Miyazaki. 1977. Changes in excitability of the cell membrane during "differentiation without cleavage" in the egg of the annelid, *Chaetopterus pergamontaceus. J. Physiol.* 272:197-216.

Holland, L., M. Gould-Somero, and M. Paul. 1984. Fertilization acid release in *Urechis* eggs. I. The nature of the acid and the dependence of acid release and egg activation on external pH. *Dev. Biol.* 103:337-342.

Holt, V. 1934. Further observations on the polarity of the eggs of *Urechis caupo. Biol. Bull.* 67:341-345.

Humphreys, W. J. 1967. The fine structure of cortical granules in eggs and gastrulae of *Mytilus edulis. J. Ultrastruct. Res.* 17:314-326.

Ii, I. and L. Rebhun. 1979. Acid release following activation of surf clam (*Spisula solidissima*) eggs. *Dev. Biol.* 72:195-200.

Ikegami, S., T. S. Okada, and S. S. Koide. 1976. On the role of calcium ions in oocyte maturation in the polychaete *Chaetopterus pergamentaceus. Dev. Growth & Differ.* 18:33-43.

Jaffe, L. A., M. Gould-Somero, and L. Z. Holland. 1979. Ionic mechanism of the fertilization potential of the marine worm, *Urechis caupo* (Echiura). *J. Gen. Physiol.* 73:469-492.

Jaffe, L. A., M. Gould-Somero, and L. Z. Holland. 1982. Studies of the mechanism of the electrical polyspermy block using voltage clamp during cross-species fertilization. *J. Cell Biol.* 92:616-621.

Jaffe, L. A., R. Kado, and L. Muncy. 1985. Propagating potassium and chloride conductances during activation and fertilization of the egg of the frog, *Rana pipiens. J. Physiol.* 368:227-242.

Jaffe, L. F. 1983. Sources of calcium in egg activation: a review and hypothesis. *Dev. Biol.* 99:265-276.

Jaffe, L. F. 1985. The role of calcium explosions, waves, and pulses in activating eggs. p. 127-165. *In: Biology of Fertilization,* Vol. 3. C. B. Metz and A. Monroy (Eds.). Academic Press, Orlando.

Johnston, R. and M. Paul. 1977. Calcium influx following fertilization of *Urechis caupo* eggs. *Dev. Biol.* 57:364-374.

Kline, D. and R. Nuccitelli. 1985. The wave of activation current in the *Xenopus* egg. *Dev. Biol.* 111:471-487.

Meech, R. and R. C. Thomas. 1987. Voltage-dependent intracellular pH in *Helix aspersa* neurones. *J. Physiol.* 390:433-452.

Ohe, Y., H. Hayashi, and K. Iwai. 1979. Human spleen histone H2B. Isolation and amino acid sequence. *J. Biochem. (Tokyo)*.85:615-624.

Osanai, K. 1976. Parthenogenetic activation of Japanese Palolo eggs with sperm extract. *Bull. Mar. Biol. Stn. Asamushi.* 15:157-163.

Pasteels, J. 1966. La reaction corticale de fecondation de l'oeuf de *Nereis diversicolor*, etudiee au microscope electronique. *Acta. Embryol. & Morphol. Exp.* 6:166-163.

Pasteels, J. J. and E. deHarven. 1962. Etude au microscope electronique du cortex de l'oeuf de *Barnea candida* (mollusque bivalve) et son evolution au moment de la fecondation, de la maturation et de la segmentation. *Arch. Biol.* 73:465-490.

Paul, M. 1975. Release of acid and changes in light scattering properties following fertilization of *Urechis caupo* eggs. *Dev. Biol.* 43:299-312.

Paul, M. and M. Gould-Somero. 1976. Evidence for a polyspermy block at the level of sperm-egg plasma membrane fusion in *Urechis caupo. J. Exp. Zool.* 196:105-112.

Peaucellier, G. 1977a. Mise en evidence du role du calcium dans la reinitiation de la meiose des ovocytes de *Sabellaria alveolata* (L.) (annelide polychete). *C. R. Hebd. Seances Acad. Sci. Ser. D Sci. Nat.* 285:913-915.

Peaucellier, G. 1977b. Initiation of meiotic maturation by specific proteases in oocytes of the polychaete annelid *Sabellaria alveolata. Exp. Cell Res.* 106:1-14.

Peaucellier, G. 1978. Acid release at meiotic maturation of oocytes in the polychaete annelid *Sabellaria alveolata. Experientia* 34:789-790.

Rebhun, L. I. 1962. Electron microscope studies on the vitelline membrane of the surf clam, *Spisula solidissima. J. Ultrastruct. Res.* 6:107-122.

Robertson, T. B. 1912. On the extraction of a substance from the sperm of a sea urchin (*Stronglyocentrotus purpuratus*) which will fertilize eggs of that species. *Univ. Calif. Publ. Physiol.* 4:103-105.

Schmidt, T., C. Patton, and D. Epel. 1982. Is there a role for the Ca^{2+} influx during fertilization of the sea urchin egg? *Dev. Biol.* 90:284-290.

Suzuki, K. and T. Ando. 1972. The complete amino acid sequence of clupeine YI. *J. Biochem. (Tokyo)* 72:1419-1432.

Tamaki, H. and K. Osanai. 1985. Re-initiation of meiosis in *Mytilus oocytes* with acrosome reaction product of sperm. *Bull. Mar. Biol. Stn. Asamushi* 18:11-23.

Tyler, A. 1932. Changes in volume and surface of *Urechis* eggs upon fertilization. *J. Exp. Zool.* 63:155-173.

Tyler, A. 1965. The biology and chemistry of fertilization. *Am. Nat.* 97:309-334.

Tyler, A. and J. Schultz. 1932. Inhibition and reversal of fertilization in eggs of the echinoid worm, *Urechis caupo. J. Exp. Zool.* 63:509-531.

Vacquier, V. and G. Moy. 1977. Isolation of bindin: the protein responsible for adhesion of sperm to sea urchin eggs. *Proc. Natl. Acad. Sci. USA* 74:2456-2460.

THE ROLE OF THE PHOSPHATIDYLINOSITOL CYCLE IN THE ACTIVATION OF THE FROG EGG

Richard Nuccitelli, James Ferguson and Jin-Kwan Han

Department of Zoology
University of California, Davis
Davis, CA 95616

Over the past five years we have filled in many steps in the sequence of events responsible for the activation of both vertebrate and invertebrate eggs. While the egg of the sea urchin has been the most popular system for the study of gamete activation, several investigators have been studying vertebrate eggs as well. The amphibian egg has emerged as the best studied vertebrate model to date. We now know that the activation events following normal fertilization in this egg include the phosphatidylinositol lipid cascade. We will review some of these recent observations here and will then concentrate on current studies from our laboratory that indicate a possible role for at least two of the phosphoinositols in the cascade.

Intracellular ionic activity changes play a central role in the activation of the metabolic processes in early development. In the early decades of this century numerous investigators demonstrated that manipulations now known to alter either pH_i (Loeb 1913) or $[Ca^{2+}]_i$ (Dalcq 1928; Mazia 1937; Pasteels 1938) could stimulate parthenogenetic development of invertebrate eggs, and recent studies have confirmed and amplified these findings. The Ca^{2+}-H^+ ionophore, A23187, is a universal activator of eggs, apparently due to its ability to elevate $[Ca^{2+}]_i$ (Steinhardt and Epel 1974; Chambers *et al.* 1974; Steinhardt *et al.* 1974). Ammonia and other weak bases, which elevate pH_i, activate numerous events of early development in the sea urchin egg such as polyadenylation, protein synthesis, increases in the K^+ conductance and fluidity of the plasma membrane (reviewed in Busa and Nuccitelli 1984). Treatments which elevate pH_i also activate sand dollar eggs (Hamaguchi 1982), the spermatozoa of numerous species, both invertebrate (Nishioka and Cross 1978; Christen *et al.* 1982; Lee *et al.* 1983;

Lambert and Epel 1979) and vertebrate (Acott and Carr 1984), and reinitiate development in dormant gastrula-stage embryos of the brine shrimp (Busa and Crowe 1983).

Thus, pH_i and $[Ca^{2+}]_i$ changes are potent metabolic and developmental effectors in numerous species, cell types and developmental stages. Their physiological significance is indicated by increasing evidence that both pH_i and $[Ca^{2+}]_i$ changes accompany normal fertilization in many eggs. Large, transient increases in $[Ca^{2+}]_i$ occur in the first minute following fertilization in the eggs of *Xenopus* (Busa and Nuccitelli 1985; Kubota *et al.* 1987), *Discoglossus pictus* (painted frog) (Nuccitelli *et al.* 1988), medaka (Ridgway *et al.* 1977; Gilkey *et al* 1978), sea urchins (Eisen and Reynolds 1984; Swann and Whitaker 1986), starfish (Eisen *et al* 1984), mouse (Cuthbertson *et al* 1981; Cuthbertson and Cobbold 1985) and hamster (Miyazaki *et al.* 1986 and this volume). Permanent increases in pH_i (of 0.3 - 0.4 U) follow fertilization in the eggs of marine invertebrates (Shen and Steinhardt 1978; Johnson and Epel 1981; Winkler *et al.* 1980; Hamaguchi 1982) and a freshwater vertebrate (Nuccitelli *et al* 1981; Webb and Nuccitelli 1982). In the sea urchin spermatozoon, the pH_i and $[Ca^{2+}]_i$ changes accompanying the acrosome reaction are thought to be central to its regulation (Tilney 1976; Tilney *et al* 1978; Schackmann and Shapiro 1981).

How does sperm-egg interaction lead to these critical changes in $[Ca^{2+}]_i$ and pH_i? The biochemical mechanisms responsible for the rise in calcium during the activation of many cell types remained a mystery until the recent discovery of the phosphatidylinositol cycle. This cycle meditates many signal transduction events in a wide variety of cells (Berridge 1987). When certain cell surface receptors bind their ligand, a G protein-regulated phospholipase C is activated which cleaves phosphatidylinositol-4,5-bisphosphate (PIP_2). PIP_2 hydrolysis results in the release of diacylglycerol (DAG) into the plasma membrane and inositol-1,4,5-trisphosphate ($Ins(1,4,5)P_3$) into the cytoplasm. $Ins(1,4,5)P_3$ causes the release of calcium from internal stores and is metabolized by two different pathways. In one pathway, $Ins(1,4,5)P_3$ can be directly dephosphorylated to $InsP_2$, which is then dephosphorylated to $InsP_1$. Alternately $Ins(1,4,5)P_3$ can be directly phosphorylated to $Ins(1,3,4,5)P_4$ which in turn is dephosphorylated to $Ins(1,3,4)P_3$ (Balla *et al* 1987). $Ins(1,3,4)P_3$ can be phosphorylated to $Ins(1,3,4,6)P_4$ (Shears *et al* 1987a), or it can be dephosphorylated to $InsP_2$ (Shears *et al* 1987b). Signal transduction therefore involves the production of many different types of polyphosphoinositols, however the exact function and interrelationships of the phosphoinositols is unknown.

Over the past few years several investigators have found that this phosphatidylinositol cycle is also involved in the mechanism of egg activation. To briefly summarize, we now know that activation involves the stimulation of a GTP-binding protein which activates phospholipase C, cleaving $Ins(1,4,5)P_3$ off of PIP_2 (Kline and Jaffe 1987). The $Ins(1,4,5)P_3$ releases intracellular Ca^{2+} and triggers a wave of increased free Ca^{2+} that spreads throughout the cytoplasm of the frog egg (Busa and Nuccitelli 1984; Busa *et al* 1985; Kubota *et al* 1987). As the $[Ca^{2+}]_i$ recovers to normal levels, pH_i increases by 0.3 U and remains at this elevated level throughout early development (Webb and Nuccitelli 1981 1982; Nuccitelli *et al* 1981). The increase in Ca^{2+} is critical for egg activation and if blocked by the injection of Ca^{2+} buffers, the fertilization potential is inhibited as well as the exocytosis of cortical granules, the elevation of the fertilization envelope, sperm decondensation and pronuclear formation (Kline 1988).

MATERIALS AND METHODS

Double-barreled Ca^{2+}-specific Microelectrodes

In order to study the localized effect of $Ins(1,4,5)P_3$ and $Ins(1,3,4,5)P_4$, double-barreled Ca^{2+}-sensitive microelectrodes were constructed by two different techniques which have been previously described (Busa 1986; Levy and Fein 1985). The first involves pulling double-barreled glass on a standard vertical puller and separating the barrels. One barrel is made into a Ca^{2+}-sensitive microelectrode and the other is back-filled with either $Ins(1,4,5)P_3$

or Ins(1,3,4,5)P_4. The barrels are then rejoined with adhesive so the tips are 10 μm apart. The other method involves the use of theta tubing which is formed into a double-barreled microelectrode on a vertical electrode puller. The tips are closer together with this method, since they were formed in essence from one piece of glass. Both styles of these electrodes allow one to examine the immediate effect of injected substances on the intracellular calcium in a very localized region of the cell.

Fura-2 Imaging

Intracellular calcium was also measured using a fura-2 imaging system similar to one previously described (Poenie *et al* 1986) with two differences. In the present study, the system was operated with only one 150 W xenon arc lamp, and the alternating 350 nm and 385 nm excitation light was generated by rotating a pair of appropriate interference filters in the light path of the xenon lamp. This arrangement gave better uniformity of intensity over the illuminated field than using two light sources into two monochrometers, and reduced the likelihood of independent intensity fluctuations in two xenon lamps giving rise to spurious changes in the apparent $[Ca^{2+}]_i$. Although such spurious changes tend to be very small, in the present experiments they might have been problematical since subtle effects were being examined.

Prior to being placed onto the microscope for imaging studies, the cells were first injected with a solution consisting of 1 mM fura-2. The fura-2 was made in up as a 1 mM stock dissolved in a buffer that resembles the cytoplasmic milieu: 19 mM NaCl, 52 mM KCl, and 10 mM HEPES, adjusted to pH 7.3 with KOH. 0.5 mM $CaCl_2$ was added so that the fura-2 was half saturated, generating a free Ca^{2+} concentration of 240 nM. The final intracellular concentration of fura-2 was approximately 100 μM. The oocytes were allowed to recover from the fura-2 injection for approximately 1h at 20°C.

Control experiments to test for Ca^{2+} contamination of the polyphosphoinositols were performed on the imaging apparatus using the procedure outlined in the Electrophysiology section below. We found no observable Ca^{2+} contamination as revealed by the Ca^{2+} imaging technique. We also injected, as a control, fructose 1, 6-bisphosphate (1.0 mM in 0.1 mM HEPES, pH 7.8) into oocytes while monitoring the cells by imaging. No intracellular $[Ca^{2+}]$ changes were observed upon injecting fructose 1, 6-bisphosphate by iontophoresis or pressure injection, unless pressures greater than 20 psi were used, which caused rapid movement of the injection pipette.

Electrophysiology

Cells were voltage-clamped with a standard, two-electrode clamp (Dagan model 8500, Minneapolis, Minn.) and the current was measured through a virtual ground circuit. The current electrode had a resistance of 1 to 2 megohms and was filled with either 3 M KCl or 3 M K^+-acetate. The voltage electrode had a resistance of 10 to 20 megohms and was filled with 3 M KCl. All cells were clamped at a potential of -60 mV.

Cells were injected either by pressure or iontophoresis. All injections were made in the animal hemisphere since the plasma membrane in that hemisphere contains more Cl^- channels (Robinson 1979; Kline and Nuccitelli 1985; Miledi and Parker 1984) and the animal cortex may contain more calcium storage sites than the vegetal cortex (Campanella and Andreuccetti 1977; Gardiner and Grey 1983; Campanella *et al* 1984; Charbonneau and Grey 1984). When pressure injections were used the tip of the injection pipet was broken back to a width of 3 μm. When iontophoresis was used, the amount injected was determined by the formula $q=-nI/zF$. The transport number was assumed to be 0.5, although it should be noted that the transport number for anions can be less than or equal to 0.1 (Purves 1981). Each phosphate on the inositol was assumed to have a charge of -2. In order to determine if the inositol solutions were contaminated with calcium, positive charge was first injected since calcium would activate Cl^- channels. None of the samples were contaminated with calcium. Throughout the text the amounts of polyphosphoinositols are reported in coulombs and the theoretical calculated amounts delivered in moles are in parentheses. As a further control, 1.0 mM fructose 1, 6-bisphosphate was made in 0.1 mM Hepes, pH 7.8, and injected into oocytes. This sugar had no effect on Cl^- currents or calcium release.

The polyphosphoinositols were a gift from Dr. Robin Irvine and were all prepared as previously described (Irvine *et al* 1984, 1986). In some experiments commercial Ins(1,4,5)P_3 was used (Sigma Co. St. Louis) but it was not used in the experiments comparing Ins(1,4,5)P_3 to Ins(2,4,5)P_3 because the Ins(1,4,5)P_3 from Sigma contains high levels of Ins(2,4,5)P_3. In all cases the inositols were prepared as 1.0 mM stock solutions in 0.1 mM Hepes and adjusted to pH 7.8 with KOH.

Centrifugation of Eggs

Xenopus eggs were obtained by squeezing females induced to ovulate via injection of 800 IU of human chorionic gonadotropin. The eggs were dejellied by gentle agitation in 2% L-cysteine in F1 solution with the pH adjusted to 7.8 with NaOH. After washing extensively with F1, eggs were gently placed in 30 ml centrifuge tubes containing 30% Ficoll-F1 and F1 solution. Ten mM chlorobutanol was added to the F1 solution to prevent possible activation during centrifugation. Eggs were placed on the Ficoll-F1 interface and centrifuged at 4080XG for 1h at 18°C. The stratified eggs were removed from the Ficoll-F1 cushion interface and transferred into the chamber filled with 30% Ficoll-F1 (with no chlorobutanol) for electrophysiological studies.

Electron Microscopy

Centrifuged eggs were fixed in 3% glutaraldehyde for 3h, washed with F1 four times then post-fixed in 1% OsO_4 for 1h. After washing with double-distilled water two times, eggs were stained with uranylacetate for 1.5h. After serial acetone dehydration steps, eggs were imbedded in plastic and sectioned at a thickness of about 100 nm.

RESULTS AND DISCUSSION

Ins(1,4,5)P_3 Triggers the Wave of Ca^{2+} Release in Mature Eggs

We have shown that Ins(1,4,5)P_3 will activate the mature eggs of the frogs, *Xenopus laevis* and *Discoglossus pictus* and those of the medaka fish, *Oryzias latipes*. When the *Xenopus laevis* egg is activated there is a permanent increase in the pH_i and a transient rise in $[Ca^{2+}]_i$. The calcium initially increases at the point of egg activation and a wave of release from intracellular stores then proceeds throughout the cytoplasm. The wave of calcium release in *Xenopus* was first measured using Ca^{2+}-sensitive microelectrode pairs and it takes approximately 6 min for the wave of Ca^{2+} to travel from the site of activation in the animal hemisphere to the opposite pole as illustrated in Figure 1 (Busa and Nuccitelli 1985). Kubota *et al* (1987) recently used the aequorin technique which indicates the Ca^{2+} levels near the cell surface. They found that the $[Ca^{2+}]_i$ increase near the surface exhibits a ring-shaped pattern that is initiated at the activation site and spreads over the egg at a rate of about 8 μm/sec. This cortical ring of increased Ca^{2+} travels at twice the speed of the $[Ca^{2+}]_i$ wave deeper in the cytoplasm. This ring-shaped wave is very similar to that found in the egg of the fresh water killifish, medaka (Gilkey *et al* 1978), which is also easily activated by Ins(1,4,5)P_3 injection (Nuccitelli 1987). The egg of the painted frog, *Discoglossus pictus*, also exhibits an increase in $[Ca^{2+}]_i$ at activation and is activated by Ins(1,4,5)P_3 injection (Nuccitelli *et al* 1988), although the wave-like nature of this Ca^{2+} increase has not been documented in the detail of the latter two vertebrates. While Ins(1,4,5)P_3 injection clearly initiates the Ca^{2+} activation wave in all three vertebrate eggs studied, it is not clear if the propagation of the wave requires a wave of Ins(1,4,5)P_3 production or is due instead to Ca^{2+}-induced-Ca^{2+} release.

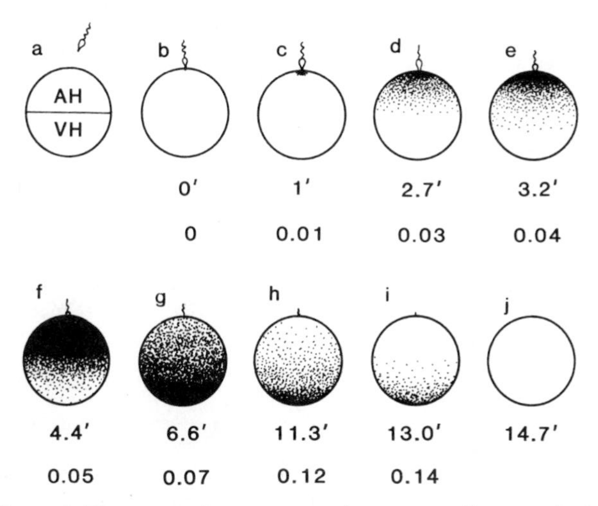

Figure 1. The approximate temporal sequence of events during the Ca^{2+} wave in a fertilized *Xenopus* egg, based on data from pairs of Ca^{2+}-specific microelectrodes. The approximate spacial extent and magnitude of increased $[Ca^{2+}]_i$ are indicated by the distribution and intensity of stippling, respectively. The first number beneath each drawing is the time, in minutes, from onset of the fertilization potential; the second number is the decimal fraction of the interval from fertilization potential to first cleavage at 22 °C. Timing of sperm incorporation is arbitrary. **(a)** Orientation of egg, with animal pole at top. Fertilization is shown occurring there for simplicity. **(b)** Onset of fertilization potential. **(c)** Ca^{2+} wave begins at the sperm entry site. **(d)** Front of $[Ca^{2+}]_i$ wave reaches equator. **(e)** $[Ca^{2+}]_i$ peaks at sperm entry site. **(f)** Front of Ca^{2+} wave reached antipode. **(g)** $[Ca^{2+}]_i$ peaks at antipode. **(h)** Recovery complete at sperm entry site. **(i)** Recovery complete in animal hemisphere. **(j)** Recovery complete throughout cell.

A Wave of Activation Current Accompanies the Ca^{2+} Wave in Some Eggs

We have used the extracellular vibrating probe to map the spatial distribution of the activation current in these three eggs, and in the case of *Xenopus* and medaka, Ca^{2+}-dependent ion channels open in a wave-like manner with a time course similar to the Ca^{2+} wave. In *Xenopus*, Ca^{2+}-sensitive Cl^- channels open which causes the membrane potential to depolarize and generate the fast, electrical block to polyspermy (Grey *et al* 1982). This activation current wave traverses the surface of the egg in 2.5-4 min. The rate of propagation in the animal and vegetal hemispheres is 17 μm/s and 14 μm/s, respectively. The propagation velocity of the calcium wave as measured with Ca^{2+}-specific microelectrodes or with aequorin appears to be somewhat slower than that of the current wave, and the explanation for this will probably require using both techniques simultaneously on the same egg. In the medaka egg, Ca^{2+}-sensitive cation channels open, allowing Na^+ and Ca^{2+} to enter the egg in a ring-shaped region that traverses the surface of the egg in 3 min. This coincides both spatially and temporally with the wave of increased $[Ca^{2+}]_i$ detected with aequorin (Gilkey *et al* 1978).

The egg of the painted frog, *Discoglossus pictus*, also exhibits an increase in $[Ca^{2+}]_i$ at activation which apparently spreads in a wave-like manner. However, there is no corresponding current wave in this egg (Nuccitelli *et al* 1988). Another difference between this egg and the *Xenopus* egg is that fertilization is localized to a 200 μm-wide region called the "animal dimple". For unknown reasons, plasma membrane ion channels are either not present or do not open in regions other than the animal dimple where there is a very high density of Cl^- channels.

Identifying the Storage Organelle that Releases Ca^{2+} Upon Binding $Ins(1,4,5)P_3$

One of the advantages of the large *Xenopus* egg is that its organelles can be easily stratified by centrifugation without damaging the cell (Figure 2). This allowed us to study the Ca^{2+} release characteristics of each organelle layer using the double-barreled Ca^{2+}-specific microelectrodes in which the $[Ca^{2+}]_i$ is measured 10 μm away from the site of $Ins(1,4,5)P_3$ injection. A major advantage of this technique is that the organelles are still in the physiological environment of the cell and possible perturbations resulting from standard, *in vitro* isolation techniques can be avoided. Stratified eggs exhibit unique organelle layers: lipid, endoplasmic reticulum, mitochondria, pigment, and yolk (Figures 2 and 3). The intracellular calcium concentration in each layer is directly measured using calcium-specific microelectrodes. We developed a double-barreled calcium electrode from theta-tubing so that the second barrel could be filled with $Ins(1,4,5)P_3$. The advantage of this technique is that the two tips of the theta-tubing electrode are very close to each other (approximately 3 to 5 μm) and this makes it possible to measure the $[Ca^{2+}]_i$ very close to the site of $Ins(1,4,5)P_3$ injection. Intracellular Ca^{2+} is very strongly buffered so that the Ca^{2+} increase due to iontophoresis of $Ins(1,4,5)P_3$ falls off very sharply with distance from the injecting electrode. We find that the ER layer releases the largest amount of Ca^{2+} in response to $Ins(1,4,5)P_3$ microinjection. As

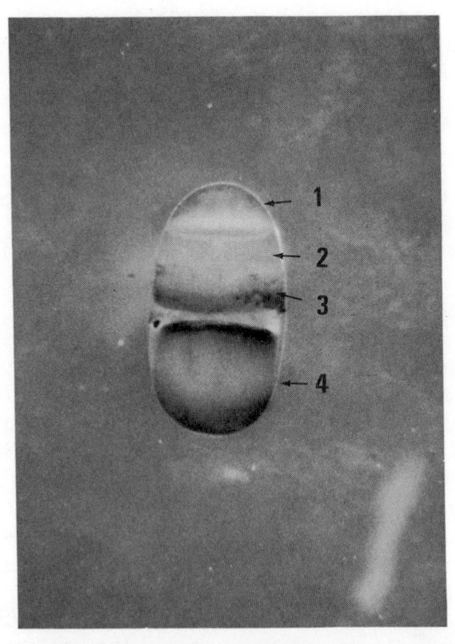

Figure 2. Photomicrograph of stratified *Xenopus* egg which had been centrifuged on Ficoll at 4080Xg for 1h. The four main organelle layers are indicated by the arrows: 1: lipid layer; 2: ER layer; 3: mitochondrial layer and; 4: yolk layer.

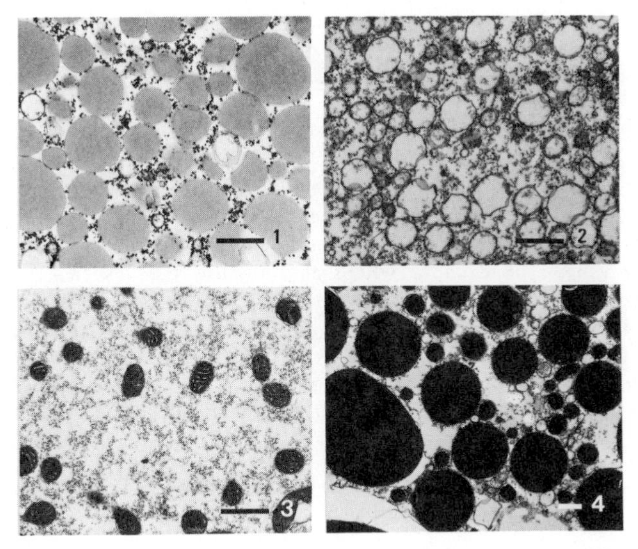

Figure 3. Electron micrographs of the four major stratified layers. **1)** lipid; **2)** Endoplasmic Reticulum; **3)** mitochondria; **4)** yolk (1, 2, and 3: 7,100X magnification; 4: 3,000X).

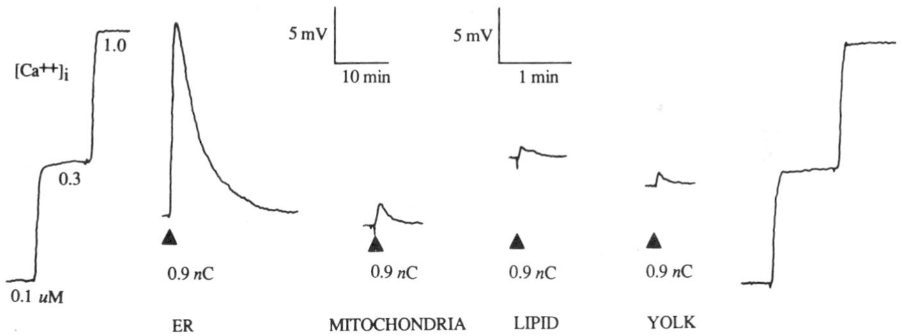

Figure 4. Tracings of double-barreled Ca^{2+}-specific microelectrode recordings during iontophoresis with the same amount of Ins(1,4,5)P_3 (0.9 nC) into different layers of a stratified egg. Trace at left and right in this figure show calcium-electrode calibrations (0.1, 0.3, 1.0 μM) before and after impalement. Each layer exhibited a slightly different level of resting $[Ca^{2+}]_i$. When the same amount of Ins(1,4,5)P_3 iontophoretic current was injected, the ER layer released the most calcium whereas the other layers released little. One possibility is that a small release of calcium from layers other than ER is due to a small contamination of ER in these layers. In all tracings in this figure, the electrical artifact due to the iontophoretic current injection was corrected by subtracting the response recorded when Ins(1,4,5)P_3 was injected into the calibration buffer outside of the egg from the calcium response recorded in each layer.

shown in Figure 4, iontophoresis of Ins(1,4,5)P_3 into the ER layer releases calcium almost immediately and the amount released is much greater than that observed in the other organelle layers. The 0.9 nC of Ins(1,4,5)P_3 iontophoresis releases sufficient calcium to reach a local concentration of approximately 800 nM which is almost the same as that accompanying fertilization (Figure 5). This result does not necessarily mean that the ER is the main Ca^{2+} release site during activation since we have never observed an activation potential in these stratified eggs. However, we have observed intact cortical granules distributed along the entire stratified egg cortex, so these eggs may still be unactivated.

The Role of Polyphosphoinositides in Ca^{2+} Release

In order to gain a better understanding of the roles of the different polyphosphoinosides in the inositol cycle during cell activation, each one must be studied in the absence of the others. It is difficult to do these experiments in the mature egg because activation involves a single explosive release of calcium which can obscure the response to the particular phosphoinositol under study. However, the immature oocyte is an ideal model system since multiple releases of calcium can be elicited from the cell without activating a regenerative wave of Ca^{2+} release. Furthermore, these cells have Ca^{2+}-dependent Cl^- channels which can be used to indirectly monitor changes in intracellular Ca^{2+} levels. When these cells are still enveloped by their follicle cells, they have muscarinic acetylcholine (ACh) receptors which activate the PI cycle upon binding ACh (Oron *et al* 1985). Acetylcholine sets off a complex

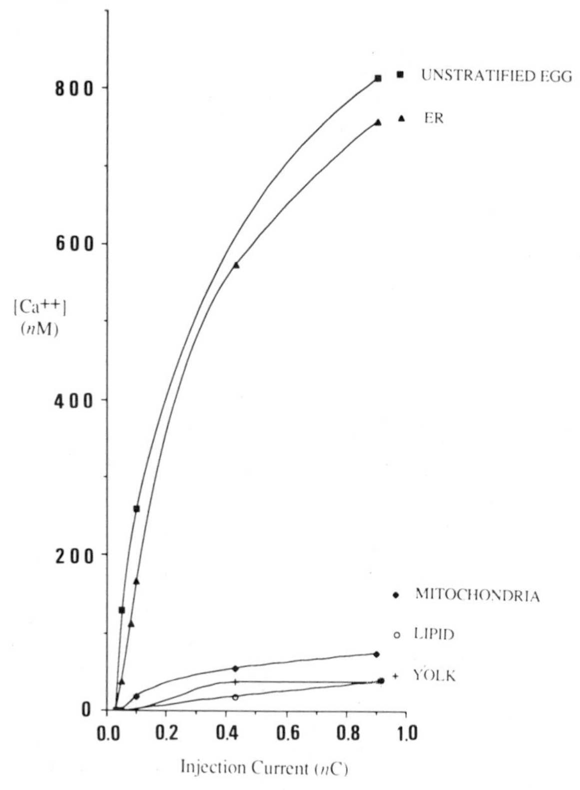

Figure 5. Increase in $[Ca^{2+}]_i$ in each of four layers in a stratified *Xenopus* egg as a function of amount of Ins(1,4,5)P_3 iontophoresis current through the other barrel of a double-barreled Ca^{2+}-specific microelectrode. For comparison, the response measured in an unstratified egg is included.

opening and closing of Cl^- channels and the injection of large amounts of Ins(1,4,5)P_3 have been reported to mimic the effects of external ACh application. Therefore we indirectly examined the ability of several different polyphosphoinositols to elicit changes in $[Ca^{2+}]_i$ by monitoring the Cl^- currents in voltage-clamped immature oocytes, and in some cases directly measured $[Ca^{2+}]_i$ using Ca^{2+}-specific microelectrodes and the Ca^{2+}-sensitive fluorescent probe, fura-2.

The Effects of Four Different Polyphosphoinositols on Cl^- Current Patterns

We examined the effects of different concentrations of Ins(1,4,5)P_3 on the pattern of the Cl^- current because under different physiological conditions varying amounts of Ins(1,4,5)P_3 may be released. It is therefore important to investigate the effects of a wide range of concentrations on Cl^- conductance. The temporal current pattern and magnitude was dependent on the amount of Ins(1,4,5)P_3 injected into the oocyte (Figure 6). Low concentrations caused a single, immediate Cl^- current due to a single release of calcium, while intermediate amounts caused a single immediate current, followed by a quiescent period, followed by oscillating Cl^- currents. When large amounts were injected, the oscillating currents increased in magnitude and were superimposed on a second, slow Cl^- conductance. The quiescent period is long enough to suggest that some biochemical event may be occurring which is responsible for the oscillations in current. These currents are totally independent of extracellular calcium, but are dependent on intracellular calcium release since they can be blocked by microinjection of BAPTA and by external application of verapamil. Ins(2,4,5)P_3 is a non-physiological form of $InsP_3$ which is not normally converted to $InsP_4$ by cellular

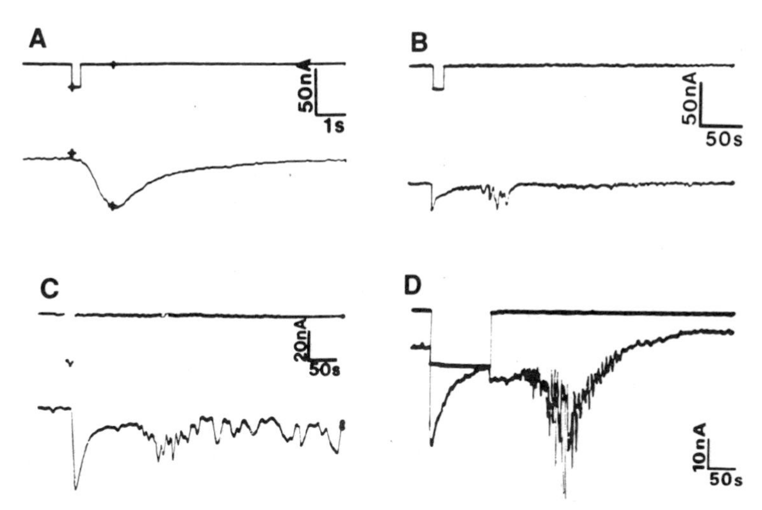

Figure 6. Response of voltage-clamped, immature stage VI oocytes of *Xenopus laevis* to iontophoresis of Ins(1,4,5)P_3. The upper trace of each figure is the iontophoresis current and the lower trace represents the current crossing the oocyte membrane. Inward current is downward and the scale bar is for this current only. In all cases the oocytes were clamped to -60 mV. **(A)** Current response to iontophoresis of 33 nC (2.9 x 10^{-14} moles) of Ins(1,4,5)P_3. **(B)** Current response to 0.5 μC (4.2 x 10^{-13} moles) of Ins(1,4,5)P_3. $InsP_3$ induces the initial Cl^- current followed by a quiescent period and oscillating Cl^- currents. **(C)** Current response to 1.2 μC (1.0 x 10^{-12} moles) of Ins(1,4,5)P_3. $InsP_3$ induces an immediate Cl^- current, followed by a quiescent period, which in turn is followed by oscillating Cl^- currents which stay close to the base line. **(D)** Current response to 3.0 μC (2.6 x 10^{-12} moles) of Ins(1,4,5)P_3. Ins(1,4,5)P_3 induces the initial Cl^- current, followed by oscillating Cl^- currents which are superimposed on a second Cl^- conductance, which is finally followed by an outward K^+ current.

kinases. We had hoped to use it to distinguish between the effects of Ins(1,4,5)P_3 and Ins(1,3,4,5)P_4. However, Ins(2,4,5)P_3 gave current responses that were very similar to those of Ins(1,4,5)P_3, and the only observed difference was that Ins(2,4,5)P_3 was 4-fold less effective in initiating the fast Cl^- current. One possible explanation for this observation is that the calcium released by Ins(2,4,5)P_3 causes an endogenous release of Ins(1,4,5)P_3 (via the activation of phospholipase C) which is then phosphorylated to produce Ins(1,3,4,5)P_4. Ins(1,3,4)P_3, which is the metabolic breakdown product of Ins(1,3,4,5)P_4 had no effect on Cl^- conductance.

Ins(1,3,4,5)P_4 injections generate a current pattern which is different from that caused by Ins(1,4,5)P_3. Ins(1,3,4,5)P_4 triggers oscillating Cl^- currents which are independent of extracellular Ca^{2+} (Figure 7). This observation has also been made by another group (Parker and Miledi 1987) and does not agree with recent controversial reports that Ins(1,3,4,5)P_4 can open plasma membrane Ca^{2+} channels in sea urchin eggs. (Irvine and Moor 1986 1987; Crossley 1988). There is often a lag time of up to 3 min between the injection of Ins(1,3,4,5)P_4 and the appearance of the currents and the pattern looks exactly like the oscillating phase of the Ins(1,4,5)P_3 and the Ins(2,4,5)P_3 response. Ins(1,3,4,5)P_4 never triggers an immediate Cl^- current. This similarity to the second phase of the Ins(1,4,5)P_3-induced current pattern suggests the possibility that during the quiescent period Ins(1,4,5)P_3 is converted to Ins(1,3,4,5)P_4 and eventually enough Ins(1,3,4,5)$_4$ may be generated to produce the oscillating Cl^- currents. The problem with this hypothesis is that there is also a quiescent period between the injection of Ins(1,3,4,5)P_4 and the appearance of any currents. Obviously there are still some steps in the pathway of the formation of these oscillating currents which are yet to be identified. It may be that calcium only activates some regulatory molecule which in turn activates the Cl^- channels. Nonetheless, this is the first evidence that different polyphosphoinositols may have distinct roles in the release of Ca^{2+}.

Intracellular Ca^{2+} Measurements

In order to study the localized changes in $[Ca^{2+}]_i$ in immature oocytes resulting from the injection of Ins(1,4,5)P_3 and Ins(1,3,4,5)P_4, double-barreled, Ca^{2+}-sensitive microelectrodes

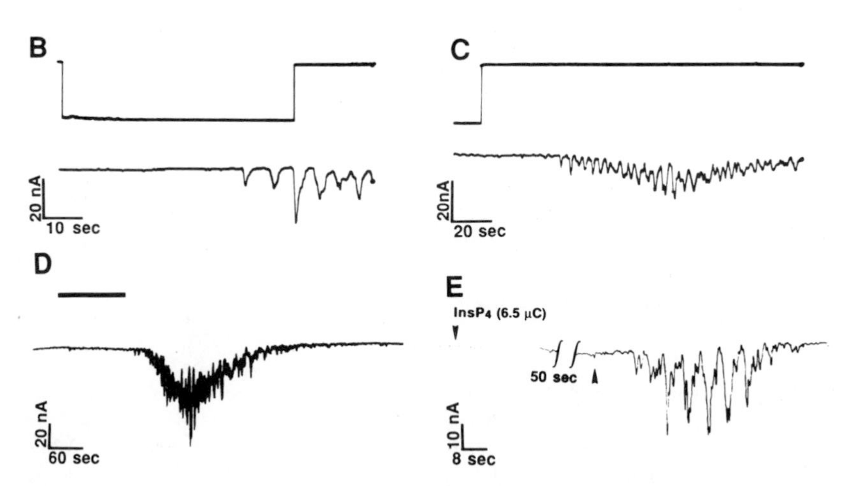

Figure 7. Oscillatory Cl^- currents triggered by the iontophoresis of Ins(1,3,4,5)P_4 into immature, stage VI oocytes. In B and C the upper trace represents the iontophoresis current, current on during downward deflection **(B)** 6.45 μC (4.2 x 10^{-12} moles). **(C)** 6.3 μC (4.1 x 10^{-12} moles). **(D)** 11.4 μC (7.4 x 10^{-12} moles). The bar represents the length of time that the iontophoresis current was on. **(E)** Currents in response to the injection of 6.5 μC (4.2 x 10^{-12} moles) of Ins(1,3,4,5)P_4 in 0-Ca^{2+}-high Mg^{2+} buffer. The first arrow indicates the time that the iontophoresis current was turned on and second arrow indicates the time when this current was turned off. There is a 50s break in the current trace.

were constructed in two different ways which have been previously described (Busa 1986; Levy and Fein 1985). These electrodes allow one to examine the immediate effect of a substance on the $[Ca^{2+}]_i$ in a very localized area of the cell. Intracellular calcium was also measured using a fura-2 imaging system similar that previously described (Poenie *et al* 1986). Among the inositol-trisphosphates studied, Ins(1,4,5)P_3 releases calcium most efficiently from intracellular stores as measured by both of these techniques (Figure 9), and Ins(1,3,4)P_3 was totally ineffective at releasing any calcium as monitored with the fura-2 imaging system. In contrast, we found that Ins(1,3,4,5)P_4 does release calcium from intracellular stores. As illustrated in Figures 8 and 9, Ins(1,3,4,5)P_4 causes an immediate release of calcium which is

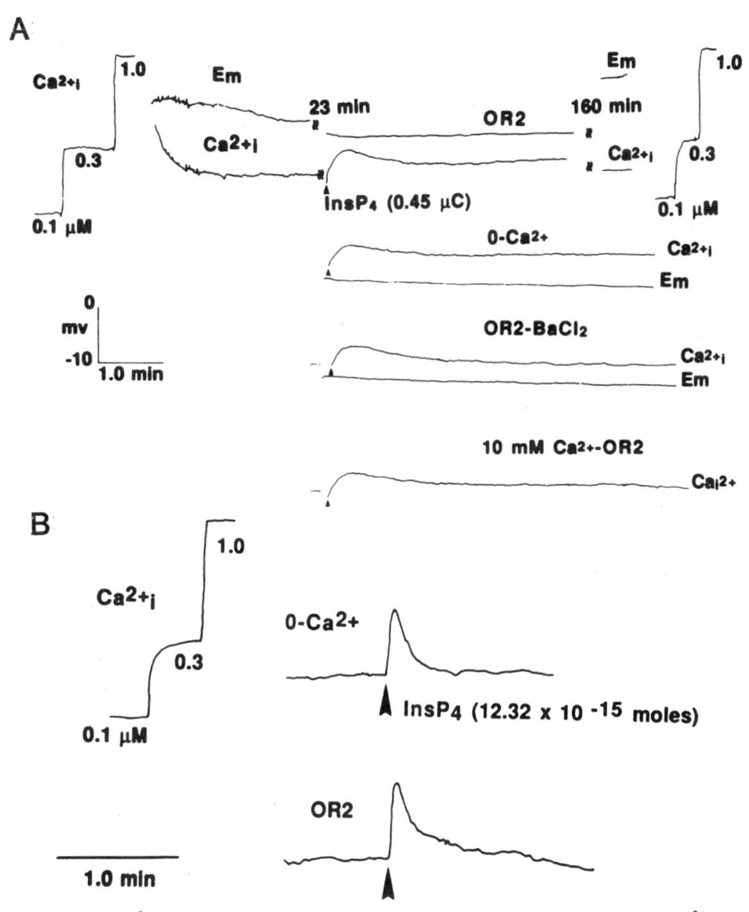

Figure 8. The $[Ca^{2+}]_i$ in an immature, stage VI oocyte measured with a Ca^{2+}-sensitive microelectrode in an oocyte which had been previously injected with Ins(1,3,4,5)P_4 under different ionic conditions. **(A)** Response of an oocyte injected with 0.45 μC (2.9×10^{-13} moles) of Ins(1,3,4,5)P_4 under the various ionic conditions indicated. The arrows indicate the injection of Ins(1,3,4,5)P_4. The calibration of the Ca^{2+}-sensitive electrode, prior to insertion and after removal is shown on the left and right of the figure, respectively. The electrode was made by the double-barreled glass technique (see Materials and Methods section). **(B)** Response of an oocyte injected with 12.3 $\times 10^{-15}$ moles of Ins(1,3,4,5)P_4 in the presence and absence of extracellular calcium. The arrows indicate the time of injection of Ins(1,3,4,5)P_4. The electrode was made by the 2theta glass technique. The calibration of the Ca^{2+}-sensitive electrode, prior to insertion, is shown on the left of the figure.

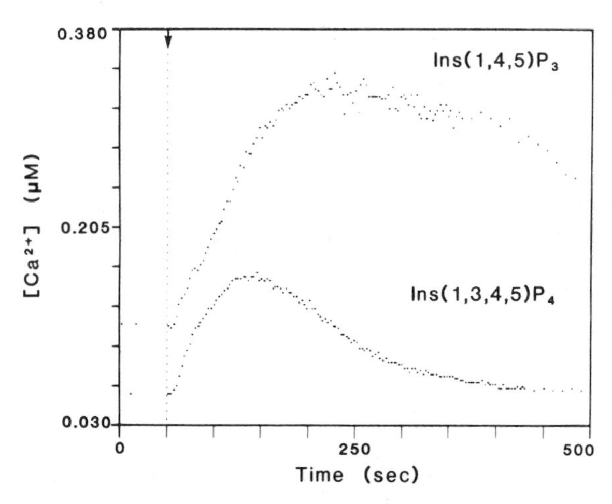

Figure 9. Time course of Ca^{2+} release in an immature, stage VI oocyte measured with the fura-2 technique in response to Ins(1,4,5)P_3 and Ins(1,3,4,5)P_4. The same oocyte was injected first with Ins(1,3,4,5)P_4 (bottom trace) and then Ins(1,4,5)P_3 (top trace). In each case the same amount of current (1.9 μC) was injected, which corresponds to 1.2 x 10^{-12} moles of Ins(1,3,4,5)P_4 and 1.7 x 10^{-12} moles of Ins(1,4,5)P_3. Injections start at the time marked by the arrow. The Ca^{2+} release was monitored by imaging the oocyte, which had been injected with fura-2 before the experiment. A circular spot approximately 24 μm in diameter was analyzed in each image collected. The spot analyzed corresponds to a region directly beneath the surface of the cell, near the site of injection. Each data point represents the average Ca^{2+} level in the circular spot during a 0.5s interval.

independent of extracellular calcium. The fact that the calcium release is immediate suggests that the Ins(1,3,4,5)P_4 is not being converted to another phosphoinositol species prior to releasing Ca^{2+}. We did not observe any oscillations in intracellular calcium in response to Ins(1,3,4,5)P_4 injection. These calcium increases induced by Ins(1,3,4,5)P_4 were not dependent on extracellular calcium. In order to compare the relative effectiveness of both of these phosphoinositols in releasing Ca^{2+}, similar amounts of each was injected into the same egg being imaged using fura-2. Comparing the traces for Ins(1,4,5)P_3 and Ins(1,3,4,5)P_4 (Figure 9) we see that the initial rate of the Ca^{2+} release is comparable for the two polyphosphoinositols; only the extent and duration of the Ca^{2+} change are different in response to the two different compounds. These results demonstrate that Ins(1,3,4,5)P_4 might be a second messenger and thus may play a role in sustaining calcium elevations during cell activation.

Ins(1,3,4,5)P_4 causes a single release of calcium but we had expected to see oscillations in intracellular calcium based on the fact that it causes oscillations in Ca^{2+}-dependent Cl^- channels. One explanation for this observation is that the Ins(1,3,4,5)P_4 triggers the release of calcium which in turn binds to and activates a regulatory molecule responsible for the oscillations. The requirement for an increase in intracellular calcium has been demonstrated by the intracellular injection of the Ca^{2+} chelator, EGTA (Parker and Miledi 1987). Furthermore we have found that verapamil completely inhibits any response to Ins(1,3,4,5)P_4. This hypothesis is reinforced by the observation that there is often a time delay between the injection of Ins(1,3,4,5)P_4 and the appearance of currents, but there is never a delay in the release of intracellular Ca^{2+}. These observations also suggest the possibility that there may be two different classes of Cl^- channels in the oocyte plasma membrane: one directly gated by Ca^{2+} and one indirectly gated by Ca^{2+}.

The PI cycle involves the generation of many different types of polyphosphoinositols. The present data now suggest that the different polyphosphoinositols may all have distinct roles in the regulation of calcium during signal transduction. Ins(1,4,5)P_3 which is short lived, probably causes the initial release of calcium from the internal stores. Ins(1,4,5)P_3 can either be converted to $InsP_2$, whose role in calcium release in oocytes has yet to be studied, or it can be converted to Ins(1,3,4,5)P_4, which can also release calcium from internal stores. It is not known if IP_4 itself is released from the membrane lipid, however, it was recently reported that activated neutrophils contain an inositol tetrakis-phosphate-containing phospholipid (Traynor-Kaplan *et al.* 1988). Therefore Ins(1,3,4,5)P_4 might be released directly from a lipid source. As inhibitors and antibodies to the phosphatases and kinases involved in the PI cycle are developed it should be possible to dissect out the unique roles of each of the phosphoinositols in cell activation.

ACKNOWLEDGEMENTS

We wish to thank Chiara Campanella for the protocol used to stratify the *Xenopus* egg. This work was supported by grant HD19966 from the NIH to R.N.

REFERENCES

Acott, R. S. and D. W. Carr. 1984. Inhibition of bovine spermatozoa by caudal epididymal fluid 2. Interaction of pH and a quiescent factor. *Biol. Reprod.* 30:926-935.

Balla, T., G. Guillemette, A. J. Baukal, and K. J. Catt. 1987. Metabolism of inositol 1,3,4-trisphosphate to a new tetrakisphosphate isomer in angiotensin-stimulated adrenal glomerulosa cells. *J. Biol. Chem.* 263:9952-9955.

Berridge, M. J. 1987. Inositol trisphosphate and diacylglycerol: Two interacting second messengers. *Annu. Rev. Biochem.* 56:159-193.

Busa, W. B. 1986. Measuring intracellular free Ca^{2+} with single- and double-barreled ion-specific microelectrodes. *Prog. Clin. Biol. Res.* 210:57-70.

Busa, W. B. and J. H. Crowe. 1983. Intracellular pH regulates transitions between dormancy and development of brine shrimp (*Artemia salina*) embryos. *Science* 221:366-368.

Busa, W. B. and R. Nuccitelli. 1984. Metabolic regulation via intracellular pH. *Am. J. Physiol.* 246:R409-438.

Busa, W. B. and R. Nuccitelli. 1985. An elevated free cytosolic Ca^{2+} wave follows fertilization in eggs of the frog, *Xenopus laevis*. *J. Cell Biol.* 100:1325-1329.

Busa, W. B., J. E. Ferguson, S. K. Joseph, J. R. Williamson, and R. Nuccitelli. 1985. Activation of frog (*Xenopus laevis*) eggs by inositol trisphosphate. I. Characterization of Ca^{2+} release from intracellular stores. *J. Cell Biol.* 101:677-682.

Campanella, C. and P. Andreuccetti. 1977. Ultrastructural observation on cortical endoplasmic reticulum and on residual cortical granules in the egg of *Xenopus laevis*. *Dev. Biol.* 56:1-10.

Campanella, C., P. Andreuccetti, C. Taddei, and R. Talevi. 1984. The modifications of cortical endoplasmic reticulum during *in vitro* maturation of *Xenopus laevis* oocytes and its involvement in cortical granule exocytosis. *J. Exp. Zool.* 229:283-293.

Chambers, E. L., B. C. Pressman, and B. Rose. 1974. The activation of sea urchin eggs by the divalent ionophores A23187 and X-537A. *Biochem. Biophys. Res. Commun.* 60:126-132.

Charbonneau, M. and R. D. Grey. 1984. The onset of activation responsiveness during maturation coincides with the formation of the cortical endoplasmic reticulum in oocytes of *Xenopus laevis*. *Dev. Biol.* 102:90-97.

Christen, R., R. W. Schackmann, and B. M. Shapiro. 1982. Elevation of the intracellular pH activates respiration and motility of sperm of the sea urchin. *J. Biol. Chem.* 91:174a.

Crossley, I., K. Swann, E. Chambers, and M. Whitaker. 1988. Activation of sea urchin eggs by inositol phosphates is independent of external calcium. *Biochem. J.* 252:257-262.

Cuthbertson, K. S. R., and P. H. Cobbold. 1985. Phorbol ester and sperm activate mouse oocytes by inducing sustained oscillations in cell Ca^{2+}. *Nature (Lond.)* 316:541-542.

Cuthbertson, K. S. R., D. G. Whittingham, and P. H. Cobbold. 1981. Free Ca^{2+} increases in exponential phases during mouse oocyte activation. *Nature (Lond.)* 294:754-757.

Dalcq, A. 1928. *Les Bases Physiologiques de la Fecondation et de la Parthenogenese.* Preses Univ. de France, Paris.

Eisen, A. and G. T. Reynolds. 1984. Calcium transients during early development in single starfish (*Asterias forbesi*) oocytes. *J. Cell Biol.* 99:1878-1882.

Eisen, A., D. P. Kiehart, S. J. Wieland, and G. T. Reynolds. 1984. Temporal sequence and spatial distribution of early events of fertilization in single sea urchin eggs. *J. Cell Biol.* 99:1647-1654.

Gardiner, D. M. and R. D. Grey. 1983. Membrane junctions in *Xenopus* eggs: Their distribution suggests a role in calcium regulation. *J. Cell Biol.* 96:1159-1163.

Gilkey, J. C., L. F. Jaffe, E. B. Ridgway, and G. T. Reynolds. 1978. A free calcium wave traverses the activating egg of the medaka, *Oryzias latipes. J. Cell Biol.* 76:448-466.

Grey, R. D., M. J. Bastiani, D. J. Webb, and E. R. Schertel. 1982. An electrical block is required to prevent polyspermy in eggs fertilized by natural mating of *Xenopus laevis. Dev. Biol.* 89:475-484.

Hamaguchi, M. S. 1982. The role of intracellular pH in fertilization of sand dollar eggs analyzed by microinjection method. *Dev. Growth & Differ.* 24:443-451.

Irvine, R. F. and R. M. Moor. 1986. Microinjection of inositol 1,3,4,5-tetrakisphosphate activates sea urchin eggs by a mechanism dependent on external Ca^{2+}. *Biochem. J.* 240:917-920.

Irvine, R. F. and R. M. Moor. 1987. Inositol (1,3,4,5) tetrakisphosphate-induced activation of sea urchin eggs requires the presence of inositol trisphosphate. *Biochem. Biophys. Res. Commun.* 146:284-290.

Irvine, R. F., K. D. Brown, and M. J. Berridge. 1984. Specificity of inositol trisphosphate-induced calcium release from permeabilized Swiss 3T3 cells. *Biochem. J.* 222:269-272.

Irvine, R. F., A. J. Letcher, J. P. Heslop, and M. J. Berridge. 1986. The inositol tris/tetrakisphosphate pathway-demonstration of Ins(1,4,5)P_3 3-kinase activity in animal tissues. *Nature (Lond.)* 320:631-634.

Johnson, C. H. and D. Epel. 1981. Intracellular pH of sea urchin eggs measured by the dimethyloxazolidinedione (DMO) method. *J. Cell Biol.* 89:284-291.

Kline, D. 1988. Calcium-dependent events at fertilization of the frog egg: injection of a calcium buffer blocks ion channel opening, exocytosis, and formation of pronuclei. *Dev. Biol.* 126:346-361.

Kline, D. and L. A. Jaffe. 1987. The fertilization potential of the *Xenopus* egg is blocked by injection of a calcium buffer and is mimicked by injection of a GTP analog. *Biophys. J.* 51:398a.

Kline, D. and R. Nuccitelli. 1985. The wave of activation current in the *Xenopus* egg. *Dev. Biol.* 111:471-487.

Kubota, H. Y., Y. Yoshimoto, M. Yoneda, and Y. Hiramoto. 1987. Free calcium wave upon activation in *Xenopus* eggs. *Dev. Biol.* 119:129-136.

Lee, H. C., C. Johnson, and D. Epel. 1983. Changes in internal pH associated with initiation of motility and acrosome reaction of sea urchin sperm. *Dev. Biol.* 95:31-45.

Levy, S. and A. Fein. 1985. Relationship between light sensitivity and intracellular free Ca concentration in Limulus ventral photoreceptors: A quantitative study using Ca-selective microelectrodes. *J. Gen. Physiol.* 85:805-841.

Loeb, J. 1913. *Artificial Parthenogenesis and Fertilization.* Univ. of Chicago Press, Chicago.

Mazia, D. 1937. The release of calcium in *Arbacia* eggs on fertilization. *J. Cell. Comp. Phys.* 10:291-304.

Miledi, R. and I. Parker. 1984. Chloride current induced by injection of calcium into *Xenopus* oocytes. *J. Physiol.* 357:173-183.

Miyazaki, S., N. Hashimoto, Y. Yoshimoto, T. Kishimoto, Y. Igusa, and Y. Hiramoto. 1986. Temporal and spatial dynamics of the periodic increase in intracellular free calcium at fertilization of golden hamster eggs. *Dev. Biol.* 118:259-267.

Nishioka, D. and N. Cross. 1978. The role of external sodium in sea urchin fertilization. p.403-413. *In: Cell Reproduction: In Honor of Daniel Mazia.* E. R. Dirkson, D. M. Prescott and C. F. Fox (Eds.). Academic Press, New York.

Nuccitelli, R. 1987. The wave of activation current in the egg of the medaka fish. *Dev. Biol.* 122:522-534.

Nuccitelli, R., D. J. Webb, S. T. Lagier, and G. B. Matson. 1981. ^{31}P NMR reveals increased intracellular pH after fertilization in *Xenopus* eggs. *Proc. Natl. Acad. Sci. USA* 78:4421-4425.

Nuccitelli, R., D. Kline, W. B. Busa, R. Talevi, and C. Campanella. 1988. A highly localized activation current yet widespread intracellular calcium increase in the egg of the frog, *Discoglossus pictus*. *Dev. Biol.* 130 (In press).

Oron, Y., N. Dascal, E. Nadler, and M. Lupu. 1985. Inositol 1,4,5-trisphosphate mimics muscarinic response in *Xenopus* oocytes. *Nature (Lond.)* 313:141-143.

Parker, I. and R. Miledi. 1987. Injection of inositol 1,3,4,5-tetrakisphosphate into *Xenopus* oöcytes generates a chloride current dependent upon intracellular calcium. *Proc. R. Soc. Lond. B.* 232:59-70.

Pasteels, J. 1938. The role of calcium in the activation of the egg of the mollusk. *Wimereux Stn. Zool. Travaux.* 13:515-530.

Poenie, M., J. Alderton, R. Steinhardt, and R. Tsien. 1986. Calcium rises abruptly and briefly throughout the cell at the onset of anaphase. *Science* 233:886-889.

Purves, R. D. 1981. *Microelectrode methods for intracellular recording and ionotophoresis.* Academic Press, London.

Ridgway, E. B., J. C. Gilkey, and L. F. Jaffe. 1977. Free calcium increases explosively in activating medaka eggs. *Proc. Natl. Acad. Sci. USA* 74:623-627.

Robinson, K. R. 1979. Electrical currents through full-grown and maturing *Xenopus* oocytes. *Proc. Natl. Acad. Sci. USA* 76:837-841.

Schackmann, R. W. and B. M. Shapiro. 1981. A partial sequence of ionic changes associated with the acrosome reaction of *Strongylocentrotus purpuratus*. *Dev. Biol.* 81:145-154.

Shears, S. B., J. B. Parry, E. K. Y. Tang, R. F. Irvine, R. H. Michell, and C. J. Kirk. 1987a. Metabolism of D-myo-inositol 1,3,4,5-tetrakisphosphate by rat liver, including the synthesis of a novel isomer of myo-inositol tetrakisphosphate. *Biochem. J.* 246:139-147.

Shears, S. B., C. J. Kirk, and R. H. Michell. 1987b. The pathway of myo-inositol 1,3,4-trisphosphate dephosphorylation in liver. *Biochem. J.* 248:977-980.

Shen, S. S. and R. A. Steinhardt. 1978. Direct measurement of intracellular pH during metabolic derepression of the sea urchin egg. *Nature (Lond.)* 272:253-254.

Steinhardt, R. A. and D. Epel. 1974. Activation of sea urchin eggs by a calcium ionophore. *Proc. Natl. Acad. Sci. USA* 71:1915-1919.

Steinhardt, R. A., D. Epel, and E. J. Carroll Jr. 1974. Is calcium ionophore a universal activator for unfertilized eggs? *Nature (Lond.)* 252:41-43.

Swann, K. and M. Whitaker 1986. The part played by inositol trisphosphate and calcium in the propagation of the fertilization wave in sea urchin eggs. *J. Cell Biol.* 103:2333-2342.

Tilney, L. G. 1976. The polymerization of actin. III. Aggregates of nonfilamentous actin and its associated proteins: a storage form of actin. *J. Cell Biol.* 69:73-89.

Tilney, L. G., D. P. Kiehart, C. Sardet, and M. Tilney. 1978. Polymerization of actin. IV. Role of Ca^{2+} and H^+ in the assembly of actin and in membrane fusion in the acrosomal reaction of echinoderm sperm. *J. Cell. Biol.* 77:536-550.

Traynor-Kaplan, A. E., A. L. Harris, B. L. Thompson, P. Taylor, and L. A. Skylar. 1988. An inositol tetrakisphosphate-containing phospholipid in activated neutrophils. *Nature (Lond.)* 334:353-356.

Webb, D. J. and R. Nuccitelli. 1981. Direct measurement of intracellular pH changes in *Xenopus* eggs at fertilization and cleavage. *J. Cell Biol.* 91:562-567.

Webb, D. J. and R. Nuccitelli. 1982. Intracellular pH changes accompany the activation of development in frog eggs: comparison of pH microelectrode and ^{31}P-NMR measurements. p. 293-324. *In: Intracellular pH, Its Measurement, Regulation, and Utilization in Cellular Functions*. R. Nuccitelli and D. W. Deamer (Eds.). Alan R. Liss, New York.

Winkler, M. M., R. A. Steinhardt, J. L. Grainger, and L. Minning. 1980. Dual ionic controls for the activation of protein synthesis at fertilization. *Nature (Lond.)* 287:558-560.

SIGNAL TRANSDUCTION OF SPERM-EGG INTERACTION CAUSING PERIODIC CALCIUM TRANSIENTS IN HAMSTER EGGS

Shun-ichi Miyazaki

Department of Physiology
Tokyo Women's Medical College
Kawada-cho, Shinjuku-ku, Tokyo, 162, Japan

The study of fertilization or egg activation has a long history, over a century, and experiments have been done mainly using sea urchin gametes, as shown in this book. The study of mammals was much delayed because of the difficulty in maintaining and fertilizing the gametes *in vitro*. In these last 30 years, these problems have been overcome through advances in cell culture techniques, and extensive studies have now been completed on mammalian fertilization, based on microscopic and morphological observations. However, because of the limitation in the number of collectable eggs, some difficulties still remain for further studies based on techniques such as biochemical analysis.

Early events in egg activation involve changes in the ionic permeability of the plasma membrane and alterations in intracellular ion concentrations. These ionic events can be analyzed with electrophysiological and optical methods in single cells. Therefore, physiological studies are possible in mammalian eggs, but reports on these events have appeared only since 1981.

Here, I will first describe the Ca^{2+} transients during activation of golden hamster eggs and in keeping with the editors' wishes, electrophysiological and optical methods will be emphasized. It is well known that a dramatic, transient increase in the intracellular Ca^{2+} concentration ($[Ca^{2+}]_i$) occurs at the early stage of fertilization in various eggs; this Ca^{2+} transient causes cortical granule exocytosis (to prevent polyspermy) and also triggers, though indirectly, the metabolic activation of inseminated eggs (see reviews by Epel 1978 and Jaffe 1985). Golden hamster eggs exhibit an interesting feature of transient but periodic rises in

$[Ca^{2+}]_i$, as demonstrated with Ca^{2+}- sensitive microelectrodes (Igusa and Miyazaki 1986) and with the Ca^{2+}-dependent luminescent protein, aequorin (Miyazaki *et al* 1986). The first two to three Ca^{2+} transients take the form of a propagating wave starting from the sperm attachment site (Miyazaki *et al* 1986). Since the series of Ca^{2+} transients begins when sperm-egg fusion is likely to occur, signalling in the plasma membrane is suggested, activated by the sperm as an external stimulus. I describe the findings that are related to signal transduction of sperm-egg interaction causing periodic Ca^{2+} rises in golden hamster eggs. Experiments are based on the microinjection of chemicals into the egg while Ca^{2+} transients are monitored by aequorin luminescence and/or by the hyperpolarization in membrane potential which is due to a Ca^{2+}-activated K^+ current. It will be shown that the signal transduction involves the activation of GTP-binding protein (G-protein) and polyphosphoinositide turnover (Miyazaki 1988a). Finally, I will propose a possible model for periodic Ca^{2+} transients, based on a linkage of continuous Ca^{2+} influx across the plasma membrane to Ca^{2+} release from intracellular stores. Then I discuss possible mechanisms of transmembrane signalling in relation to the proposed model. A mini review has appeared elsewhere (Miyazaki 1988b).

MATERIALS AND METHODS

Preparation of Eggs and Spermatozoa

Eggs. Two- to four- month-old females are superovulated by intraperitoneal injection of pregnant mare's serum gonadotropin (PMSG), followed 48h later by human chorionic gonadotropin (HCG) (Yanagimachi 1969). Mature eggs are collected from the oviducts 15.5h after the HCG injection. Usually, 10-30 eggs are obtained from a female. The diameter of the egg is 72 μM. The eggs are freed from cumulus cells and the zona pellucida by sequential treatment with 0.05% hyaluronidase for 30-60s at 23-26°C and 0.07% trypsin for 1.5 min. Zona-free eggs are useful, because 1) microelectrode impalement is easier, 2) sperm directly attach to the egg vitellus, and sperm-egg fusion begins 10-30s later, indicating the initiation of sperm-egg interaction, 3) upon changing external medium, or adding drugs, their effects on the zona pellucida are excluded. A disadvantage of zona-free eggs is that they are always polyspermic unless the number of applied sperm is specially limited.

A modified Krebs-Ringer solution (BWW medium; Biggers *et al* 1971) is used as standard medium. Bovine serum albumin is added before use (4 mg/ml). Zona-free eggs are transferred to a 0.4 ml drop of the medium in a plastic petri dish and covered with paraffin oil. To make eggs stick to the bottom more easily, the dish is pretreated with poly-L-lysine (50 μg/ml). The dish is placed on the stage of an inverted, phase-contrast microscope, surrounded by a ring heater. Experiments are performed at 31-32°C. Higher temperatures cause an increased leakage conductance in the plasma membrane when penetrated by electrode(s), and lower temperatures tend to inactivate the movement of sperm.

Sperm. Spermatozoa obtained from the cauda epididymis do not have the ability to fertilize eggs; they first must undergo capacitation and the acrosome reaction. Factors inducing these changes have been extensively studied (see review by Yanagimachi 1988). Our method is as follows: approximately 15 μl of spermatozoa taken with a toothpick is suspended in 0.4 ml of modified Tyrode's solution containing 10% bovine adrenal gland extract; 30 μl of this sperm suspension is then added to 0.4 ml of the solution containing 20% human serum (Yanagimachi 1970). Sperm incubated for 4-5h at 37°C (gas phase, 5% CO_2 in air) are then available for the insemination of eggs.

Egg Activation During Electrophysiological Recording

To record the membrane potential, a glass microelectrode filled with 4M K acetate or 3M KCl is inserted into the egg with the aid of current by oscillation of the preamplifier. The resistance of the electrode is 60-100 MΩ. To monitor a change in the membrane conductance, constant current pulses are applied through the intracellular electrode by means of a bridge circuit. Since the current-voltage relation is linear at potentials between -10 and -150 mV

(Miyazaki and Igusa 1982), changes in the amplitude of the potential step in response to the current pulse indicate changes in membrane conductance.

After microelectrode impalement, 5-15 μl of cultured sperm suspension is added to the medium surrounding the egg, using a microsyringe. Sperm application is controlled so that sperm come to attach to the egg surface one by one (1-10 sperm per egg in 10 min.). Active flagellar motion of the attached sperm decreases gradually and then stops within 10-30s. This is considered to be the time when the sperm and egg membranes begin to fuse (Yanagimachi 1978). According to electron microscopic studies, the sperm plasma membrane above the equatorial segment fuses first with the egg plasma membrane (Bedford *et al* 1979). The sperm head is then covered with the protruded egg surface and its numerous microvilli. The sperm is then incorporated into the egg cytoplasm in about 15 min (for details, see Yanagimachi 1978, 1988).

Freshly ovulated eggs are arrested at metaphase of the second meiotic division (Austin 1961). The earliest indications of egg activation by sperm are the breakdown of the cortical granules and the resumption of the second meiotic division (Austin 1961). These phenomena are not visible during electrical recording, but after withdrawing the electrode the penetration of sperm is recognized histologically as the swelling of the head and the resumption of the second meiosis is identified as the separation of egg chromosomes (Igusa *et al* 1983). In our experiments in the hamster egg, activation is defined by the resumption of the second meiosis. However, the identification is difficult in the experiment with two electrodes, because of damage of the egg upon withdrawing the electrodes.

Methods for Recording Ca^{2+} Transients

Ca^{2+}-sensitive microelectrodes. One of the methods for measurement of $[Ca^{2+}]_i$ is to use a Ca^{2+}-sensitive microelectrode. Since the egg is a relatively large cell, this method is applicable (Igusa and Miyazaki 1986). A bevelled glass micropipette is back-filled under pressure with 10^{-7}M Ca^{2+} (pCa 7) standard saline and then filled with the liquid neutral Ca^{2+} ionophore ETH 1001 (Ca^{2+} cocktail, Fluka) by suction up to 200-600 μM from the tip. Electrodes are calibrated before or after the experiment with standard Ca^{2+} solutions (Igusa and Miyazaki 1986). The e.m.f. of the Ca^{2+}-electrode changes with the Nernstian slope (30 mV per pCa unit at 31°C) down to pCa 6.5 (3.2×10^{-7} M), but the slope between pCa 6 and 7 is usually 20-22 mV (28 mV even in the best electrode). The $[Ca^{2+}]_i$ is estimated from the standard curve. The disadvantage of the Ca^{2+}-electrode is the slow response time: 90% response time was 2.7 or 3.7s for increasing or decreasing $[Ca^{2+}]$ between pCa 7 and 6, respectively (Igusa and Miyazaki 1986).

The $[Ca^{2+}]_i$ is measured by the potential difference between Ca^{2+}-electrode and KCl electrode, both of which are inserted into an egg. This cancellation of the membrane potential (V_m) is incomplete if a Ca^{2+} rise is accompanied by a fast V_m change as in fertilized hamster eggs (see below), because of extremely high resistance of the Ca^{2+}-electrode (18-35 GΩ). Therefore, a third electrode (filled with 3M KCl) is introduced for passing current and thereby clamping V_m (see Figure 2a). Application of the voltage clamp method enabled us to record Ca^{2+} transients uncontaminated with V_m changes (Igusa and Miyazaki 1986).

Aequorin luminescence. Another method for recording Ca^{2+} transients is the use of the Ca^{2+}-binding photoprotein, aequorin. An unfertilized egg is injected with an aequorin solution composed of 9.6 mg/ml purified aequorin, 100 μM EGTA and 10 mM PIPES at pH 7.0, using volume-controlled microinjection techniques (Hiramoto 1961; Kishimoto 1986). After injection and withdrawing the pipette, the egg is transferred to the experimental dish or test tube and then inseminated.

Total luminescence generated by the intracellular Ca^{2+}-aequorin reaction in a single egg is recorded with the Bioluminescence Reader (Aloka BLR-101) on a pen recorder (see Figure 1a). During the measurement the zona-free hamster egg in a test tube becomes polyspermic because of uncontrolled application of sperm.

To investigate the spatial distribution of the Ca^{2+} transients, the photon counting imaging method with a supersensitive video camera system is used (Hamamatsu Photonics, C1966-20). Photon-limited events are intensified with two-dimensionally arranged photon-counting tubes and visualized as light spots appearing momentarily on the video screen. The temporal integration of these spots clarifies the temporal distribution of a Ca^{2+} rise (Eisen *et al* 1984; Miyazaki *et al* 1986). The video-recorded raw image of the light spots is processed by the attached image processor using facilities such as, continuous accumulation (accumulation of all light spots during Ca^{2+} transient, see Figure 3) or sequential accumulation (accumulation at every 1/8 to 1s interval). The video-recorded, processed image is played back and video frames are stopped one by one and then photographed at the desired moment.

RESULTS AND DISCUSSION

Periodic Ca^{2+} Transients

Poenie *et al* (1985) have demonstrated Ca^{2+} transients in single sea urchin eggs at fertilization. The $[Ca^{2+}]_i$ increases from the resting level (ca. 0.15 μM) to the peak of 2-2.5 μM

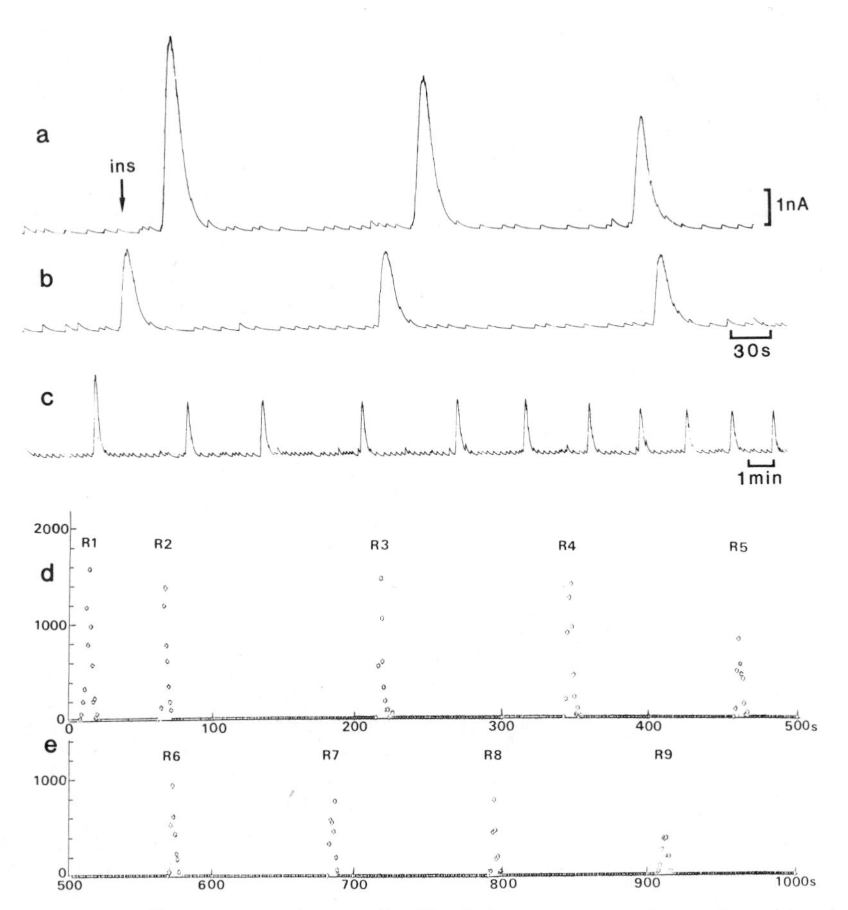

Figure 1. Ca^{2+} transients in the fertilized hamster egg, shown by aequorin luminescence. **(a-c)** Continuous record from an egg with the Bioluminescence Reader. Insemination is indicated by "ins". The luminescence intensity is indicated by the photomultiplier current (vertical bar). The egg was polyspermic. **(d and e)** Record from an egg with a single sperm attached. Light intensity in a restricted area of the egg on the TV screen was measured and presented by arbitrary unit on the ordinate. Abscissa: time after sperm attachment. Rl-R9: the first to ninth Ca^{2+} transient. (From Miyazaki *et al* 1986).

in 10-15s and then declines slowly over several minutes. They have also shown succeeding small oscillations in $[Ca^{2+}]_i$ during progression to the first cell division. In hamster eggs, the Ca^{2+} rise occurs repeatedly at intervals of 40-120s (Miyazaki *et al* 1986). Each Ca^{2+} transient lasts for 12-18s and it appears in an all-or-none fashion, although the amplitude decreases gradually to a constant value. These features are shown in Figures 1a-1c where aequorin luminescence was measured with the Bioluminescence Reader. The egg shown in Figures la-lc was polyspermic because of the technical limitation. When only a single sperm was applied to the egg under microscope through a fine glass capillary as in the case of Figures 1d-1e, it was shown that even a single sperm is capable of inducing repeated responses. For a single sperm, however, the interval between Ca^{2+} transients later than the third response is prolonged (sometimes longer than 3 min) while multiple sperm produce more frequent Ca^{2+} rises at fairly constant intervals of 40-60s.

Figure 2a shows Ca^{2+} rises measured with a Ca^{2+}-electrode in combination with the voltage clamp method. The basal $[Ca^{2+}]_i$ before insemination was 0.2-0.4 μM (Igusa and Miyazaki 1986). The value seems to be a little more than the actual value, probably because of leakage of Ca^{2+} into the egg due to damage caused by electrode impalement. In the first three responses after insemination, the $[Ca^{2+}]_i$ reaches 1-2 μM, which is roughly close to the estimated value of 2.5-4.5 μM (Steinhardt *et al* 1977) or about 2 μM (Poenie *et al* 1985) at fertilization of the sea urchin egg. For later responses in the fertilizing hamster egg the peak $[Ca^{2+}]_i$ decreases to 0.6-0.7 μM. The basal $[Ca^{2+}]_i$ slightly increases during the series of Ca^{2+} transients (Igusa and Miyazaki 1986).

Considering that the Ca^{2+}-electrode measures $[Ca^{2+}]_i$ at its tip, the depth of the electrode was varied in the cytoplasm between 3-30 μM, but no significant difference was found in the peak and approximate time course of each Ca^{2+} transient (Igusa and Miyazaki 1986). This finding indicates that the $[Ca^{2+}]_i$ increases in the deep in the cytoplasm, and it is not restricted to the egg cortex.

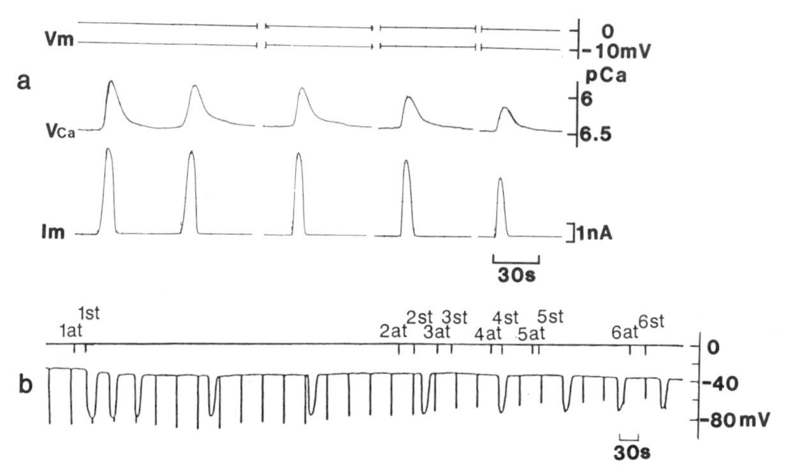

Figure 2. Ca^{2+} transients in the fertilized hamster egg. **(a)** $[Ca^{2+}]_i$ measurement with a Ca^{2+}-electrode. V_m: the membrane potential clamped at -10 mV. V_{Ca}: the potential of Ca^{2+}-electrode subtracted by V_m. I_m: membrane current under voltage clamp. The first to fifth responses are shown. There are three interruptions in the presented record (From Igusa and Miyazaki 1986). **(b)** Membrane potential change during fertilization. 1-6 at: attachment of the first to sixth sperm. 1-6 st: stop of flagellar motion of the first to sixth sperm. Constant current pulses of 0.5 nA and 300 ms duration were applied continuously. (Miyazaki, unpublished).

Hyperpolarizing Responses

As seen in Figure 2a, each Ca^{2+} rise always coincides with a transient, outward membrane current. Consistently, the membrane potential measurement without voltage clamp shows repeated, transient hyperpolarizations in fertilizing hamster eggs (see Figure 2b) (Miyazaki and Igusa 1981a). The hyperpolarizing response (HR) is due to a K^+ current activated by an increase in $[Ca^{2+}]_i$, because 1) its reversal potential (-83 mV in BWW medium; $[K^+]_o$ = 5.5 mM) shifts with the Nernstian slope for K^+ ions when $[K^+]_o$ is changed, 2) the HR is blocked by intracellular injection of EGTA, and 3) injection of Ca^{2+} into an egg induces a hyperpolarization similar to the HR (Miyazaki and Igusa 1982). Here, two interesting things can be noted. The first is that the fertilization potential in the golden hamster egg consists of recurring hyperpolarizations, quite different from those in eggs of other species (see review by Hagiwara and Jaffe 1979). In other mammals, the mouse (Jaffe *et al* 1983; Igusa *et al* 1983) and rabbit (McCulloh *et al* 1983), the fertilization potential consists of, or at least includes, repeated hyperpolarizations, although they are much smaller than those in the hamster. The second is that the hamster egg seems to be an interesting model for the study of the movement of Ca^{2+} ions, because there is an easy way of monitoring a change in $[Ca^{2+}]_i$ by a hyperpoarization, which is recorded with a conventional single electrode.

With precise observation of HRs at fertilization of hamster eggs, the first HR (i.e., the first Ca^{2+} transient) occurs at about the time when flagellar motion of the first attached sperm stops (1 "st" in Figure 2b), i.e., when the sperm and egg membranes are likely to fuse (for detailed analysis, see Miyazaki and Igusa 1981a). Subsequent HRs don't always coincide with the stopping of flagellar motion of additional sperm (see "sts" in Figure 2b). The amplitude of HRs decreases gradually, associated with a decrease in membrane resistance. The series of HRs persists for at least 2h when multiple sperm fused with the egg. The resting potential is between -20 and -25 mV before insemination and it shifts gradually in the hyperpolarizing direction to about -40 mV during a series of HRs (Figure 2b).

The occurrence of the series of Ca^{2+} transients has been shown to be related to activation of hamster eggs (Igusa *et al* 1983). Repeated Ca^{2+} transients have also been demonstrated in aequorin-injected mouse eggs during artificial activation and fertilization (Cuthbertson *et al* 1981; Cuthbertson and Cobbold 1985). Compared with hamster eggs, the zona-free mouse egg shows a much longer delay between insemination and the first Ca^{2+} rise (45-90 min) and a much longer duration of each Ca^{2+} rise (about 1 min or more). The frequency of Ca^{2+} transients in mouse eggs is 1-6 per 10 min.

Spreading Ca^{2+} Rise

Gilkey *et al* (1978) first demonstrated a Ca^{2+} wave traversing the activating egg of the medaka fish by aequorin luminescence. This kind of observation was impossible in a single, smaller egg until a system using photon counting imaging with a supersensitive TV camera was developed. This method has revealed that a propagating Ca^{2+} wave also takes place in the first two to three Ca^{2+} transients in the inseminated hamster egg (Miyazaki *et al* 1986). Figure 3 is a sample record of the first Ca^{2+} transient, shown by continuous accumulation of light spots (on the TV monitor screen) of aequorin luminescence. In the first response the increase in $[Ca^{2+}]_i$ begins near the sperm attachment site, and the Ca^{2+} rise spreads over the entire egg within 4-7s (6.5s in Figure 3). The Ca^{2+} rise attains its peak in 5-8s and declines with almost even distribution (Miyazaki *et al* 1986). In the second and sometimes the third response, the spreading Ca^{2+} rise is also seen, starting from the same focus, but spreading more rapidly (~2s). The succeeding Ca^{2+} rises occur synchronously in the whole egg within 1s, without any detectable foci. It seems that the first 2-3 responses are triggered by the attached sperm, while later responses are built up and repeated as the characteristics of the inseminated egg. The frequency of the responses, however, is somehow dependent on the number of additional sperm, as already mentioned.

The spreading velocity of the Ca^{2+} rise in the hamster egg is 16-28 μM/s, if the Ca^{2+} wave is assumed to travel along the egg surface. This is a little faster than 12 μM/s in the medaka fish (Gilkey *et al* 1978) and 8-10 μM/s in *Xenopus* (Busa and Nuccitelli 1985; Kubota *et al* 1986).

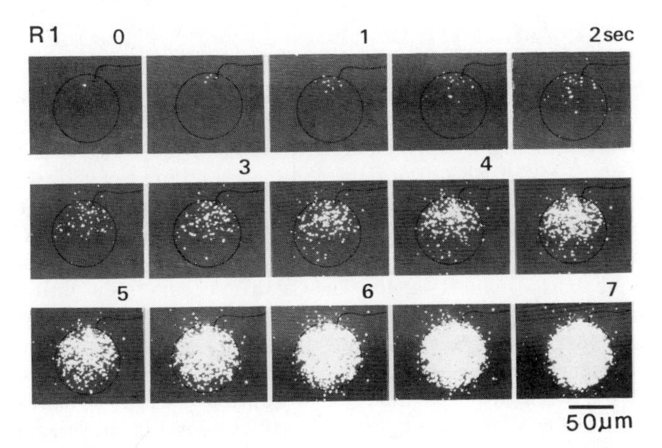

Figure 3. Propagating Ca^{2+} rise in the first response, demonstrated by aequorin luminescence with a supersensitive TV camera system. All light spots on the TV screen were accumulated from zero time (when the first spot appeared inside the egg) and photographed at every 0.5s (from 0.5 to 7s). The record was obtained from the egg shown in Figure 1d-e, which was monospermic. A single sperm attached at the top. (From Miyazaki *et al* 1986).

Microinjection of Ions and Drugs

To examine the mechanism involved in the periodic Ca^{2+} rises and in the signal transduction activated by sperm, microinjection of ions and drugs into an egg is useful. Two ways for microinjection were used. One is iontophoretic injection for charged substances. The solution containing the substance at high concentration is loaded in a glass microelectrode with slightly broken tip and is injected with constant current pulses. With this method, repeated injections are possible, but the amount injected is unknown, although it can be roughly estimated from knowing the total injected charge.

Another way is microinjection by pressure. The method has been described in detail previously (Miyazaki 1988a). The desired volume of the solution with a known concentration of a substance was injected under pressure in 1-2s near the center of the egg cytoplasm. When two or three kinds of solutions are injected, each solution is separated by silicon oil in a single pipette. The volume injected was 2-8 pl for each solution, which is 1-4% of the total egg volume of 200 pl. Even with this pressure injection, the exact concentration of injected substance at its site of action is unknown. The only way is to tentatively calculate the intracellular concentration under the assumption of even distribution in the cytoplasm. Hereafter, the calculated final concentration of, for instance, substance A will be expressed as [A].

Upon microinjection, the Ca^{2+} transients were monitored by the hyperpolarizing response (HR) and/or aequorin luminescence. Since a HR has an exact one-to-one correspondence to a Ca^{2+} transient (see Figures 2a, 4a and 4c), most records were obtained simply by membrane potential measurement with a conventional electrode. The conductance increase at the peak of an HR is considered to be a relatively quantitative indicator of an increase in $[Ca^{2+}]_i$ (Igusa and Miyazaki 1983).

Ca^{2+} Release Induced by Injection of $InsP_3$ and Ca^{2+}

Generally, an increase in $[Ca^{2+}]_i$ is due either to Ca^{2+} influx across the plasma membrane or to Ca^{2+} release from intracellular stores or binding sites. The Ca^{2+} transient in activating eggs is thought to be due to Ca^{2+} release, based on findings in sea urchin eggs (Steinhardt *et al* 1977; Epel 1978; Jaffe 1985). Of possible factors mediating Ca^{2+} release, inositol 1,4,5-trisphosphate ($InsP_3$) is known to be a second messenger to mobilize Ca^{2+} from intracellular stores in variety of cells (Berridge and Irvine 1984). Whitaker and Irvine (1984) first showed that injection of $InsP_3$ caused elevation of the fertilization membrane in the sea urchin egg: this phenomenon is based on cortical granule exocytosis due to Ca^{2+} release. The $InsP_3$-induced Ca^{2+} release has been confirmed in the sea urchin egg (Crossley *et al* 1988) and has also shown in the *Xenopus* (Busa *et al* 1985) and hamster eggs (Miyazaki 1988a).

When $InsP_3$ is injected into an unfertilized hamster egg by pressure, a Ca^{2+} transient is induced with no measurable delay, as demonstrated by a hyperpolarization and an increase in aequorin luminescence (Figures 4a, 4c). The Ca^{2+} rise occurs in the entire egg and lasts for 13-18s, similar to the Ca^{2+} transients seen upon insemination. Experiments with varied concentrations of $InsP_3$ have shown that the $InsP_3$-induced Ca^{2+} rise is of an all-or-none nature: the conductance increase at the peak of the hyperpolarizing response (HR) is altered only about twofold while the final $InsP_3$ concentration ([$InsP_3$]) is increased 3,000 times (Miyazaki 1988a). The critical concentration of $InsP_3$ is 80 nM in the injection pipette (2 nM in the egg, assuming uniform distribution). The $InsP_3$-induced Ca^{2+} rise occurs even in Ca^{2+}-free medium (Figure 4b) or in Ca^{2+}-free medium containing 0.5 mM EGTA, indicating that the Ca^{2+} rise is due to intracellular Ca^{2+} release.

A single injection of $InsP_3$ under pressure usually produces only one Ca^{2+} release. A series of small Ca^{2+} rises are seen only when an extremely high concentration of $InsP_3$ is used. The injected $InsP_3$ may be immediately turned over. Repeated injections of $InsP_3$ into the same egg are possible if we use the iontophoretic injection method instead of pressure injection. When injection current of $InsP_3$ (negative pulse) is increased little by little, a Ca^{2+} release is induced in an all-or-none fashion (Miyazaki *et al* 1988). Once a Ca^{2+} transient is produced, the same injection current fails to generate a second Ca^{2+} release for 100-200s; that is, there is a refractory period for the induction of Ca^{2+} release. $InsP_3$ can be applied continuously using the

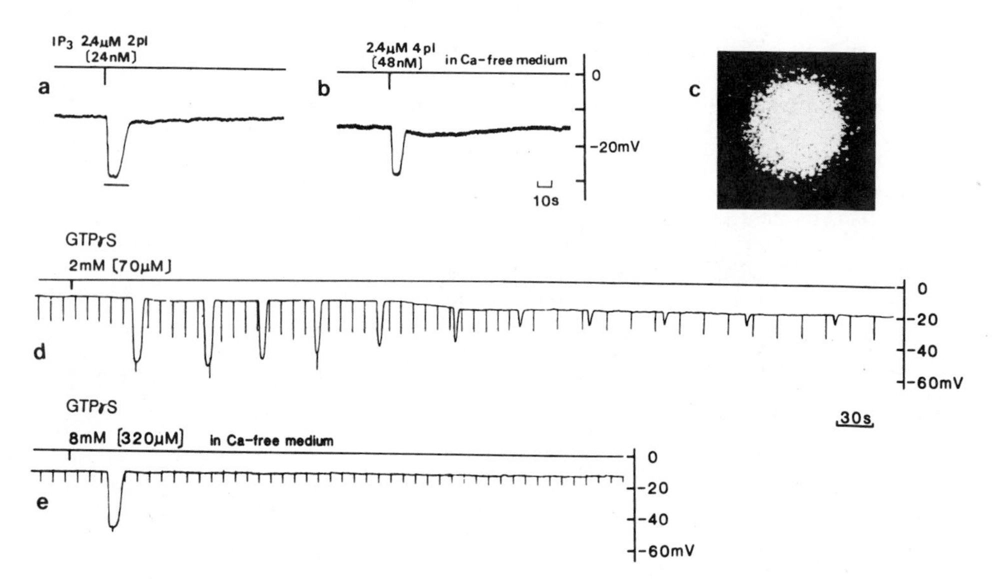

Figure 4. Ca^{2+} transient(s) induced by injection of $InsP_3$ or GTPγS. **(a)** A hyperpolarizing response (HR) induced by $InsP_3$ (2.4 µM is the pipette or 24 nM in the egg). $InsP_3$ was injected at the moment indicated by the vertical bar on the top trace. **(b)** HR induced by $InsP_3$ in Ca^{2+}-free medium. **(c)** Aequorin luminescence during the HR shown in a. All light spots are accumulated during the time indicated by the horizontal bar under the HR in a. **(d and e)** Periodic HRs induced by injection of GTPγS in standard medium and in Ca^{2+}-free medium. (From Miyazaki 1988a).

injection pipette with a blunt tip: $InsP_3$ leaks out from the pipette to the egg cytoplasm. The $[Ca^{2+}]_i$ did not increase persistently but increased periodically at intervals of 40-60s, despite continuous application of $InsP_3$ (Miyazaki *et al* 1988). The periodic Ca^{2+} rises disappeared when leakage of $InsP_3$ was stopped by applying a backing current (positive d.c. current) to the injection pipette.

Another factor inducing a Ca^{2+} release is Ca^{2+} itself. The mechanism of Ca^{2+}-induced Ca^{2+} release has been found to exist in the sarcoplasmic reticulum of skeletal muscles (see review by Endo 1977). This mechanism has been postulated to work in activating eggs, since it explains well the propagating Ca^{2+} release (Gilkey *et al* 1978).

When Ca^{2+} ions are injected iontophoretically into an unfertilized hamster egg while the membrane potential is recorded, a regenerative hyperpolarizing response (HR) is induced with an apparent threshold (Figure 5a) (Igusa and Miyazaki 1983). This response was induced even in Ca^{2+}-free external medium. The regenerative HR is followed by a refractory period of 60-120s (see Figure 5a). Of course, the injected Ca^{2+} *per se* causes a small hyperpolarization based on a Ca^{2+}-activated K^+ conductance, but conductance increase at the peak of the hyperpolarization is roughly proportional to the injection current of Ca^{2+}, when measured at the refractory period after a regenerative HR. Thus, it is evident that there exists a mechanism for non-linear enhancement of the Ca^{2+} rise in response to Ca^{2+} itself.

GTP-binding Protein-mediated Signaling

$InsP_3$ is one of the products derived from breakdown of polyphosphoinositides of the plasma membrane. There are a number of reports that phosphodiesterase, which mediates the cleavage of phosphatidylinositol 4,5-bisphosphate (PIP_2) into $InsP_3$ and diacylglycerol, is activated by ligand-receptor binding by way of a GTP-binding protein (see review by Stryer and Bourne 1986). Involvement of G-protein in the signal transduction of sperm-egg inter-

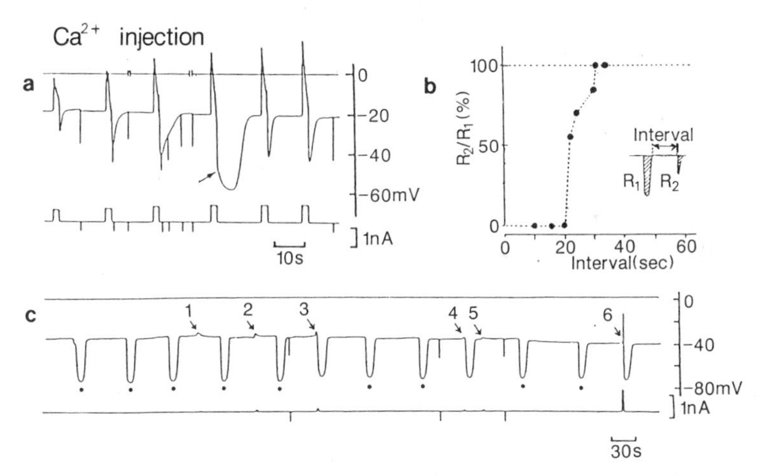

Figure 5. Iontophoretic injection of Ca^{2+} into the hamster egg. **(a)** A regenerative HR (arrow) in an unfertilized egg, induced by incremental current pulses. The bottom trace is injection current of Ca^{2+} **(b)** Refractory period of the regenerative HR in a fertilized egg. A regenerative response (R_1) and the next response (R_2) are quantified by the area in the record (see inset), and the ratio R_2/R_1 is plotted as a function of the interval of Ca^{2+} injections. **(c)** Ca^{2+} injections in an inseminated egg. Sperm-mediated periodic HRs are indicated by dots and the timing of Ca^{2+} injections is indicated by arrows. Arrows 1 and 2: subthreshold injection; 3: critical injection generating a HR; 4 and 5: injections showing a HR followed by the refractory period; 6: Ca^{2+} was injected with large suprathreshold current, but it causes only a similar HR to other HRs (From Igusa and Miyazaki 1983).

action causing Ca^{2+} release has been demonstrated in the sea urchin (Turner *et al* 1986) and hamster (Miyazaki 1988a). To activate a G-protein artificially, guanosine-5´-O-(3-thiotriphosphate) (GTPγS), the hydrolysis-resistant analog of GTP, is injected into the egg.

Injection of GTPγS (as well as GTP) into an unfertilized hamster egg by pressure, produced Ca^{2+} transient(s). The minimum effective [GTPγS] was 12 μM (330 μM in the pipette) (Miyazaki 1988a). For injection of GTP, 200 μM [GTP] was apparently comparable to 12 μM [GTPγS]. A remarkable feature of the GTPγS-induced Ca^{2+} rise is the delay after injection. It is 160-200s at lower [GTPγS] and is shortened with increasing [GTPγS]. However, the delay of 25s still remained, even when GTPγS was injected close to the cell's cortex with high concentrations. This time lag is probably due to a G-protein-mediated process causing a Ca^{2+} rise and not simply to diffusion of GTPγS to the plasma membrane.

A [GTPγS] more than 50 μM produces periodic HRs with fairly constant intervals of 40-60s, as shown in Figure 4d. This pattern of repeated HRs is similar to that induced by sperm (Figure 2b), except for the remarkable attenuation beginning from the third or fourth HR. Aequorin luminescence has shown that the first 2-3 responses are associated with the Ca^{2+} rise in the entire egg whereas succeeding Ca^{2+} rises occur at a localized area of the cytoplasm (Miyazaki 1988a).

The activation of G-protein by GTPγS should be antagonized by GDPβS [guanosine-5´-O-(2-thiodiphosphate)], the hydrolysis-resistant analog of GDP. Preinjection of GDPβS inhibited the occurrence of Ca^{2+} transient(s) in response to subsequent injection of GTPγS, in a dose-dependent manner (Miyazaki 1988a). For 70-80 μM [GTPγS], 1 mM [GDPβS] reduced the frequency of HRs and 7 mM [GDPβS] completely blocked the occurrence of HR. GDPβS did not affect the $InsP_3$-induced Ca^{2+} rise, indicating that the blocking site of GDPβS precedes $InsP_3$-induced Ca^{2+} release.

As described above, there is a G-protein-mediated process causing Ca^{2+} transient(s) in the hamster egg. Since GTPγS-induced periodic Ca^{2+} rises are interrupted by an interposed Ca^{2+} release induced by $InsP_3$ injection (Miyazaki 1988a), the G-protein-mediated process is likely to be the production of $InsP_3$ by way of activation of phosphodiesterase. Involvement of this process in formation of repeated Ca^{2+} transients during fertilization has been demonstrated by inhibitory effect of GDPγS on sperm-induced Ca^{2+} rises (Miyazaki 1988a). When GDPβS was preinjected, only the first 2-3 HRs were produced by sperm at 5 mM [GDPβS] in the egg, and no HR occurred at 8 mM [GDPβS] while more than 5 sperm attached to the egg and stopped their flagellar motion. Thus, GDPβS inhibits the occurrence of both GTPγS-induced and sperm-mediated Ca^{2+} transients at similar concentration ranges. Injection of GDPβS during the series of periodic HRs after insemination blocked subsequent HRs. Therefore, the series of Ca^{2+} rises upon fertilization requires persistent activation, not transient activation as a trigger, of the G-protein-mediated process.

The supposed G-protein is not identified in terms of pertussis toxin or cholera toxin sensitivity. Both toxins (injected into the egg) neither stimulated nor inhibited the occurrence of both GTP-induced and sperm-mediated Ca^{2+} rises (Miyazaki 1988a). Pertussis toxin- and cholera toxin-insensitive G-proteins which stimulate phosphodiesterase have been reported in other cells (Martin *et al* 1986; Merritt *et al* 1986).

Contribution of Ca^{2+} Influx

The series of HRs in fertilized hamster eggs is reduced in frequency and eventually stops after perfusion of Ca^{2+}-free medium (Igusa and Miyazaki 1983), indicating that Ca^{2+} influx contributes to the series of Ca^{2+} rises. A simple explanation is to postulate that the periodic Ca^{2+} rise is due to periodically increased Ca^{2+} influx. However, there are some findings against this hypothesis: attempts to change Ca^{2+} influx did not affect the amplitude of HRs but affected the frequency of HRs (Igusa and Miyazaki 1983). The frequency of HRs was reduced by lowering the external Ca^{2+} concentration, $[Ca^{2+}]_o$, or by adding Mn^{2+} or Co^{2+} to the external medium, without substantial effect on the amplitude, K^+ conductance increase and reversal potential of each HR. The frequency of HRs was increased by raising $[Ca^{2+}]_o$ in a time-and concentration-dependent manner, associated with a rather slight decrease in the size of HRs.

These findings lead to the conclusion that the series of periodic Ca^{2+} rises is dependent on Ca^{2+} influx but each Ca^{2+} rise is due to intracellular release of Ca^{2+}.

The findings described above suggest that continuous Ca^{2+} influx occurs, possibly due to an increase in Ca^{2+} permeability of the plasma membrane caused by the fused sperm. The hyperpolarizing shift of the membrane potential (HS) during the series of periodic HRs (Figure 2b) is likely to reflect this increased Ca^{2+} permeability. Usually, an increased Ca^{2+} permeability causes a depolarization, but in hamster eggs, it is able to cause a hyperpolarization via Ca^{2+}-induced K^+ conductance. In fact, the HS was blocked by injection of EGTA (Miyazaki and Igusa 1982) and it was decreased by lowering $[Ca^{2+}]_o$ or by adding Mn^{2+} or Co^{2+} to the external medium (Igusa and Miyazaki 1983). With a Ca^{2+}-electrode, the basal $[Ca^{2+}]_i$ has also been shown to increase slightly during the series of Ca^{2+} transients (Igusa and Miyazaki 1986).

GTPγS-produced multiple Ca^{2+} transients superimposed on a slight hyperpolarizing shift of the membrane potential (HS) (see Figure 4d). Upon removal of external Ca^{2+}, only the first one to two Ca^{2+} transients occurred but succeeding ones as well as HS were absent (see Figure 4e). The HS was not observed either, when Ca^{2+} transients were blocked by GDPβS (Miyazaki 1988a).

Linkage of Ca^{2+} Influx to Ca^{2+} Release

We have proposed a model based on a linkage of continuous Ca^{2+} influx to intracellular Ca^{2+} release for periodic Ca^{2+} transients in fertilized hamster eggs (Igusa and Miyazaki 1983; Miyazaki 1983). In the findings described above, the condition for reducing Ca^{2+} influx always reduces the HS as well as the frequency of HRs. Thus, Ca^{2+} influx seems to be a factor determining the interval between Ca^{2+} releases.

Some interesting findings were obtained by injection of Ca^{2+} into fertilized eggs at the pause between sperm-mediated HRs (Igusa and Miyazaki 1983). 1) The threshold injection current of Ca^{2+} for inducing a regenerative HR is about one-tenth of that necessary in unfertilized eggs (arrow 3 in Figure 5c; compare with Figure 5a): sensitivity of Ca^{2+}-induced Ca^{2+} rise becomes much higher after insemination. 2) Sperm-mediated HRs are interrupted by interposed HR(s) induced by Ca^{2+} injection(s) and the periodicity is reset by an injected Ca^{2+}-induced HR (Figure 5c): sperm-mediated Ca^{2+} rise and Ca^{2+}-induced Ca^{2+} rise are based on process(es) at common sites. 3) The refractory period of the regenerative HR is 40-50s in fertilized eggs (Figure 5b): the refractory period is comparable to the interval between sperm-mediated Ca^{2+} rises. 4) Continuous, repetitive injections of Ca^{2+} with constant pulses produce periodic HRs, the frequency of which is dependent on the injection current: extrinsic, continuous supply of Ca^{2+} can be converted to transient, periodic increase in $[Ca^{2+}]_i$ and the rate of the former relates to the frequency of the latter.

In summary, the proposed model predicts that continuous Ca^{2+} influx supplies Ca^{2+} for reloading the intracellular stores, which become ready to release Ca^{2+} stimulated by factor(s) such as $InsP_3$ or Ca^{2+} itself. The interval between Ca^{2+} transients is considered to be the time required for reloading the stores, which is revealed by the refractory period upon injection of $InsP_3$ or Ca^{2+}

Possible Mechanism of Signal Transduction

Possible mechanism involved in the signal transduction of sperm-egg interaction in the hamster egg is schematized in Figure 6. It should be noted that this scheme is tentative and speculative, and includes many questions. But it may be helpful for understanding the following discussion.

Activation of G-protein. The first Ca^{2+} transient in the hamster egg seems to be analogous to that in the sea urchin egg, although the time course is quite different. A possible mechanism of transmembrane signaling is activation of polyphosphoinositide turnover by way of activation of a G-protein, as proposed in the sea urchin egg where there is biochemical as well as physiological evidence (see review by Turner and Jaffe 1989). The problem is how the sperm activates the G-protein. If we assume ligand-receptor binding for activation of a G-

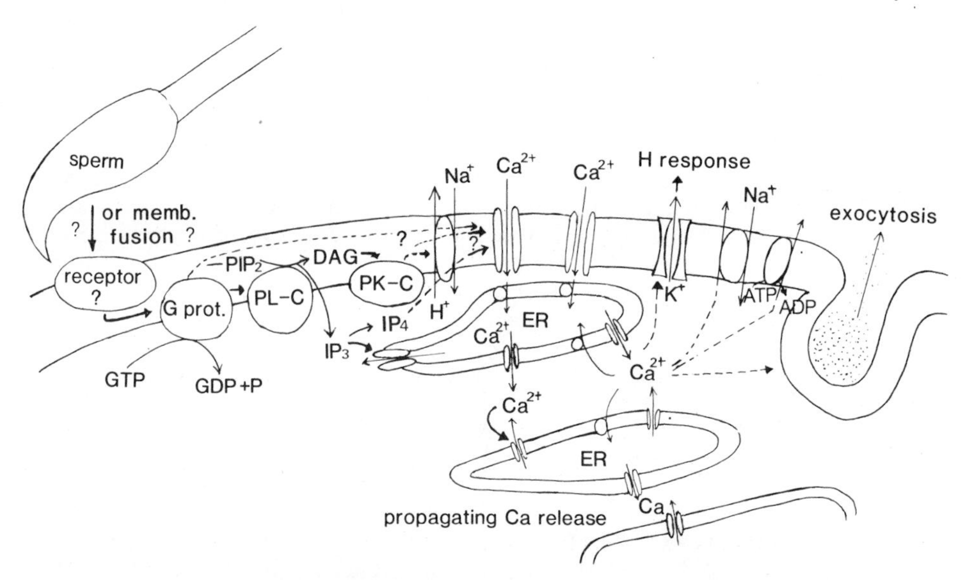

Figure 6. A schematic drawing of possible mechanism of signalling in the activating hamster egg. Unknown pathways are indicated by question marks. Ca^{2+} stores are attributed to endoplasmic reticulum (ER), which has been suggested to be the site of $InsP_3$- induced Ca^{2+} release (see Berridge and Irvine 1984). A H^+-Na^+ exchange system has been shown to be activated by protein kinase C (PK-C) in the sea urchin egg (see Swann *et al* 1987). This system raises intracellular pH, which is an important factor for metabolic activation of sea urchin eggs (see Epel 1978). The system is unknown in the hamster egg. PL-C: phospholipase C; DAG: diacylglycerol.

protein, as is the usual case in other cells, a receptor protein for sperm or a substance derived from sperm is postulated to exist in the egg plasma membrane. Such a receptor(s) has not been identified so far. Fusion of the lipid bilayer membranes between sperm and egg may activate a G-protein or breakdown of phosphatidylinositol; or a substance(s) in the sperm plasma membrane may be a direct activator of a G-protein. A different mechanism has been proposed by Swann *et al* (1987) in the sea urchin: the sperm brings $InsP_3$ into the egg. This mechanism cannot be ruled out in the hamster egg, considering that the frequency of Ca^{2+} transients is somehow dependent on the number of attached sperm. On the other hand, the block of the first Ca^{2+} transient by preinjected GDPβS argues against this hypothesis.

$InsP_3$ and Ca^{2+}-wave. Whatever the source is, $InsP_3$ is likely to be the initial inducer of intracellular Ca^{2+} release near the sperm attachment site. For a propagating Ca^{2+} release, a local increase in $[Ca^{2+}]_i$ is thought to cause release of Ca^{2+} from neighboring stores. At present, it remains unknown in the hamster egg whether the propagation is attributed to Ca^{2+}-induced Ca^{2+} release alone or to a recycling process between Ca^{2+}-stimulated production of $InsP_3$ and $InsP_3$-induced Ca^{2+} release as proposed in the sea urchin by Swann and Whitaker (1986).

The second and third Ca^{2+} responses are similar to the first one in the peak $[Ca^{2+}]_i$ (Figure 2a) and propagating increase of $[Ca^{2+}]_i$. They occur with short intervals after the preceding response (often shorter than 40s; the minimum interval in established periodicity of later responses). There may be reserve Ca^{2+} stores with an apparently high threshold for a stimulus to release Ca^{2+}.

Activation of Ca^{2+} permeability. Later Ca^{2+} transients seem to occur repeatedly with a rhythm determined by characteristics of the inseminated egg. A model based on reloading of Ca^{2+} stores by continuous Ca^{2+} influx has already been described. It is a general phenomenon that the sperm induces an increase in membrane permeabilities of the egg to various ions (Na^+, K^+, Ca^{2+} or Cl^-), generating the fertilization potential (Hagiwara and Jaffe 1979). The dominant ionic species are different from animal to animal. In the hamster egg, Ca^{2+} permeability is increased.

How is Ca^{2+} permeability increased? It seems probable that a G-protein-mediated process is responsible for elevation of Ca^{2+} permeability, as this process is involved in formation of periodic Ca^{2+} rises. A pathway of Ca^{2+} across the plasma membrane is Ca channel. In hamster eggs there exist voltage-dependent Ca channels, which mediate a Ca^{2+}-dependent action potential (Miyazaki and Igusa 1981b). These or perhaps other kinds of Ca channels may be activated upon insemination. There are reports that G-proteins activate K channels (see review by Neer and Clapham 1988). There is a possibility that G-protein itself may activates Ca channels. Another possibility is activation of Ca channels through diacylglycerol and protein kinase C. The C kinase has been shown to recruit covert Ca channels in *Aplysia* neurons (Strong *et al.* 1987). However, our recent studies have shown that the application of the phorbol ester, TPA (12-0-tetradecanoyl phorbol acetate) is inhibitory and sphingosine, a C kinase inhibitor, facilitates GTPγS-induced Ca^{2+} transients (Swann, Igusa and Miyazaki, unpublished). Another candidate for activation of Ca^{2+} channels is inositol 1,3,4,5-tetrakisphosphate (IP_4). Irvine and Moore (1986, 1987) have proposed a model in the sea urchin egg that IP_4 causes an influx of Ca^{2+} to be sequestered into intracellular Ca^{2+} stores apposed to the plasma membrane; and $InsP_3$ induces release of Ca^{2+} from these stores. However, there is a report arguing against such an effect by IP_4 (Crossley *et al.* 1988).

Periodic Ca^{2+} transients. What stimulates release of Ca^{2+} from reloaded stores? $InsP_3$ is capable of inducing repeated Ca^{2+} releases, if it is supplied continuously. Actually, activation of a G-protein-mediated process has been suggested to persist during fertilization (Miyazaki, 1988a). On the other hand, the finding that GTPγS-induced Ca^{2+} transients attenuate a local $[Ca^{2+}]_i$ suggests additional factor(s) responsible for causing a synchronous Ca^{2+} release throughout the egg (within 1s) in later responses of fertilized eggs. It is interesting that Ca^{2+} stores become much sensitive to injected Ca^{2+} to release Ca^{2+} in inseminated eggs (Figure 5c).

The Ca^{2+} extrusion system is also important for formation of repetitive Ca^{2+} rises. The recovery of increased $[Ca^{2+}]_i$ is about 10s or less. A Ca^{2+} pump has been suggested to exist because of the prolongation of the recovery phase of HRs, caused by the application of La^{3+}, high $[Ca^{2+}]_o$ or quercetin (Georgiou et al 1987) or by raising external pH (Georgiou *et al* 1988). A Na^+-Ca^{2+} exchange system has been suggested as well, by the prolonged recovery phase upon removal of external Na^+ (Igusa and Miyazaki 1983; Georgiou *et al* 1988).

Summary. Although there are still many unknown respects, I propose, at present, two kinds of pathways for the G-protein-mediated process: one is the production of $InsP_3$ for inducing Ca^{2+} release and the other is elevation of Ca^{2+} permeability for maintaining periodic Ca^{2+} releases. Further studies are needed to elucidate the unknown mechanisms and complete the scheme in Figure 6, and they are also needed to elucidate the subsequent processes following the Ca^{2+} rises which lead to egg activation and cell proliferation.

ACKNOWLEDGEMENTS

Most of the experiments on hamster eggs described in this article were done in the Department of Physiology, Jichi Medical School. The author is grateful to Professor Kyoji Maekawa for his support and to Dr. Yukio Igusa for collaboration in most of the studies and for discussion. He also thanks Dr. Karl Swann for discussion during the preparation of this article and Misses Yumiko Hodota, Satomi Shinozaki and Midori Okada for technical assistance.

REFERENCES

Austin, C. R. 1961. Fertilization of mammalian eggs *in vitro*. *Int. Rev. Cytol.* 12:337-359.

Bedford, J. M., H. D. M. Moore, and L. E. Franklin. 1979. Significance of the equatorial segment of the acrosome of the spermatozoa in eutherian mammals. *Exp. Cell Res.* 119: 119-126.

Berridge, M. J. and R. F. Irvine. 1984. Inositol trisphosphate, a novel second messenger in cellular signal transduction. *Nature (Lond.)* 312:315-321.

Biggers, J. D., W. K. Whitten, and D. G. Whittingham. 1971. The culture of mouse embryo *in vitro*. p. 86-116. *In: Methods in Mammalian Embryology*. J. C. Daniel (Ed.). Freeman, San Francisco.

Busa, W. B., J. E. Ferguson, S. K. Joseph, J. R. Williamson, and R. Nuccitelli. 1985. Activation of frog (*Xenopus laevis*) egg by inositol trisphosphate. I. Characterization of Ca^{2+} release from intracellular stores. *J. Cell Biol.* 101:677-682.

Busa, W. B. and R. Nuccitelli. 1985. An elevated free cytosolic Ca^{2+} wave follows fertilization in egg of the frog, *Xenopus laevis*. *J. Cell Biol.* 100:1325-1329.

Crossley, I., K. Swann, E. Chambers, and M. Whitaker. 1988. Activation of sea urchin eggs by inositol phosphates is independent of external calcium. *Biochem. J.* 252:257-262.

Cuthbertson, K. S. R. and P. H. Cobbold. 1985. Phorbol ester and sperm activate mouse oocytes by inducing sustained oscillations in cell Ca^{2+}. *Nature (Lond.)*. 316:541-542.

Cuthbertson, K. S. R., D. G. Whittingham, and P. H. Cobbold. 1981. Free Ca^{2+} increases in exponential phases during mouse oocyte activation. *Nature (Lond.)* 294:754-757.

Eisen, A., D. P. Kiehart, S. J. Wieland, and G. T. Reynolds. 1984. Temporal sequence and spatial distribution of early events of fertilization in single sea urchin egg. *J. Cell Biol.* 99:1647-1654.

Endo, M. 1977. Calcium release from the sarcoplasmic reticulum. *Physiol. Rev.* 57:71-108.

Epel, D. 1978. Mechanisms of activation of sperm and egg during fertilization of sea urchin gametes. *Curr. Top. Dev. Biol.* 12:185-246.

Georgiou, P., C. Bountra, A. McNiven, and C. R. House. 1987. The effect of lanthanum, quercetin and dinitrophenol on calcium-evoked electrical responses in hamster eggs. *Q. J. Exp. Physiol.* 72:227-241.

Georgiou, P., C. R. House, A. I. McNiven, and S. Yoshida. 1988. On the mechanism of a pH-induced rise in membrane potassium conductance in hamster eggs. *J. Physiol. (Lond.)* 402:121-138.

Gilkey, J. C., L. F. Jaffe, E. B. Ridgeway, and G. T. Reynolds. 1978. A free calcium wave traverses the activating egg of the medaka, *Oryzias latipes*. *J. Cell Biol.* 76:448-466.

Hagiwara, S. and L. A. Jaffe. 1979. Electrical properties of egg cell membranes. *Annu. Rev. Biophys. Bioeng.* 8:385-416.

Hiramoto, Y. 1961. Microinjection of the live spermatozoa into sea urchin eggs. *Exp. Cell Res.* 27:416-426.

Igusa, Y. and S. Miyazaki. 1983. Effects of altered extracellular and intracellular calcium concentration on hyperpolarizing responses of the hamster egg. J.Physiol. (Lond.) 340: 611-632.

Igusa, Y. and S. Miyazaki. 1986. Periodic increase of cytoplasmic calcium in fertilized hamster eggs measured with calcium-sensitive electrodes. *J. Physiol. (Lond.)* 377:193-205.

Igusa, Y., S. Miyazaki and N. Yamashita. 1983. Periodic hyperpolarizing responses in hamster and mouse eggs fertilized with mouse sperm. *J. Physiol. (Lond.)* 340:633-647.

Irvine, R. F. and R. M. Moor. 1986. Micro-injection of inositol 1,3,4,5-tetrakisphosphate activates sea urchin eggs by a mechanism dependent on external Ca^{2+}. *Biochem. J.* 240:917-920.

Irvine, R. F. and R. M. Moor. 1987. Inositol (1,3,4,5) tetrakisphosphate-induced activation of sea urchin eggs requires the presence of inositol trisphosphate. *Biochem. Biophys. Res. Commun.* 146:284-290.

Jaffe, L. A., A. P. Sharp, and D. P. Wolf. 1983. Absence of an electrical polyspermy block in the mouse. *Dev. Biol.* 96:317-323.

Jaffe, L. F. 1985. The role of calcium explosions, waves and pulses in activating eggs. p. 127-165. *In: Biology of Fertilization*. C. B. Metz and A. Monroy (Eds.). Academic Press, New York.

Kishimoto, T. 1986. Microinjection and cytoplasmic transfer in starfish oocytes. *Methods Cell Biol.* 27:379-394.

Kubota, H. Y., Y. Yoshimoto, M. Yoneda, and Y. Hiramoto. 1986. Free calcium wave upon activation in *Xenopus* eggs. *Dev. Biol.* 119:129-136.

Martin, T. F., D. O. Lucas, S. M. Bajjalieh, and J. A. Kowalchyk. 1986. Thyrotropin-releasing hormone activates a Ca^{2+}-dependent polyphosphoinositide phosphodiesterase in permeable GH_3 cells. *J. Biol. Chem.* 261:2918-2927.

McCulloh, D. H., C. E. Rexroad, Jr., and H. Levitan. 1983. Insemination of rabbit eggs is associated with slow depolarization and repetitive diphasic membrane potentials. *Dev. Biol.* 95:372-377.

Merritt, J. E., C. W. Taylor, R. P. Rubin, and J. W. Putney, Jr. 1986. Evidence suggesting that a novel guanine nucleotide regulatory protein couples receptors to phospholipase C in exocrine pancreas. *Biochem. J.* 236:337-343.

Miyazaki, S. 1983. Periodic hyperpolarizations in fertilized hamster eggs: possible linkage of Ca influx to intracellular Ca release. p. 219-231. *In: The Physiology of Excitable Cells.* A. Grinnell and W. J. Moody, Jr. (Eds.). Alan R. Liss Inc., New York.

Miyazaki, S. 1988a. Inositol 1,4,5-trisphosphate-induced calcium release and guanine nucleotide-binding protein-mediated periodic calcium rises in golden hamster eggs. *J. Cell Biol.* 106:345-354.

Miyazaki, S. 1988b. Fertilization potential and calcium transients in mammalian eggs. *Dev. Growth & Differ.* 30:603-610.

Miyazaki, S., N. Hashimoto, Y. Yoshimoto, T. Kishimoto, Y. Igusa, and Y. Hiramoto. 1986. Temporal and spatial dynamics of the periodic increase in intracellular free calcium at fertilization of golden hamster eggs. *Dev. Biol.* 118:259-267.

Miyazaki, S. and Y. Igusa. 1981a. Fertilization potential in golden hamster eggs consists of recurring hyperpolarizations. *Nature (Lond.)* 290:702-704.

Miyazaki, S. and Y. Igusa. 1981b. Ca-dependent action potential and Ca-induced fertilization potential in golden hamster eggs. p. 305-311. *In: The Mechanism of Gated Calcium Transport Across Biological Membranes.* S. T. Ohnishi and M. Endo (Eds.). Academic Press, New York.

Miyazaki, S. and Y. Igusa. 1982. Ca-mediated activation of a K current at fertilization of golden hamster eggs. *Proc. Natl. Acad. Sci. USA.* 79:931-935.

Miyazaki, S., Y. Igusa and K. Swann. 1988. Involvement of GTP-binding protein in signal transduction of sperm-egg interaction at fertilization of hamster eggs. *J. Physiol. Soc. Japan* 50:390.

Neer, E. J. and D. E. Clapham. 1988. Roles of G protein subunits in transmembrane signalling. *Nature (Lond.)* 333:129-134.

Poenie, M., J. Aldertson, R. Y. Tsien, and R. A. Steinhardt. 1985. Changes of free calcium levels with stages of the cell division cycle. *Nature (Lond.)* 315:147-149.

Steinhardt, R., R. Zucker, and G. Schatten. 1977. Intracellular calcium release at fertilization of the sea urchin egg. *Dev. Biol.* 58:185-196.

Strong, J. A., A. P. Fox, R. W. Tsien, and L. K. Kaczmare. 1987. Stimulation of protein kinase C recruits covert calcium channels in *Aplysia* bag cell neurons. *Nature (Lond.)* 325: 714-717.

Stryer, L. and H. R. Bourne. 1986. G proteins: a family of signal transducers. *Annu. Rev. Cell Biol.* 2:391-419.

Swann, K., B. Ciapa, and M. J. Whitaker. 1987. Cellular messengers and sea urchin egg activation. p. 45-69. *In: Morecular Biology of Invertebrate Development.* A. O'Conner (Ed.). Alan R. Liss, New York.

Swann, K. and W. Whitaker. 1986. The part played by inositol trisphosphate and calcium in the propagation of the fertilization wave in sea urchin eggs. *J. Cell Biol.* 103:2333-2342.

Turner, P. R. and L. A. Jaffe. 1989. G proteins and the regulation of oocyte maturation and fertilization. p. 297-318. *In: The Cell Biology of Fertilization.* G. Schatten and H. Schatten (Eds.). Academic Press, Orlando. (In Press).

Turner, P. R., L. A. Jaffe, and A. Fein. 1986. Regulation of cortical vesicle exocytosis in sea urchin egg by inositol 1,4,5- trisphosphate and GTP-binding protein. *J. Cell Biol.* 102: 70-76.

Whitaker, M. J. and R. F. Irvine. 1984. Inositol 1,4,5-trisphosphate microinjection activates sea urchin eggs. *Nature (Lond.)* 312:636-639.

Yanagimachi, R. 1969. *In vitro* capacitation of hamster spermatozoa by follicular fluid. *J. Reprod. Fertil.* 18:275-286.

Yanagimachi, R. 1970. *In vitro* capacitation of golden hamster spermatozoa by homologous and heterologus blood sera. *Biol. Reprod.* 3:147-153.

Yanagimachi, R. 1978. Sperm-egg association in mammals. *Curr. Top. Dev. Biol.* 12:83-105.

Yanagimachi, R. 1988. Mammarian fertilization. p.135-185. *In: The Physiology of Reproduction.* E. Knobil and J. Neill (Eds.). Raven Press, New York.

EGG-INDUCED MODIFICATIONS OF THE MURINE ZONA PELLUCIDA

Gregory S. Kopf[1], Yoshihiro Endo[1], Peter Mattei[2], Shigeaki Kurasawa[1], and Richard M. Schultz[2]

Division of Reproductive Biology[1]
Department of Obstetrics and Gynecology
School of Medicine, and Department of Biology[2]
University of Pennsylvania, Philadelphia, PA 19104-6080

ABSTRACT

We are interested in the biochemical mechanisms involved in the early events of mammalian egg activation, with special reference to the mechanism(s) by which the block to polyspermy at the level of the zona pellucida (ZP) is established. Since some of the initial events of sperm-induced egg activation in lower species are accompanied by the hydrolysis of polyphosphoinositides, we examined the effects of protein kinase C (PKC) activators (phorbol diesters and diacylglycerols) and microinjected inositol 1,4,5-trisphosphate ($InsP_3$)

on sperm-ZP binding, the ZP-induced acrosome reaction, fertilization, and egg induced modifications of the ZP.

Sperm penetration of the ZP and fertilization are inhibited in mouse eggs treated with biologically active phorbol diesters and the diacylglycerol, sn-1,2,-dioctanoyl glycerol (diC_8). The effect is mediated by the ZP, since ZP-free eggs treated with these compounds are fertilized to the same extent as untreated eggs; these compounds also exert no direct inhibitory effect on the sperm's ability to fertilize an egg. The inhibitory effect of these compounds on fertilization is due to a step subsequent to sperm binding, since the ability of untreated and treated eggs to bind sperm is similar. Two-dimensional reduction gel electrophoresis of iodinated ZP obtained from phorbol diester or diC_8-treated eggs demonstrated that ZP2 is modified to $ZP2_f$; such a modification is similar to that observed in ZP isolated from 2-cell embryos. Analysis of the sperm receptor and acrosome reaction-inducing activities of the ZP (*i.e.*, properties of ZP3) from these treated eggs demonstrated that receptor activity remained but that the acrosome reaction-inducing activity was altered; sperm could initiate but not complete the acrosome reaction. Similarly, purified ZP3 prepared from phorbol diester-treated eggs possesses full sperm receptor activity, and can initiate but not complete the acrosome reaction. This is in contrast to ZP3 from 2-cell embryos, which cannot even initiate an acrosome reaction. Thus, we have been able to elicit an egg-induced dissociation of the receptor and acrosome reaction-inducing activities of the ZP, which should provide a means to study the components of ZP3 involved in the acrosome reaction. Moreover, ZP obtained from phorbol diester-treated eggs may serve as a probe to delineate the sequence of events comprising the acrosome reaction.

Since the previous experiments utilizing PKC activators represent the examination of one-half of a second messenger cascade generated upon the hydrolysis of phosphatidylinositol-1,4,-bisphosphate (PIP_2) by a PIP_2-specific phospholipase C, we investigated the effects of microinjecting $InsP_3$ into mouse eggs on egg activation and egg-induced modifications of the ZP.

A concentration-dependent increase in the percentage of eggs displaying an egg-induced modification of ZP2 to $ZP2_f$ was observed in cells microinjected with increasing concentrations of $InsP_3$ (0.1 nM to 4 μM final concentration). $InsP_3$ concentrations greater than 1 nM elicited an egg-induced modification of ZP2 to $ZP2_f$; EC_{50}=5 nM. These $InsP_3$ effects also occurred in the absence of extracellular Ca^{2+}. Microinjection of inositol 1,3,4,5-tetrakisphosphate, which has been implicated in activating plasma membrane associated Ca^{2+} channels in eggs of lower species, did not result in an egg-induced modification of ZP2. Similarly, microinjection of $I(1,4)P_2$, $I(2,4,5)P_3$, or $I(1,3,4)P_3$, which do not release intracellular Ca^{2+}, failed to cause an egg-induced modification of ZP2. Analysis of sperm binding to the ZP of $InsP_3$ microinjected eggs demonstrated that binding was decreased when compared to vehicle injected eggs. These binding studies suggest that at least one of the biological activities of the ZP (*i.e.*, sperm receptor activity) may be modified by $InsP_3$ microinjection. It remains to be determined whether the acrosome reaction-inducing properties of the ZP are modified by such treatments. Fluorograms of ^{35}S-methionine labeled mouse eggs that had been injected with $InsP_3$ revealed labeling patterns similar to that of vehicle injected eggs; these patterns contrasted with those seen with 1-cell embryos. This is consistent with a low level of egg activation (17%), as demonstrated by emission of a second polar body. $InsP_3$ injection, therefore, did not generate those global changes in protein synthesis normally associated with fertilization.

INTRODUCTION

It is now appreciated that sperm-egg interaction in both invertebrate and vertebrate species consists of a series of carefully orchestrated and tightly regulated events that initially results in an egg-induced activation of the sperm and ultimately results in a sperm-induced activation of the egg. It is becoming apparent that specific aspects of both egg-induced sperm activation and sperm-induced egg activation have elements analogous to ligand-receptor-second messenger systems known to be important in the mediation of intercellular communication between somatic cells. In the mammalian egg, and the mouse egg in particular, species-specific sperm-egg recognition and interaction, sperm activation (*i.e.*, the acrosome reaction), and an egg-induced block to polyspermy all appear mediated by the extracellular matrix surrounding the egg called the zona pellucida (ZP).

Composition and Function of the Mouse Egg Zona Pellucida

The ZP of the mouse egg is composed of three sulfated glycoproteins designated as ZP1, ZP2, and ZP3 (Bleil and Wassarman 1980a; Shimizu *et al.* 1983; Wassarman 1988), which are present in molar ratios of approximately 1:10:10, respectively. These proteins are the major biosynthetic product of the growing oocyte, and are secreted throughout the period of oocyte growth (Bleil and Wassarman 1980b). ZP1 (M_r=200,000) (Bleil and Wassarman 1980a) is a dimer connected by intermolecular disulfide bonds and may function to maintain the three dimensional structure of the ZP by cross linking filaments that are composed of repeating structures of ZP2/ZP3 heterodimers (Greve and Wassarman 1985).

ZP2 (M_r=120,000 under nonreducing and reducing conditions) (Bleil and Wassarman 1980a), which is present in both oocytes and unfertilized eggs, may mediate the binding of acrosome reacted sperm to the ZP (Bleil and Wassarman 1986; Bleil *et al.* 1988). Upon fertilization ZP2 is modified to a form called $ZP2_f$ (Bleil and Wassarman 1981), and this modification is effected by the egg. $ZP2_f$ has a M_r=120,000 under nonreducing conditions, but under reducing conditions the M_r=90,000. This egg-induced modification probably results from proteolysis of ZP2, generating fragments which are held together by disulfide bonds. The biological consequence of the ZP2 to $ZP2_f$ shift is that $ZP2_f$ no longer will bind to acrosome reacted sperm (Bleil and Wassarman 1986; Bleil *et al.* 1988).

ZP3 (M_r=83,000) accounts for both the sperm receptor and the acrosome reaction-inducing activities of the ZP of unfertilized eggs (Bleil and Wassarman 1980c; 1983). The sperm receptor activity of ZP3 appears to be conferred by O-linked carbohydrate moieties and not by its polypeptide chain (Florman *et al.* 1984; Florman and Wassarman 1985); an α-linked terminal galactose residue may be involved in this binding activity. On the other hand, the acrosome reaction-inducing activity of ZP3 may be conferred by both the carbohydrate and protein portions of the molecule (Florman *et al.* 1984; Wassarman *et al.* 1985), although the exact nature of the interaction between the protein and carbohydrate required for biological activity is not clear at this time. Fertilization results in the loss of both the sperm receptor and acrosome reaction-inducing activities of the ZP3 molecule (Bleil and Wassarman 1980c 1983); the loss of these activities appears to be associated with a relatively minor biochemical modification of the ZP3 molecule, since the electrophoretic mobility of ZP3 from fertilized eggs is similar to that of ZP3 from unfertilized eggs.

Sperm-Zona Pellucida Interaction Prior to and After Sperm-Egg Fusion and the Zona Pellucida Block to Polyspermy in the Mouse

Based on the above mentioned properties of the ZP glycoproteins from unfertilized and fertilized eggs, and based on the fact that only acrosome intact sperm bind to the ZP of unfertilized eggs (Saling *et al.* 1979), the following sequence of events are thought to comprise sperm-ZP interaction prior to and after sperm-egg fusion in the mouse. Acrosome intact sperm bind to the ZP (Saling *et al.* 1979; Florman and Storey 1982) in a species specific manner by interaction with ZP3 and, once this binding occurs, ZP3 then mediates the acrosome reaction. These initial interactions occur via interactions between the ZP and the plasma membrane overlying the head of the sperm. Secondary interactions of the acrosome reacted sperm with ZP2 then occur through the interaction of the exposed inner acrosomal membrane of the sperm (Bleil and Wassarman 1986; Bleil *et al.* 1988). These acrosome reacted sperm penetrate the ZP, traverse the perivitelline space, and then bind and fuse with the plasma membrane of the egg. Once sperm-egg fusion has occurred the egg undergoes the cortical granule reaction (Szollsi 1967; Barros and Yanagimachi 1971,1972; Wolf and Hamada 1977; Lee *et al.* 1988). This exocytotic event results in the release of cortical granule-associated enzymes into the perivitelline space, which probably convert ZP2 to $ZP2_f$ and modify ZP3 such that it loses both its sperm receptor and acrosome reaction inducing activities. Although the component(s) of the cortical granule exudate responsible for these ZP modifications have not been isolated, a trypsin-like protease may be involved (Gwatkin *et al.* 1973; Wolf and Hamada 1977). As a consequence of these ZP modifications, acrosome intact sperm no longer bind to the ZP, and sperm that are bound to the ZP and have acrosome reacted can no longer interact and penetrate the ZP since they are unable to interact with $ZP2_f$ (Bleil and Wassarman 1986; Bleil *et al.* 1988). Such egg induced modifications of the ZP constitute the ZP block to polyspermy. Although there is evidence to support an additional plasma membrane polyspermy block in mouse (Wolf 1978) and other mammalian eggs (Stewart-Savage and Bavister 1988), there is no evidence to support the existence of a rapid electrical polyspermy block in mammalian eggs (Jaffe *et al.* 1983).

Biochemical Basis of the Sperm-Induced Activation of Nonmammalian and Mammalian Eggs

The nature of the sperm-induced biochemical signal(s) that ultimately results in cortical granule exocytosis, the ZP and plasma membrane polyspermy blocks, and metabolic activation of the mammalian egg are presently not known. Parthenogenetic activation of mouse eggs by several agents is accompanied by a large rise in the intracellular Ca^{2+} concentration (Cuthbertson *et al.* 1981). The importance of Ca^{2+} as an ionic signal in egg activation has been well documented in invertebrate, amphibian and teleost eggs (Jaffe 1985). In nonmammalian species it has been shown that the cortical granule reaction is triggered by a rise in free intracellular Ca^{2+} that appears to originate from an intracellular reticulum-like complex in the cortex of the egg (Poenie *et al.* 1982; Poenie and Epel 1987), and spreads in a wave from the point of sperm-egg fusion around the cortex of the egg in a centripetal fashion (Jaffe 1985). Those factors controlling the release of Ca^{2+} from this store, its re-sequestration, and the mechanism by which this cation causes exocytosis are only starting to be investigated.

Polyphosphatidylinositide turnover appears to represent at least one response that may operate to control some of the early events of egg activation in both nonmammalian and mammalian species. Increased diphosphoinositide and triphosphoinositide levels accompany fertilization in sea urchin eggs; these changes precede the cortical granule reaction (Turner *et al.* 1984). Microinjection of $InsP_3$ also activates eggs and produces a cortical reaction similar to that observed during fertilization (Whitaker and Irvine 1984). This $InsP_3$-induced Ca^{2+} release is thought to occur through the activation of a guanine nucleotide binding regulatory protein(s) (G-protein) (Turner *et al.* 1986 1987), proteins known to act as transmembrane signal transduction elements distal to the binding of cell surface receptors by their respective ligands (Gilman 1987). The nature of such a G-protein and the mechanism by which it couples sperm fusion events at the egg plasma membrane to the generation of intracellular second messengers in the egg is not known. One possibility is that fertilization is accompanied by a G-protein mediated stimulation of phospholipase C, which generates $InsP_3$ and 1,2, diacylglycerol (DAG). The resultant $InsP_3$-induced intracellular Ca^{2+} release may then play a role in cortical granule exocytosis. An additional role for this $InsP_3$-mediated Ca^{2+} release may also be to propagate the fertilization wave observed in this species through the feedback release of additional $InsP_3$ through Ca^{2+} activation of the phospholipase C (Swann and Whitaker 1986; Ciapa and Whitaker 1986). The DAG that is generated may also play a role in the cortical granule reaction since recent attention has focused on the role(s) of protein phosphorylations catalyzed by protein kinase C (PKC) in the regulation of exocytosis in other systems (Kikkawa and Nishizuka 1986); however, there is no current evidence for such a mode of action in the sea urchin egg. The fertilization-induced activation of the egg plasma membrane Na^+/H^+ exchanger, which brings about the alkalinization of the intracellular pH required for DNA synthesis, may be regulated by DAG induced PKC activation, since tumor promoting phorbol esters appear to activate this ion exchanger (Swann and Whitaker 1985). Since physiological Ca^{2+} concentrations stimulate the hydrolysis of phosphatidylinositol-1,4,-bisphosphate (PIP_2) in egg plasma membranes (Whitaker and Steinhardt 1985), it is possible that the resultant generation of both $InsP_3$ and DAG serve to couple the increased intracellular Ca^{2+} transient and the cortical reaction with the resultant alkalinization of intracellular pH via the Na^+/H^+ exchanger.

Fertilization induced increases in intracellular Ca^{2+} levels also appear to be a primary ionic signal in the activation of amphibian eggs. Both the fertilization potential (e.g., chloride conductance) and cortical granule exocytosis appear to be dependent on a rise in intracellular Ca^{2+}, since both of these events are blocked when the intracellular free Ca^{2+} is buffered with [1,2-bis(o-aminopenoxy)ethane-N,N,N1N1-tetraceticacid] (BAPTA) (Kline 1988). Such changes in Ca^{2+} concentrations may result from the activation of a G-protein with the subsequent generation of $InsP_3$ through the activation of a phospholipase C (Kline and Jaffe 1987). Iontophoretic injection of $InsP_3$, in fact, brings about a series of early activation events including membrane depolarization (fertilization potential), release of Ca^{2+} from intracellular stores, cortical contraction and cortical granule exocytosis (Busa *et al.* 1985). As in the sea urchin, these responses appear to be identical to those events observed during the sperm-induced activation of these eggs.

We are just beginning to understand the biochemical mechanisms by which sperm-induced activation of mammalian eggs occur. In the hamster, sperm-egg interaction produces transient but periodic rises in intracellular Ca^{2+} (Igusa and Miyazaki 1986; Miyazaki *et al.* 1986). Such Ca^{2+} transients can be generated when $InsP_3$ is microinjected, and these effects are mimicked by the injection of GTP or GTPγS (Miyazaki 1988). Both the sperm-induced and GTPγS-induced Ca^{2+} transients are inhibited by preinjection of GDPβS. These data support the idea that sperm-induced Ca^{2+} transients in mammalian eggs might act through a similar G-protein - polyphosphoinositide signaling system. The involvement of PKC in generating these Ca^{2+} transients has been suggested by experiments demonstrating that tumor promoting phorbol diesters induce transient oscillations in intracellular Ca^{2+} and parthenogenetic activation of ZP-free mouse eggs (Cuthbertson and Cobbold 1985); these effects appear similar to those changes in Ca^{2+} concentrations observed during fertilization (Cuthbertson *et al.* 1981). The mechanism by which these effects occur are not clear since extremely high concentrations of phorbol diesters were required to bring about these effects.

It can be concluded, therefore, that common mechanisms of sperm-induced egg activation probably exist in both nonmammalian and mammalian eggs. In attempts to understand further mammalian sperm-ZP interactions prior to and after fertilization, we have investigated the potential role of polyphosphoinositide turnover in the mouse egg with regard to the egg-induced ZP polyspermy block. We demonstrate that activators of PKC (a diacylglycerol or phorbol diesters) and $InsP_3$ cause egg-induced modifications in the ZP that, at least in one case, mimic only part of the egg-induced ZP modification that normally occurs after fertilization of mouse eggs. The possible mechanisms by which these agents effect these ZP modifications and the role of such a intracellular signaling system in the ZP polyspermy block will be discussed.

MATERIALS AND METHODS

Collection of Gametes

Gametes were obtained from 12-24 week old male and 8-12 week old female Swiss Webster mice. Superovulated eggs were incubated in modified Krebs Ringer bicarbonate buffer (m-KRB) (Toyoda and Chang 1974). This medium was supplemented with either 4 mg/ml polyvinylpyrrolidone (PVP) (for phorbol diester studies) or 4 mg/ml BSA (for inositol phosphate studies) (Endo *et al.* 1987a,b; Kurasawa *et al.* 1989). Sperm were capacitated in m-KRB containing 4 mg/ml BSA for 90 min (Endo *et al.* 1987a)

Assessment of Effects of Phorbol Diesters and DiC_8 on *In Vitro* Fertilization

The methods of Endo *et al.* (1987a) were used for *in vitro* fertilization, as well as the preparation and assessment of effects of agents on *in vitro* penetration and fertilization of cumulus intact, cumulus-free and cumulus-free ZP-free mouse eggs.

Microinjection of Mouse Eggs

$InsP_3$ (Amersham), structural analogs of $InsP_3$ (Calbiochem), and vehicle (distilled H_2O) were injected into mouse eggs using a Narishige micromanipulator and Medical Systems Corp.

picoinjector. Each egg was injected with 10 pl, which represents approximately 5% of the total volume of a mature mouse egg.

Isolation of Structurally-Intact Zonae Pellucidae from Control and Experimentally Treated Eggs

ZP were removed mechanically with a mouth operated micropipet with a bore size approximately 2/3 to 3/4 that of the diameter of the egg. To minimize proteolysis caused by disrupting the eggs during this step, the ZP were removed in medium (Earle's minimal essential medium containing 100 μg/ml pyruvate, 10 μg/ml gentamicin, 3 mg/ml PVP, 10 mM Hepes, pH 7.2) containing 1 mg/ml EDTA, 10 μg/ml lima bean trypsin inhibitor, and 0.1 mM phenylmethylsulfonylfluoride. The ZP were then extensively washed in either m-KRB (for subsequent incubation with sperm) or 1 mM benzamidine, followed by distilled water (for subsequent iodination). These methods are described in detail by Endo *et al.* (1987a).

Preparation of Solubilized Zonae Pellucidae

ZP were solubilized from control and experimentally treated mouse eggs, and two-cell embryos by incubation in m-KRB, pH 4 (adjusted with 1 N HCl) for approximately 10 min at 37°C. When the ZP were solubilized, the medium was withdrawn and neutralized with an appropriate amount of 1N NaOH (Endo *et al.* 1987b).

Iodination of Zona Pellucida Proteins

Structurally intact ZP, isolated from control and experimentally treated eggs as described above, were washed four times in 1 mM benzamidine and once in distilled water. The ZP were then iodinated using ^{125}I Bolton-Hunter reagent and processed according to the method of Endo *et al.* (1987a). Chloramine-T was also used to iodinate either pools of ZP from various treatments or the ZP from individual eggs, and was used in conjunction with the microinjection experiments. Briefly, after the benzamidine washing step individual structurally intact ZP were transferred to a 200 μl drop of 1 M NaCl containing 1% Triton X-100 and 1mM benzamidine for 5 min. The ZP were then washed three times in 0.1% Triton X-100 containing 10 mM sodium phosphate, pH 7.2 and transferred to a 0.5 ml polypropylene conical vial in a volume less than 5 μl. To this tube was added 25 μl 0.5 M sodium phosphate, pH 7.2 and 50 μCi Na^{125}I. The reaction was started by the addition of 5 μl chloramine T (1 mg/ml in 0.5 M sodium phosphate, pH 7.2), and after 1 min 50 μl sodium metabisulfite (2.4 mg/ml in 0.05 M sodium phosphate, pH 7.2) was added to stop the reaction. The contents of the reaction vial were then transferred to a 35 mm culture dish and the iodinated ZP (either single ZP or groups of ZP) were washed through six 200 μl drops of bicarbonate-free Earle's minimal essential medium containing 3 mg/ml PVP to remove unincorporated radioactivity. ZP were then counted and prepared for electrophoresis as described below.

Purification of ZP3

ZP3 was purified from control and experimentally treated eggs, or 2-cell embryos by the method of Bleil and Wassarman (1980c). ^{125}I-labeled ZP3 was used to monitor the position of ZP3 on polyacrylamide gels and to determine the overall recovery, which was between 50-60%. The tracer ZP were iodinated as previously described (Endo *et al.* 1987a).

One-Dimensional Gel Electrophoresis, Two-Dimensional Reduction Gel Electrophoresis and Autoradiography

Iodinated ZP were subjected to either one-dimensional gel electrophoresis or two-dimensional reduction electrophoresis as previously described (Endo *et al.* 1987a). Autoradiography was performed at -80°C using Kodak X-AR X-ray film with Cronex® intensifying screens.

[^{35}S]Methionine Radiolabeling of Egg Proteins and Two-Dimensional Gel Electrophoresis

Eggs and embryos were incubated in medium containing 1 mCi/ml of [^{35}S]methionine for 2h and then prepared for and subjected to two-dimensional gel electrophoresis as previously described (Bornslaeger *et al.* 1986). Proteins were detected by fluorography.

Sperm Binding Assays

Sperm binding to structurally-intact isolated ZP. Isolated ZP from control and experimentally treated eggs were incubated with capacitated sperm (2 x 10^5 cells/ml) for 15 min. After this incubation loosely associated sperm were removed from the ZP by careful washing with a wide bore micropipet (this time point is considered the 0 time of the experiment). The number of sperm bound to the ZP were then assessed over time.

Sperm binding to control and experimentally-treated ZP-intact eggs. Five to eight microinjected eggs and two 2-cell embryos were incubated for 5 or 30 min in m-KRB containing 2 x 10^6/ml of capacitated sperm. In some experiments sperm were capacitated in the presence of 100 ng/ml pertussis toxin to prevent the ZP-induced acrosome reaction; sperm treated in this manner still retain their ability to bind to the ZP (Endo *et al.* 1987c). The eggs were then removed, washed through three drops of m-KRB and then placed in individual droplets of medium; at this stage essentially no sperm bound to the 2-cell embryo. Individual eggs were placed under coverslips and the number of sperm bound to the eggs counted; this corresponded to time 0. After 15, 30, and 60 min relative to the time 0 point, the eggs were monitored for the number of sperm bound.

Competition sperm binding assay for the assessment of the sperm receptor activity of the ZP. A modification of the assay for sperm receptor activity (Bleil and Wassarman 1980c) was used. In this assay five to eight ovulated eggs and two two-cell embryos were incubated for 1h at 37°C in a 10 μl drop of m-KRB containing 2000 capacitated sperm and increasing concentrations of either solubilized ZP or purified ZP3 isolated from control eggs, experimentally treated eggs, or two-cell embryos. The two-cell embryos served as an internal negative control for nonspecific binding. The eggs and embryos were then removed with a mouth-operated micropipet and washed through three 100 μl drops of m-KRB. The eggs were then placed under a coverslip and the number of sperm heads counted by focusing through several optical planes using Nomarski optics.

Chlortetracycline Fluorescence Assay

The fluorescent probe chlortetracycline (CTC) was used to monitor the acrosomal status of mouse sperm bound to structurally intact ZP, sperm incubated with solubilized ZP, or sperm incubated with purified ZP3 (Saling and Storey 1979; Ward and Storey 1984; Lee and Storey 1985). The different patterns of CTC fluorescence are characterized by specific distributions of the dye on various regions of the cell. The pattern representing the capacitated state (B-pattern) is characterized by a bright epifluorescene over the anterior portion of the head and midpiece; the post-acrosomal region displays no fluorescence. The appearance of this fluorescent pattern is correlated with acrosome intact sperm, as assessed by transmission electron microscopy (Florman and Storey 1982). The S-pattern, which represents an intermediate stage that appears prior to the completion of the acrosome reaction, is characterized by bright fluorescence over the midpiece and punctate fluorescence over the anterior portion of the head; the postacrosomal region displays an irregular pattern of fluorescence. The appearance of the S-pattern correlates closely with the loss of ionic gradients, as determined with pH gradient sensitive probes (Lee and Storey 1985). Sperm that have completed the acrosome reaction (AR-pattern) are characterized by a lack of fluorescence over the anterior portion of the head with the midpiece still displaying fluorescence. The appearance of this pattern is correlated with acrosome reacted sperm, as assessed by transmission electron microscopy (Saling *et al.* 1979). Examples of these different CTC patterns are shown in Figure 1. In experiments using structurally intact ZP five to ten ZP with bound sperm were examined at each time point for each experimental treatment, and the percentage of the different CTC patterns calculated as a raw percentage. In experiments using solubilized ZP or purified ZP3 at least 100 sperm were scored for the percentage of the different CTC patterns at each time point.

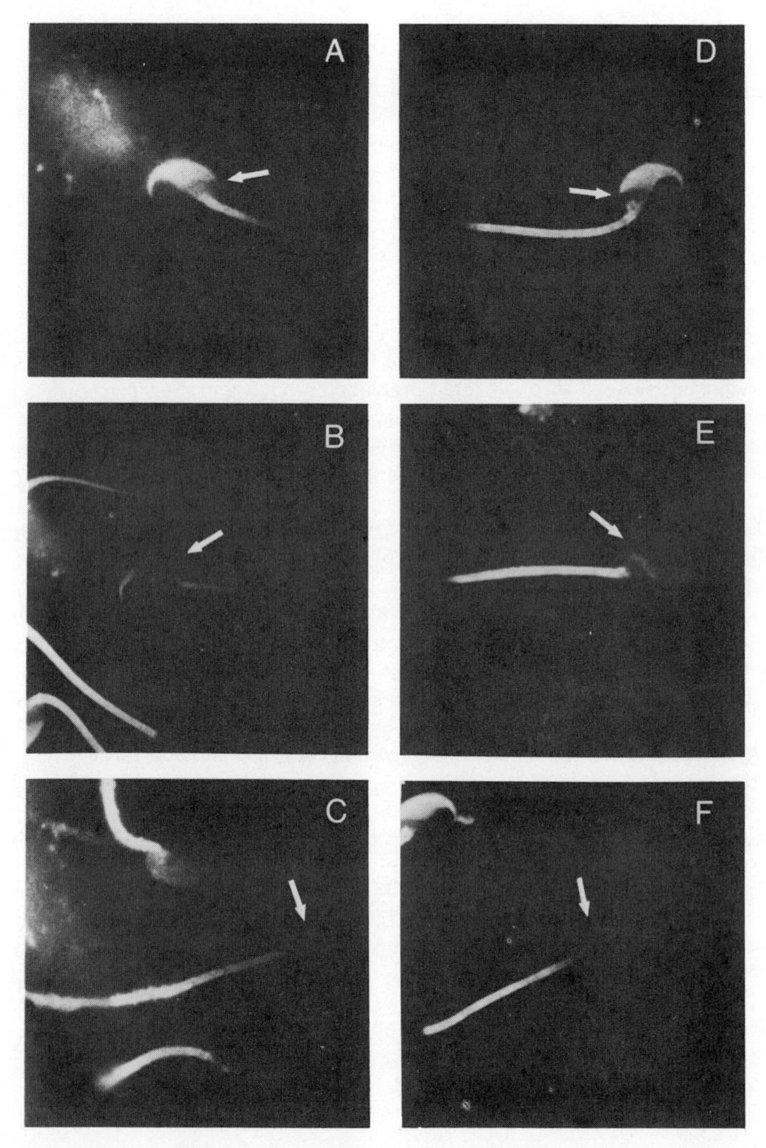

Figure 1. Epifluorescent photomicrographs of capacitated mouse sperm bound to mechanically isolated ZP after an initial incubation of cumulus-free mouse eggs in either the absence (A-C) or presence of 10 ng/ml TPA (D-F). (A,D) B-pattern. (B,E) S-pattern. (C,F) AR-pattern. Reprinted from Endo *et al.* (1987a) with permission of Academic Press, Inc.

RESULTS AND DISCUSSION

Phorbol Diesters and a Diacylglycerol Inhibit *In Vitro* Fertilization

When cumulus-free mouse eggs were pretreated with biologically active phorbol diesters such as 4β-phorbol 12,13 didecanoate (4β-PDD) prior to the addition of capacitated sperm, sperm penetration and fertilization were inhibited in a concentration dependent manner (Figure 2A). This inhibitory effect was not observed when eggs were initially treated with 4α-phorbol 12,13 didecanoate (4α-PDD), a steroisomer of 4β-PDD which does not activate PKC (Niedel *et al.* 1983) (Figure 2A). Similar results were obtained when these experiments were carried out using cumulus-enclosed eggs (data not shown). The inhibitory effects observed with 4β-PDD were observed with 12-O-tetradecanoyl phorbol 13-acetate (TPA), as well as the more natural activator of PKC, sn-1,2-dioctanoyl glycerol (diC_8). Therefore, although an effect of these compounds to activate PKC was not demonstrated, the stereospecificity and selectivity of these agents on sperm penetration and fertilization most likely reflects the ability of these compounds to stimulate PKC.

The Inhibitory Effects of Phorbol Diesters and a Diacylglycerol on Fertilization are Due to an Egg-Induced Modification of the Zona Pellucida

It is interesting to note that the percent fertilization correlated closely with the percent penetration of the eggs at all phorbol diester and diC_8 concentrations tested (Figure 2). Thus, if sperm penetration of the ZP occurred, subsequent events leading to fertilization (at least to the two pronuclei stage) appeared unaffected. Such an observation would argue against a toxic effect of these compounds on the egg (at least up to the pronuclear stage). This was confirmed by experiments demonstrating that sperm entry into ZP-free eggs, which had been treated with phorbol diesters, was not inhibited when compared to penetration of ZP-free eggs that had not been initially treated with these compounds (data not shown) (In a given population of capacitated mouse sperm about 10-20% of the sperm have undergone a spontaneous acrosome reaction; it is these acrosome reacted sperm that fertilize a ZP-free egg).

Although eggs treated with either the phorbol esters or diC_8 were washed prior to incubation with capacitated sperm, it was possible that residual levels of these PKC activators could exert their inhibitory effects on fertilization by adversely affecting sperm function (e.g., inhibiting sperm binding to the ZP or the ZP-induced acrosome reaction). In order to test this possibility, capacitated sperm were incubated with either TPA or 4β-PDD, and then chal-

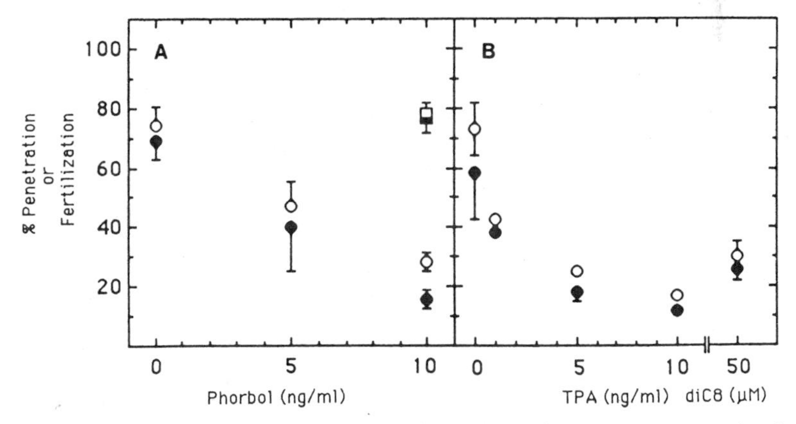

Figure 2. Effects of phorbol diesters and a diacylglycerol on sperm penetration (open symbols) and fertilization (closed symbols) of cumulus cell-free mouse eggs. **(A)** 4β-phorbol-12,13-didecanoate, circles; 4α-phorbol-12,13-didecanoate, squares. **(B)** TPA and diC_8. The data are expressed as mean +/- SEM, or the range. Reprinted from Endo *et al.* (1987a) with permission of Academic Press, Inc.

lenged with mechanically removed intact ZP from untreated eggs. As shown in Table 1, sperm binding to these ZP, as well as the sequential changes in sperm CTC fluorescence, indicative of capacitation (B-pattern), intermediate stages prior to the completion of the acrosome reaction (S-pattern) and the acrosome reaction (AR-pattern) were unaffected by the phorbol diester treatment (see Figure 1 for representative fluorescence patterns).

Likewise, the inhibitory effects of the phorbol diesters were not due to direct effects on the ZP. Intact mechanically removed ZP incubated for 1h in the presence of 10 ng/ml TPA, washed, and then inseminated with capacitated sperm bound sperm and induced the acrosome reaction to the same extent as untreated ZP (data not shown).

Taken together, these data suggest that, 1) the inhibitory effects of the phorbol diesters and diC_8 on fertilization was at the level of the ZP, and 2) the ZP-mediated inhibition of sperm penetration and fertilization required the presence of the egg during the phorbol diester or diC_8 treatment, implying an egg-induced modification of the ZP.

It is known that fertilization brings about an egg-induced modification of the ZP, such that ZP2 (M_r=120,000) (Bleil and Wassarman 1980a) is modified to $ZP2_f$ (M_r=90,000 under reducing conditions) (Bleil and Wassarman 1981) and both the sperm receptor and acrosome reaction inducing activities of ZP3 are abolished (Bleil and Wassarman 1980c; 1983). In order to determine whether the phorbol diesters and diC_8 are exerting their inhibitory effects by causing an egg-induced modification of the ZP, intact ZP from control and treated ZP-intact eggs were isolated, radioiodinated with Bolton Hunter reagent, and subjected to two-dimensional reduction gel electrophoresis. Figure 3 demonstrates that phorbol diester treatment of ZP-intact eggs results in a modification of ZP2 to $ZP2_f$, when compared to untreated eggs (compare Figure 3A with 3B). This ZP2 to $ZP2_f$ shift appears similar to that observed in ZP obtained from 2-cell embryos (compare Figure 3B with 3C). In experiments not shown, densitometric analysis of autoradiograms from one dimensional gels demonstrated that the ZP2 of eggs treated with either TPA or diC_8 underwent a shift in electrophoretic mobility (to $ZP2_f$) that represented a conversion of 84% and 98%, respectively, when compared to the ZP2 shift that occurred in 2-cell embryos. Therefore, at the gross electrophoretic level, the egg-induced ZP2 modification which occurred in eggs treated with phorbol diesters and diC_8 appeared similar to that observed in 2-cell embryos. Although there were no significant differences in

Table 1. Effect of Treatment of Sperm with PKC Activators on Sperm Binding and the Acrosome Reaction[a]

			% Fluorescent Pattern[b]		
Expt.	Time (h)	No. bound sperm[c]	B	S	AR
Control	1	34	92 (90/98)	8 (8/98)	0 (0/98)
	2	11	62 (34/55)	38 (21/55)	0 (0/55)
	3	8	0 (0/39)	10 (4/39)	90 (35/39)
TPA	1	35	95 (101/106)	5 (5/106)	0 (0/106)
	2	13	60 (39/65)	40 (26/65)	0 (0/65)
	3	8	0 (0/40)	10 (4/40)	90 (36/40)
Control	1	19	100 (125/125)	0 (0/125)	0 (0/125)
	2	3	67 (10/15)	33 (5/15)	0 (0/15)
	3	2	0 (0/10)	10 (1/10)	90 (9/10)
4β-PDD	1	13	100 (66/66)	0 (0/66)	0 (0/66)
	2	5	56 (15/27)	44 (12/27)	0 (0/27)
	3	2	0 (0/12)	8 (1/12)	92 (11/12)

[a] Reprinted from Endo *et al.* (1987a) with permission of Academic Press, Inc.
[b] The numbers in parenthesis represent the ratio of the number of sperm exhibiting the characteristic fluorescence pattern to the total number of sperm analyzed.
[c] The numbers represent the average number of sperm bound per zona in which five zonae were examined.

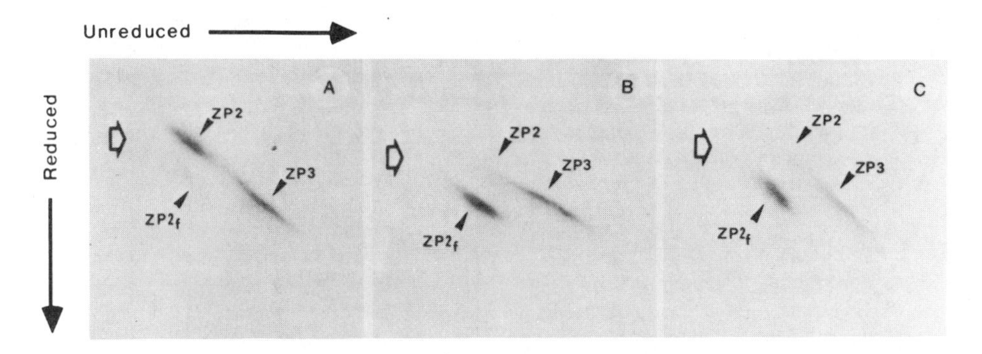

Figure 3. Autoradiograms of ZP isolated from **(A)** untreated eggs, **(B)** TPA-treated eggs, and **(C)** 2-cell embryos. Two dimensional "reduction electrophoresis" was performed on the iodinated ZP isolated from each of the three groups. The arrowheads indicate the positions of ZP2, $ZP2_f$, and ZP3. The open arrowhead indicates the position of ZP1 that was visible on the original autoradiograms but did not reproduce well. Reprinted from Endo *et al.* (1987a) with permission of Academic Press, Inc.

the mobility of ZP1 and ZP3 between control and treated eggs using either one-dimensional or two dimensional reduction gel electrophoretic techniques, minor differences might not be resolved in such systems. It was critical, therefore, that the sperm receptor and acrosome reaction-inducing activities of the ZP (activities associated with ZP3) from phorbol diester and diC_8 treated eggs be examined to determine whether such treatment affected these important biological activities of the ZP.

Phorbol Diester and diC_8 Treatment of ZP-Intact Eggs Does Not Affect the Sperm Binding Activity but Modifies the Acrosome Reaction-Inducing Activity of the Zona Pellucida

Although the above results indicated that PKC activators inhibit fertilization by stimulating an egg-induced modification of the ZP (as seen by a ZP2 to $ZP2_f$ shift), changes in the properties of ZP3 were of interest since this molecule mediates sperm binding and the acrosome reaction. In the previous experiments the phorbol diester- and diC_8-induced inhibition of fertilization, which is at the level of the ZP, could be due to either an inhibition of sperm binding to the ZP or an inhibition of the ZP-induced acrosome reaction. When ZP were isolated from control, phorbol diester- or diC_8-treated eggs and challenged with sperm, both initial sperm binding to ZP (not shown) and the number of sperm bound to ZP over time were similar (Table 2). There were no changes in sperm binding to ZP from eggs treated with concentrations of TPA as high as 200 ng/ml, a concentration previously shown to activate mouse eggs (Cuthbertson and Cobbold 1985). Phorbol diester or diC_8 treatment of the eggs, therefore, did not appear to affect the sperm binding activity of the ZP.

A possible explanation for the phorbol diester- and diC_8-induced inhibition of sperm penetration of the ZP and resultant fertilization is that these bound sperm fail to undergo the acrosome reaction. This possibility was next examined. Sperm bound to ZP from eggs treated with biologically active phorbol diesters or diC_8 displayed differences in the time course of appearance of the CTC fluorescent patterns. Although sperm bound to the ZP from phorbol diester- and diC_8-treated eggs initiate an acrosome reaction (*i.e.*, B to S transition) they fail to complete it, and accumulate in the S-pattern (Figure 4). Such sperm are not irreversibly stuck in this intermediate stage of the acrosome reaction, since further addition of either solubilized ZP from untreated eggs or A23187 to S-pattern sperm will cause them to undergo an S to AR transition (Kligman *et al.* 1988). [A direct comparison of the kinetics of the

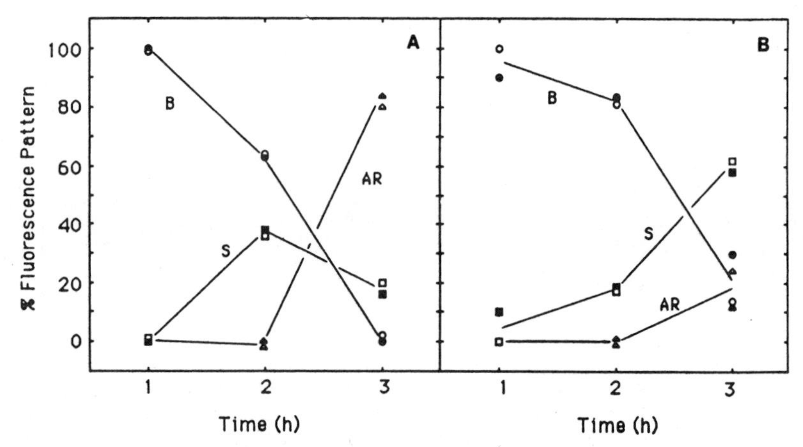

Figure 4. Effects of phorbol diester and diC_8 treatment of ZP-intact eggs on the time couse of the acrosome reaction of sperm bound to the ZP. **(A)** ZP from untreated (open symbols) or 4α-phorbol-12,13-didecanoate-treated (closed symbols) eggs. **(B)** ZP from 4β-phorbol-12,13- didecanoate-treated (open symbols) or diC_8-treated (closed symbols) eggs. Reprinted from Endo *et al.* (1987a) with permission of Academic Press, Inc.

acrosome reaction between control and experimental groups was possible in these experiments since, 1) the patterns of CTC fluorescence of sperm incubated with ZP from either control or treated eggs were similar (Figure 1; compare panels A-C with D-F) and 2) sperm treated with phorbol diesters display the same series of changes in CTC fluorescence as do untreated sperm (Lee *et al.* 1987)].

These data demonstrate that treatment of eggs with phorbol diesters or diC_8 results in an apparent dissociation, in intact ZP, of the sperm receptor activity from the acrosome reaction-inducing activity. This is in contrast to ZP modifications that normally occur after fertilization, in which both biological activities of the ZP (and of ZP3) have been shown to

Table 2. Effect of Treatment of Eggs with PKC Activators on Sperm Binding[a]

		# Sperm Bound/Zona[b]					
Experiment	Time (h)	Control	4a-PDD	4b-PDD	TPA	diC_8	2PN
I	1	19	11	13			
	2	3	7	5			
	3	2	2	2			
II	1	12	20	22			
	2	2	10	6			
	3	4	11	10			
III	1	13			14	15	
	2	12			11	13	
	3	5			4	4	
IV[c]	1	10			8		1
	2	8			10		2
	3	4			3		1

[a] Reprinted from Endo *et al.* (1987a) with permission of Academic Press, Inc.
[b] Data are expressed as the average number of sperm bound per zona, where five zonae were analyzed.
[c] Fertilized eggs with two pronuclei (2-PN) were challenged with capacitated sperm and binding to zonae determined.

be lost. Although this experimental dissociation clearly has interesting implications with regard to the biochemical mechanism(s) by which the egg modifies its ZP, it was possible that the structural integrity of the ZP under these experimental conditions, rather than a unique modification of the ZP3 molecule, was the underlying reason for these observations. It was necessary, therefore, to examine the sperm binding and acrosome reaction-inducing properties of ZP3 purified from untreated and phorbol-diester treated eggs to answer this question.

Zona competition sperm binding assays confirmed, as suggested by previous binding assays (Table 2), that ZP3 from phorbol diester-treated eggs possessed similar amounts of sperm receptor activity as compared to that of ZP3 isolated from untreated eggs (Figure 5). Specificity of competition by ZP3 for sperm binding to ZP-intact eggs in these assays was confirmed by demonstrating that ZP3 from 2-cell embryos, which possesses no sperm receptor activity (Bleil and Wassarman 1980c), did not compete for sperm binding to ZP-intact eggs (Figure 5). When the acrosome reaction was examined, ZP3 from untreated eggs was shown to induce a complete acrosome reaction (B to S to AR transitions), as monitored by the CTC assay (Figure 6). ZP3 isolated from phorbol diester-treated eggs, in contrast, did not induce a complete acrosome reaction as a consequence of the inability of the sperm to undergo an S to AR transition (Figure 6); this is consistent with observations using intact ZP (Figure 4). An additional observation was also evident from these experiments. Although previous studies indicated that ZP3 isolated from 2-cell embryos did not induce the acrosome reaction, the methods employed assayed an end point (*i.e.*, the completion of the acrosome reaction), and would not detect intermediates in this process (Bleil and Wassarman 1983). When ZP3 from 2-cell embryos were tested in the CTC assay, the sperm did not complete the acrosome reaction as a consequence of their inability to undergo a B to S transition (Figure 6). Therefore, the inability of ZP isolated from phorbol diester treated eggs to induce a complete acrosome reaction represents an egg-induced modification of the ZP3 that may be different from that which occurs after fertilization.

In summary, treatment of ZP-intact mouse eggs with activators of PKC results in an egg-induced modification of the ZP such that there is a dissociation of the sperm receptor activity from the acrosome reaction-inducing activity of ZP3. In contrast, fertilization results in the loss of both of these biological activities of ZP3. It is presumed that differences in ZP3 from untreated, phorbol diester-treated and fertilized eggs reflect those portions of the mole-

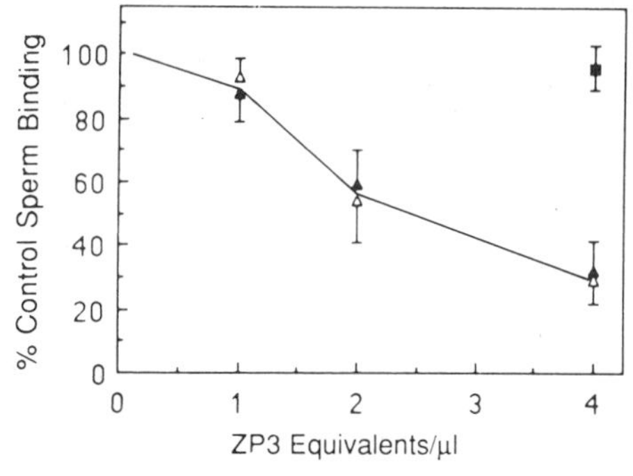

Figure 5. Concentration-dependent inhibition of sperm binding to ZP-intact eggs by ZP3 isolated from untreated eggs (▲), TPA-treated eggs (Δ), and two-cell embryos (■). Sperm competition assays were performed in which increasing amounts of ZP3 from the aforementioned eggs was added and its subsequent effect on the binding of sperm to ZP-intact eggs measured. Data are expressed as percentage of binding of sperm to ZP-intact eggs in the absence of added ZP3 and represent the means +/- SEM, where n > 3. Reprinted from Endo *et al.* (1987b) with permission of Academic Press, Inc.

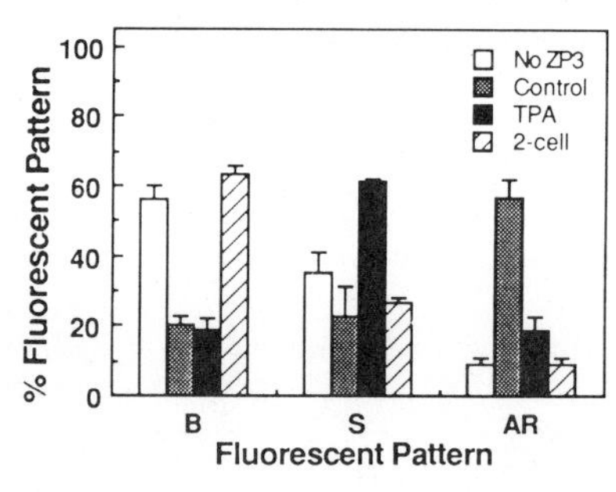

Figure 6. Distribution of chlortetracycline (CTC) fluorescent patterns of sperm incubated with no ZP3, or ZP3 isolated from untreated eggs, TPA-treated eggs, or 2-cell embryos. It should be noted that a fraction of sperm undergo a spontaneous B to S transition (Lee and Storey 1985). Moreover, in a given population of capacitated sperm, there exists a subpopulation of about 10-20% of cells that have undergone an acrosome reaction. The data are expressed as means +/- SEM, where n=3. The differences in the distribution of CTC patterns between sperm incubated in the absence of ZP3 and ZP3 isolated from 2-cell embryos are not significant ($P > 0.14$, t-test). Differences in the distribution of both the S and AR patterns of sperm incubated in the presence of ZP3 isolated from either untreated or TPA-treated eggs are significant ($P < 0.003$, t-test). Reprinted from Endo *et al.* (1987b) with permission of Academic Press, Inc.

cule that participate in the acrosome reaction. Studies examining such biochemical differences, therefore, should facilitate the analysis of those portions of ZP3 that are involved in inducing the acrosome reaction. The use of ZP from phorbol diester-treated eggs should also aid in studies of the mechanism(s) of the acrosome reaction, since it is now possible to use these ZP to study the B to S transition independently from the S to AR transition. Determination of the biochemical correlates of the S-pattern, which has characteristics of an intermediate stage prior to the completion of the acrosome reaction, is one such area of study that would benefit from such an experimental system.

Models to Explain the Mechanism(s) by which Phorbol Diesters and diC$_8$ Cause an Egg-Induced Dissociation of the Sperm Receptor Activity from the Acrosome Reaction-Inducing Activity of ZP3

A number of models could account for this phorbol diester-induced dissociation of the two biological properties of ZP3. As shown in Figure 7, one model (case I) would predict that a single component of ZP3 is responsible for both the B to S and S to AR transitions of the acrosome reaction. Phorbol diester treatment of the eggs results in a modification of this component such that the S to AR transition is inhibited. In case II (Figure 7), one component of ZP3 is responsible for the B to S transition and another component is responsible for the S to AR transition; phorbol diester treatment results in a modification of the latter component. At present neither of these possibilities can be distinguished.

Some interesting possibilities arise from the observation that phorbol diester or diC$_8$ treatment of eggs elicits ZP modifications that represent a "partial" or "incomplete" ZP response when compared to the response observed at fertilization. As described earlier, a cortical granule reaction may be involved in the egg-induced modification of the ZP, presumably through the release of granule-associated enzymes that specifically modify the ZP glycoproteins. A possible explanation for the "partial" ZP response of eggs treated with the

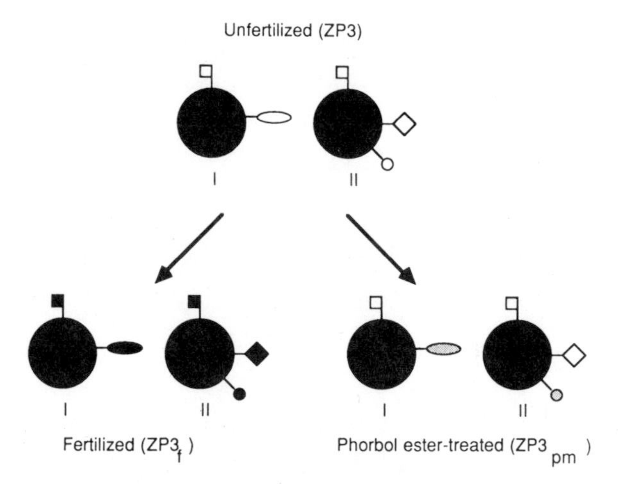

Figure 7. Schematic representation of changes in ZP3 following fertilization or treatment of ZP-intact eggs with phorbol diesters. The figure depicts two possibilities that currently cannot be distinguished. Case I represents a model in which one component (elipse) of ZP3 is responsible for both the B to S and S to AR transitions of the acrosome reaction. Case II represents a model in which one component of ZP3 is responsible for the B to S transition (diamond) and another component is responsible for the S to AR transition (circle). In both cases, the square denotes the sperm receptor component of ZP3. For case I, fertilization results in a modification of both the sperm receptor and acrosome reaction-inducing components, as represented by the solid square and elipse; these modifications result in a loss of both biological activities of ZP3. Phorbol diester treatment of eggs results in retention of sperm receptor activity but also a partial modification of the component involved in inducing the acrosome reaction, as represented by the stippled elipse; this modification results in retention of sperm receptor activity but also a loss in the ability to induce a complete acrosome reaction. For case II, fertilization results in a modfication of both the sperm receptor and the two components involved in the acrosome reaction, although it is possible that only the component involved in the B to S transition is modified. Phorbol diester treatment does not cause a modification of the sperm receptor component, but that component involved in the S to AR transition is modified, as represented by the stippled circle. $ZP3_f$ represents ZP3 isolated from fertilized eggs. $ZP3_{pm}$ represents ZP3 isolated from phorbol diester treated eggs that displays a partial modification (pm) of the acrosome reaction inducing activity. Reprinted from Endo *et al.* (1987b) with permission of Academic Press, Inc.

phorbol diesters or diC_8 may be a consequence of heterogeneity in the cortical granule population. Electron microscopy reveals the existence of both light and dark staining populations of cortical granules (Nicosia *et al.* 1977). These populations may reflect granules at different stages of maturity but whose contents are the same, or may reflect heterogeneity of mature granules that have different protein compositions. In sea urchins, for example, although the cortical granules appear homogeneous as assessed by electron microscopy, only about 20% of the granules contain a specific cortical granule antigen, as determined by immunocytochemistry (Anstrom *et al.* 1988). Such observations provide biochemical evidence that there is heterogeneity in cortical granule populations. Perhaps phorbol diester or diC_8 treatment of the mouse eggs result in the selective release of the contents of a subpopulation of cortical granules that are capable of only performing certain aspects of the ZP modifications, which *in toto* constitute a complete ZP reaction. Consistent with this hypothesis is the observation that eggs treated with biologically active phorbol diesters, but not inactive phorbol diesters, display only a partial reduction in the number of *Lens culinaris* agglutinin-staining granules (Cherr *et al.* 1988), presumed to be cortical granules, when compared to A2187-

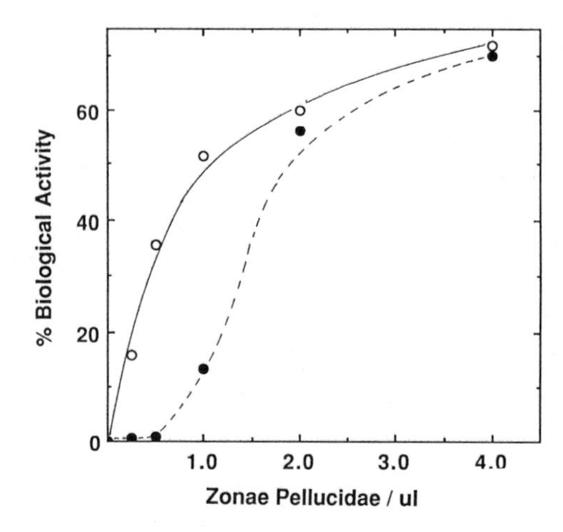

Figure 8. Schematic representation of the relationship between ZP3 concentration and the expression of sperm binding activity (open circles) and the induction of the acrosome reaction (closed circles). The hyperbolic "kinetics" for sperm binding activity are consistent with single interactions mediating binding, whereas the sigmoidal "kinetics" of the acrosome reaction are indicative of a cooperative interaction mediated by several ZP3 components. Adapted from Bleil and Wassarman (1983) and reprinted with permission of Academic Press, Inc.

activated or fertilized eggs (Ducibella, Kopf and Schultz, unpublished results). It is clear that additional studies will be required to verify such a hypothesis.

An alternative model may be proposed to explain the dissociation of sperm receptor and acrosome reaction inducing activities without postulating a heterogeneous population of cortical granules possessing specific ZP modifying functions. This model is based on the fact that multiple interactions between ZP3 molecules and sperm surface sites may be required *in toto* for binding and the induction of the acrosome reaction. The interaction of acrosome intact sperm with the sperm receptor activity component of the ZP3 molecule is described by a hyperbolic function, whereas the interaction of acrosome intact sperm with the acrosome reaction inducing component(s) of the ZP3 molecule is described by a sigmoidal function (Bleil and Wassarman 1983) (Figure 8). Thus, the hyperbolic "kinetics" of sperm receptor activity are consistent with single interactions mediating binding, whereas the sigmoidal "kinetics" of the acrosome reaction are indicative of a cooperative interaction mediated by several ZP3 components. Normally, after sperm-egg fusion has occurred, the majority of cortical granules undergo exoctyosis and presumably release enzymes that specifically abolish the ZP3 interaction sites required for both sperm binding and the induction of the acrosome reaction; these two biological properties of the ZP3 molecule are, therefore, lost (Figure 9). Phorbol diester or diC_8 treatment of the eggs may result in the exocytosis of only a limited number of these homogeneous cortical granules; the reduced amount of the released ZP3 modifying enzymes abolishes only a portion of the ZP3 interaction sites (Figure 9). The limited destruction of these sites still allows normal binding to occur and allows the sperm to initiate the acrosome reaction (B to S transition), but abolishes the ability of the sperm to complete the acrosome reaction (S to AR transition). This model, although consistent with our data, will require further testing to determine its validity.

Microinjected Inositol 1,4,5,-Trisphosphate Causes an Egg-Induced ZP2 to $ZP2_f$ Conversion of the Zona Pellucida

As reviewed earlier, polyphosphatidylinositide turnover, with the subsequent generation of two intracellular second messengers (DAG and $InsP_3$), may represent an early fertilization

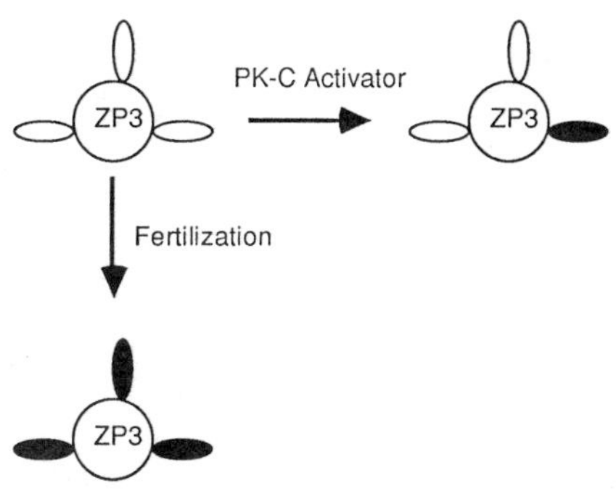

Figure 9. Schematic representation of a model depicting a mechanism by which sperm binding and the induction of the acrosome reaction are mediated by multiple ZP3 interaction sites (open elipses). Successful sperm binding requires the interaction of sperm with one of these sites, which gives rise to the hyperbolic "kinetics" depicted in Figure 8. Once this interaction has occurred additional interactions with the other ZP3 sites are required for the complete acrosome reaction; this accounts for the sigmoidal "kinetics" depicted in Figure 8. Fertilization results in an egg-induced modification of all of these interaction sites (denoted by the dark elipses), which results in a ZP3 molecule that no longer has sperm binding or acrosome reaction-inducing activity. On the other hand, phorbol diester-treatment of eggs results in a limited modifi-cation of these sites. The consequence of such a limited modification is that ZP3 still possesses sperm receptor activity. ZP3, however, does not possess sufficient numbers of interaction sites to allow the cooperative interactions necessary to induce a complete acrosome reaction. Rather, ZP3 possesses a sub-optimal number of sites, which permits the sperm to initiate an acrosome reaction (*i.e.*, B to S transition), but not to complete it.

response of the egg in both invertebrate and vertebrate species, and may constitute at least one mechanism by which some of the early events of egg activation are mediated. Our previous experiments using activators of PKC would, in principle, only activate one-half of such a bifurcating second messenger cascade system, and the "partial" ZP responses obtained under these experimental conditions is consistent with this idea. Therefore, we examined the effects of microinjecting $InsP_3$ and its structural analogs into mouse eggs on egg activation, electrophoretic modifications of the ZP, and the sperm receptor and acrosome reaction inducing activities of the ZP (Kurasawa *et al.* 1989).

When ZP-intact mouse eggs were microinjected with 4 μM $InsP_3$ (final concentration), an unphysiologically high concentration of this second messenger, a low percentage (15%) of the eggs activated, as evidenced by second polar body emission within 1.5h. Vehicle-injected (H_20) eggs, in contrast, did not activate. The ZP from $InsP_3$ injected eggs that activated always displayed an electrophoretic mobility shift of ZP2 to $ZP2_f$, characteristic of an activated egg (Figure 10). When the ZP from $InsP_3$ microinjected eggs that did not show evidence of activation were analyzed, approximately 85% of the ZP displayed a ZP2 to $ZP2_f$ shift (Figure 10); the loss of ZP2 and appearance of $ZP2_f$ were confirmed by two-dimensional reduction electrophoresis and yielded a pattern similar to that seen in Figure 3B,C. The conversion of ZP2 to $ZP2_f$ was total in approximately 70% and partial (50% conversion) in 30% of the ZP from $InsP_3$ injected eggs. When the ZP from unactivated vehicle injected eggs were examined only about 30% of the ZP displayed a ZP2 to $ZP2_f$ conversion; in most cases the conversion was partial.

Since high concentrations of microinjected $InsP_3$ resulted in an egg-induced ZP2 to $ZP2_f$ conversion, we examined the concentration-response relationship between $InsP_3$ and the ZP2 to $ZP2_f$ shift. A concentration-dependent increase in the percentage of ZP-intact eggs displaying an egg-induced conversion of ZP2 to $ZP2_f$ was observed over an $InsP_3$ concent-

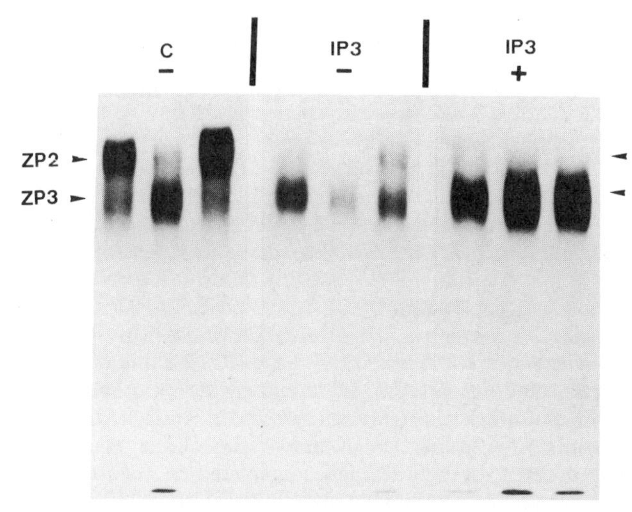

Figure 10. Autoradiograms of individual ZP isolated from either vehicle-injected or $InsP_3$-injected eggs. (C) Vehicle-injected eggs. ($InsP_3$-) $InsP_3$ injected eggs that did not activate. ($InsP_3$ +) $InsP_3$-injected eggs that did activate. Note the consistent loss of ZP2 in the $InsP_3$ injected eggs that did and did not activate.

ration range (final) of 0.1 nM to 4 μM. $InsP_3$ concentrations greater than 1 nM elicited the ZP2 to $ZP2_f$ conversion with an EC_{50}=5 nM. This effective concentration of $InsP_3$ compares favorably to that necessary to induce the cortical granule-mediated elevation of the fertilization envelope in sea urchin eggs, as well as Ca^{2+} release from intracellular stores in both sea urchin (Whitaker and Irvine 1984; Swann and Whitaker 1986) and hamster (Miyazaki 1988) eggs. These $InsP_3$ effects also occurred in eggs microinjected in medium depleted of extracellular Ca^{2+}, which is consistent with the known effects of $InsP_3$ in mediating Ca^{2+} release from intracellular stores.

Specificity of the Inositol 1,4,5,-Trisphosphate Effect to Induce a ZP2 to $ZP2_f$ Conversion of the Zona Pellucida

A variety of structural analogs of $InsP_3$ were also tested to determine the specificity of the $InsP_3$ effect to bring about an egg-induced ZP2 to $ZP2_f$ conversion. Microinjection of 100 nM to 1 μM inositol 1,3,4,5-tetrakisphosphate [I(1,3,4,5)P_4] (final concentration), which has been implicated in activating plasma membrane associated Ca^{2+} channels in sea urchin eggs (Irvine and Moor 1986, 1987), did not result in an egg-induced ZP2 to $ZP2_f$ conversion. Similarly, microinjection of 1 μM I(1,4)P2, I(2,4,5)P3 or I(1,3,4)P3 (final concentration), which have been shown in other systems not to release intracellular Ca^{2+}, did not effect a ZP2 to $ZP2_f$ conversion. It appears, therefore, that the ZP2 to $ZP2_f$ conversion brought about by the injection of inositol phosphates is specific for $InsP_3$, an agent known to release Ca^{2+} from intracellular stores. Whether $InsP_3$ is, in fact, working through such a mechanism in mouse eggs is not known at this time, although it has been demonstrated that $InsP_3$ microinjection of hamster eggs results in the elevation of free intracellular Ca^{2+} (Miyazaki 1988).

Effects of Inositol 1,4,5,-Trisphosphate Microinjection on Egg-Induced Changes in the Sperm Binding and Acrosome Reaction-Inducing Properties of the Zona Pellucida

Studies with $InsP_3$ to date have focused on the effects of this second messenger on an egg-induced ZP2 to $ZP2_f$ conversion. $InsP_3$ microinjection of mouse eggs may also modify the sperm receptor and acrosome reaction-inducing properties of the ZP (those activities associated with ZP3). In order to address this question we examined ZP isolated from vehicle and $InsP_3$-injected eggs on sperm binding and the acrosome reaction. In initial experiments we examined the number of sperm bound at 1, 2, and 3h time points. As shown in Table 3 sperm binding, although lower in the ZP from $InsP_3$ injected eggs versus vehicle injected eggs at the 1h time point, was similar in both groups at the 2 and 3h time points. Those sperm that were bound to both the ZP from vehicle and $InsP_3$-injected eggs underwent the acrosome reaction with similar kinetics. The lower extent of binding could represent the inability of acrosome reacted bound sperm to establish a secondary binding with ZP2, for the following reason: If ZP2 mediates the binding of acrosome reacted sperm and $ZP2_f$ cannot interact with acrosome reacted sperm, then although the sperm can bind to $InsP_3$-injected eggs and undergo the acrosome reaction, they cannot establish the secondary binding with ZP2, which has been modified to $ZP2_f$. Thus, the bound acrosome reacted sperm would dissociate from the ZP. The design of this experiment would, therefore, not be amenable to testing whether $InsP_3$ microinjection affects the sperm receptor and acrosome reaction-inducing activities of the ZP. Lower sperm binding could still occur in the presence of a full complement of sperm receptor activity of the ZP.

To resolve this question, sperm should be allowed to establish primary binding to the ZP but be prevented from undergoing an acrosome reaction and establishing secondary binding. Furthermore, these experiments should be done with individual eggs so that once the sperm binding assay was performed on that given egg, the ZP from that egg could be analyzed electrophoretically to determine whether there was an associated ZP2 to $ZP2_f$ conversion. Such a system would enable us to determine whether, 1) the sperm receptor activity of the ZP is, indeed, affected by $InsP_3$ microinjection and 2) whether there is a correlation between a modification of sperm receptor activity and a ZP2 to $ZP2_f$ conversion.

Pertussis toxin treatment of sperm was used to examine primary binding in the absence of secondary binding. Pertussis toxin treatment of mouse sperm does not affect their ability to become capacitated or bind to the ZP (Endo *et al.* 1987c). Pertussis toxin does inhibit, however, the acrosome reaction by preventing the ZP3-induced B to S transition (Endo *et al.* 1988). Sperm, capacitated in the presence of pertussis toxin, were challenged with ZP from vehicle and $InsP_3$-injected eggs, and sperm binding at 0, 15, 30, and 60 min was monitored. It was demonstrated that microinjection of mouse eggs with $InsP_3$ results in lower sperm binding to the ZP than the respective vehicle injected eggs at all of the time points examined (not shown). This reduction is not due to the inability of acrosome reacted sperm to establish secondary binding to $ZP2_f$, since the pertussis toxin pretreatment of the sperm inhibits the ZP-induced acrosome reaction. $InsP_3$ injected eggs, therefore, have a reduced ability to bind sperm which may indicate that the sperm receptor component of ZP3 is modified by the $InsP_3$ microinjection.

These studies also establish a close correlation between the egg-induced ZP2 to $ZP2_f$ conversion and a decreased ability of the ZP to bind acrosome-intact sperm. We have now analyzed the data from 60 vehicle and $InsP_3$-injected eggs, and in 83% of the cases there is a

Table 3. Effect of Microinjection of Inositol 1,4,5-Trisphosphate into Mouse Eggs on Sperm Binding and the Acrosome Reaction[a]

			% Fluorescent Pattern[b]		
Experiment	Time (h)	# Sperm Bound[c]	B	S	AR
Control	1	12.1 +/- 3.1	91	9	0
	2	3.9 +/- 2.5	61	32	7
	3	5.7 +/- 1.1	2	35	63
$InsP_3$	1	5.5 +/- 4.0	84	12	0
	2	4.4 +/- 2.1	42	44	14
	3	4.1 +/- 3.6	13	45	43

[a] Data represent either the mean +/- S.D. or mean of five individual experiments.
[b] Patterns analyzed by the CTC assay.
[c] Data are expressed as the average number of sperm bound per zona, where five zonae were analyzed.

positive correlation between a ZP2 to $ZP2_f$ conversion and reduced binding of acrosome intact sperm, which we infer to represent a modification of ZP3. These data suggest that $InsP_3$ microinjection does not result in a complete dissociation between the ZP2 to $ZP2_f$ conversion and changes in the biological activities of ZP3 (e.g., sperm receptor or acrosome reaction inducing activities). It should be noted, however, that in about 20% of the cases this correlation was not observed. Although this may represent a dissociation in egg-induced changes of ZP2 and ZP3, it is premature to conclude that this is actually the case.

Effects of Inositol 1,4,5,-Trisphosphate Microinjection on the Patterns of Protein Synthesis in Mouse Eggs

Fertilization is associated with a characteristic set of changes in the pattern of protein synthesis. Although $InsP_3$-injected eggs display changes in the ZP characteristic of fertilization (*i.e.*, a ZP2 to $ZP2_f$ conversion and an apparent modification of the sperm receptor activity of ZP3), they do not reveal the changes in the patterns of protein synthesis associated with fertilization. This is consistent with the low level of egg activation, as assessed by pronuclear formation. Thus, although $InsP_3$ can bring about modifications of the ZP known to occur at fertilization, it does not elicit a full egg activation response, but rather a subprogram of events that occurs during egg activation.

CONCLUSIONS AND FUTURE DIRECTIONS

Events in mammalian fertilization comprising sperm-egg recognition, sperm-egg fusion, early events of egg activation, and the ZP block to polyspermy are still not well defined at the present time. Our laboratories are focusing on the role of polyphosphoinositide turnover in regulating the early events of egg activation and the ZP block to polyspermy. Although it is likely that this process is mediated by a G-protein(s), the biochemical nature of the receptor for the sperm on the egg surface, the molecular identity of the G-proteins involved, and the coupling mechanisms of the putative egg-associated sperm receptor with this G-protein(s), as well as the consequences of this coupling, are still unknown. Biochemical and molecular approaches should enable us to answer questions such as whether sperm-egg fusion results in the generation of DAG and $InsP_3$ through a G-protein modulated phospholipase C. The ultimate targets of DAG and $InsP_3$ action within the egg are also of great interest. Does $InsP_3$ serve to modulate intracellular Ca^{2+} in a manner similar to that seen in hamster eggs? Does PKC play a role in cortical granule exocytosis and, if so, what is the role of protein phosphorylation in this important exocytotic event?

The role of cortical granules in mediating early events in mammalian egg activation has yet to be resolved. The biochemical basis for fertilization-induced cortical granule exocytosis is only starting to be determined in both invertebrate and mammalian species. Virtually nothing is known about the contents of mammalian egg cortical granules and the role that specific granule components play in modifying the ZP. The answers to such questions should provide insight into the molecular mechanisms by which ZP modifications alter the receptivity of the egg for sperm. The question of heterogeneity in the cortical granule population is still to be addressed; the answers to such questions might provide insight into the apparent dissociations of ZP functions that we have observed experimentally. Work directed at isolating and characterizing mammalian cortical granules will help to resolve many of the above mentioned issues.

Finally, work directed towards defining the biochemical nature of the determinants in both ZP3 and ZP2, as well as the corresponding sperm-associated ZP3 and ZP2 binding site(s), involved in primary sperm binding, the induction of the acrosome reaction, and secondary sperm binding should aid in the development of a unifying hypothesis regarding sperm-egg recognition, sperm activation, and sperm-egg fusion.

ACKNOWLEDGMENTS

Research performed by the authors was supported by grants from the National Institutes of Health (HD 19096 to G.S.K., HD 18604 to R.M.S., and HD 22732 to G.S.K. and R.M.S.), Mellon Foundation (G.S.K.), University Research Fund (R.M.S.), and Rockefeller Foundation (Y.E. and S.K.). G.S.K. and R.M.S. would like to thank Philip Hugo for assistance with some of the experiments described above, Jeff Bleil for discussions regarding the role of ZP2 in sperm binding, and Paul Wassarman for discussions regarding modes of sperm-ZP3 interactions.

REFERENCES

Anstrom, J. A., J. E. Chin, D. S. Leaf, A. L. Parks, and R. A. Raff. 1988. Immunocytochemical evidence suggesting heterogeneity in the population of sea urchin egg cortical granules. *Dev. Biol.* 125:1-7.

Barros, C. and R. Yanagimachi. 1971. Induction of the zona reaction in golden hamster eggs by cortical granule material. *Nature (Lond.)* 233:268-269.

Barros, C. and R. Yanagimachi. 1972. Polyspermy-preventing mechanisms in the golden hamster egg. *J. Exp. Zool.* 180:251-266.

Bleil, J. D. and P. M. Wassarman. 1980a. Structure and function of the zona pellucida: Identification and characterization of the proteins of the mouse oocyte's zona pellucida. *Dev. Biol.* 76:185-202.

Bleil, J. D. and P. M. Wassarman. 1980b. Synthesis of zona pellucida proteins by denuded and follicle-enclosed mouse oocytes during culture *in vitro*. *Proc. Natl. Acad. Sci. USA* 77: 1029-1033.

Bleil, J. D. and P. M. Wassarman. 1980c. Mammalian sperm-egg interaction: Identification of a glycoprotein in mouse egg zonae pellucidae possessing receptor activity for sperm. *Cell* 20:873-882.

Bleil, J. D. and P. M. Wassarman. 1981. Mammalian sperm-egg interaction: Fertilization of mouse eggs triggers modification of the major zona pellucida glycoprotein, ZP2. *Dev. Biol.* 86:189-197.

Bleil, J. D. and P. M. Wassarman. 1983. Sperm-egg interactions in the mouse: Sequence of events and induction of the acrosome reaction by a zona pellucida glycoprotein. *Dev. Biol.* 95:317-3

Bleil, J. D. and P. M. Wassarman. 1986. Autoradiographic visualization of the mouse egg's sperm receptor bound to sperm. *J. Cell Biol.* 102:1363-1371.

Bleil, J. D., J. M. Greve, and P. M. Wassarman. 1988. Identification of a secondary sperm receptor in the mouse egg zona pellucida: Role in maintenance of binding of acrosome-reacted sperm to eggs. *Dev. Biol.* 128:376-385.

Bornslaeger, E. A., P. Mattei, and R. M. Schultz. 1986. Involvement of cAMP-dependent protein kinase A and protein phosphorylation in regulation of mouse oocyte maturation. *Dev. Biol.* 114:453-462.

Busa, W. B., J. E. Ferguson, S. K. Joseph, J. R. Williamson, and R. Nuccitelli. 1985. Activation of frog (*Xenopus laevis*) eggs by inositol trisphosphate. I. Characterization of Ca^{2+} release from intracellular stores. *J. Cell Biol.* 101:677-682.

Cherr, G. N., E. Z. Drobnis, and D. R. Katz. 1988. Localization of cortical granule constituents before and after exocytosis in the hamster egg. *J. Exp. Zool.* 246:81-93.

Ciapa, B. and M. Whitaker. 1986. Two phases of inositol polyphosphate and diacylglycerol production at fertilization. *FEBS. Lett.* 195:347-351.

Cuthbertson, K. S. R., D. G. Whittingham, and P. H. Cobbold. 1981. Free Ca^{2+} increases in exponential phases during mouse oocyte activation. *Nature (Lond.)* 294:754-757.

Cuthbertson, K. S. R. and P. H. Cobbold. 1985. Phorbol ester and sperm activate mouse oocytes by inducing sustained oscillations in cell Ca^{2+}. *Nature (Lond.)* 316:541-542.

Endo, Y., R. M. Schultz, and G. S. Kopf. 1987a. Effects of phorbol esters and a diacylglycerol on mouse eggs: Inhibition of fertilization and modification of the zona pellucida. *Dev. Biol.* 119:199-209.

Endo, Y., P. Mattei, G. S. Kopf, and R. M. Schultz. 1987b. Effects of a phorbol ester on mouse eggs: Dissociation of sperm receptor activity from acrosome reaction-inducing activity of the mouse zona pellucida protein, ZP3. *Dev. Biol.* 123:574-577.

Endo, Y., M. A. Lee, and G. S. Kopf. 1987c. Evidence for the role of a guanine nucleotide-binding regulatory protein in the zona pellucida-induced mouse sperm acrosome reaction. *Dev. Biol.* 119:210-216.

Endo, Y., M. A. Lee, and G. S. Kopf. 1988. Characterization of an islet activating protein--sensitive site in mouse sperm that is involved in the zona pellucida-induced acrosome reaction. *Dev. Biol.* 129:12-24.

Florman, H. M. and B. T. Storey. 1982. Mouse gamete interactions: The zona pellucida is the site of the acrosome reaction leading to fertilization *in vitro*. *Dev. Biol.* 91:121-130.

Florman, H. M. and P. M. Wassarman. 1985. O-linked oligosaccharides of mouse egg ZP3 account for its sperm receptor activity. *Cell* 41:313-324.

Florman, H. M., K. B. Bechtol, and P. M. Wassarman. 1984. Enzymatic dissection of the functions of the mouse egg's receptor for sperm. *Dev. Biol.* 106:243-255.

Gilman, A. G. 1987. G proteins: Transducers of receptor-generated signals. *Annu. Rev. Biochem.* 56:615-649.

Greve, J. M. and P. M. Wassarman. 1985. Mouse egg extracellular coat is a matrix of interconnected filaments possessing a structural repeat. *J. Mol. Biol.* 181:253-264.

Gwatkin, R. B. L., D. T. Williams, J. F. Hartmann, and M. Kniazuk. 1973. The zona reaction of hamster and mouse eggs: Production *in vitro* by a trypsin-like protease from cortical granules. *J. Reprod. Fertil.* 32:259-265.

Irvine, R. F. and R. M. Moor. 1986. Micro-injection of inositol 1,3,4,5-tetrakisphosphate activates sea urchin eggs by a mechanism dependent on external Ca^{2+}. *Biochem. J.* 240: 917-920.

Irvine, R. F. and R. M. Moor. 1987. Inositol (1,3,4,5) tetrakisphosphate-induced activation of sea urchin eggs requires the presence of inositol trisphosphate. *Biochem. Biophys. Res. Commun.* 146:284-290.

Igusa, Y. and S. Miyazaki. 1986. Periodic increase of cytoplasmic free calcium in fertilized hamster eggs measured with calcium sensitive electrodes. *J. Physiol. (Lond.)* 377:193-205.

Jaffe, L. A., A. P. Sharp, and D. P. Wolf. 1983. Absence of an electrical polyspermy block in the mouse. *Dev. Biol.* 96:317-323.

Jaffe, L. 1985. The role of calcium explosions, waves and pulses in activating eggs. p. 127-166. *In: Biology of Fertilization, vol. 3.* C. B. Metz and A. Monroy (Eds.) Academic Press, Orlando.

Kikkawa, U. and Y. Nishizuka. 1986. The role of protein kinase C in transmembrane signalling. *Annu. Rev. Cell Biol.* 2:149-178.

Kligman, I., G. S. Kopf, and B. T. Storey. 1988. Characterization of an intermediate stage of the zona pellucida-mediated acrosome reaction in mouse sperm. *Fertil. Steril.* 50:(Suppl.), 530-531.

Kline, D. 1988. Calcium-dependent events at fertilization of the frog egg: Injection of a calcium buffer blocks ion channel opening, exocytosis, and formation of pronuclei. *Dev. Biol.* 126:346-361.

Kline, D. and L. A. Jaffe. 1987. The fertilization potential of the *Xenopus* egg is blocked by injection of a calcium buffer and is mimicked by injection of a GTP analog. *Biophys. J.* 51:398a.

Kurasawa, S., R. M. Schultz, and G. S. Kopf. 1988. Inositol 1,4,5-trisphosphate-induced changes in the mouse zona pellucida. *Biol. Reprod.* 38(Suppl. 1):74.

Kurasawa, S., M. Schultz, and G. S. Kopf. 1989. Egg-induced modifications of the zona pellucida of mouse eggs: Effects of microinjected inositol 1,4,5-trisphosphate. *Dev. Biol.* (In press).

Lee, M. A. and B. T. Storey. 1985. Evidence for plasma membrane impermeability to small ions in acrosome-intact mouse spermatozoa bound to mouse zonae pellucidae, using an aminoacridine fluorescent probe: time course of the zona-induced acrosome reaction monitored by both chlorteracycline and pH probe fluorescence. *Biol. Reprod.* 33:235-246.

Lee, M. A., G. S. Kopf, and B. T. Storey. 1987. Effects of phorbol esters and a diacylglycerol on the mouse sperm acrosome reaction induced by the zona pellucida. *Biol. Reprod.* 36:617-627.

Lee, S. H., K. K. Ahuja, D. J. Gilburt, and D. G. Whittingham. 1988. The appearance of glycoconjugates associated with cortical granule release during mouse fertilization. *Development* 102:595-604.

Miyazaki, S., N. Hashimoto, Y. Yoshimoto, T. Kishimoto, Y. Igusa, and Y. Hiramoto. 1986. Temporal and spatial dynamics of the periodic increase in intracellular free calcium at fertilization of golden hamster eggs. *Dev. Biol.* 118:259-267.

Miyazaki, S. 1988. Inositol 1,4,5-trisphosphate-induced calcium release and guanine nucleotide-binding protein-mediated periodic calcium rises in golden hamster eggs. *J. Cell Biol.* 106:345-353.

Nicosia, S. V., D. P. Wolf, and I. Masato. 1977. Cortical granule distribution and cell surface characteristics in mouse eggs. *Dev. Biol.* 57:56-74.

Niedel, J. E., L. J. Kuhn, and G. R. Vandenbark. 1983. Phorbol diester receptor copurifies with protein kinase C. *Proc. Natl. Acad. Sci. USA* 80:36-40.

Poenie, M., C. Patton, and D. Epel. 1982. Use of calcium precipitants during fixation of sea urchin eggs for electron microscopy: Search for the calcium store. *J. Cell Biol.* 95:161a.

Poenie, M. and D. Epel. 1987. Ultrastructural localization of intracellular calcium stores by a new cytochemical method. *J. Histochem. Cytochem.* 35:939-956.

Saling, P. M. and B. T. Storey. 1979. Mouse gamte interactions during fertilization *in vitro*: Chlortetracycline as fluorescent probe for the mouse sperm acrosome reaction. *J. Cell Biol.* 33:544-555.

Saling, P. M., J. Sowinski, and B. T. Storey. 1979. An ultrasturctual study of epididymal mouse spermatozoa binding to zonae pellucidae *in vitro*: Sequential relationship to the acrosome reaction. *J. Exp. Zool.* 209:229-238.

Shimizu, S., M. Tsuji, and J. Dean. 1983. *In vitro* biosynthesis of three sulfated glycoproteins of murine zonae pellucidae by oocytes grown in follicle culture. *J. Biol. Chem.* 258: 5858-5863.

Stewart-Savage, J. and B. D. Bavister. 1988. A cell surface block to polyspermy occurs in golden hamster eggs. *Dev. Biol.* 128:150-157.

Swann, K. and M. Whitaker. 1985. Stimulation of the Na/H exchanger of sea urchin eggs by phorbol ester. *Nature (Lond.)* 314:274-277.

Swann, K. and M. Whitaker. 1986. The part played by inositol trisphosphate and calcium in the propagation of the fertilization wave in sea urchin eggs. *J. Cell Biol.* 103:2333-2342.

Szollsi, D. 1967. Development of cortical granules and the cortical reaction in rat and hamster eggs. *Anat. Rec.* 159:431-446.

Toyoda, Y. and M. C. Chang. 1974. Fertilization of rat eggs *in vitro* by epididymal spermatozoa and the development of eggs following transfer. *J. Reprod. Fertil.* 36:9-22.

Turner, P. R., M. P. Sheetz, and L. A. Jaffe. 1984. Fertilization increases the polyphosphoinositide content of sea urchin eggs. *Nature (Lond.)* 310:414-415.

Turner, P. R., L. A. Jaffe, and A. Fein. 1986 Regulation of cortical granule exocytosis in sea urchin eggs by inositol 1,4,5-trisphosphate and GTP-binding protein. *J. Cell Biol.* 102:70-76.

Turner, P. R., L. A. Jaffe, and P. Primakoff. 1987. A cholera toxin-sensitive G-protein stimulates exocytosis in sea urchin eggs. *Dev. Biol.* 120:577-583.

Ward, C. R. and B. T. Storey. 1984. Determination of the time course of capacitation in mouse spermatozoa using a chlortetracycline fluorescence assay. *Dev. Biol.* 104:287-296.

Wassarman, P. M., J. D. Bleil, H. M. Florman, J. M. Greve, and R. J. Roller. 1985. The mouse egg's receptor for sperm: What is it and how does it work? *Cold Spring Harbor Symp. Quant. Biol.* 50:11-19.

Wassarman, P. M. 1988. Zona pellucida glycoproteins. *Annu. Rev. Biochem.* 57:414-442.

Whitaker, M. and R. F. Irvine. 1984. Inositol 1,4,5-trisphosphate microinjection activates sea urchin eggs. *Nature (Lond.)* 312:636-639.

Whitaker, M. and R. A. Steinhardt. 1985. Ionic signaling in the sea urchin egg at fertilization. *Biol. Fertil.* 3:167-221.
Wolf, D. P. 1978. The block to sperm penetration in zona-free mouse eggs. *Dev. Biol.* 64:1-10.
Wolf, D. P. and M. Hamada. 1977. Induction of zonal and oolemal blocks to sperm penetration in mouse eggs with cortical granule exudate. *Biol. Reprod.* 17:350-354.

AN ODE TO EDWARD CHAMBERS: LINKAGES OF TRANSPORT, CALCIUM AND pH TO SEA URCHIN EGG AROUSAL AT FERTILIZATION

David Epel

Department of Biological Sciences
Hopkins Marine Station of Stanford University
Pacific Grove, CA 93950

AN INTRODUCTION, A DEFINITION AND A DEDICATION

The events which surround fertilization and ultimately result in the beginning of the developmental program are referred to as "egg activation", as attested by the title of this chapter. However, the definition of activation is "to make active" and the first item I discuss is whether this term is appropriate for what occurs at fertilization. I do not ask this to be pedantic or pugnacious, but because a search for a more appropriate term might lead to a better definition/description and understanding of fertilization.

First, I submit that the egg does not simply become "more active." To be sure, there is an increase in rate of some synthetic processes such as protein synthesis and in anabolic processes such as respiration. But many of the changes are not simply step-ups in ongoing processes, as implied by the term "activation". For example, there is specific initiation of new protein synthesis (e.g., the cyclins) and specific initiation of nucleic acid synthesis (e.g., DNA). There is also the remodeling of the egg cortex as a consequence of the cortical reaction. Then there are patent structural alterations associated with the movement of the sperm and egg pronucleus, along with an orchestrated movement of organelles from the cortex to the cell surface. There then follows a precise series of cell divisions in which the blastomeres are initially equal, but in later cleavages become unequal. A polarity is also present such that the blastomeres at the animal and vegetal poles of the embryo behave differently, with the cells at the vegetal end migrating into the blastocoel and experiencing a different fate than the cells at the animal end.

Thus, the changes that result from the sperm-egg interaction are indeed more profound than simply encompassed in the idea of becoming more active. It is in reality an unfolding of a developmental program, preset/differentiated/ arranged in the oocyte and then unleashed/initiated/turned on/"activated" at fertilization.

The challenge to the student of fertilization, then, is to first understand the unfolding of this developmental program at insemination and ultimately to describe its differentiation or set-up/ordering/manufacturing in the oocyte during oogenesis. The unfolding at fertilization has so far has been more tractable since this event can be synchronously initiated by addition of sperm to a suspension of eggs and the consequences then followed in either suspension or in single cells.

The comprehension of the differentiation of the oocyte, which results in the egg being "ready" to be fertilized, has however been much more difficult. Such comprehension will first require the cataloging of the events of fertilization. Once such a catalog is available, one would need a good experimental organism which permits facile examination of oogenesis so as to describe and understand the nature of the gene action which results in the setting up of this developmental program.

A major craftsman in understanding the unfolding of fertilization has been Edward Chambers. His initial work on ion transport and then ionic fluxes has stimulated numerous researchers to enter the field of fertilization and his research has resulted in some of our most important insights. In this article I have chosen to address Ted Chamber's contributions in a personal vein, noting how his work has impacted my own thinking and where it has led my own research. Hopefully this will indicate how much his work and thought has meant to me and to the understanding of the remarkable consequences that result from sperm-egg interaction.

I shall consider three events surrounding the turning on of the egg's developmental program. The first will be the initiation/activation/unfolding of new transport systems as a consequence of fertilization, an area begun with the pioneering work of Chambers in the late 40's. I shall then consider our current understanding on the role of Na^+-H^+ exchange and the resultant rise in intracellular pH (pH_i) that accompanies fertilization. (Again, an area in which important initial insights were provided by Chambers). Finally, I will present a progress report on our current work examining the role of calcium in egg activation. In this I will describe our studies on enzymes in permeabilized cells, aimed at understanding the arousal of the metabolism of the egg and which suggests that a Ca^{2+}-mediated global change is involved in enzyme regulation.

TRANSPORT CHANGES AT FERTILIZATION

Students of the sea urchin embryo have long appreciated the changes in permeability that occur at insemination. Early research, especially in the 30's, focused on changes in permeation of molecules affecting the osmotic properties of the egg and evidenced as changes in egg volume (see, e.g., Davson 1959). In the 40's, the advent of isotopes led to the pioneering work of Chambers showing a remarkable change in phosphate permeability at insemination (Brooks and Chambers 1948, 1954; Chambers and White 1949, 1954; Chambers and Chambers, 1949). Then in collaboration with Arthur Whiteley, they described the kinetics of the process and the energy dependence of the initial setting up of this transport system (Chambers and Whiteley 1966; Whiteley and Chambers 1961). Later work revealed that other transport systems were also activated, as for nucleosides and amino acids, and showed that the properties of these different transport systems were remarkably similar (Epel 1972; Piatigorsky and Whiteley 1965; Schneider 1985). For example, nucleoside, phosphate and amino acid transport are all sodium-dependent, they all appear with similar temporal kinetics after fertilization and the expression of the transport is prevented by inhibitors of energy metabolism when applied during the early phase but with no affect if these same inhibitors are applied later (Schneider 1985). This suggests that the expression of the transporters all depend on some common property/event.

The process is indeed a remarkable one. Looking at amino acid transport, with which I am best acquainted, one sees that the aliphatic amino acids such as valine, leucine, or isoleucine are transported by a sodium-independent mechanism before fertilization. Following insemination, a very large increase in transport activity for the same amino acids begins to appear, but now this transport is sodium-dependent (Epel 1972). Remarkably, the transport of neutral amino acids such as glycine, serine and alanine, which was previously non-existent in the unfertilized egg, now begins to appear and at a very rapid rate. An example of this is depicted for the transport of glycine in Figure 1. The establishment of this transport is not immediate, is not a step function, but appears over a period of 60 min (Epel 1972) and as noted, similar kinetics are seen for the transport of nucleosides and phosphate (Piatigorsky and Whiteley 1965; Chambers and Whiteley 1966; Schneider 1985).

The establishment of these transport systems is profoundly affected by low concentrations of cytochalasin B (Epel and Johnson 1976; Allemand *et al.* 1987a) or application of high hydrostatic pressure (Swezey *et al.* 1987). The period of sensitivity to cytochalasin and high pressure occurs soon after fertilization; in the first few minutes. If cytochalasin or high pressure are applied after the transport is fully established (as 40-60 min after fertilization) however, there is little affect on transport (Epel and Johnson 1976; Allemand *et al.* 1987a; Swezey *et al.* 1987). Thus, a pressure-sensitive and cytochalasin-sensitive process associated with the early events of fertilization is necessary for transport to be established. The process is not the cortical granule exocytosis; cytochalasin does not affect the cortical reaction at concentrations which prevent the evolution of transport activity (Allemand *et al.* 1987a). High hydrostatic pressure can affect the cortical reaction, but if a "pressure titration" is done, it is seen that pressures between 2,000-4,000 psi will prevent the expression of transport but not affect the cortical reaction (Swezey *et al.* 1987). Thus, the pressure-sensitive and cytochalasin-sensitive process is clearly not the cortical exocytosis.

One possibility is that pressure and cytochalasin are both affecting the insertion of a transporter in the plasma membrane, such as preventing fusion of a vesicle containing the transporter (Swezey *et al.* 1987). Another possibility is that a post-fertilization establishment of a steeper sodium gradient is necessary. Thus, Ciapa *et al.* (1984) have shown that sodium-potassium ATPase activity increases after fertilization and causes a 30-40% drop in intracellular sodium. Related to this, Allemand *et al.* (1986) have presented interesting circumstantial arguments that the activation of the sodium-potassium ATPase and resultant establishment of this gradient may be essential to the operation (activation?) of

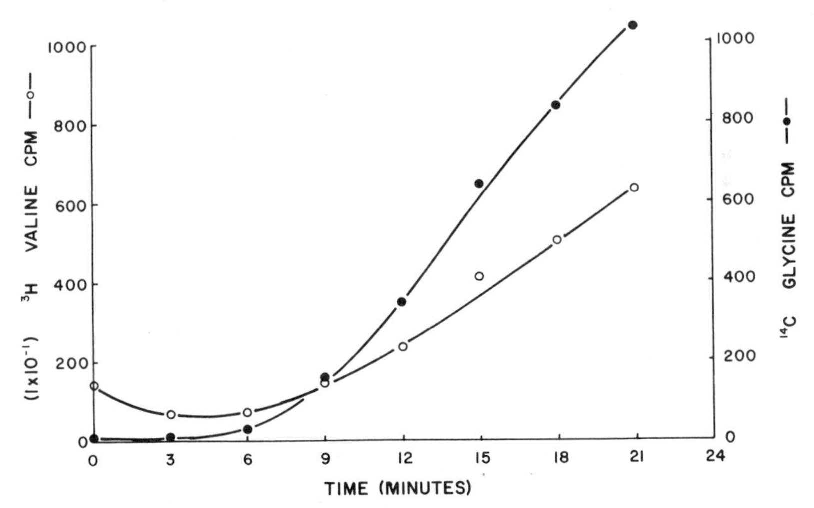

Figure 1. Rate of glycine and valine transport following fertilization of S. purpuratus eggs. Eggs or embryos were pulsed for three minutes with H^3-valine and C^{14}-glycine and uptake determined as described in Epel (1972). Note that valine transport already occurs in the unfertilized egg, that it decreases after fertilization and then increases. Glycine transport is virtually non-existent in the unfertilized egg and then transport activity appears and increases to a very high rate.

sodium-dependent transport. Thus, cytochalasin inhibits the increased sodium-potassium ATPase activity after fertilization (Allemand *et al.* 1987a) and a number of experimental treatments which modify the sodium influx or modify the sodium pumping similarly modify amino acid transport (Allemand *et al.* 1986). These two hypotheses (insertion of the transporter *vs* Na^+-K^+ ATPase) are not necessarily mutually exclusive and suggest that either the sodium-potassium ATPase, the amino acid transporters, or both are inserted/activated into the plasma membrane in a programmed fashion after fertilization.

THE ROLE OF pH_i IN EGG ACTIVATION

An Historical Overview

It is now well established that fertilization in sea urchin eggs is followed by a brief bout of Na^+-H^+ exchange with resultant rise in pH_i (see recent review by Epel 1988). The concept did not come easily and an experiment of Ted Chambers provided the critical information.

The fact that eggs release acid had been known for years and the prevailing wisdom until the mid-70's was that this acid release came from the cortical granule exocytosis (Ishihara 1968). In 1970 Nakazawa *et al.* proposed another possibility – namely that the acid release might be a Ca^{2+}-H^+ exchange as based on experiments with EGTA. The hypothesis of Nakazawa *et al.* (1970) was not tested in calcium-free seawater since eggs cannot be easily fertilized in this medium. The subsequent finding (in 1974) that one can activate eggs with calcium ionophores (Steinhardt and Epel 1974) provided a means for Miles Paul and myself to test this Ca^{2+}-H^+ exchange hypothesis (Paul and Epel 1975). To our surprise, we found that normal acid release occurred in calcium-free media and that indeed acid release could take place in every media tried except sodium-free media. In the absence of sodium, there was a 70% inhibition of acid release but there was also some inhibition of cortical granule exocytosis. Applying the prevailing wisdom of the time, we interpreted this inhibition of acid release as ensuing from the apparent inhibition of the cortical reaction. (Later we realized that this inhibition of the cortical reaction in Na^+-free media was a pathological consequence of prolonged incubation in the absence of sodium; if activated soon after suspension in Na^+-free media, the inhibition was less pronounced).

Meanwhile, the ammonia-activation of eggs had been discovered by Steinhardt and Mazia (1973). Paul, James Johnson and I subsequently looked at the effects of ammonia on acid release and were surprised to find that there was an acidification of the medium when ammonia was added, although there was no cortical reaction. If these eggs were subsequently fertilized, there was now only a small amount of acid release. This finding clearly indicated that the cortical reaction was not the major source of acid release at fertilization (Paul *et al.* 1976). James Johnson, Paul and myself began to re-examine our earlier data on sodium requirements for acid release and quickly realized that there was in fact a Na^+-H^+ exchange occurring, in which the acid release did indeed depend on sodium and that the stoichiometry of sodium for hydrogen was close to 1:1 (Johnson *et al.* 1976).

What did these results mean in terms of egg activation? At exactly this time came the abstract and then the paper from Chamber's lab, showing a sodium requirement for the first ten minutes after fertilization and not thereafter (Chambers 1975; 1976). The time limits of the sodium requirement corresponded almost exactly to the period of Na^+-H^+ exchange. Putting our observations together, we quickly realized that Na^+-H^+ exchange was essential for egg activation, that it was not the sodium influx *per se* that was important but rather the efflux of hydrogen ions with a resultant rise in pH_i. The fact the ammonia substituted for Na^+, combined with crude measurements of pH_i by looking at the pH_i of lysates, led us to advance the idea of this Na^+-H^+ exchange causing a pH_i rise at fertilization which was essential for egg activation (Johnson *et al.* 1976).

This pH_i rise has since been confirmed by measurements with pH microelectrodes (Shen and Steinhardt 1978), NMR (Winkler *et al.* 1982), weak acid distribution (Johnson and Epel 1981) and fluorescent probe methods (Whitaker and Steinhardt 1981). Chamber's initial observations on the inhibition of development by deletion of sodium (Chambers 1975; 1976) have been confirmed and considerably extended in terms of the effects of Na-free media or fertilization in amiloride (and derivatives) on protein synthesis (Winkler *et al.* 1980; Dubé *et al.* 1985), DNA synthesis (Dubé *et al.* 1985), microtubule migration (Schatten *et al.* 1985) and even some aspects of the actin changes at fertilization (Begg *et al.* 1982; Carron and Longo (1982).

The other line of evidence is that this arrest of development in Na-free media or amiloride can be reversed by treatments which directly raise pH_i, as by addition of weak bases such as ammonia (Johnson *et al.* 1976). Also, incubation of unfertilized eggs in these weak bases directly raises pH_i (Shen and Steinhardt 1978, 1979) and also induces part of the normal post-fertilization response. For example, the rate of protein synthesis is increased and DNA synthesis and chromosome condensation are initiated (Epel *et al.* 1974). Cytoskeletal events also take place; for example, fertilization normally involves a movement of the egg pronucleus towards the center of the egg and this centering occurs when the pH_i is raised with ammonia (Mar 1980).

A Simplified View of pH_i as Regulator

It has therefore been postulated, on the basis of the above types of studies, that there are two ionic phases to the fertilization response. An initial calcium rise is deemed responsible for the cortical reaction, activation of the calcium-dependent NAD kinase and probably other events and the subsequent pH_i rise is deemed responsible for increased rate of protein synthesis, initiation of DNA synthesis and induction of the aforementioned microtubule-mediated events (Reviews by Epel 1980; Whitaker and Steinhardt 1985). This dual ionic signal concept is also consistent with the work of Michael Whitaker, Laurinda Jaffe and their colleagues (reviewed in this book) showing that the signal transduction events at fertilization involve polyphosphoinositide hydrolysis, with the initial inositol trisphosphate increase causing calcium release and the secondary bout of diacylglycerol production presumably activating sodium-hydrogen exchange (but note the article by Shen in this volume which suggests calcium may be acting directly).

The above evidence has seemed convincing and the general assumption is that this pH_i increase is critical for initiating the later sequelae of fertilization. The assumption received further generalization by the finding that a similar bout of Na^+-H^+ exchange and resultant pH_i rise follows the application of growth factors to cells in culture (see review by Epel and Dubé 1987). The concept widely promulgated is therefore that Na^+-H^+ exchange and resultant pH_i increases are critical for promoting development in the case of fertilization and for promoting growth and mitosis in the case of tissue culture cells.

Problems with this Simplistic View of pH_i

There are, however, several disturbing observations which indicate that the pH_i rise *by itself* is an inadequate stimulus. The first indication of problems came from the experiments of Nishioka and Cross (1978). Prior to their work, assessment of the effects of Na^+-free media was accomplished by activation of eggs with calcium ionophore or by insemination of eggs in regular seawater and thence rapidly washing the eggs into Na^+-free media (e.g., Chambers 1976). Direct insemination in Na^+-free media was not possible since the sperm were immotile. Nishioka and Cross (1978), however, observed that sperm motility – and hence fertilization – could occur if the pH_o of the Na^+-free media was raised to 9.0.

They found that in such eggs a cortical reaction ensued, but that development did not progress. This could mean that sodium was important; however, these arrested eggs could be rescued by the addition of ammonia. So, the pH_i was presumably important and perhaps the reason the eggs would not develop unless ammonia was added was that the pH_i rise was not adequate at an external pH of 9.0. However, subsequent work of Johnson and Epel (1981) showed that eggs and embryos were not perfect pH regulators, and that when the pH_o around unfertilized eggs was raised to 9, the pH_i should have gone to the level of the fertilized egg. Yet Nishioka and Cross (1978) found that eggs fertilized in sodium-free seawater at pH 9 did not develop.

A more serious problem with the pH_i idea came from experiments of Whitaker and Steinhardt (1981), based on some observations of Chambers and Hinkley (1979) on partially-activated eggs. Chambers and Hinkley (1979) found that partial activation of eggs could be achieved by touching eggs with A23187-coated rods. Two classes of eggs – and of activation – were seen. In eggs with less than 50% cortical reaction, nuclear events as evidenced by nuclear membrane breakdown did not occur; eggs with greater than 50% cortical granule activation did undergo nuclear events.

Whitaker and Steinhardt (1981) then asked about the pH_i rise in these partially activated eggs. Did it rise in eggs where nuclear events took place? Did it not rise as much in the cases where nuclear events did not occur? Using an intracellular fluorescent dye to monitor pH_i, they found that the pH_i rose in both cases! pH_i was apparently an insufficient stimulus.

What other changes might be occurring in these eggs? Whitaker and Steinhardt (1981) also looked at a calcium-induced reaction, the activation of NAD kinase, and found that eggs which underwent nuclear events had undergone a full activation of NAD kinase but eggs which were insufficiently activated did not evidence the NAD kinase activation. This could mean that the NAD kinase change was important; or that the calcium rise was critical and in cases where an adequate calcium rise did not occur, the eggs consequently did not activate. These workers went on to look at changes in pyridine nucleotides in the case of ammonia activation and found that although NAD kinase was not activated by ammonia, an alteration in redox ratio occurred mimicking the change that normally follows NAD kinase activation. They therefore postulated that perhaps the redox change mimicked by ammonia – and normally induced by fertilization – might be critical for nuclear activation.

Whether this specific hypothesis is correct or not, the important observation they made relative to this discussion is that pH_i can increase but eggs are not necessarily activated. So, to reiterate, the pH_i increase by itself is not an adequate stimulus. Yet, ammonia can activate eggs, so this may be by providing something more than just a pH_i rise.

The aforementioned work of Nishioka and Cross (1978) and Johnson and Epel (1981) similarly implied this, since raising pH_o around unfertilized eggs to a pH of 9.0 yielded a pH_i similar to that caused by fertilization, yet the eggs did not overtly activate. This was studied further by Dubé and Epel (1986), who compared the increase in protein synthesis when pH_i was raised by ammonia as opposed to raising pH_i by raising pH_o. They observed a marked difference in the efficacy of these two modes of raising pH_i ; when one raised the pH_i to similar levels with ammonia or pH_o, the rates of protein synthesis in ammonia were two times greater. More importantly, DNA synthesis was never initiated by high pH_o whereas it was initiated by ammonia.

A third example indicating a difference between raising pH_i by incubation in ammonia *versus* raising pH_i by incubation in high pH_o comes from work of Allemand *et al.* (1986), who examined increased amino acid transport after fertilization. They found that this transport activity is not initiated if one activates eggs in sodium-free seawater. If one adds ammonia to these inhibited eggs, transport is initiated; but, if one raises pH_i by raising pH_o, the transport is not initiated.

In conclusion, in a number of cases it appears that directly raising pH_i by raising pH_o is very different than when one raises pH_i with ammonia. The critical question then relates to what ammonia is doing? Is it the alteration of the redox ratio, a fortuitous mimicking of postfertilization events by ammonia that is important (Whitaker and Steinhardt 1981)? Or is it the discharging of intracellular acidic vesicles, which are known to be discharged by ammonia (Lee and Epel 1983; Allemand *et al.* 1987b)? Or is it a direct action of ammonia on some enzyme system in the egg? Or could it be a requirement for a change in sodium levels, which can be stimulated by ammonia and which normally decreases after fertilization and ensues from activation of the Na^+/K^+ ATPase?

TARGETS OF pH_i RISE

The previous paragraphs have documented some of the evidence that the post-fertilization pH_i rise is important for turning on of egg metabolism. However, I have also pointed out caveats in terms of whether the effect is really due to the H^+ decrease, or whether it results from sodium changes, from alkalinization of cytoplasmic vesicles or a yet undescribed target. Irrespective of the exact mechanisms, however, it is clear that some consequence of Na^+-H^+ exchange activity is important for turning on the egg. In this section I consider what this might be.

I describe here a phenomonological study by Dubé *et al.* (1985) on the Na^+-H^+ exchange and resultant pH_i rise. The question asked is whether pH_i, which increases after fertilization, must be maintained in this new higher position (*i.e.*, is this a new permissive pH_i level necessary for new egg activity?) or whether the change in pH_i is acting in a trigger mode (in which case the pH_i must only be raised for some brief period after insemination). The approach we have used (Dubé *et al.* 1985) to test these alternatives (presented diagrammatically in Figure 2) is to allow the pH_i to increase after fertilization and then drive the pH_i back to the prefertilization level by incubating the egg in weak acids at a lower extracellular pH_o (see e.g., Grainger *et al.* 1979). The weak acids used are acetate and DMO with pK's respectively of 4.7 and 6.3. At lower pH's these weak acids will be driven into the egg and cause the pH_i to decrease (and the resultant pH_i is then assessed by incubation of the eggs in radioactive DMO, using defined procedures).

The results of these experiments (Dubé *et al.* 1985) indicate that the pH_i need only be elevated for 10-20 min after insemination. Preventing the rise, or driving pH_i down during the first 20 min after insemination, prevents activation. After this time, one can drive the pH_i back to the unfertilized level and still get progression through mitosis. (this progression, however, is much slower than normal and cytokinesis is impaired).

These results indicate that whatever the nature of the Na^+-H^+ exchange (i.e., for increased alkalinity of the cytoplasm, or for discharging cytoplasmic vesicles or for altering sodium or for ?), the Na^+-H^+ exchange activity and pH_i rise is only needed for turning on egg metabolism during the first 20 min. Thereafter the affect is on rate with some consequences for cytokinesis. A challenge for the future is to determine the targets of Na^+-H^+ exchange. Is it one of the above, such as vesicle alkalinization? Or is it acting on some critical master enzyme, such as a kinase affecting many enzymes in a global situation? If so, could this be an enzyme related to the factor which propels cells through mitosis, the Mitosis Promoting Factor

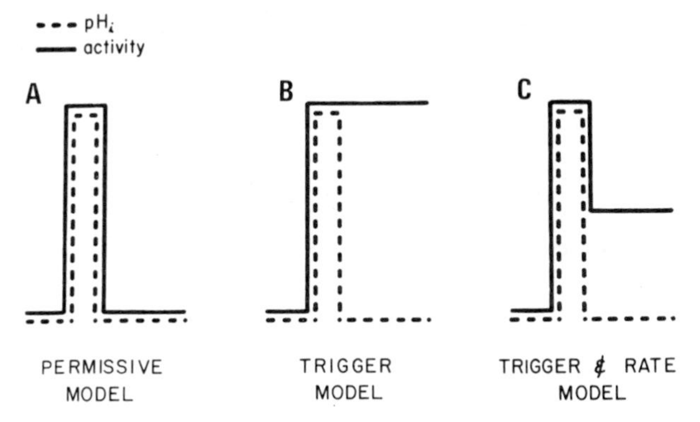

Figure 2. Consequences of raising and then dropping pH_i on metabolic activity following fertilization under three different models. In all cases, elevating pH_i causes an increase in activity. **A.** If pH_i is permissive, decreasing the pH_i to the unfertilized value should result in a decrease in activity back to the level of the unfertilized egg. **B.** If pH_i acts as a trigger, dropping pH_i back to the unfertilized value should have no consequences for activity. **C.** If pH_i has both a trigger and rate effect, then dropping pH_i will decrease activity but activity will remain above the unfertilized level.

(MPF, see review by Dunphy and Newport 1988)? or some other type of kinase perhaps solely involved in the developmental initiation? Given the parallels to growth factor action and their common affect in raising pH_i, it may be that there is a general pH sensitive target, with MPF as an intriguing candidate.

In concluding this section on pH_i, it is disappointing to note that the sites sensitive to the pH_i increase are still not clear. Indeed, as noted above, one is not even sure whether it is the pH_i change itself that is important, whether a sodium influx is critical which can be mimicked by ammonia, or whether it is some combination of sodium and pH_i changes which are important. Clearly, the experiments showing that raising pH_i by itself is not an adequate stimulus indicate that the pH_i situation is more complex.

A final possibility to consider is that the Na^+-H^+ exchange activity somehow starts a "calcium clock" and that it is this Ca^{2+} clock that it critical for forwarding cell activity. We know that following insemination a calcium cycle ensues (Poenie *et al.* 1985) which is also initiated by incubation in ammonia.

My current interpretation is that there are at least two events related to Na^+-H^+ exchange activity which are necessary for egg activation. One is the pH_i rise, which can be artificially induced by simply raising pH_o. The other event is one which can be mimicked by ammonia. Whether this is provision of a counterion, stimulation of the sodium-potassium ATPase and subsequent changes in sodium ion or some direct affect of ammonia, will await further studies. The answer to this will be important not only for understanding egg activation but perhaps mitogenesis in general.

ROLE OF THE CALCIUM INCREASE

The other change that occurs at fertilization is a transient rise and fall of cytoplasmic calcium (Steinhardt *et al.* 1977). Evidence that this is critical comes from experiments in which the calcium level is experimentally manipulated; raising calcium levels as with ionophores (Steinhardt and Epel 1974, Chambers *et al.* 1974) or calcium buffers (Hamaguchi and Hiramoto 1981) will activate eggs whereas preventing the calcium rise with high concentrations of calcium chelators prevents activation (Zucker and Steinhardt 1978; Hamaguchi and Hiramoto 1981).

It is generally assumed that the increased calcium is directly responsible for metabolic activation, but studies on polyphosphoinositide metabolism by Whitaker and colleagues could also be interpreted as showing that the raise of calcium is needed only for the autocatalytic propagation of the cortical reaction and production of diacylglycerol through the action of the calcium-activated phospholipase C (see excellent review by Swann *et al.* 1987 and also Ciapa and Whitaker 1986; Swann and Whitaker 1986). The "contrarian" idea here would be that the diacylglycerol would then activate protein kinase C, with the resultant protein phosphorylations being critical for the subsequent steps. The calcium rise would have no direct role other than in propagating the cortical reaction and in producing diacylglycerol.

Arguing against this idea is the fact that there are a number of events which are regulated by calcium and which occur at the time of the calcium rise. However, almost all of these are related to the cortical reaction and cortex modifications and these Ca^{2+}-mediated changes are not essential for activating DNA and protein synthesis, chromosome cycles etc. So, perhaps calcium is only needed for events associated with the initial aspects of fertilization but not for the subsequent metabolic turn-on.

One line of evidence in favor of the idea that Ca^{2+} is an essential link between early events and late events comes from the work of Shen (this colloquium), which suggests that in fact the turn-on of the Na^+-H^+ exchanger is not mediated by protein kinase C but rather by calcium-calmodulin. If so, we have an important connection between the early calcium rise and early events, i.e., calcium induces cortical exocytosis and other events associated with sperm incorporation and also activates the Na^+-H^+ exchanger whose activity is necessary for further development.

As noted, however, most of the described calcium-mediated events center around the cortical reaction and cortical reorganization. For example, the cortical exocytosis is directly mediated by calcium, through some type of calmodulin-type interaction (see Whitaker and Steinhardt 1985 for review). Similarly, the changes in actin polymerization at fertilization may be calcium-mediated (Spudich *et al.* 1988). However, neither the cortical reaction nor the actin polymerization are essential for initiation of new metabolic activity since one can prevent the cortical exocytosis with high hydrostatic pressure (e.g., Swezey *et al.* 1987) and actin polymerization with cytochalasin (Longo 1978; Byrd and Perry 1980) and the initiation of new metabolic activity still takes place.

One of the earliest calcium-mediated changes at fertilization is the activation of the enzyme NAD kinase, which promotes the synthesis of new NADP and NADPH (Epel *et al.* 1981). This change may have a major role in generating hydrogen peroxide for the hardening process of the fertilization membrane (see, e.g., Turner *et al.* 1988). Whether the pyridine nucleotide changes have a role in subsequent turn-on of new synthesis is unclear, although as noted above there is the interesting correlation between redox state and DNA synthesis (Whitaker and Steinhardt 1981). So, perhaps this is an important thread linking calcium to egg arousal.

TRACING A POST-FERTILIZATION CHANGE

Glucose-6-Phosphate Dehydrogenase

An approach we have been utilizing in the last few years to examine the turn-on of egg metabolism has been to follow a post-fertilization change and determine how it is regulated. We have been especially concentrating on the regulation of the enzyme glucose-6-phosphate dehydrogenase (G6PD). Our evidence suggests that the changes in this enzyme are calcium-mediated and that a global calcium-mediated change in protein-protein interactions (perhaps a change in cell structure) occurs which influences many enzymes.

The origin of this work comes from an old observation of Isono and colleagues in Japan (Isono 1963; Isono and Yasumasu 1968). They found that this enzyme, the first enzyme of the pentose-phosphate shunt, undergoes a dramatic change in location after fertilization; the enzyme activity was associated with a pellet fraction in the unfertilized egg but shortly after fertilization the activity became soluble. The evidence for this was suspect, however, since the change could only be seen at low ionic strength; in high salt media more representative of the egg cytoplasm the enzyme was always soluble (Isono 1963).

Robert Swezey and I decided to re-examine the reality of this phenomena and observed that the association of the enzyme with the cytomatrix could be maintained in physiological salt solutions only if high concentrations of protein were present (Swezey and Epel 1986). This association was lost when the eggs were fertilized. Thus, at the high protein and high salt levels typical of the *in vivo* situation the enzyme remained associated with the pellet in unfertilized egg homogenates but was soluble in homogenized fertilized egg homogenates. We also found that the activity of the enzyme was considerably reduced in the bound state but was much more active when soluble. So, if this change indeed occurred *in vivo* there could be significant consequences for enzyme activity (Swezey and Epel 1986).

Permeabilized Cell Studies

We became attracted to a permeabilized cell system developed by Suprynowicz and Mazia (1985) which utilizes high voltage electric discharge to permeabilize the cells. When such cells are maintained in a low-calcium media the "pores" remain open and one can then percolate substrates into the cells and examine enzyme activity *in situ*. Our findings (Swezey and Epel 1988) indicated that in these permeabilized cells the activity of G6PD was severely repressed in the unfertilized egg. When the same cells were permeabilized after fertilization, however, there was a 10- to 15-fold increase in activity. We examined the activity of a number of other enzymes and found that of the six enzymes examined, activity of five of them were similarly repressed in the unfertilized state but activity was enhanced after insemination.

Importantly, this modulation of enzyme activity was not seen in cell-free homogenates. Rather, in the cell-free homogenates the activity of the enzymes was much higher and there were no differences between fertilized and unfertilized eggs. Thus, a modulation of activity is only seen in the permeabilized cells and this is considerably relieved at fertilization.

What causes this change in enzyme activity, which can only be seen in the permeabilized cells? We looked at the role of pH_i and calcium changes in this action. Raising pH_i with ammonia had no immediate affect on G6PD activity as assessed in the permeabilized cells (but see later). Activation with calcium ionophore-induced a change similar to fertilization. However, ionophore activation in sodium-free media (which prevents the pH_i rise) resulted in about a 40-50% inhibition. So, a pH_i increase cannot induce the change by itself; increasing calcium does, but the calcium-induced change also requires some aspect of the Na^+-H^+ exchange activity.

Better evidence that calcium is involved in regulating G6PD activity has come from studies in which we examined the effects of prolonged incubation in ammonia. Poenie *et al.*. (1985), in their initial study of calcium changes during the cell cycle of the sea urchin embryo, noted that there is an increase in calcium just before nuclear membrane breakdown during the first mitotic cycle. When one activates eggs with ammonia, there is no initial calcium change (as seen at fertilization) but a calcium change is observed just before nuclear membrane breakdown.

What happens to G6PD activity when eggs are incubated in ammonia? As noted, there is no initial change in G6PD activity following ammonia addition and when the pH_i has risen. However, during a prolonged incubation, sometime between 40-60 min, there is an abrupt shift in G6PD activity almost identical to that seen at fertilization. The period of change corresponds in time to the calcium rise that occurs during incubation in ammonia and lends support to the idea that the G6PD change is calcium-mediated. Whether this is a direct or indirect affect of calcium is not yet known.

What is the nature of the change that results in this large increase of G6PD activity? A clue comes from studies in which we have looked at the pH optima of the enzyme. As shown in Figure 3, the enzyme in homogenates shows the characteristic behavior of G6PD; a monotonic increase in activity as pH is raised. This same behavior, however, is not seen in the permeabilized cells. Indeed, there is a slight decrease in activity as the pH is raised. As this difference in pH profile is characteristic of enzymes bound to surfaces, our working hypothesis is that the enzyme in the permeabilized cells (and presumably *in vivo*) is associated with cell structures and that this association changes as a consequence of fertilization.

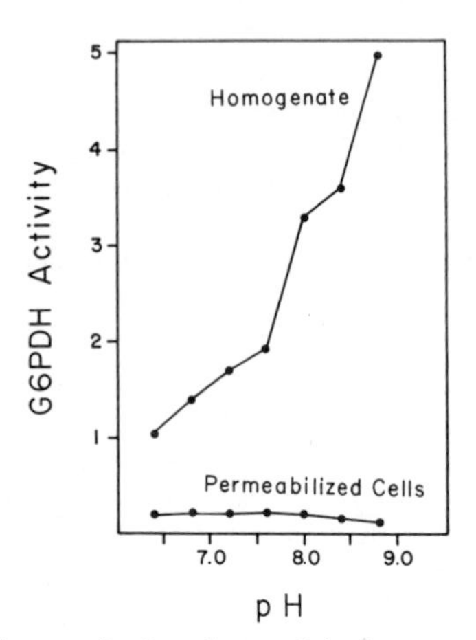

Figure 3. Glucose-6-phosphate dehydrogenase (G6PDH) activity as a function of pH in homogenates or permeabilized cells of S. purpuratus embryos. As seen, activity in these two states is radically different as the pH changes. Conditions of assay as described in Swezey and Epel (1988).

Where do these results leave us? First, I note that all of our data so far has been on permeabilized cells and that it is critical to ascertain whether the same change occurs *in vivo*. The data of Isono and Yasumasu (1968) suggest that the pentose shunt is activated (G6PD activity?) but more definitive evidence needs to be generated.

We are approaching this by the development of a system to transiently permeabilize eggs and our intent is to introduce radioactive 6-phosphogluconic acid into them. This enzyme can only be metabolized in one direction and will allow us to assess the activity of 6-phosphogluconic acid dehydrogenase *in vivo* (this enzyme undergoes a similar shift as G6PD). We should then be able to tell whether the activity changes *in vivo* correspond to what we see in the permeabilized cells. Is activity operating at only a fraction of potential in the unfertilized egg and is there a large increase of activity after fertilization?

Assuming the enzyme change seen in the permeabilized case is real, what does this tell us about how the egg is activated/turned on/aroused at fertilization? If structural changes are critical, we will need to expand our view of major structural changes at fertilization. We know that the cortex is reorganized (e.g., Vacquier 1981); perhaps there is also cytoplasmic reorganization which results in changes and alterations in enzyme activity. If this idea is correct, it would appear to be modulated by the calcium increase.

Clearly more work is necessary to look at the mechanism of egg arousal at fertilization. As the contributions to this monograph indicate, great advances in understanding the initial aspects of sperm-egg interaction have been made. The quest of the 1990's will be to understand how this interaction leads to the arousal and initiation of new development. Ted Chamber's contributions from the 1940's to the present provide us with the tools and knowledge to answer this quest.

REFERENCES

Allemand, D., G. De Renzis, P. Payan, and J-P. Girard. 1986. Regulatory and energetic role of Na^+ in amino acid uptake by fertilized sea urchin eggs. *Dev. Biol.* 118:19-27.

Allemand, D., B. Ciapa, and G. De Renzis. 1987a. Effect of cytochalasin B on the development of membrane transports in sea urchin eggs after fertilization. *Dev. Growth & Differ.* 29:333-340.

Allemand, D., G. De Renzis, J-P. Girard, and P. Payan. 1987b. Activation of amino acid uptake at fertilization in the sea urchin egg. Requirement for proton compartmentalization during cytosolic alkalosis. *Exp. Cell Res.* 169:169-177.

Begg, D., L. J. Rebhun, and H. Hyatt. 1982. Structural organization of actin in the sea urchin egg cortex: microvillar elongation in the absence of actin bundle formation. *J. Cell Biol.* 93:24-32.

Brooks, S. C. and E. L. Chambers. 1948. Penetration of radioactive phosphate into the eggs of *Strongylocentrotus purpuratus*, *S. franciscanus* and *Urechis caupo*. *Biol. Bull.* 95:262-263.

Brooks, S. C. and E. L. Chambers. 1954. The penetration of radioactive phosphate into marine eggs. *Biol. Bull.* 106:279-296.

Byrd, W. and G. Perry. 1980. Cytochalasin B blocks sperm incorporation but allows activation of the sea urchin egg. *Exp. Cell Res.* 126:333-342.

Carron, C. P. and F. J. Longo. 1982. Relation of cytoplasmic alkalinization to microvillar elongation and microfilament formation in the sea urchin egg. *Dev. Biol.* 89:128-137.

Chambers, E. L. 1975. Na^+ is required for nuclear and cytoplasmic activation of sea urchin eggs by sperm and divalent ionophores. *J. Cell Biol.* 67:60a.

Chambers, E. L. 1976. Na^+ is essential for activation of the inseminated sea urchin egg. *J. Exp. Zool.* 197:149-154.

Chambers, E. L. and R. Chambers. 1949. Ion exchanges and fertilization in echinoderm eggs. *Am. Nat.* 83:269-284.

Chambers, E. L. and R. E. Henkley. 1979. Non-propagated cortical reaction induced by the divalent ionophore A23187 in eggs of the sea urchin, *Lytechinus variegatus*. *Exp. Cell Res.* 124:441-446.

Chambers, E. L., B. C. Pressman, and B. Rose. 1974. The activation of sea urchin eggs by divalent ionophores A23187 and X-537A. *Biochem. Biophys. Res. Commun.* 60:126-132.

Chambers, E. L. and W. E. White. 1949. The accumulation of phosphate and evidence for synthesis of adenosine triphosphate in the fertilized sea urchin egg. *Biol. Bull.* 97: 225-226.

Chambers, E. L. and W. E. White. 1954. The accumulation of phosphate by fertilized sea urchin eggs. *Biol. Bull.* 106:297-307.

Chambers, E. L. and A. H. Whiteley. 1966. Phosphate transport in fertilized sea urchin eggs. I. Kinetic aspects. *J. Cell. Physiol.* 68:289-308.

Ciapa, B. and M. Whitaker. 1986. Two phases of inositol polyphosphate and diacylglycerol production at fertilization. *FEBS Lett.* 195:247-351.

Ciapa, B., D. Allemand, P. Payan, and J. P. Girard. 1984. Sodium-potassium exchange in sea urchin egg. I. Kinetic and biochemical characterization at fertilization. *J. Cell. Physiol.* 121:235-242.

Davson, H. 1959. A Textbook of General Physiology. Little, Brown and Co., Boston.

Dubé, F. and D. Epel. 1986. The relation between intracellular pH and rate of protein synthesis in sea urchin eggs and the existence of a pH-independent event triggered by ammonia. *Exp. Cell Res.* 162:191-204.

Dubé, F., T. Schmidt, C. H. Johnson, and D. Epel. 1985. The hierarchy of requirements for an elevated intracellular pH during early development of sea urchin embryos. *Cell* 40: 657-666.

Dunphy, W. G. and J. W. Newport. 1988. Unraveling of mitotic control mechanisms. *Cell* 55: 925-928

Epel, D. 1972. Activation of an Na^+-dependent amino acid transport system upon fertilization of sea urchin eggs. *Exp. Cell Res.* 72:74-89.

Epel, D. 1980. Experimental analysis of the role of intracellular calcium in the activation of the sea urchin egg at fertilization p. 169-186. *In: The Cell Surface: Mediator of Developmental Processes.* (S. Subtelny and N. K. Wessells (Eds.). Academic Press, New York.

Epel, D. 1988. The role of Na^+-H^+ exchange and intracellular pH changes in fertilization. *In: Na^+-H^+ Exchange.* S. Grinstein (Ed.) CRC Press, Boca Raton. (In press).

Epel, D. and F. Dubé. 1987. Intracellular pH and cell proliferation p. 364-394. *In: Control of Animal Cell Proliferation.* A. Boynton and H. L. Leffert (Eds.). Academic Press, Orlando.

Epel, D. and J. D. Johnson. 1976. Reorganization of the sea urchin egg surface at fertilization and its relevance to the activation of development. p. 105-120. *In: Biogenesis and Turnover of Membrane Molecules.* J. S. Cook (Ed.) Raven Press, New York.

Epel, D., C. Patton, R. W. Wallace, and W. Y. Cheung. 1981. Calmodulin activates NAD kinase of sea urchin eggs; an early event of fertilization. *Cell* 23:543-549.

Epel, D., R. Steinhardt, T. Humphreys, and D. Mazia. 1974. An analysis of the partial metabolic derepression of sea urchin eggs by ammonia. The existence of independent pathways. *Dev. Biol.* 40:245-255.

Grainger, J. L., M. M. Winkler, S. S. Shen, and R. A. Steinhardt. 1979. Intracellular pH controls protein synthesis rate in sea urchin egg and early embryo. *Dev. Biol.* 68:396-406.

Hamaguchi, Y. and Y. Hiramoto. 1981. Activation of sea urchin eggs by microinjection of calcium buffers. *Exp. Cell Res.* 134:171-179.

Ishihara, K. 1968. An analysis of acid polysaccharides produced at fertilization of sea urchin eggs. *Exp. Cell Res.* 51:473-484.

Isono, N. 1963. Carbohydrate metabolism in sea urchin eggs IV. Intracellular localization of enzymes of the pentose phosphate cycle in unfertilized and fertilized eggs. *J. Fac. Sci. Univ. Tokyo* 10:37-53.

Isono, N. and I. Yasumasu. 1968. Pathways of carbohydrate breakdown in sea urchin eggs. *Exp. Cell Res.* 50:616-626.

Johnson, C. H. and D. Epel. 1981. Intracellular pH of sea urchin eggs measured by the DMO method. *J. Cell Biol.* 89:284-291.

Johnson, J. D., D. Epel, and M. Paul. 1976. Na^+-H^+ exchange is required for activation of sea urchin eggs after fertilization. *Nature (Lond.)* 262:661-664.

Lee, H. C. and D. Epel. 1983. Changes in intracellular acidic compartments in sea urchin eggs after activation. *Dev. Biol.* 98:446-454.

Longo, F. J. 1978. Effects of cytochalasin B on sperm-egg interactions. *Dev. Biol.* 67:249-265.

Mar, H. 1980. Radial cortical fibers and pronuclear migration in fertilized and artificially activated eggs in *Lytechinus pictus*. *Dev. Biol.* 78:1-13.

Nakazawa, T., K. Asami, R. Shoger, A. Fujiwara, and I. Yasumasu. 1970. Ca^{+2} uptake, H^{+} ejection and respiration in sea urchin eggs on fertilization. *Exp. Cell Res.* 65:143-146.

Nishioka, D. and N. Cross. 1978. The role of external sodium in sea urchin fertilization. p. 403-414. *In: Cell Reproduction*. E. R. Dirksen, D. M. Prescott, and C. F. Fox (Eds.) Academic Press, New York.

Paul, M. and D. Epel. 1975. Formation of fertilization acid by sea urchin eggs does not require specific cations. *Exp. Cell Res.* 94:1-6.

Paul, M., J. D. Johnson, and D. Epel. 1976. Fertilization acid of sea urchin eggs is not a consequence of cortical granule exocytosis. *J. Exp. Zool.* 197:127-133.

Piatigorsky, J. and A. H. Whiteley. 1965. A change in permeability and uptake of C^{14}-uridine in response to fertilization in *Strongylocentrotus purpuratus* eggs. *Biochim. Biophys. Acta* 108:404-418.

Poenie, M., J. Alderton, R. Tsien, and R. Steinhardt. 1985. Changes of free calcium levels with stages of the cell division cycle. *Nature (Lond.)* 315:147-149.

Schatten, G., T. Bestor, R. Balczon, J. Henson, and H. Schatten. 1985. Intracellular pH shift leads to microtubule assembly and microtubule-mediated motility during sea urchin fertilization: correlations between elevated intracellular pH and microtubule disassembly. *Eur. J. Cell Biol.* 36:116-127.

Schneider, E. G. 1985. Activation of Na^{+}-dependent transport at fertilization in the sea urchin: requirements of both an early event associated with exocytosis and a later event involving increased energy metabolism. *Dev. Biol.* 108:152-163.

Shen, S. S. and R. A. Steinhardt. 1978. Direct measurement of intracellular pH during metabolic derepression of the sea urchin egg. *Nature (Lond.)* 272:253-254.

Shen, S. S. and R. A. Steinhardt. 1979. Intracellular pH and the sodium requirement at fertilization. *Nature (Lond.)* 282:87-89.

Spudich, A., J. T. Wrenn, and N. K. Wessells. 1988. Unfertilized sea urchin eggs contain a discrete cortical shell of actin that is subdivided into two organizational states. *Cell Motil. Cytoskeleton* 9:85-96.

Steinhardt, R. A. and D. Epel. 1974. Activation of sea urchin eggs by a calcium ionophore. *Proc. Natl. Acad. Sci. USA* 71:1915-1919.

Steinhardt, R. A. and D. Mazia. 1973. Development of K^{+}-conductance and membrane potentials in unfertilized sea urchin eggs after exposure to NH_4OH. *Nature (Lond.)* 241: 400-401.

Steinhardt, R. A., R. Zucker, and G. Schatten. 1977. Intracellular calcium release at fertilization in the sea urchin egg. *Dev. Biol.* 58:185-196.

Suprynowicz, F. A. and D. Mazia. 1985. Fluctuation of the Ca^{+2}-sequestering activity of permeabilized sea urchin embryos during the cell cycle. *Proc. Natl. Acad. Sci. USA* 82: 2389-2393.

Swann, K. and M. J. Whitaker. 1986. The part played by inositol trisphosphate and calcium in the propagation of the fertilization wave in sea urchin eggs. *J. Cell Biol.* 103: 2333-2342.

Swann, K., B. Ciapa, and M. Whitaker. 1987. Cellular messengers and sea urchin egg activation. p. 45-69. *In: Molecular Biology of Invertebrate Development*. J. D. O'Connor (Ed.). Alan R. Liss, New York.

Swezey, R. R. and D. Epel. 1986. Regulation of glucose-6-phosphate dehydrogenase activity in sea urchin eggs by reversible association with cell structural elements. *J. Cell Biol.* 103: 1509-1515.

Swezey, R. R. and D. Epel. 1988. Enzyme stimulation upon fertilization is revealed in electrically permeabilized sea urchin eggs. *Proc. Natl. Acad. Sci.* 85:812-816.

Swezey, R. R., T. Schmidt, and D. Epel. 1987. Effects of hydrostatic pressure on actin assembly and initiation of amino acid transport upon fertilization of sea urchin eggs. p.

95-111. *In: Current Perspectives in High Pressure Biology*. H. W. Jannasch, R. E. Marquis, and A. M. Zimmerman (Eds.). Academic Press, London.

Turner, E. L., J. Hager, and B. M. Shapiro. 1988. Ovothinol replaces glutathione peroxidase as a hydrogen peroxide scavenger in sea urchin eggs. *Science* 242:939-941.

Vacquier, V. D. 1981. Dynamic changes of the egg cortex. *Dev. Biol.* 84:1-26.

Whitaker, M. J. and R. A. Steinhardt. 1981. The relation between the increase in reduced nicotinamide nucleotides and the initiation of DNA synthesis in sea urchin eggs. *Cell* 25: 95-103.

Whitaker, M. J. and R. A. Steinhardt. 1985. Ionic signaling in the sea urchin egg at fertilization. p. 168-222. *In: Biology of Fertilization*, Vol. 3. C. B. Metz and A. Monroy (Eds.). Academic Press, Orlando.

Whiteley, A. H. and E. L. Chambers. 1961. The differentiation of a phosphate transport mechanism in the fertilized egg of the sea urchin. p. 387-401. *In: Symposium on Germ Cells and Development*. Institut Intern. d'Embryologie and Fondazione A. Baselli.

Winkler, M. M., E. Nelson, C. Lashbrook, and J. W. B. Hershey. 1982. ^{31}P- NMR study of the activation of the sea urchin egg. *Exp. Cell Res.* 139:217-222.

Winkler, M. M., R. A. Steinhardt, J. L. Grainger, and L. Minning. 1980. Dual ionic controls for the activation of protein synthesis at fertilization. *Nature (Lond.)* 287:558-560.

Zucker, R. S. and R. A. Steinhardt. 1978. Prevention of the cortical reaction in fertilized sea urchin eggs by injecting calcium-chelating ligands. *Biochim. Biophys. Acta* 541:459-466.

Index